IET POWER AND ENERGY SERIES 81

Cyber-Physical-Social Systems and Constructs in Electric Power Engineering

Other related titles in this series:

Volume 73 **Wide Area Monitoring, Protection and Control Systems: The enabler for Smarter Grids** A. Vaccaro and A. Zobaa (Editors)
Volume 77 **Wide Area Monitoring of Interconnected Power Systems** A.R. Messina
Volume 79 **Vehicle-to-Grid: Linking electric vehicles to the smart grid** J. Lu and J. Hossain (Editors)

Cyber-Physical-Social Systems and Constructs in Electric Power Engineering

Edited by
Siddharth Suryanarayanan,
Robin Roche and Timothy M. Hansen

The Institution of Engineering and Technology

Published by The Institution of Engineering and Technology, London, United Kingdom

The Institution of Engineering and Technology is registered as a Charity in England & Wales (no. 211014) and Scotland (no. SC038698).

First published 2016

The Institution of Engineering and Technology
Michael Faraday House
Six Hills Way, Stevenage
Herts, SG1 2AY, United Kingdom

www.theiet.org

British Library Cataloguing in Publication Data
A catalogue record for this product is available from the British Library

ISBN 978-1-84919-936-0 (hardback)
ISBN 978-1-84919-937-7 (PDF)

Typeset in India by MPS Limited
Printed in the UK by CPI Group (UK) Ltd, Croydon

Contents

Foreword

Recent years have seen explosive growth in computing and communications. Just in the last two decades, internet and mobile phones have had great impact on all aspects of our lives at home and work: e-mail, social networks, e-commerce, entertainment, transportation, to name just a few aspects.

Increasingly, our physical surroundings and systems are getting infused with computing and communications. For example, a modern automobile contains several computer networks and processors. And, smart meters are being deployed in our homes and workplaces to make our electric energy consumption and management more efficient. Aircraft engines are being fitted with sensors, computing and communications devices that are transmitting data on the state of the engine for proactive and efficient maintenance. Internet-of-things, or the industrial internet, offers powerful and compelling visions of the rapid evolution of this cyber-physical reality.

It is likely that within the next decade, this infusion of sensors, communications, computing and control into our physical systems and surroundings will have the same transformative impact that internet and mobile phones have had in the last two decades.

Will this transformation benefit humans in the pursuit of more satisfying, happier, and productive lives?

A positive answer to this question is at the center of our vision of future of cyber-physical-social systems. This is particularly important in view of the contemporary commentary in mainstream discussions leading to fears of job losses, privacy, and security. We must make all possible efforts to avoid these dystopian visions and achieve much more positive outcomes in the coming decades.

The essential idea is that human behavior, as individuals and as members of various groups, should be at the center of our thinking in the design and development of internet-of-things and cyber-physical systems. Deep understandings from the fields of psychology, sociology, economics, and anthropology should be combined with cyber-physical systems analysis and design frameworks and tools for the creation of the future cyber-physical-social systems.

The potential for benefits is, indeed, very great. One of the major concerns at the current time is the significant slowdown in total factor productivity, particularly in the developed nations. One can imagine a future where well-designed cyber-physical-social systems will make our work lives more satisfying and safer while increasing productivity and quality. Imagine how our education processes might become more personalized, effective, and efficient by the deployment of cyber-physical-social education systems that leverage the deep understandings from science of learning and powerful technologies in the classrooms and home while also make the work of teachers more satisfying and productive. Can cyber-physical-social medical systems in hospitals and at home help us lead healthier lives, reduce healthcare cost burdens while also make the work of health care workers more satisfying and productive?

Can we make our transportation systems more efficient, resilient, safer, and sustainable by the development of suitable cyber-physical-social automobiles, trains, airplanes, and systems management and operations? Can the demanding work in mining and construction become safer and more productive by the development of appropriate cyber-physical-social devices and systems?

To realize these promising and attractive futures will require engineers, physical scientists, biomedical scientists, computer scientists, and social-behavioral scientists to deeply integrate knowledge, tools, techniques, and modes of thinking to create new ways of conceiving, designing, and implementing future cyber-physical-social systems. This, however, is far from easy. The challenge lies in the academic disciplinary traditions and separations that make such integration very difficult.

I am delighted that work along this line has started to appear. This volume is an example of such an amalgam of ideas from different fields pertaining to the cyber-physical-social aspects of electric power systems engineering. The coeditors of this book, who are actively engaged in this area of emerging research, have organized the chapters in sections apropos to the ongoing modernization of the electricity infrastructure, namely, stability and security; controls and economics in the end-user realm; social aspects; and, testbeds for validation. The contributing authors are experts in their field and bring a wealth of knowledge to this book via their contributions.

It is often stated that we live in a time of accelerating technological change. Indeed, the rates of internet and smartphone technologies penetration are remarkable by all historical standards. Therefore, we might anticipate rapid development and penetration of internet-of-things and design and development of cyber-physical-social systems. Indeed, as Bill Gates has observed, "We always overestimate the change that will occur in the next two years and underestimate the change that will occur in the next ten. Don't let yourself be lulled into inaction." I am confident that the research community, far from being lulled into inaction, will seize this incredible opportunity to improve the human condition.

Pramod P. Khargonekar
Eckis Professor of Electrical and Computer Engineering,
Dean Emeritus, Herbert Wertheim College of Engineering,
University of Florida,
Gainesville, FL.
April 7, 2016.

Acknowledgments

With this book, we hope to provide the interested reader with an overview and some current research in the fields of cyber-physical systems and cyber-physical-social systems for electric power system applications. The book is primarily intended as a reference for researchers in the field of electric power engineering. It may also be used as supplementary reading material in an advanced graduate course on emerging research topics in electric power engineering.

We would like to thank all the authors who have contributed to this book. We appreciate the quality of their contributions, their commitment to meeting the deadlines, and their patience toward our multiple requests. We also acknowledge Dr Pramod P. Khargonekar, Eckis Professor of Electrical and Computer Engineering and Dean Emeritus of the Herbert Wertheim College of Engineering at the University of Florida (Gainsville, Florida, USA), for writing a foreword on the importance of cyber-physical-social systems to the changing landscape of technological advancements.

Ms Jennifer Grace, Mr Paul Deards, and Ms Olivia Wilkins of The IET are thanked for their dedication to guiding this book toward completion and for enduring the numerous questions we posed at every stage.

Finally, we would like to thank our respective academic institutions – Colorado State University (Fort Collins, Colorado, USA), Université de Technologie de Belfort-Montbéliard (Belfort, France), and South Dakota State University (Bookings, South Dakota, USA) – for the support and encouragement offered to this effort.

We thoroughly enjoyed reviewing and editing these chapters and have learned a lot in doing so. We hope the reader shares this feeling.

Sincerely,
-Sid Suryanarayanan, Fort Collins, CO
Robin Roche, Belfort, France
Tim Hansen, Brookings, SD

ACKNOWLEDGEMENTS

Part 1

Introduction

Chapter 1

Introduction to cyber-physical-social systems and their applications in power systems engineering

Siddharth Suryanarayanan[1], *Robin Roche*[2] *and Timothy M. Hansen*[3]

1.1 What is a cyber-physical system?

Recent technological progress has made possible significant advances in the computation and communication fields, and enabled the emergence of large, networked infrastructures. Example applications may be found in the transportation, health care, manufacturing, and energy domains.

Such large systems already include what are known as embedded systems, i.e., computational systems designed to control and/or monitor a physical system. Although the concept of embedded systems has been in use for many years, the growing trend of interconnecting many physical and computational components to form large networks presents new challenges and requires novel approaches to design and control.

The concept of cyber-physical systems (CPS) was introduced in 2006 to fill this gap in knowledge [1]. A CPS may be described as a typically large networked system, made of tightly interconnected physical and computational components, operating in a networked fashion. The history of CPS may be traced to the seminal article titled "As we may think" by Vannevar Bush in 1945, where the author challenged physicists returning from the war to develop the "memex," a computer device for storing and accessing large sets of information like books and encyclopedias by humans [2,3]. Significant advances in the fields of communications and network engineering; computation, control and systems theory and engineering; information systems; Internet engineering; and sensor systems have led to the progression of the human–machine experience, thus paving the way for the evolution of the theory and hardware of CPS. Kim and Kumar present a detailed description of the history of CPS in Reference 4.

[1]Department of Electrical and Computer Engineering, Colorado State University, Fort Collins, Colorado, USA
[2]IRTES Laboratory (EA 7274) and FCLAB Research Federation (FR CNRS 3539), UTBM, Université Bourgogne Franche-Comté, Belfort, France
[3]Electrical Engineering and Computer Science Department, South Dakota State University, Brookings, South Dakota, USA

CPS may be seen as similar to the Internet, but applied to the physical world. For example, the Internet-of-Things (IoT) [5], may be considered as an enabler for CPS, in that CPS applications require efficient sensing and communication infrastructures. CPS are at the frontiers of the engineering and computer science fields, in that they aim to combine the most recent advances in both disciplines. The systems integration approach is therefore a central element to the CPS concept. Current challenges at the frontier of both fields include architectures, interoperability, networked control, standards and test procedures, verification and validation, and security. CPS-enabled critical infrastructures (e.g., for energy, transportation, or smart cities) may possess the promise of solutions to the grand challenges facing the engineering community in the twenty-first century [6].

Only considering technical aspects is insufficient, especially when humans are expected to use and be impacted by the designed CPS. A system can operate perfectly from a technical point-of-view, but if its users are unable to understand how to interact with it or are not convinced of its usefulness, the system may never reach the point of fulfilling the intended purpose. Thus, the *social* aspects of such systems must be considered in the design process.

1.2 What is a cyber-physical-social system?

An emerging, yet challenging, frontier for CPS applications is the inclusion of the *social* aspect in engineering. The end-user of a critical engineering infrastructure determines the utility of that domain, and any advance should also increase the quality of life of the end-user. According to Reference 7, a "smart" infrastructure includes an *active* end-user (sometimes described as a "prosumer") with a notably different role in participation than in the past. The active end-user, enabled with information in real-time or near real-time and the ability to make decisions, is no longer a passive participant in the control and operation of the critical infrastructure. An empowered end-user, with a hitherto unprecedented level of information and control, is a paradigm shifting concept such as residents of "smart" homes in the electricity domain controlling their energy usage to save money and provide grid ancillary services. To understand, model, simulate, develop, build, test, analyze, and enhance these futuristic manifestations imposes a fundamental requirement of the consideration of the social (and societal) aspect of CPS. The human-centric CPS, which marks the next generation of CPS, is called the cyber-physical-*social* system (CPSS) [8].

A CPSS integrates computation, physical components, and human cognition to achieve socially aware advancement in the operation of critical infrastructures and their interdependencies. Figure 1.1 illustrates the difference between the CPS and CPSS concepts. Each circle represents a domain: cyber, physical, and social. Areas 1–4 represent the intersections of the different domains into the system-level concept. Area 1 represents CPS, while area 2 represents cyber-social systems (e.g., the Internet). Area 4 is at the intersection of all three domains, and corresponds to the proposed concept of CPSS.

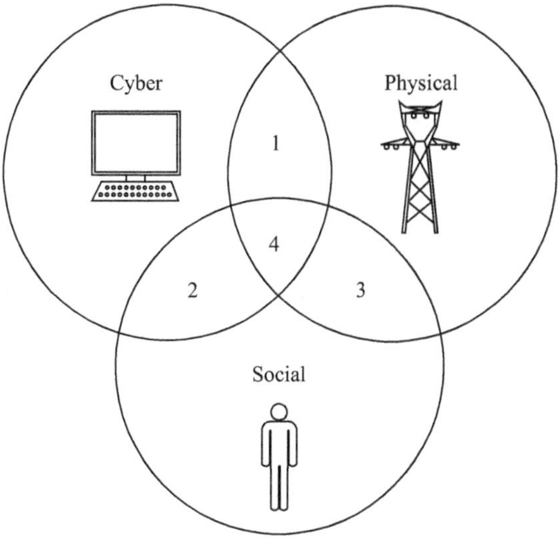

Figure 1.1 *The system-level concept of CPSS versus other domain-intersecting concepts. Area 1 corresponds to the existing idea of CPS; Area 4 adds a human-cognition component to extend the idea to CPSS*

1.3 Applications of CPSS in power engineering

Electric power systems traditionally rely on sensor, communication, and control infrastructures known as supervisory control and data acquisition (SCADA) systems. SCADA is used to transfer measurement data to central transmission and/or distribution (T&D) system operators so the best operational decisions can be made to ensure high power delivery quality while minimizing costs, usually through dedicated software. Therefore, power systems are a CPS, with the interrelated physical electric equipment (e.g., lines, transformers, loads, generators) and cyber SCADA systems.

This situation is now changing, as T&D systems are evolving toward what is now known as the "smart(er) grid." Figure 1.2 illustrates some of the ongoing changes, including a higher share of renewable energy generation and energy storage at the bulk and distribution levels, and the emergence of new entities and systems such as microgrids [9] and energy service providers.

While transmission systems have long been instrumented, new equipments, such as phasor measurement units (PMUs) [10], are being installed to provide much more accurate and timely data. These devices are useful to monitor large transmission systems, where instabilities can propagate and, in the worst case, lead to cascading outages. Disturbances can originate from electrical faults or failures, but also from malicious cyber-security exploits and attacks. Such topics intrinsically require considering both cyber and physical aspects of the grid, as weaknesses in either lead to weakness in the whole system. For example, communication delays or

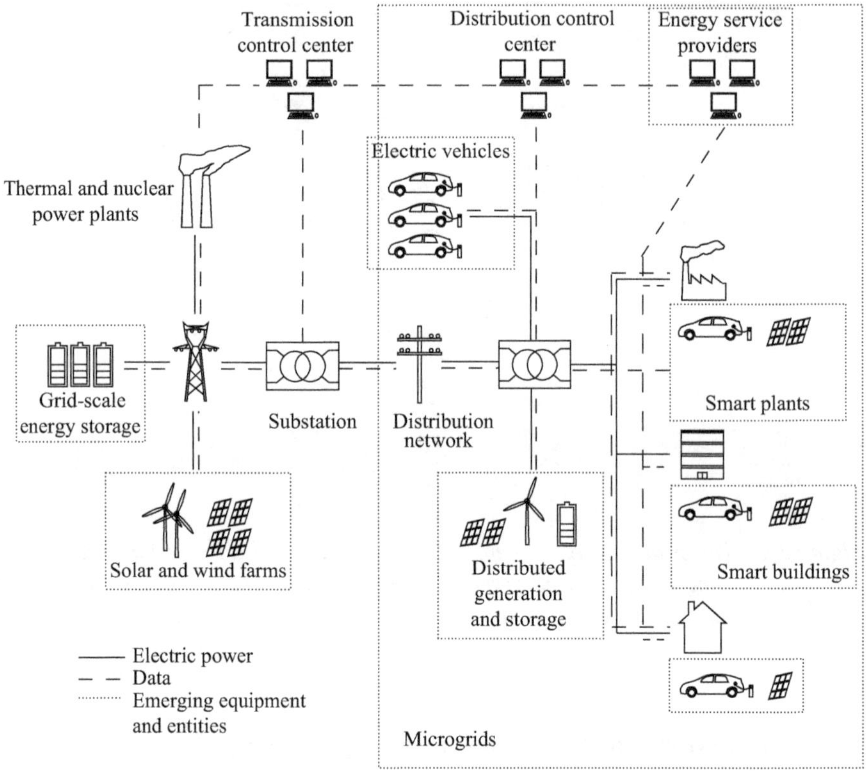

Figure 1.2 Illustrations of some of the changes expected in the future smart electric grid

manipulated information may lead to improper or delayed control actions, which may in turn worsen grid condition. A first series of challenges thus lies in the close integration and interdependence between the cyber and physical infrastructures, and in ensuring the stability of the system under all conditions.

Similar to transmission, distribution is becoming "smarter," i.e., better instrumented and coupled with dedicated equipment for monitoring and control. Instrumentation is being deployed on the distribution system, e.g., for substation automation, but also at end-user premises through advanced metering infrastructure (AMI) and smart meters. Current challenges interlink technical, economic, and social aspects. Technical challenges result from the growing penetration of distributed generation, as well as distributed storage and electric vehicles, which are changing traditional power flows and require decentralized control due to the increasing scale of the system. This new equipment, together with AMI and smart meters, is also creating significant amounts of data that must be collected, stored, and analyzed [11]. Economic challenges are related to how costs and benefits are shared among the many players (e.g., end-users, aggregators, and utilities), and how markets are structured as a result of the growing share of resources located at the edge of the grid. Finally, social challenges result from the active behavior of end-users, modeled as prosumers. End-user decisions depend on multiple criteria, such as cost savings, comfort, ease of use, and privacy, and in turn impact power system operations. In many distribution studies, the social aspect can no longer be ignored, and adopting the CPSS perspective becomes necessary.

To reduce costs and shorten development duration, extensive simulation and validation studies now usually precede the deployment of new technologies. However, traditional simulation tools are often not suitable for the CPS or the CPSS perspective, in that they typically focus on a specific discipline, such as the cyber or physical domains. New testbeds are thus needed to tackle current challenges that lie at the interface of multiple domains, e.g., power systems, communication systems, and cyber-security. Real-time simulation, coupled with hardware-in-the-loop (HIL) capabilities, is an enabler for faster production development and validation. Over the past few years, the research community has lead an effort to create such new tools, which are necessary to validate CPS- and CPSS-based concepts.

The smart grid is therefore a profound paradigm shift from conventional bulk power generation feeding passive loads to a system where electricity is also generated and stored at the distribution level, and where massive information flows are available. As a consequence, new concepts and tools are essential to facing the challenges of today and the future in electric power systems viewed as CPS and CPSS.

1.4 Organization of the book

This book presents a collection of some recent research in the areas of CPS and CPSS applications to electric power engineering. The constituent chapters are arranged in the following parts of the book:

Part 1. Introduction;
Part 2. Stability and security of the grid;
Part 3. New philosophies of control and economics in distribution systems;
Part 4. Social aspects and implementations; and
Part 5. Testbeds for validation of the research concepts.

The motivation to organize the book thusly is due to the importance of the abovementioned topics to modernization of the electric grid, as described in Section 1.3. While the grid is undergoing transformation in the various dimensions of the cyber, physical, social, and systems engineering aspects, we have chosen the above classification for the contents of our book because of the heightened recent interest in these areas. Table 1.1 lists the information of title and authors for the constituent chapters.

For instance, the security issues of the grid were brought to the forefront via numerous blackout incidents worldwide in 2015. These incidents include malicious causes that were both physical (e.g., Pakistan in January 2015 [12] and Crimea region in November 2015 [13]) and cyber (e.g., Western Ukraine in December 2015 [14]). The impact of such events may be contained by implementing advanced control technologies in the grid that are enabled with high computational and communication capabilities, which is the focus of the chapters in Part 2. In Chapter 2, the authors present a framework of distributed optimal control for increasing the transfer capability and maintaining stability in electric grids, while considering a predefined communication architecture. Chapter 3 presents three different communication and computational architectures as solutions for translating traditional centralized strategies for wide-area monitoring and control to distributed strategies. Both Chapters 2 and 3 utilize the alternating direction method of multipliers (ADMM) as a tool. The authors of Chapter 4 describe applications of control-based techniques that use distributed energy resources (DERs) for maintaining generator coherency during malicious cyber attacks on the power system. They describe two approaches: (a) a biologically inspired flocking method for modeling generator coherency; and (b) a generator desynchronization scenario due to a maliciously intended subset of DERs. The uniqueness of the latter approach is a scenario where the uncorrupted DERs take on the role of a protection system to bring generators to synchronism. The last chapter in Part 2 deals with a method of transient stability analysis based on change in energy functions, while describing several options for attaining post-fault security in an electric power system.

While the modernization efforts in the transmission sector of the electricity grid are expected to be *evolutionary*, those in the distribution realm are expected to be *revolutionary*. According to Reference 15, the distribution system of the future will possess transformative features enabled by pervasive and new technologies such as plug-in hybrid electric vehicles, automated metering infrastructures, networked connections, and two-way power flow and communication services. Further, distribution system users are expected to be active participants with access to information (e.g., real-time pricing) and asset controllability (e.g., resources participating in grid support through demand response and other ancillary services). The chapters in Part 3 introduce the prospect of new control methods, new participants in electricity markets, and new philosophies of customer management in the electricity grid. Chapter 6 presents the concept of an aggregator entity that delivers a competitive real-time pricing program for participating end-users to utilize their assets in demand response. Some mathematical models of end-user behavior are also presented. In Chapter 7, the authors present an overview of flexible operation of DERs with multiple levels of control using an agent-based approach. In particular, they address the interoperability of the built environment with electric grid operation to achieve flexibility in managing resources. Chapter 8 describes a literature review of the state-of-the-art of integrating electric and hybrid electric vehicles to the power grid. Chapter 9 addresses multiple algorithms based on agent technology for increasing the use of renewable energy resources in *smart* neighborhoods through electricity trading. Microgrids, which refer to fully operational and self-contained subsets of an electric power system that supplies its end-users electricity of required reliability and price, are increasing in their implementation and acceptance. A microgrid is viewed as an option to ride through system events seamlessly from the perspective of the end-user, thus positively impacting reliability. A microgrid offers self-sufficiency to its loads using local generation; thus, offering the advantage of deferring potentially high electricity prices during certain times of peak load in the system. In that regard, Chapter 10 describes the architecture of a DC-powered microgrid and presents the need and a method for the compensation of voltage errors in microgrids. A review of the literature on low-bandwidth communication is presented, and a framework for a compensation scheme is proposed, where the deviation in voltage and unequal shares of loads are compensated separately. Part 3 concludes with Chapter 11, where a study of the integration of electric vehicles into the grid considering technical, economic, environmental, and social aspects is presented. The proposed strategy focuses on the cooperation and coordination of responsive plug-in electric vehicles via a modified economic dispatch algorithm that considers the social component through a responsiveness model.

The chapters in Part 4 address the social aspects and implementations of some of the technologies presented in the previous parts of the book. In Chapter 12, the authors present some barriers to consumer adoption of *smart* grid technologies from the perspective of social science. They emphasize the need for interdisciplinary analysis for the successful design and implementation of such technical systems. The chapter starts with a historical introduction to the concept of *smart* homes with a focus on thermostats and appliances. Basic concepts on user-centered design and the diffusion of innovations are then presented. The second part deals with *smart* meters, and analyzes the shortcomings of a utility's deployment as a case study. Several keys to successful community engagement and lessons learned are then identified. The third part introduces community-based social marketing and then focuses on distributed generation with interviews to analyze the motivations and barriers to development. Chapter 13 addresses the risks, threats, and mitigation strategies related to cyber-security issues in SCADA systems, covering both technological and social aspects. In Chapter 14, the authors describe a demand response system for commercial buildings that considers occupant behaviors and present the results of a field experiment to show the effectiveness of the proposed approach.

Testbeds are vital for validating and verifying the results of research prior to deploying them in a critical infrastructure like the electricity grid. Part 5 of the book deals with some examples of existing and emerging testbeds relevant to CPSS studies for electric power engineering. Chapter 15 presents an overview of a cyber-physical security testbed and its layers and operation. The testbed has a unique computer network simulation layer that provides advanced cyber-attack and defense simulation capabilities. Chapter 16 describes the design objectives and challenges for CPS security experimentation and testbed development, while addressing some of the limitations of current efforts. Chapter 16 presents various case studies on cyber-security using a testbed at Iowa State University. The final chapter in Part 5 describes the concept of geographically distributed real-time digital simulators, including topics on HIL testing, aimed at utilizing resources and infrastructure that may be available at disperse locations. Examples included are the development of a testbed for geographically distributed real-time simulations between two national laboratories in the United States and another venture between two European academic institutions.

Table 1.1 Contents of the book

Chapter	Title	Authors
Part 2: Stability and security of the grid		
2	Distributed control design for damping inter-area oscillations in cyber-physical power networks	M. Mahmoudi and K. Tomsovic
3	Distributed algorithms for wide-area oscillation monitoring in power systems: a cyber-physical perspective	S. Nabavi, J. Zhang, and A. Chakrabortty
4	Cyber-physical strategies for generator coherency in the face of malicious attack	P. Srikantha, J. Kocsis, and D. Kundur
5	Dynamic contingency analysis and remedial action tools for secure electric cyber-physical systems	J. Mitra, M. Benidris, and N. Nguyen
Part 3: New philosophies of control and economics in distribution systems		
6	New philosophies of control and economics in distribution systems	T. M. Hansen, R. Roche, S. Suryanarayanan, *et al.*
7	Application of the agent-based technology to invoke demand flexibility under the SG-BEAMS integration framework	L. A. Hurtado, P. H. Nguyen, and W. L. Kling
8	Grid integration of electric and hybrid electric vehicles in cyber-physical-social systems	W. Su, R. Ma, and S. Xu
9	Increasing local renewable energy use in smart neighborhoods through coordinated trading	B. Celik, R. Roche, D. Bouquain, and A. Miraoui
10	Compensation of droop control in DC microgrid with multiple distributed generators	N. Yang, D. Paire, F. Gao, and A. Miraoui
11	Co-operative responsive electric vehicles for social-economic dispatch	M. Rahmani-andebili and G. K. Venayagamoorthy
Part 4: Social aspects and implementations		
12	If you build it, will they come? Getting consumers on board with the future of the smart grid	P. A. Aloise-Young, J. E. Cross, and P. K. Sandoval
13	Risks, threats and mitigation strategies for SCADA systems	H. Janicke, A. Cook, A. Nicholson, and K. Jones
14	Occupant engaged fast demand response for commercial buildings	Z. Song, S. Zheng, and S. Srivastava
Part 5: Testbeds for validation of the research concepts		
15	A testbed for closed-loop cyber-physical-social system simulation and security analysis	R. Liu, R. Goodfellow, and A. K. Srivastava
16	Cyber-physical-social system security testbeds for an attack-resilient smart grid	A. Ashok, P. Wang, and M. Govindarasu
17	Distributed real-time simulations for electric power engineering	M. Mohanpurkar, M. Panwar, S. Chanda, *et al.*

References

[1] E. A. Lee, "Cyber-physical systems – are computing foundations adequate?," *NSF Workshop on Cyber-Physical Systems: Research Motivation, Techniques, and Roadmap*, October 2006, 9 pp.

[2] V. Bush, "As we may think," *The Atlantic*, July 1945, vol. 176, pp. 101–108.

[3] J. Geelan. "Why I argue that 'the IoT era' began already in 1945." Linkedln, April 25, 2015. [Online] Available: *https://goo.gl/sS3nKj*. Accessed on: March 7, 2016.

[4] K-D. Kim and P. R. Kumar, "Cyber-physical systems: a perspective at the centennial," *Proceedings of the IEEE*, May 2012, vol. 100, pp. 1287–1308.

[5] N. Gershenfeld, R. Krikorian, and D. Cohen, "The Internet of things," *Scientific American*, October 2004, vol. 291, no. 4, pp. 76–81.

[6] National Academy of Engineering. "Grand Challenges – 14 Grand challenges for Engineering." [Online] Available: *http://www.engineeringchallenges.org/challenges.aspx*. Accessed on: March 13, 2016.

[7] A. Annaswamy, A. Hussain, A. Chakrabortty, and M. Cvetković, "Foundations of infrastructure CPS," in *Proceedings of IEEE American Control Conference*, July 2016, 14 pp.

[8] P. Khargonekar, "Cyber-physical social systems and resilient infrastructure – A perspective from NSF," *Keynote lecture, KAUST Conference on Human-Machine Networks and Intelligent Infrastructure*, Saudi Arabia, October 2015.

[9] N. Hatziargyriou, H. Asano, R. Iravani, and C. Marnay, "Microgrids," *IEEE Power and Energy Magazine*, July–August 2007, vol. 5, no. 4, pp. 78–94.

[10] A. G. Phadke, B. Pickett, M. Adamiak *et al.*, "Synchronized sampling and phasor measurements for relaying and control," *IEEE Transactions on Power Delivery*, January 1994, vol. 9, no. 1, pp. 442–452.

[11] G. R. Gray, J. Simmins, G. Rajappan, G. Ravikumar, and S. A. Khaparde, "Making Distribution Automation Work: Smart Data Is Imperative for Growth," *IEEE Power and Energy Magazine*, January–February 2016, vol. 14, no. 1, pp. 58–67.

[12] S. Masood. "Rebels tied to blackout across most of Pakistan." New York Times, January 25, 2015. [Online] Available: *http://goo.gl/fQYGbk*. Accessed on: March 5, 2016.

[13] I. Nechepurenko. "Crimea to face power shortages for months, officials say." New York Times, January 8, 2016. [Online] Available: *http://goo.gl/jqe3Pm*. Accessed on: March 5, 2016.

[14] K. Zetter. "Inside the cunning, unprecedented hack of Ukraine's power grid." Wired Magazine, March 3, 2016. [Online] Available: *http://goo.gl/WW64zN*. Accessed on: March 5, 2016.

[15] H. E. Brown, S. Suryanarayanan, and G. T. Heydt, "Some characteristics of emerging distribution systems considering the smart grid initiative," *The Electricity Journal*, June 2010, vol. 23, no. 5, pp. 64–75.

Part 2

Stability and security of the grid

Chapter 2

Distributed control design for damping inter-area oscillations in cyber-physical power networks

Meimanat Mahmoudi[1] *and Kevin Tomsovic*[1]

Abstract

The next generation of the electricity grid, also known as 'Smart Grid', is one of the most complex cyber-physical systems (CPS) due to its extreme dimension, geographic reach, and high reliability requirements. One of the main concerns for secure and reliable operation of power systems is the small signal stability problems caused by inter-area oscillations. In the future grid, enhancing the transfer capability while maintaining system stability requires damping these oscillations. In this chapter, we proposed a distributed optimal control framework using group sparse regularization functions. The proposed control aims to optimize a standard cost criterion while penalizing the number of communication links. The group sparse regularization approach is used to induce a desired communication structure and encode prior information about the underlying system into the control design. We present two applications of the proposed algorithm for damping inter-area modes in power networks. Our results suggest that the proposed method provides flexibility in designing wide-area type controls by allowing for a predefined communication structure. It can also act as an alternative approach to modal analysis methods in finding effective measurement-control loops in the system. The ability to encode system constraints in the control design objective is another major advantage of the method.

2.1 Introduction

The next generation of the electricity grid, also known as 'Smart Grid', is one of the most complex cyber-physical systems (CPS) due to its extreme dimension,

[1]Department of Electrical Engineering and Computer Science, University of Tennessee, Knoxville, TN USA

geographic reach, and high reliability requirements. Recent trends toward deregulation and deployment of distributed generation on one hand, and the technical trend of rapidly advancing communication technologies on the other hand, suggest that the complexity and interactivity of power networks will continue to grow [1]. Thus, the future power grid is characterized by a substantial increase in uncertainty and complexity that it was not designed to manage. Without careful crafting of new control schemes, power systems may suffer from a decrease of reliability, which has always been the foundation of its design and operation. In the future grid, integration of large quantities of variable energy sources, emergence of power transfers in unprecedented magnitudes and directions, and operation of the grid closer to its limits, will make maintaining high reliability challenging [2].

One of the main concerns for the secure and reliable operation of power systems is the small signal stability problem. In the highly interconnected structure of power systems, an unstable mode can have immediate impacts over a wide area and may even lead to cascading failures and blackouts [3]. Most of the previous blackout events exhibited sustained low-frequency oscillations [4]. The notable breakup of the Western Electricity Coordinating Council (WECC) system on August 10, 1996, was due to undamped oscillations [5]. Other major oscillation events include a generation loss on October 9, 2003, multiple line trippings on June 4, 2003 [6], and the BC-Alberta separation event on August 4, 2000 [7]. These events raised great concerns about the adverse effect of oscillations on power system operation.

Local and inter-area oscillations are two distinct types of oscillations which have presented problems in power systems. Local mode oscillations are caused by a generator or group of generators swinging against the rest of the system. Local plant modes are usually induced by the action of Automatic Voltage Regulators (AVRs) operating at high output [8]. Local modes generally have natural frequencies in the range of 1–2 Hz. Their characteristics are fairly well understood and can be analyzed adequately using a detailed representation only in the vicinity of the plant [9]. Power system stabilizers (PSS) can be used to achieve a satisfactory damping for local modes.

Inter-area oscillations involve groups of machines in one part of the system oscillating against groups of machines in other parts of the system. They may be caused by either high-gain exciters or groups of machines closely coupled but interconnected by weak ties [10]. Inter-area modes have frequencies typically in the range of 0.1–1 Hz. Characteristics of these modes and factors influencing them are not fully understood which makes them far more complex to study and to control. A detailed model of the entire system is required to study inter-area modes [11].

Large power systems typically exhibit multiple dominant inter-area modes. Insufficient damping of this type of oscillation may lead to a limitation of power transfer capability or, worse than that, a growth in amplitude of the oscillations which could even cause a system to collapse. With highly increasing power exchange between utilities over the existing transmission network, power systems are driven closer to their limits, especially those of transmission capacity. Therefore, enhancing the transfer capability, while maintaining system stability, greatly relies on damping these inter-area oscillations.

The traditional approach for damping inter-area oscillations is by installing PSS to provide supplementary control action through the excitation of the generators. The increased use of high-response exciters, while improving transient stability, adversely affected the damping of local plant modes and shifted the focus for PSS tuning to local mode stability [12]. In addition, due to the lack of observability of certain inter-area modes in local measurements, conventional PSS with local measurements as input, cannot always effectively damp these type of oscillations. For instance, under certain operating conditions an inter-area mode may be controllable from one area and observable from another [13]. Moreover, the performance of conventional PSS is sensitive to operating point changes in the system which presents significant challenges in the design [14]. The practical feasibility of PSS is also limited due to the fact that power system models have been found inadequate for describing real-time operating conditions [5].

Introduction of Phasor Measurements Units (PMUs) offers the possibility of control based on Wide-area Measurements (WAMs) and provides great potential to overcome the shortcomings of conventional local controllers. Researchers achieved promising results by applying WAMs to Wide-area Damping Control (WADC) of inter-area oscillations. A Linear Matrix Inequality (LMI) based centralized WADC was proposed in Reference 15 that provides supplementary damping signals to AVRs of selected generators in addition to their local PSS signal. In the proposed design procedure, only a few measurements with high observability of inter-area modes are selected and only a few generators with the high controllability of those modes are chosen as control locations. A novel control inversion framework was proposed in Reference 16 which can invert PMU-based control designs developed for reduced-order power systems to controllers in actual higher-order systems via optimization methods. In practice, to enhance the damping performance of inter-area oscillations, a WADC framework was designed to modulate multiple HVDC systems and was implemented in China Southern Power Grid (CSG) [17].

Although several other studies also confirm that the use of remote measurement signals could increase damping of inter-area modes beyond that attainable by local signals [18–20], WADC is still in its infancy due to many difficulties associated with its design and implementation. Centralized framework for wide-area control suggested by most of the previous studies is one of the main challenges. As power networks are large-scale systems, a centralized controller is practically difficult to implement. While local controllers are blind to the widespread effects of disturbances and centralized controllers are practically difficult to implement, distributed controllers can achieve a trade-off between performance and required communications. In a distributed control scheme, limited communication between different controllers is used to improve the control performance. In a distributed architecture, the controller is composed of several interconnected units. Each local control agent can only access a subset of the global information (e.g., sensor measurements, model parameters) and actuate on a subset of the inputs, perhaps in its vicinity. Distributed controllers have received considerable attention since the early 2000s [21–25]. Motivation for designing localized controllers comes from Reference 21, where it was shown that optimal controllers for spatially invariant systems, although centralized, have an inherent spatial-decay property. For these systems, dependence of the controller agent on remote measurements decays exponentially with increasing distance from the controller [26].

In this chapter, inspired by recent advances in the field of compressive sensing and the sparsity-promoting optimal control design in Reference 27, we propose a distributed linear quadratic optimal control framework using group sparse regularization function. The proposed control aims to optimize a standard cost criterion while penalizing the number of communication links. The group sparse regularization approach is used to induce a desired sparsity pattern and encode prior information about the underlying system into the control design. We then present two applications of the proposed algorithm for damping inter-area modes in power networks:

1. Inducing a desired communication network based on system constraints and
2. Control and communication codesign to find efficient measurement and control signal pairs that improve the damping of inter-area modes.

The proposed optimal control design is then evaluated using the two-area four-machine test system. Our results suggest that the proposed method provides increased flexibility in designing wide-area type controls by allowing for a predefined communication structure. It also can act as an alternative approach to modal analysis methods in finding effective measurement-control loops in the system. The ability to encode system constraints in the control design objective is another major advantage of the method.

The remainder of the chapter is organized as follows. Section 2.2 describes the power system electromechanical model with the proposed distributed control input. The distributed control design algorithm using group sparse regularization functions is presented in Section 2.3. Sections 2.4 and 2.5 illustrate two applications for the proposed method in designing control for damping inter-area modes in power networks. Section 2.6 concludes the chapter.

2.2 Power system electromechanical model with distributed control input

To capture the electromechanical oscillations in power systems, we initially use the detailed, nonlinear, standard Differential Algebraic Equation (DAE) model to describe the power system dynamic behavior. Let $x(t) \in \mathbb{R}^n$ denote a vector that contains the dynamic states of synchronous generators; $y(t) \in \mathbb{R}^q$ denote the system algebraic states, including bus voltage magnitudes and angles; and $u(t) \in \mathbb{R}^m$ denote the control action. Then, power system dynamic behavior can be described by a set of DAEs of the form:

$$\dot{x}(t) = f(x(t), y(t), u(t)) \tag{2.1a}$$

$$0 = g(x(t), y(t), u(t)) \tag{2.1b}$$

Here, the dynamic equation (2.1a) accounts for the electromechanical dynamics of the synchronous generators and their excitation control. The algebraic equation (2.1b) accounts for load flow and generator stator equations. The control action $u(t)$ can represent the input to generator excitation, governor, or power electronic devices. Each generator is modeled using a third-order flux-decay model [28] along with its excitation:

$$\begin{aligned}
\frac{d}{dt}w_i(t) &= \frac{\pi f_s}{H_i}(P_{m_i} - P_{e_i} - P_{D_i}) \\
\frac{d}{dt}\delta_i(t) &= w_i(t) - w_s(t) \\
\frac{d}{dt}E'_{q_i}(t) &= \frac{1}{T'_{d0_i}}[E_{fd_i}(t) - E'_{q_i}(t) - (X_{d_i} - X'_{d_i})I_{d_i}(t)] \\
\frac{d}{dt}E_{fd_i}(t) &= \frac{1}{T_{A_i}}[-E_{fd_i}(t) + K_{A_i}(V_{ref_i} - E_{t_i}(t) + V_{DLQR_i}(t))]
\end{aligned} \tag{2.2}$$

In (2.2), $V_{DLQR_i}(t)$ is the supplementary damping control signal designed by the proposed distributed controller. The PSS are modeled as illustrated in Figure 2.1. In this chapter, without loss of generality, we assumed that the distributed control design will provide a supplementary signal to the excitation of selected generators. It should be noted, however, that model (2.1) and the proposed control design are sufficiently general to include different control inputs, load models, and renewable generation. Figure 2.2 shows the proposed distributed control agent and its information exchange pattern at k^{th} generator. Next, we linearize system (2.1) at a stationary operating point and derive the linear state-space model:

$$\dot{x}(t) = Ax(t) + Bu(t) \tag{2.3}$$

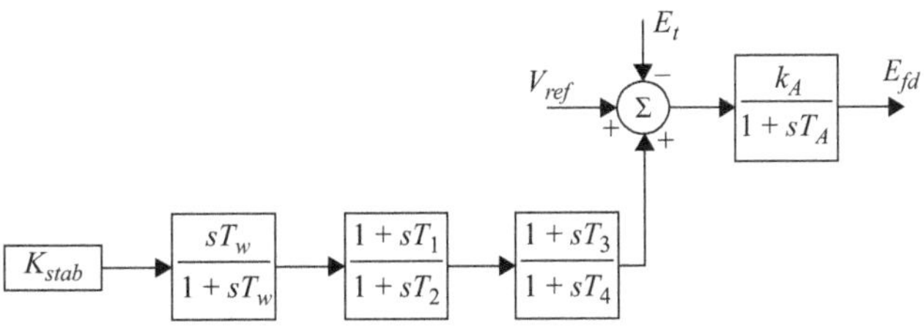

Figure 2.1 Two-stage power system stabilizer model

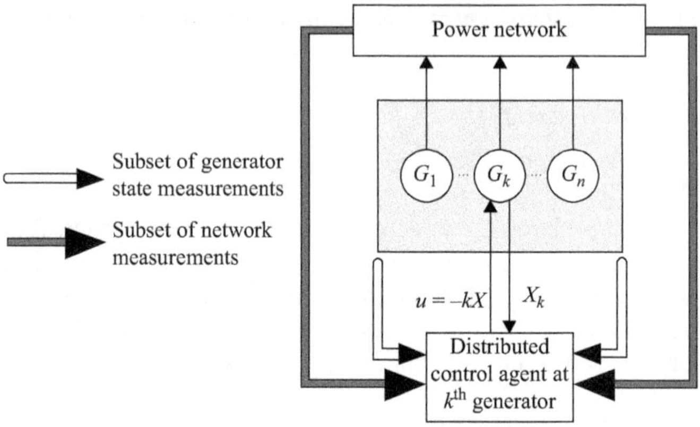

Figure 2.2 Proposed distributed control agent and its information exchange pattern at k^{th} generator

where $A \in \mathbb{R}^{n \times n}$ and $B \in \mathbb{R}^{n \times m}$. This linearized full model of the system will be used to design the controller. Practically, a large-scale system may have a large number of states, inputs, and outputs and the detailed model (2.3) of a large power network will be of a large scale. Thus, one of the challenges for designing wide-area controls for large-scale systems arises from the absence of mathematical models that can capture the electromechanical dynamic couplings in the system [29]. Aggregated and equivalent models have provided a good alternative for wide-area analysis of large-scale power systems since the 1980s [30]. Recently proposed aggregation methods can take advantage of PMU measurements and construct simplified inter-area models of large power systems by using dynamic measurements available from limited points on the transmission lines [31]. System identification techniques can also be used to derive low-order state-space models of a power system [32].

2.3 Distributed control design using group sparse regularization functions

In modeling and control, in networks of dynamical systems, it is crucial to understand the interactions between control network structure and the underlying dynamic properties. Constraints on communication network can limit the control performance. A trade-off exists between the achievable performance and the communication cost. If all controllers can communicate with each other, and decide on the control actions to be applied to each subsystem, the best possible performance is achievable. However, this requires prohibitively excessive communication and computation. On the other hand, in a fully decentralized scenario, every controller acts independently and applies control actions to its corresponding subsystem solely based on the subsystem's output measurements. Although decentralized control has minimal communication and

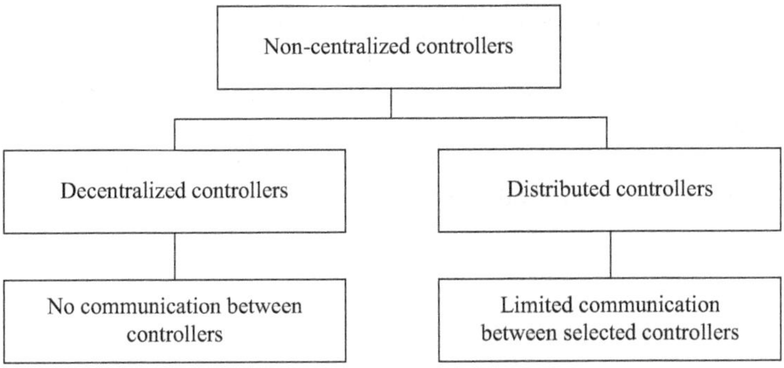

Figure 2.3 Categories of non-centralized controllers

computation requirement, it may result in poor performance. A desired scenario is the limited exchange of information in a distributed control architecture [33].

Distributed control is a widely used, but not well-defined term [34]. Here, distributed control refers to a control structure where each local control agent can only access a subset of the global information and actuate on a subset of the inputs. Figure 2.3 categorizes non-centralized control structures as they are defined in this chapter. The conventional controller design problem assumes that all the controllers in the system have access to the same information. Thus, the controller design problem is to design a controller K for a plant P such that K stabilizes the plant and meets a certain performance criteria like min $\|f(P,K)\|$. A distributed control problem can then be defined generally in the form:

$$\begin{aligned}&\text{minimize } \|f(P,K)\|\\&\text{subject to } K \text{ stabilizes } P\\&K \in S\end{aligned} \tag{2.4}$$

where S is a subspace. For a general linear time-invariant system P and subspace S, there is no known tractable algorithm for computing the optimal K. The unique feature that simplifies the conventional control problem is the information pattern. The information pattern represents the information set that each decision maker has access to when it makes the decision and calculates the control input. Since in a distributed control problem the information set for different controllers is different, the problem is difficult to solve. To overcome this design complexity, new approaches have recently been developed that initially design a centralized controller and apply compressed sensing algorithms, [35], to sparsify the control structure while preserving stability [26,36,37]. These methods open new possibilities in wide-area control design with limited communication and provide an alternative solution for the major challenge in distributed control design, namely the information pattern constraint. In this chapter, we propose a novel distributed control design for damping inter-area oscillations. The newly developed approach provides the ability to encode prior knowledge about the underlying system into the control design objective using group sparse regularization functions. In power systems, this prior knowledge can be a predefined communication structure, measurement availabilities or network variables and dynamic features that can observe or affect the inter-area modes. The grouping approach will greatly expand the application domain of the algorithm and the flexibility of the control design. Details of the proposed method and its applications are described in the following subsection.

2.3.1 Distributed linear quadratic control design using group sparse regularization functions

As discussed in Section 2.1, one of the main challenges in designing a practical WADC is selecting the measurements that each controller agent in a distributed architecture needs to have access to. One approach for addressing this issue is the sparsity-promoting linear quadratic control proposed in Reference 36 which simultaneously identifies the optimal control structure and optimizes the closed-loop performance. This method was applied to WADC design in Reference 27. In Reference 36, to induce a control architecture with limited communication, the standard quadratic performance index was regularized with an ℓ_1-norm penalty on the feedback matrix. However, use of ℓ_1-norm regularization encourages sparsity among individual elements of the controller matrix, without considering the potential structural relationships existing in the system. When dealing with a large-scale system with a sensor-rich structure and communication constraints, using readily available information or historic data about the system in the control design can greatly enhance the controllers performance, flexibility, and practicality. A considerable body of research in the field of compressive sensing has recently been devoted to design regularizations capable of encoding more prior information about the problem in searching for patterns of nonzero elements [38–40]. Inspired by the applications in compressive sensing, we propose a distributed linear quadratic control design using group sparse regularization functions. Group sparse regularizations consider sums of norms of appropriate subsets of measurements, or groups of states, based on the system characteristics. These groupings can influence the sparsity patterns of the resulting control and improve the feedback control performance.

The group-sparse linear quadratic optimal control problem is defined as follows:

$$
\begin{aligned}
&\min_{K} \int_0^{\infty} (x(\tau)'Qx(\tau) + u(\tau)'Ru(\tau))d(\tau) + \lambda\Omega(K) \\
&\text{subject to} \\
&\text{dynamics:} \qquad \dot{x}(t) = Ax(t) + Bu(t) \\
&\text{feedback control:} \quad u(t) = Kx(t) \\
&\text{stability:} \qquad A - BK \ \text{ is Hurwitz} \\
&\text{initial condition:} \quad x(0) = x_0
\end{aligned}
\tag{2.5}
$$

where $\lambda \geq 0$ is a regularization parameter, and $Q \in \mathbb{R}^{n \times n}$ and $R \in \mathbb{R}^{p \times p}$ are positive semi-definite and positive definite matrices that denote the state and control weights, respectively. Ω is a convex, typically non-smooth sparsity-inducing regularization function. The stability condition in (2.5) guarantees that the closed-loop system will be stable [41]. When it is only desired that the solution of optimal control problem have limited number of nonzero elements, Ω is chosen to be ℓ_1-norm [36]:

$$\Omega(K) \triangleq \sum_{i,j} |K_{i,j}| \tag{2.6}$$

(other norms, such as ℓ_2 and ℓ_∞ can also be used to form different objectives). Regularizing by the ℓ_1-norm induces sparsity in the sense that, depending on the strength of the regularization parameter, λ, a number of elements in K, will be equal to zero. In the proposed group sparse approach, based on the prior knowledge of the system, we organize the elements of the feedback matrix, K in groups. The objective is then to select or remove all the variables forming a group. A regularization function exploiting this group structure will help in designing controllers which are more flexible to adapt to communication and measurement constraints that we may practically have in the system. The group norm used in our method is the so-called ℓ_1/ℓ_2-norm:

$$\Omega(K) \triangleq \sum_{g \in \mathcal{G}} \eta_g \left\| K_g \right\|_2 \tag{2.7}$$

where $\mathcal{G}$ is a partition of $\{1, ..., m\}$ and $m = n \times p$ is the total number of elements in K. η_g are strictly positive weights and K_g denotes the vector in $\mathbb{R}^{|g|}$ recording the coefficients of K indexed by group g in $\mathcal{G}$. As defined in (2.7), Ω is known as a mixed ℓ_1/ℓ_2-norm. It behaves like an ℓ_1-norm on the vector $(\left\| K_g \right\|_2)_{g \in \mathcal{G}}$ in $\mathbb{R}^{|g|}$, and therefore, Ω induces group sparsity. In other words, each $\left\| K_g \right\|_2$, and equivalently each K_g, is encouraged to be set to zero. On the other hand, within the groups g in $\mathcal{G}$, the ℓ_2-norm does not promote sparsity.

The proposed group-sparse regularization approach can be used to induce a desired communication structure (e.g., star, ring, hierarchical) or to encode prior information about the underlying system into the control design. In power systems, this prior knowledge can be communication costs and constraints, probability of measurement availabilities, power transfers in critical tie-lines, inertia and damping distributions, and the topology of the system.

The optimization problem (2.5) is difficult in part because of the non-smooth nature of the norm function. To solve the group-sparse optimal control problem (2.5), we use the Alternating Direction Method of Multipliers (ADMM) [42]. In each iteration of ADMM the required proximal gradient step which induces the group sparsity pattern is solved using SPArse Modeling Software package (SPAMS) [38]. We incorporated the SPAMS solver into the Sparsity-Promoting Linear Quadratic Regulator (LQRSP) toolbox [27], to solve the proposed optimization. Two applications of the proposed method for damping inter-area oscillations in power networks are described in the following sections. Below is a summary of the algorithmic approach for the proposed distributed linear quadratic control design.

Distributed linear quadratic control design algorithm using group sparse regularization functions

Step 1. Derive the state-space model of the system (2.3) analytically or by using aggregation or identification methods.
Step 2. Define grouping structures based on the desired communication network design.
Step 3. Construct the graphs representing the groups in the system and select their weights based on their grouping objective.
Step 4. Solve optimal control problem (2.5) with $\lambda = 0$ to obtain the centralized feedback matrix K.
Step 5. Set the regularization parameter λ based on the application and the desired level of sparsity.
Step 6. Solve the optimization problem (2.5) using ADMM algorithm implemented in LQRSP toolbox. Enforce grouping structures (2.7) in the proximal gradient step of ADMM using SPAMS software.
Step 7. Solve the structured optimal control problem based on the final sparsity pattern identified in Step 7.

2.4 Application I: inducing a desired communication structure for damping inter-area oscillations in power networks

Based on the report from North American Synchrophasor Initiative (NASPI) the total number of PMUs installed in the North American grid was more than 1100 by 2014, offering nearly 100 percent observability of the transmission system. PMU measurements are taken at a high speed (30 observations per second) and can be transmitted over a communication network to the control stations. Due to communication limitations and the large scale of the power grid, these sensor measurements cannot instantaneously be available to all the controllers in the system. In practice, each local control agent can only access a subset of the measurements. Limitations on the information exchange pattern in control design in power systems can come from different sources. Control actions being employed by separate utilities, lack of availability of some measurements and privacy considerations are just a few of the reasons. Therefore, it will be highly beneficial if the control design method can consider communication limitations and measurement access constraints and stabilize the

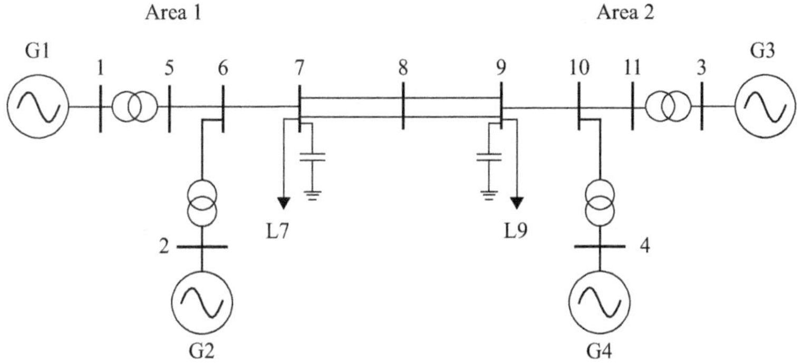

Figure 2.4 Two-area, four-machine test system [10]

system based on a predefined communication structure. In the following subsection application of our proposed method for inducing a desired communication structure is illustrated on the two-area four-machine test system.

2.4.1 Illustration on the two-area four-machine system

Consider the two-area four-machine power system in Figure 2.4 modeled using (2.2). This test system was created to exhibit different types of oscillations that occur in both large and small interconnected power systems. A detailed description of the system parameters can be found in Reference 10. We use this test system to validate the performance of the proposed WADC in inducing different predefined communication structures and explore the effect of the information patterns on the damping of the inter-area mode. In this configuration, 400 MW of active power exported from area 1 to area 2 causes the transmission lines to be stressed and the inter-area mode to be unstable. All generators are equipped with PSS to provide damping for local and inter-area modes. We modified this test system to create a very low-damped inter-area mode while local modes can be fairly damped using PSS. We doubled the amount of inertia on G_2 and G_4, setting $H_2 = 13s$ and $H_4 = 12.35s$, to increase the distance between frequencies of the local modes and the inter-area mode. The stabilizer gains of the generators were decreased to $K_{stab} = [10; 0; 10; 0]$ to allow for lightly damped modes. We linearized this system and derived the linear state-space model representation (2.3). Small signal analysis is then performed to identify system modes and the respective mode shapes. Table 2.1 shows the local and the inter-area modes of the system and their frequency and percentage damping. In power networks damping is considered adequate if all electromechanical modes have a damping ratio of around five percent. Damping ratios less than three percent are considered too weakly damped and becomes a cause of concern. Thus, this test system has a too weakly damped inter-area mode that may lead to further instabilities.

Table 2.1 Two-area system modes (base-case)

Mode no.	Eigenvalues	Frequency (Hz)	Percentage damping	Mode type
1	$-0.0500 \pm 3.1069j$	0.4945	1.61	Inter-area
2	$-0.2574 \pm 6.2966j$	1.0030	4.08	Local
3	$-0.2823 \pm 6.4472j$	1.0271	4.37	Local

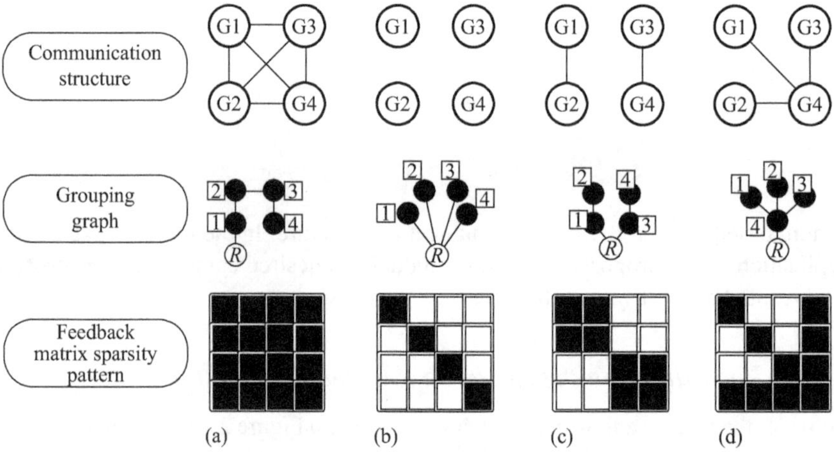

Figure 2.5 Communication structures and grouping graphs for (a) Centralized, (b) Decentralized, (c) Local, and (d) Star communication networks and the predicted sparsity pattern of the feedback matrix

Our goal is to solve the proposed optimal control problem (2.5) for a predefined communication structure using grouping regularization functions. This structure can be defined based on communication cost and constraints in the system. Figure 2.5 shows the communication structures and grouping graphs for centralized, decentralized, local and star communication networks along with the predicted sparsity patterns of their respective feedback matrices. We use graph representation to show the grouping structures. Each grouping graph contains one white vertex which represents the root of the graph depicted by R and a set of black vertices that represent the generators labeled by the generator number. A desired grouping structure is created by adding a path to the graph which starts from the root and connects the black vertices that are included in that group. The sequence of black vertices in the path can be chosen arbitrarily and will not affect the grouping definitions, but it should not create a loop. For example, the grouping graph in Figure 2.5 (a) contains only one path, i.e., $(R, 1, 2, 3, 4)$ which indicates that all the generators are grouped together. In other word, in the centralized design all the generators can communicate with each

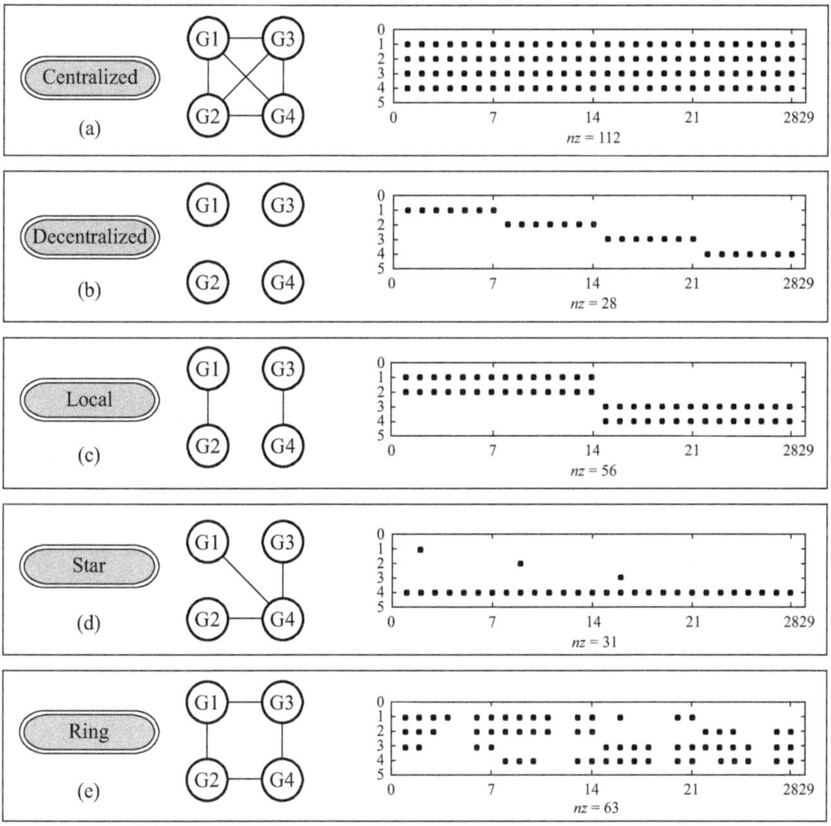

Figure 2.6 Communication structures and their respective designed feedback control matrix sparsity patterns (a) Centralized, (b) Decentralized, (c) Local, (d) Star, and (e) Ring

other. The grouping graph in Figure 2.5 (b) contains four paths, $(R, 1)$, $(R, 2)$, $(R, 3)$, and $(R, 4)$ which shows that the generators are not allowed to communicate in the decentralized design. In the matrix representation, the 16 divided blocks represent generator gains. The block diagonal elements of the matrix are the generator self-gains, and off-diagonal blocks are correlation gains between the four generators in the system. Black blocks represent vectors of nonzero elements. Figure 2.5 only shows the groupings between different generators, but it should be noted that all seven states of each generator (i.e., $[\Delta\omega_i, \Delta\delta_i, \Delta E'_{q_i}, \Delta E_{fd_i}, \Delta V_1, \Delta V_2, \Delta V_s]$) are also treated as a group.

Having group structures, the proposed optimal control problem (2.5) is then solved for the different information patterns. Figure 2.6 shows the communication structures and their resulting feedback control matrix patterns. The feedback matrix

Table 2.2 Inter-area modes for different communication structures

Case	Inter-area mode	Frequency (Hz)	Percentage damping
Base case	$-0.0500 \pm 3.1069j$	0.4945	1.61
Centralized	$-0.4603 \pm 3.0790j$	0.4900	14.78
Decentralized	$-0.1322 \pm 3.1299j$	0.4981	4.22
Local	$-0.1429 \pm 3.1383j$	0.4994	4.54
Star	$-0.2202 \pm 3.1300j$	0.4981	7.01
Ring	$-0.3755 \pm 3.1478j$	0.5009	11.84

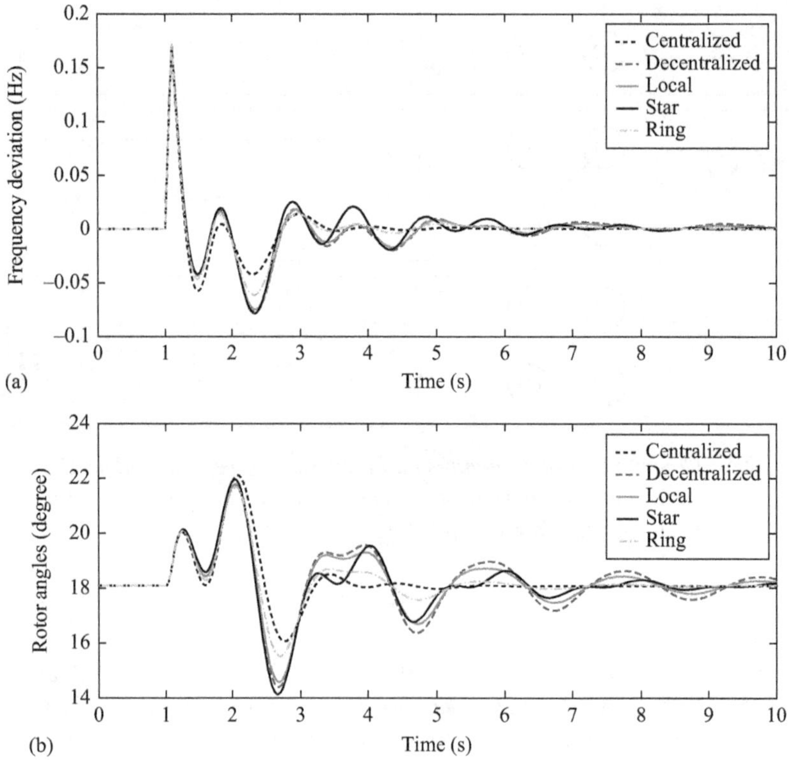

Figure 2.7 Nonlinear simulation results for different control structures: (a) Frequency deviation (Hz) and (b) Rotor Angles (degree)

K, is a N by ($N \times n$) matrix where N is the number of generators and n is the number states for each generator. For the two-area system with four generators modeled with seven states, K is a 4×28 matrix. The blue dots represent the nonzero elements of the matrix. Below each matrix, the total number of the nonzero elements is indicated by nz . As illustrated in Figure 2.6 (a), for the centralized case where all the generators are allowed to communicate the resulting feedback control matrix is a full matrix. In Figure 2.6 (b), i.e., the decentralized case, there is no communication between the generators; thus, the resulting control matrix has the block diagonal pattern with nz exactly equal to 28 nonzero elements which shows that each generator only uses its own state information for the control purpose. Table 2.2 compares the inter-area modes for different communication structures. In this test system, the ring structure has the next best damping after the optimal centralized case which was predictable due to the high level of information exchanged between the controllers. For simulation of the dynamic behavior, the disturbance scenario considered is a three-phase fault at bus 8 at time $t = 1s$ that is cleared after six cycles. The nonlinear time domain simulation of the power network is carried out using MATLAB. Figure 2.7 shows frequency deviations and rotor angles for G_1 for the five communication structures. The nonlinear simulation also confirms the results obtained in linear analysis.

Here, we provided examples of design and implementation of the distributed control with predefined communication structures to show the effectiveness of the proposed algorithm. However, it should be noted that our objective is not to find the best performing controller. The performance of these controllers cannot be compared while they have different amount of data communicated and different levels of optimality. Here, the proposed method provides a framework to find the controllers that best fit communication costs and constraints.

2.5 Application II: control and communication codesign for damping inter-area oscillations in power networks

As discussed in Section 2.1 several studies confirm that using remote measurements can improve damping of the inter-area modes beyond that obtainable by local signals. The main challenge is then finding a few pairs of measurements and control inputs among all the options that can achieve an effective inter-area oscillation damping. The choice of measurements and control input signals are typically based on modal analysis and frequency response information. Geometric measures and residues are two methods that have been used in power system studies to quantify the observability and controllability of the modes in the system [15]. Geometric measures are proven to be more accurate compared with residues in predicting the contribution of the kth mode of the system in output measurements and the mode's sensitivity to input perturbations [43]. An alternative approach is to use the proposed distributed control design to find the few measurements that when communicated can improve the damping for inter-area modes. For this purpose, the regularization parameter λ is increased gradually from zero to one to reach a desired level of sparsity in the feedback control matrix. Since some of the generator states are practically difficult to measure, the grouping feature of the algorithm provides the benefit to choose from the available sets of measurements by setting more weights, i.e., higher η_g, on those groups. To illustrate the effectiveness of the proposed method we studied two scenarios on

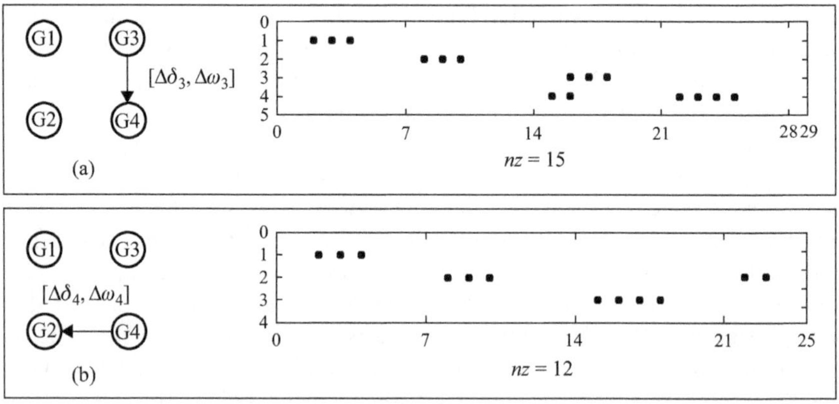

Figure 2.8 Communication structures and respective feedback control sparsity patterns for (a) Scenario 1 and (b) Scenario 2

the two-area four-machine test system introduced in Section 2.4.1 and compared the results with geometric measure approach. In the first scenario, we solved the group-sparse linear quadratic optimal control problem (2.5) for the base case and gradually increased λ from zero to obtain a sparse feedback matrix with one communication link. In grouping the states between different generators equal higher weights were selected for the set of $[\Delta\delta_i, \Delta\omega_i]$ measurements, to emphasize the availability of these measurements compared with other states. Figure 2.8 (a) shows the resulting communication structure and respective feedback control sparsity pattern. This result suggests measurements $[\Delta\delta_3, \Delta\omega_3]$ and the excitation control input G_4 as the effective measurement-control pair to improve the damping of the inter-area mode. Then the geometric measures for this scenario were calculated. Here we briefly explain the geometric measures and their formulations. The interested reader is referred to Reference 44 for the detailed explanation.

Consider the state-space model of the system described as (2.3). Eigenanalysis of matrix A provides the n distinct eigenvalues ($\lambda_i, i = 1, .., n$) and the corresponding matrices of right and left eigenvectors, $\Phi = [\Phi_1, ..., \Phi_n]$ and $\Psi = [\Psi_1, ..., \Psi_n]$, respectively. The geometric measures of controllability m_{ci} and observability m_{oi} associated with mode i are defined as follows:

$$m_{ci}(k) = \cos(\alpha(\Psi_i, b_k)) = \frac{|b_k^T \Psi_i|}{\|\Psi_i\| \, \|b_k\|} \tag{2.8}$$

$$m_{oi}(l) = \cos(\alpha(c_l, \Phi_i)) = \frac{|c_l \Phi_i|}{\|\Phi_i\| \, \|c_l\|} \tag{2.9}$$

where b_k is the kth column of input matrix B (referring to the kth input) and c_l is the lth row of output matrix C (referring to the lth output). $|z|$ and $\|z\|$ are defined as the modulus and Euclidean norm of z which sets m_{ci} and m_{oi} between 0 and 1. $\alpha(\Psi_i, b_k)$ is the geometrical angle between the input vector b_k and the ith left eigenvector, while $\alpha(c_l, \Phi_i)$ is the geometrical angle between the output vector c_l and the ith right eigenvector. Using (2.8) and (2.9), the joint controllability observability measure is defined as:

$$m_{coi}(k, l) = m_{ci}(k).m_{oi}(l) \tag{2.10}$$

It is feasible to control the mode i with the input u_k and the measured value y_l, if $m_{coi}(k)$ is different from zero. Pairs of u_k and y_l with the maximum m_{coi} will be the most efficient choice. Figure 2.9 (a)–(c) shows the observability, controllability, and joint controllability/ observability measures of speed deviations and excitation control inputs of inter-area mode for the four generators in the first scenario. Based on the calculated geometric measures the inter-area mode is best controllable and best observable from G_4. This will result in a decentralized measurement-control loop. However, if we search for a remote measurement, based on the joint controllability/observability measure, $\Delta\omega_3$ and excitation control of G_4 are the second most efficient control loop. Analysis of the geometric measures of rotor angles will also produce the same result. Thus, the control loop suggested by our distributed control design algorithm is compatible with the traditional modal analysis.

In the second scenario, to evaluate the controller ability in finding efficient wide-area control loops we modified the test system. The inertia constant of G_4 is tripled and G_4 is considered as an equivalent area without a definitive point of control. Then we solved the group-sparse linear quadratic optimal control problem (2.5) and gradually increased λ to obtain a sparse feedback matrix with one communication link. Choice of groups and their weights is similar to scenario 1. Figure 2.8 (b) shows the resulting communication structure and respective feedback control sparsity pattern. This result suggests measurements $[\Delta\delta_4, \Delta\omega_4]$ and the excitation control input G_2 as the effective measurement-control pair to improve the damping of the inter-area mode. Then the geometric measures for this scenario were calculated. Figure 2.9 (d)–(f) shows the observability, controllability, and joint controllability/observability measures of speed deviations and excitation control inputs of inter-area mode for the four generators in the second scenario. By removing the control on G_4, the inter-area mode has the best controllability from G_2 and best observability from $\Delta\omega_4$. Therefore, the results from the proposed framework are again compatible with the traditional modal analysis. The proposed distributed control design not only considers the communication costs but also eliminates the need to perform additional modal analysis computations for finding best measurement-control pairs for a given system.

The results for the two scenarios on two-area four-machine system show that the proposed algorithm can successfully find the efficient measurement-control loops and has the potential to be considered as an alternative approach to traditional modal analysis methods for this purpose. Our experience with this method on larger test

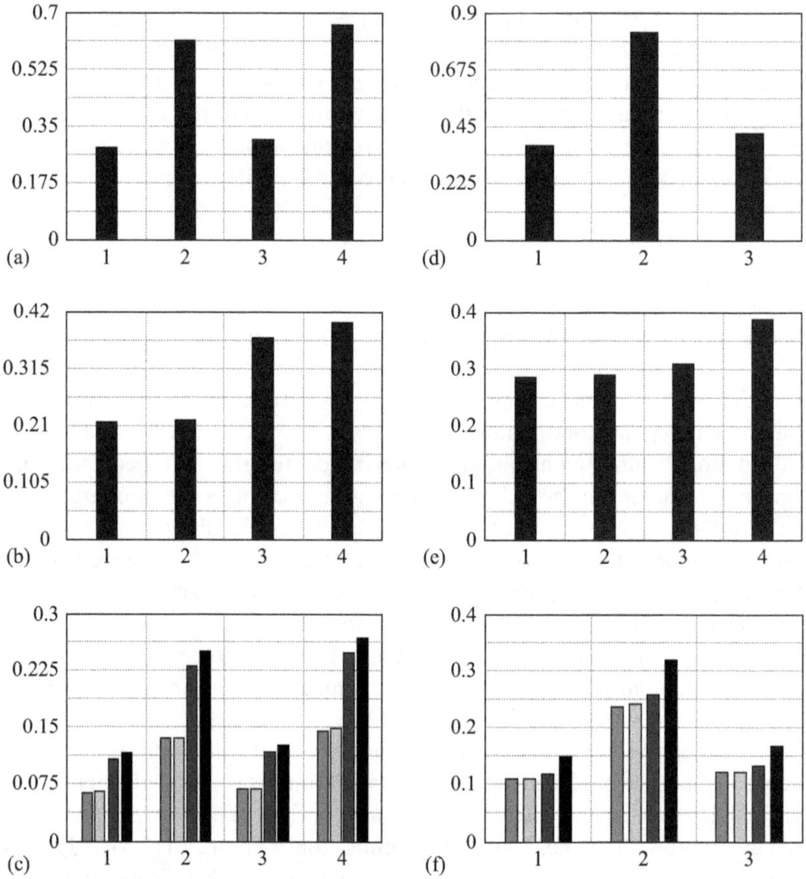

Figure 2.9 Geometric measure of speed deviation measurements and excitation control inputs for inter-area mode for two scenarios in two-area four-machine test systems, Scenario1: (a) Controllability measure, (b) Observability measure, and (c) Joint controllability/observability measure; Scenario 2: (d) Controllability measure, (e) Observability measure, and (f) Joint controllability/observability measure

systems with multiple low-damped or unstable inter-area modes reveals the presence of another important factor in the design which is the interaction between different control loops that may adversely affect each other. In that case, the objective function and group structures should be chosen based on the desired performance that might require damping a particular mode or increasing the overall damping in the system. The interaction measure introduced in Reference 43 provides a good criterion for validating the results.

2.6 Conclusion

Development of trustworthy CPS has been a challenging problem due to their complex nature, especially in the context of critical infrastructure such as the power grid. This chapter addressed one of the main stability problems in power systems and proposed a new control scheme considering the cyber-physical aspects of the grid. Low-damped inter-area oscillations are one of the major causes of stability concerns in power systems. To improve the damping of these low-frequency oscillations, a novel wide-area control scheme has been presented in this chapter. The proposed distributed optimal control framework optimizes a standard performance criterion while limiting the amount of required communication. The newly developed approach provides the ability to encode prior knowledge about the underlying system into the control design objective using group sparse regularization functions. In power systems, this prior knowledge can be a predefined communication structure, measurement availabilities or network variables and dynamic features that can observe or affect the inter-area modes. The grouping approach will greatly expand the application domain of the algorithm and the flexibility of the control design. Two applications of the proposed framework have been introduced and evaluated in this chapter. First, the grouping objective was used to select a communication structure based on system constraints. Second, a control and communication codesign problem was solved to find efficient measurement and control signal pairs to improve the damping of inter-area modes. For this application, groups were defined based on measurement availabilities. Our results suggest that the proposed method opens new possibilities in designing wide-area type controls by including knowledge of the system constraints and characteristics into the design objective. It can also act as an alternative approach to modal analysis methods in finding effective measurement-control loops in the system. Further investigations can explore the robustness of the designed controller to communication failures. Implementation of the proposed method on a large system like WECC can help to unveil interesting relations between sparsity of the physical layer and the communication layer in power systems.

Nomenclature

ω_i	Rotational speed of the rotor, in rad/s
ω_s	Base angular speed, in rad/s
δ_i	Rotor electrical angle, in rad (subscript denotes the *i*th generator)
P_{e_i},P_{m_i}	Electrical and mechanical power, in pu
H_i	Inertia constant, in seconds
E'_{q_i}	q-axis voltage, in pu
E_{fd_i}	Field voltage, in pu
E_{t_i}	Generator terminal voltage, in pu
X_{d_i},X'_{d_i}	d-axis reactance and d-axis transient reactance, in pu

K_{A_i}	Generator excitation gain
V_{ref_i}	Reference voltage, in pu
T'_{d0_i}, T_{A_i}	d-axis transient time constant and generator excitation time constant, in seconds
I_{d_i}	d-axis current, in pu
V_{DLQR_i}	Proposed Distributed LQR output, in pu
f_s	Synchronous frequency, in Hz
K_{stab}	Power system stabilizer gain
T_w	Washout filter time constant, in seconds
T_1, T_2, T_3, T_4	Phase compensation block time constants, in seconds

Acknowledgment

The authors gratefully acknowledge the support of the National Science Foundation through grant CNS-1239366 and support by the Engineering Research Center Program of the National Science Foundation and the Department of Energy under NSF Award Number EEC-1041877 and the CURENT Industry Partnership Program.

References

[1] M. Amin, "Challenges in reliability, security, efficiency, and resilience of energy infrastructure: toward smart self-healing electric power grid," in *Power and Energy Society General Meeting - Conversion and Delivery of Electrical Energy in the 21st Century, 2008 IEEE*, July 2008, pp. 1–5.

[2] K. Moslehi and R. Kumar, "A reliability perspective of the smart grid," *Smart Grid, IEEE Transactions on*, June 2010, vol. 1, no. 1, pp. 57–64.

[3] J. Paserba, "Analysis and control of power system oscillations," CIGRE special publication, *Technical Brochure no. 111*, January 1996.

[4] Z. Huang, N. Zhou, F. Tuffner, *et al.*, "Improving small signal stability through operating point adjustment," in *Power and Energy Society General Meeting, 2010 IEEE*, July 2010, pp. 1–8.

[5] D. N. Kosterev, C. Taylor, and W. Mittelstadt, "Model validation for the August 10, 1996 WSCC system outage," *Power Systems, IEEE Transactions on*, August 1999, vol. 14, no. 3, pp. 967–979.

[6] J. Hauer, H. Lee, J. Burns, and R. Baker, "Preliminary analysis of western system oscillation event on June 4, 2003: BPA and Canada," *Working Note for the WECC Disturbance Monitoring Work Group*, 2003.

[7] J. Hauer, "Preliminary examination of the Alberta trip on August 4, 2000," *Working Note for the WECC Modeling and Validation Work Group*, 2002.

[8] S. Juan, P. John, C. Charles, K. Prabha, and L. Einar, *Small-Signal Stability and Power System Oscillations*. CRC Press, Boca Raton, FL, 2007.

[9] P. Kundur, D. Lee, and H. Zein El-Din, "Power system stabilizers for thermal units: analytical techniques and on-site validation," *Power Apparatus and Systems, IEEE Transactions on*, January 1981, vol. PAS-100, no. 1, pp. 81–95.

[10] P. Kundur, N. Balu, and M. Lauby, *Power System Stability and Control*. McGraw-Hill, New York, NY, 1994.

[11] P. Kundur, M. Klein, G. Rogers, and M. Zywno, "Application of power system stabilizers for enhancement of overall system stability," *IEEE Transactions on Power Systems*, May 1989, vol. 4, no. 2, pp. 614–626.

[12] B. Pal and B. Chaudhuri, *Robust Control in Power Systems*. Springer Science & Business Media, Springer US, 2006.

[13] M. Aboul-Ela, A. Sallam, J. McCalley, and A. Fouad, "Damping controller design for power system oscillations using global signals," *Power Systems, IEEE Transactions on*, May 1996, vol. 11, no. 2, pp. 767–773.

[14] Z. Wang, C. Chung, K. Wong, and C. Tse, "Robust power system stabilizer design under multi-operating conditions using differential evolution," *IET Generation, Transmission & Distribution*, September 2008, vol. 2, pp. 690–700.

[15] Y. Zhang and A. Bose, "Design of wide-area damping controllers for interarea oscillations," *Power Systems, IEEE Transactions on*, August 2008, vol. 23, no. 3, pp. 1136–1143.

[16] A. Chakrabortty, "Wide-area damping control of power systems using dynamic clustering and TCSC-Based redesigns," *Smart Grid, IEEE Transactions on*, September 2012, vol. 3, no. 3, pp. 1503–1514.

[17] L. Peng, W. Xiaochen, L. Chao, *et al.*, "Implementation of CSG's wide-area damping control system: Overview and experience," in *Power Systems Conference and Exposition, 2009. PSCE '09. IEEE/PES*, March 2009, pp. 1–9.

[18] D. Dotta, A. e Silva, and I. Decker, "Wide-area measurements-based two-level control design considering signal transmission delay," *Power Systems, IEEE Transactions on*, February 2009, vol. 24, no. 1, pp. 208–216.

[19] H. Ni, G. Heydt, and L. Mili, "Power system stability agents using robust wide area control," *Power Systems, IEEE Transactions on*, November 2002, vol. 17, no. 4, pp. 1123–1131.

[20] I. Kamwa, R. Grondin, and Y. Hebert, "Wide-area measurement based stabilizing control of large power systems-a decentralized/hierarchical approach," *Power Systems, IEEE Transactions on*, February 2001, vol. 16, no. 1, pp. 136–153.

[21] B. Bamieh, F. Paganini, and M. Dahleh, "Distributed control of spatially invariant systems," *Automatic Control, IEEE Transactions on*, July 2002, vol. 47, no. 7, pp. 1091–1107.

[22] R. D'Andrea and G. Dullerud, "Distributed control design for spatially interconnected systems," *Automatic Control, IEEE Transactions on*, September 2003, vol. 48, no. 9, pp. 1478–1495.

[23] G. Dullerud and R. D'Andrea, "Distributed control of heterogeneous systems," *Automatic Control, IEEE Transactions on*, December 2004, vol. 49, no. 12, pp. 2113–2128.

[24] M. Mahmoudi, J. Dong, K. Tomsovic, and S. Djouadi, "Application of distributed control to mitigate disturbance propagations in large power networks," in *North American Power Symposium (NAPS), 2015*, October 2015, pp. 1–6.

[25] M. Mahmoudi and K. Tomsovic, "A distributed control design methodology for damping critical modes in power systems," in *Power and Energy Conference at Illinois (PECI), 2016 IEEE*.

[26] M. Fardad, F. Lin, and M. Jovanovic, "Sparsity-promoting optimal control for a class of distributed systems," in *American Control Conference (ACC), 2011*, June 2011, pp. 2050–2055.

[27] F. Dorfler, M. Jovanovi, M. Chertkov, and F. Bullo, "Sparsity-promoting optimal wide-area control of power networks," *Power Systems, IEEE Transactions on*, September 2014, vol. 29, no. 5, pp. 2281–2291.

[28] P. Sauer and M. Pai, *Power System Dynamics and Stability*. Prentice Hall, Upper Saddle River, NJ, 1998.

[29] A. Chakrabortty and J. Chow, "Measurement-based methods for model reduction of power systems using synchrophasors," in *Power System Coherency and Model Reduction*, J. H. Chow, Ed. Springer, New York, 2013, vol. 94, pp. 159–197.

[30] J. Chow, *Time-scale Modeling of Dynamic Networks with Applications to Power Systems*, ser. Lecture notes in control and information sciences. Springer-Verlag, Berlin, 1982.

[31] A. Chakrabortty, J. Chow, and A. Salazar, "A measurement-based framework for dynamic equivalencing of large power systems using wide-area phasor measurements," *Smart Grid, IEEE Transactions on*, March 2011, vol. 2, no. 1, pp. 68–81.

[32] R. Eriksson and L. Soder, "Wide-area measurement system-based subspace identification for obtaining linear models to centrally coordinate controllable devices," *Power Delivery, IEEE Transactions on*, April 2011, vol. 26, no. 2, pp. 988–997.

[33] F. Lin, "Structure identification and optimal design of large-scale networks of dynamical systems," PhD dissertation, The University of Minnesota, December 2012.

[34] V. Gupta. Introduction to distributed control. [Online]. Available: https:// www.cds.caltech.edu/murray/wiki/images/c/c7/Lecture2d_gupta.pdf, May 2006.

[35] D. L. Donoho, "Compressed sensing," *IEEE Transactions on Information Theory*, April 2006, vol. 52, no. 4, pp. 1289–1306.

[36] F. Lin, M. Fardad, and M. Jovanovic, "Design of optimal sparse feedback gains via the alternating direction method of multipliers," *Automatic Control, IEEE Transactions on*, September 2013, vol. 58, no. 9, pp. 2426–2431.

[37] U. Munz, M. Pfister, and P. Wolfrum, "Sensor and actuator placement for linear systems based on h_2 and h_∞ optimization," *Automatic Control, IEEE Transactions on*, November 2014, vol. 59, no. 11, pp. 2984–2989.

[38] R. Jenatton, J. Mairal, G. Obozinski, and F. Bach, "Proximal methods for hierarchical sparse coding," *Journal of Machine Learning Research*, July 2011, vol. 12, pp. 2297–2334.

[39] A. Taalimi, S. Ensafi, H. Qi, S. Lu, A. Kassim, and C. Tan, "Multimodal dictionary learning and joint sparse representation for hep-2 cell classification," in *Medical Image Computing and Computer-Assisted Intervention MICCAI 2015*, N. Navab, J. Hornegger, M. W. Wells, and F. A. Frangi, Eds. Springer International Publishing, Switzerland, 2015, vol. 9351, pp. 308–315.

[40] A. Taalimi, H. Qi, and R. Khorsandi, "Online multi-modal task-driven dictionary learning and robust joint sparse representation for visual tracking," in *Advanced Video and Signal Based Surveillance (AVSS), 2015 12th IEEE International Conference on*, August 2015, pp. 1–6.

[41] K. Zhou and J. Doyle, *Essentials of Robust Control*. vol. 104, Prentice Hall, Upper Saddle River, NJ, 1998.

[42] N. Parikh and S. Boyd, "Proximal algorithms," *Foundations and Trends in Optimization*, January 2014, vol. 1, no. 3, pp. 127–239.

[43] A. Heniche and I. Kamwa, "Control loops selection to damp inter-area oscillations of electrical networks," *Power Systems, IEEE Transactions on*, May 2002, vol. 17, no. 2, pp. 378–384.

[44] H. Hamdan and A. Hamdan, "On the coupling measures between modes and state variables and subsynchronous resonance," *Electric Power Systems Research*, 1987, vol. 13, no. 3, pp. 165–171.

Chapter 3

Distributed algorithms for wide-area monitoring of power systems: a cyber-physical perspective

Jianhua Zhang[1], *Seyedbehzad Nabavi*[2], *Aranya Chakrabortty*[1] *and Yufeng Xin*[3]

Abstract

This chapter develops a distributed algorithmic framework for executing critical transmission-level operations of power systems using Synchrophasor data. As the number of phasor measurement units continues to increase exponentially, it is rather intuitive that the current state-of-the-art centralized communication and information processing architecture of wide-area measurement system will no longer be sustainable under such data-explosion, and a completely distributed cyber-physical architecture will need to be developed. Very little attention has yet been paid to the most critical consequence of this envisioned distributed architecture – namely, distributed algorithms. Our primary task in this chapter is to describe how one can develop distributed optimization methods for solving real-time wide-area monitoring problems with investigation of their convergence, robustness, and implementation issues.

3.1 Introduction

Following the Northeast blackout of 2003, Wide-Area Measurement System (WAMS) technology using Phasor Measurement Units (PMUs) has largely matured for the North American grid [1]. However, as the number of PMUs scales up to the thousands in the next few years under the US Department of Energy's smart grid demonstration initiative, Independent System Operators (ISOs) and utility companies are struggling to understand how the resulting gigantic volumes of real-time data can be efficiently harvested, processed, and utilized to solve wide-area monitoring and control problems for any realistic power system interconnection. It is intuitive that the current

[1]Department of Electrical and Computer Engineering, North Carolina State University, Raleigh, NC, USA
[2]New York Power Authority, White Plains, NY, USA
[3]Renaissance Computing Institute, Chapel Hill, NC, USA

state-of-the-art centralized communication and information processing architecture of WAMS will no longer be sustainable under such a data explosion, and a completely distributed cyber-physical architecture will need to be developed [2]. In the Eastern Interconnection (EI) of the US grid, e.g., about 60 PMUs are currently streaming data via the Internet to a super phasor data concentrator (SPDC) which is handling about 100,000 data points per second. This architecture will no doubt become untenable as the EI scales up to 300–400 PMUs by 2015. Research is currently being carried out by the Data and Network Management Task Team (DNMTT) of North American Synchrophasor Initiative (NASPI) on the implementation of this distributed architecture with the prime research focus being protocols, Quality-of-Service (QoS), latency, bandwidth, and security [3].

However, very little attention has yet been paid to perhaps two of the most critical consequences of this envisioned distributed architecture, namely, (1) formulation of traditional centralized strategies for wide-area monitoring and control as distributed algorithms and (2) investigation of numerical stability and convergence properties of these algorithms when they are implemented in a completely asynchronous environment that is bound to arise in any practical wide-area communication network [4,5]. Partly due to a lack of a cyber-physical research infrastructure and partly due to the priorities set forth by PMU installations, the NASPI community has not yet delved into investigating how the currently used centralized algorithms for wide-area monitoring and control can be translated into a distributed computing framework once the aforementioned decentralized WAMS architecture is realized in the coming years [6]. Development of such algorithms will obviously be imperative not only for increasing reliability by eliminating single-point failures, but also for minimizing network transit. As shown in Reference 7, transmitting data across a wide-area communication network (WAN) is expensive, the links can be relatively slow, and the bandwidth-per-dollar will indeed grow slower than other computing resources leading to distributed PMU data processing as a natural choice.

Motivated by this challenge, in this chapter we first propose three different distributed communication and computational architectures to address problem (1) and secondly address problem (2) by focusing on the fundamental question on how asynchrony in a communication network can influence the performance for one of the most critical wide-area monitoring applications, namely, modal estimation of electromechanical oscillations using synchrophasors. Several centralized algorithms for solving this problem have been proposed over the past decade including the Eigenvalue Realization Algorithm (ERA) [8], Prony analysis [9], mode metering [10], and Hilbert-Huang transform [11]. However, all of these algorithms are based on offline techniques, and that too using only a handful (but, observable) set of PMUs. In contrast, we formulate the mode estimation problem as a global consensus problem for the coefficients of the characteristic polynomial of the system, and then solve it using Alternating Direction Method of Multipliers (ADMM). The communication network required to execute this distributed estimation is shown in Figure 3.1. The physical grid is assumed to be divided into multiple balancing regions or *areas*, which may or may not be coherent, but belong to different utility companies. PMUs in each area communicate their data in real-time to estimator(s) or Phasor Data Concentrators

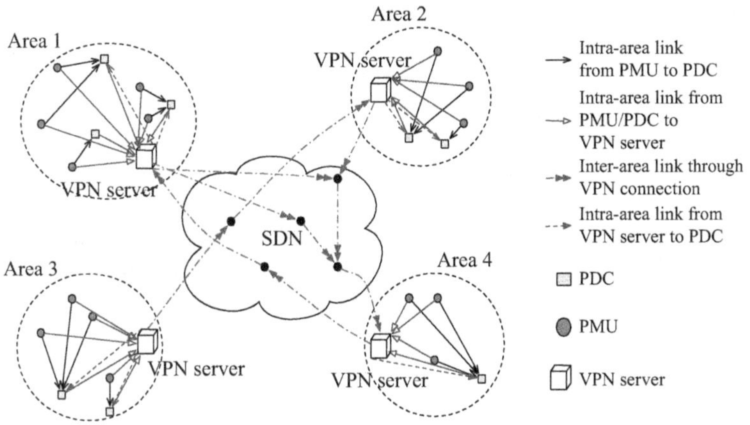

Figure 3.1 Distributed architecture for wide-area PMU-PDC communications

(PDC) located at the local control center via a Virtual Private Network (VPN). These local PDCs can then share information between each other and also with a central PDC located at the ISO through a controllable wide-area network such as a Software Defined Network (SDN). The key idea is to make use of this distributed network protocol to run local consensus at the PDCs inside each area, iteratively generate *myopic* estimates of the coefficients of the characteristic polynomial, and let the PDCs communicate either with each other or with the central PDC to reach a global solution using several variants of the ADMM [12]. Both synchronous and asynchronous communications are considered.

Our proposed framework clearly demonstrates how ADMM can be a beneficial tool for distributed mode estimation in power systems, and what types of performance bottlenecks, accuracy issues, and computation delays it may result in. The innovation of this work is, therefore, in proposing a bridge between the cyber and physical implementation of distributed WAMS. We illustrate our algorithms via offline simulations of two IEEE prototype power system models, and discuss the benefits and drawbacks of each algorithm in light of security and data privacy. Distributed consensus algorithms in power systems have been reported in recent papers such as [13–17], but mostly in the context of distributed optimal power flow, distributed generation, and demand-side management, and not for wide-area oscillation monitoring. Preliminary results on the first architecture proposed in this chapter have recently been reported in our conference papers [18,19]. However, the results outlined in this chapter are significantly more expansive than those in References 18 and 19 including two new architectures with unique sets of distributed algorithms, a discussion of their convergence properties, the pertinent issue of asynchronous communication in real-world SDNs.

Besides cyber-physical architectures for distributed estimation, another contribution of this chapter is to develop a number of heuristic variants of the ADMM algorithm for mitigating the perennial problem of asynchrony in wide-area communication networks. We consider a probabilistic traffic model for modeling delays in the wide-area networks, and study how these delays enter the process of information exchange between the PDCs. We propose four different update rules by which the convergence rate of ADMM and the accuracy of estimation can be made immune to the asynchrony resulting from the network traffic as much as possible. We carry out extensive simulations using an IEEE prototype power system model to show possible numerical instabilities and sensitivities of the ADMM convergence on our proposed strategies. Preliminary results on two of the proposed algorithms, considering only uni-directional delays, have recently been reported in our conference paper [20]. This chapter extends those initial findings to two new strategies, each with multiple sub-scenarios, and considers the combined impact of both uplink and downlink delays in asynchronous ADMM.

3.2 Problem formulation

We consider a power system consisting of n generator buses and n_l load buses. Each synchronous generator is modeled by a second-order swing equation, while each load bus is modeled by two algebraic equations for active and reactive power balance. We convert this differential-algebraic model to a completely differential model using standard techniques of Kron reduction [21] and arrive at a linearized state variable model for the n-machine system as:

$$\left[\begin{array}{c}\Delta\dot{\delta}(t)\\ \hline \Delta\dot{\omega}(t)\end{array}\right] = \underbrace{\left[\begin{array}{c|c}\mathbf{0}_{n\times n} & \omega_s I_n\\ \hline M^{-1}L & -M^{-1}D\end{array}\right]}_{A}\left[\begin{array}{c}\Delta\delta(t)\\ \hline \Delta\omega(t)\end{array}\right],$$

$$\mathbf{y}(t) = [\Delta\theta_1(t), \ldots, \Delta\theta_p(t)]^T \tag{3.1}$$

where $\Delta\delta = [\Delta\delta_1, \ldots, \Delta\delta_n]^T$, $\Delta\omega = [\Delta\omega_1, \ldots, \Delta\omega_n]^T$, $M = \mathrm{diag}(M_1, \ldots, M_n)$, and $D = \mathrm{diag}(D_1, \cdots, D_n)$, with $\Delta\delta_i$, $\Delta\omega_i$, M_i, and D_i being the small-signal angle deviation, the small-signal frequency deviation, inertia, and mechanical damping of generator i, respectively. I_n is the $(n \times n)$ identity matrix, and ω_s is the synchronous speed of the system. The Laplacian matrix represents the connectivity of the areas in the equivalent topology whose elements, neglecting the line resistances, are defined as

$$[L]_{ij} = \frac{E_i E_j}{x_{ij}} \cos(\delta_{i0} - \delta_{j0}), \quad [L]_{ii} = -\sum_k [L]_{ik} \tag{3.2}$$

where x_{ij} is the equivalent reactance between generators i and j in the Kron-reduced form, E_i is the internal voltage, and δ_{i0} is the equilibrium angle of generator i. We consider the output vector $\mathbf{y}(t)$ to be a set of phase angle measurements $\Delta\theta_i(t)$, $i = 1, \ldots, p$, measured by PMUs at p designated buses. Other outputs such as bus voltages and frequencies may also be considered but we restrict our analysis to phase angles only. The eigenvalues of A are denoted by $(-\sigma_l \pm j\Omega_l)$. Our objective is to estimate these $2n$ eigenvalues of A from $\mathbf{y}(t)$ in a distributed fashion using multiple computational resources. For this purpose, we next describe how the commonly used Prony algorithm for modal estimation can be cast as a distributed optimization problem. We first recall the centralized Prony algorithm, and thereafter reformulate it as a distributed algorithm using three different cyber-physical architectures.

3.3 Modal estimation using Prony method

A generic expression for the solution of $\Delta\theta_i(t)$ in (3.1) can be written as

$$\Delta\theta_i(t) = \sum_{l=1}^{n} \left(r_{il} e^{(-\sigma_l + j\Omega_l)t} + r_{il}^{*} e^{(-\sigma_l - j\Omega_l)t} \right) \tag{3.3}$$

Each component in the RHS of (3.3) is referred to as a *mode*, where r_{il} is the residue of mode l reflected in the ith output. Sampling $\Delta\theta_i(t)$ with a uniform sampling period of T, a generic expression for the z-transform of $\Delta\theta_i(k) \triangleq \Delta\theta_i(t)|_{t=kT}$, $(k = 0, 1, \ldots, m-1)$, with m being the total number of measured samples, can be rewritten as

$$\Delta\theta_i(z) = \frac{b_{i0} + b_{i1}z^{-1} + b_{i2}z^{-2} + \cdots + b_{i2n}z^{-2n}}{1 + a_1 z^{-1} + a_2 z^{-2} + \cdots + a_{2n}z^{-2n}} \tag{3.4}$$

The zeros of the transfer function in (3.4) are dependent on the output index i. The poles, however, are identical for every output, and, therefore, independent of i. We recall that our objective is to estimate σ_l, Ω_l, and r_{il} in (3.3). We next state three steps of the Prony algorithm by which this can be achieved from (3.4) [9].

Step 1. The first step of the Prony algorithm is to find a_1 through a_{2n} by solving

$$\underbrace{\begin{bmatrix} \Delta\theta_i(2n) \\ \Delta\theta_i(2n+1) \\ \vdots \\ \Delta\theta_i(2n+\ell) \end{bmatrix}}_{\mathbf{c}_i} = \underbrace{\begin{bmatrix} \Delta\theta_i(2n-1) & \cdots & \Delta\theta_i(0) \\ \Delta\theta_i(2n) & \cdots & \Delta\theta_i(1) \\ \vdots & & \vdots \\ \Delta\theta_i(2n+\ell-1) & \cdots & \Delta\theta_i(\ell) \end{bmatrix}}_{H_i} \underbrace{\begin{bmatrix} -a_1 \\ -a_2 \\ \vdots \\ -a_{2n} \end{bmatrix}}_{\boldsymbol{a}}, \tag{3.5}$$

where ℓ is an integer satisfying $2n + \ell \leq m - 1$. Let us concatenate $\mathbf{c}_i$ and H_i in (3.5) for PMU $i = 1, \ldots, p$. One can find $\boldsymbol{a}$ by solving a least squares (LS) problem defined as

$$\min_{\boldsymbol{a}} \frac{1}{2} \left\| \begin{bmatrix} H_1 \\ \vdots \\ H_p \end{bmatrix} \boldsymbol{a} - \begin{bmatrix} \mathbf{c}_1 \\ \vdots \\ \mathbf{c}_p \end{bmatrix} \right\|^2 \tag{3.6}$$

where $\|\cdot\|$ denotes the two-norm of a vector.

Step 2. Once $\boldsymbol{a}$ is computed, the next step is to find the roots of the discrete-time characteristic polynomial as shown in the denominator of (3.4). Let these roots be denoted by $\bar{z}_l$, $l = 1, \ldots, 2n$. Finally, the eigenvalues of A in (3.1) can be calculated as $\ln(\bar{z}_l)/T$.

Step 3. The final step is to find the residues r_{il} in (3.3). This can be done by forming the following *Vandermonde* equation, and solving it for the residues $r_{i1}, \ldots, r_{in}$:

$$\begin{bmatrix} \Delta\theta_i(0) \\ \Delta\theta_i(1) \\ \vdots \\ \Delta\theta_i(m) \end{bmatrix} = \begin{bmatrix} 1 & 1 & \cdots & 1 \\ (\overline{z}_1)^{1/T} & (\overline{z}_2)^{1/T} & \cdots & (\overline{z}_{2n})^{1/T} \\ \vdots & \vdots & & \vdots \\ (\overline{z}_1)^{m/T} & (\overline{z}_2)^{m/T} & \cdots & (\overline{z}_{2n})^{m/T} \end{bmatrix} \begin{bmatrix} r_{i1} \\ r_{i1}^* \\ \vdots \\ r_{in} \\ r_{in}^* \end{bmatrix} \tag{3.7}$$

Note 1: The method cannot estimate the order of the system $2n$. If n is not known apriori, it is usually considered to be a large number, and thereafter the modes with negligible residues are discarded. However, n is limited by the number of available measurements as well as the computational memory.

Note 2: The method can be performed in real time. That is, one may solve (3.6) while iteratively updating H_i, and $\mathbf{c}_i$ as new measurements become available.

3.4 Proposed architectures for distributed modal estimation

In this section, we show how Step 1 of the centralized approach delineated in Section 3.3 can be recast as a distributed optimization problem using intra- and inter-regional PMU-PDC architectures. We wish to clarify at the very outset that the term "PDC" in our algorithms is used in a much broader sense, and not just as a data aggregator. It essentially refers to any computing agent that can process PMU data and run algorithms on them. In other words, as long as the data from a given PMU is communicated to a computing station, whether it be a hardware PDC, software PDC, a local server, or even a *data center* – we would collectively refer to all of these processing units as a "PDC" for convenience. We assume the CPS infrastructure of the grid to be divided into five distinct layers. The lowest layer contains the stochastic variations in loads and events due to nature and human activities. The second level is the physical power system model. The third and fourth layers consist of real-time PMU data processing (and computation) at the substation level and the control center level, respectively. Between the second, third, and fourth layers, data communication architecture is configured to handle massive amounts of PMU data using the network shown in Figure 3.1. The topmost layer is the application layer, which for our purpose is the estimation of the eigenvalues of the matrix A in (3.1) using the distributed communication protocol of this network. If necessary, an internal hierarchy of multiple area-level decentralization layers can also be created (in fact, our third architecture is based on this assumption). Too many sub-layers, however, can lead to unacceptable processing delays and latency violations. Based on these intuitions, we next describe our proposed computational architectures.

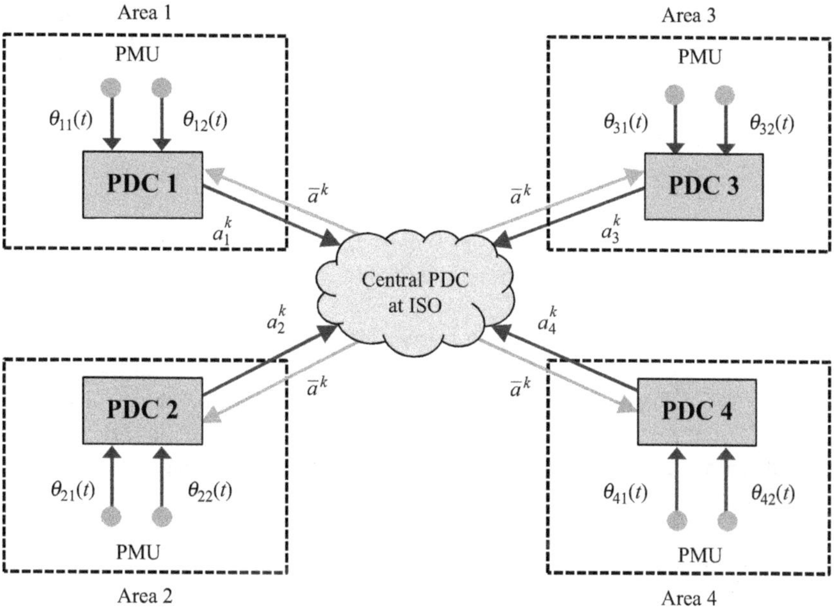

Figure 3.2 Architecture 1 using S-ADMM for a four-area network

3.4.1 Architecture 1: distributed Prony using standard ADMM (S-ADMM)

The LS problem (3.6) can be regarded as a *global consensus problem* over a network of N utility companies or areas. We assume every area to be equipped with one aggregated PDC as shown in Figure 3.2. The consensus problem can be described as

$$\underset{\boldsymbol{a}_1,\ldots,\boldsymbol{a}_N,\mathbf{z}}{\text{minimize}} \quad \sum_{j=1}^{N} \frac{1}{2} \|\hat{H}_j \boldsymbol{a}_j - \hat{\mathbf{c}}_j\|^2, \tag{3.8}$$

$$\text{subject to } \boldsymbol{a}_j - \mathbf{z} = 0, \ \text{for } j = 1, \ldots, N$$

Here, $\hat{H}_j \triangleq [H_{j,1}^T \ H_{j,2}^T \ \ldots \ H_{j,N_j}^T]^T$, $\hat{\mathbf{c}}_j \triangleq [\mathbf{c}_{j,1}^T \ \mathbf{c}_{j,2}^T \ \cdots \ \mathbf{c}_{j,N_j}^T]^T$, where N_j is the total number of PMUs in Area j, and $H_{j,i}$ and $\mathbf{c}_{j,i}$ are constructed as in (3.5) from the time samples of $\Delta\theta_{j,i}$, which is the ith PMU measurement in Area j, $i = 1, \ldots, N_j$. The global consensus solution, denoted by $\mathbf{z} \in \mathbb{R}^{2n}$, is the solution of (3.6) that is obtained when the local estimates of the N regional PDCs, denoted by $\boldsymbol{a}_j, j = 1, \ldots, N$, reach the same value.

The standard ADMM (S-ADMM) estimation method uses the Lagrangian multiplier approach to solve (3.8) in an iterative, distributed way [12]. The *augmented Lagrangian* for (3.8) is:

$$L_\rho = \sum_{j=1}^{N} \left(\frac{1}{2} \| \hat{H}_j \boldsymbol{a}_j - \hat{\mathbf{c}}_j \|^2 + \mathbf{w}_j^T (\boldsymbol{a}_j - \mathbf{z}) + \frac{\rho}{2} \| \boldsymbol{a}_j - \mathbf{z} \|^2 \right), \tag{3.9}$$

where $\boldsymbol{a}_j$ and $\mathbf{z}$ are the vectors of the *primal* variables, $\mathbf{w}_j$ is the vector of the *dual variables* or the Lagrange multipliers associated with (3.8), and $\rho > 0$ denotes a *penalty factor*.

Before the S-ADMM algorithm starts, the central PDC fixes the order of the system $2n$, and the initial height of the Hankel blocks $(\ell + 1)$ for all local PDCs as shown in (3.5). Each local PDC j (i.e., the PDC located in Area j) then waits until the $(2n + \ell)^{th}$ sample of the measurement arrives. In order to ensure the real-time nature of the algorithm, at iteration k, each local PDC j constructs the matrices $\hat{H}_j^{(k)}$ and $\hat{\mathbf{c}}_j^{(k)}$ as follows:

$$\hat{H}_j^{(k)} \triangleq [(H_{j,1}^{(k)})^T \cdots (H_{j,N_j}^{(k)})^T]^T, \; \hat{\mathbf{c}}_j^{(k)} \triangleq [(\mathbf{c}_{j,1}^{(k)})^T \cdots (\mathbf{c}_{j,N_j}^{(k)})^T]^T \tag{3.10}$$

where

$$H_{j,i}^{(k)} \triangleq \begin{bmatrix} \Delta\theta_{j,i}(2n-1) & \cdots & \Delta\theta_{j,i}(0) \\ \Delta\theta_{j,i}(2n) & \cdots & \Delta\theta_{j,i}(1) \\ \vdots & & \vdots \\ \Delta\theta_{j,i}(m^{(k-1)}) & \cdots & \Delta\theta_{j,i}(m^{(k-2n)}) \end{bmatrix} \tag{3.11a}$$

$$\mathbf{c}_{j,i}^{(k)} \triangleq \left[\Delta\theta_{j,i}(2n) \; \Delta\theta_{j,i}(2n+1) \; \cdots \; \Delta\theta_{j,i}(m^{(k)}) \right]^T \tag{3.11b}$$

for $i = 1, \ldots, N_j$. Here, $\Delta\theta_{j,i}(m^{(k)})$ is the most recent measurement sample available to PDC j from the ith PMU in its area at iteration k. Using (3.10) and (3.11), the S-ADMM algorithm can be written as:

Distributed Prony using standard ADMM

1. Each PDC j initializes $\boldsymbol{a}_j^{(0)}$, $\mathbf{z}^{(0)}$, and $\mathbf{w}_j^{(0)}$, $j = 1, \ldots, N$.
2. At iteration k:
   a. PDC j constructs $\hat{H}_j^{(k)}$ and $\hat{\mathbf{c}}_j^{(k)}$ from (3.10).
   b. PDC j updates $\boldsymbol{a}_j$ as:

$$\boldsymbol{a}_j^{(k+1)} = \arg\min_{\boldsymbol{a}_j} L_\rho = ((\hat{H}_j^{(k)})^T \hat{H}_j^{(k)} + \rho I_{2n})^{-1} ((\hat{H}_j^{(k)})^T \hat{\mathbf{c}}_j^{(k)} - \mathbf{w}_j^{(k)} + \rho \mathbf{z}^{(k)}) \tag{3.12}$$

   c. PDC j transmits $\boldsymbol{a}_j^{(k+1)}$ to the central PDC.

d. Central PDC calculates

$$\mathbf{z}^{(k+1)} \triangleq \bar{\boldsymbol{a}}^{(k+1)} = \frac{1}{N}\sum_{j=1}^{N} \boldsymbol{a}_j^{(k+1)} \tag{3.13}$$

e. Central PDC broadcasts $\mathbf{z}^{(k+1)}$ to all local PDCs.

f. PDC j updates $\mathbf{w}_j$ as:

$$\mathbf{w}_j^{(k+1)} = \mathbf{w}_j^{(k)} + \rho(\boldsymbol{a}_j^{(k+1)} - \mathbf{z}^{(k+1)}) \tag{3.14}$$

Since (3.8) is a convex optimization problem, as $k \to \infty$, $\mathbf{z}^{(k)}$ in (3.13) converges to the global minimum of (3.8), and so does each individual $\boldsymbol{a}_j^{(k)}$ due to consensus. Let these optimal values be denoted as $\mathbf{z}^*$ and $\boldsymbol{a}_j^*$. Once $\mathbf{z}^* = \boldsymbol{a}_1^* = \cdots = \boldsymbol{a}_N^*$ is calculated, every local PDC can compute the eigenvalues of A using Step 2 of the Prony algorithm described in Section 3.3. It can also compute the mode residues using Step 3.

Note: The conventional ADMM algorithm for consensus problems converges at the rate of $O(1/k)$, as shown in Reference 12. However, unlike Reference 12, where the objective function to be minimized is assumed to be time-invariant, (3.12) in our proposed S-ADMM varies over time. If we hold $\hat{H}_j^{(k)}$ and $\hat{\mathbf{c}}_j^{(k)}$ to be constant at $\hat{H}_j^{(0)}$ and $\hat{\mathbf{c}}_j^{(0)}$, respectively, for all j and k, then the $O(1/k)$ convergence rate can be guaranteed.

3.4.2 Architecture 1 with asynchronous communication (A-ADMM)

One assumption behind Architecture 1 is that all local PDCs are performing their respective optimization steps with equal speed, and the communication delays between the local PDCs and the central PDC are also equal, i.e., they are *synchronous*. However, in reality, the PDCs may not be perfectly synchronized with each other due to differences in their processing speeds as well as due to various communication delays such as routing, queuing, and transfer delays in the SDN shown in Fig. 3.1. One possible solution to overcome this asynchrony is to force the central PDC to wait until it receives data from all local PDCs. In that case, the total end-to-end delay for each iteration will be dependent on the slowest communication link, and hence the entire algorithm may become very slow. An alternative approach would be to use the recently developed method called *Asynchronous ADMM* (A-ADMM) [22,23]. In this method, the central PDC receives the updates only from a subset of the N local PDCs at every iteration k, referred to as *active* PDCs. Let this set be denoted by $\mathscr{S}_{(k)}$. It then calculates $\mathbf{z}^{(k+1)}$ using the most recent local estimates from all PDCs. Let $T^{(k+1)}$ be the time instant at which $\mathbf{z}^{(k+1)}$ is computed. The central PDC then broadcasts ($\mathbf{z}^{(k+1)}$, $T^{(k+1)}$) to every local PDC. Upon receiving $T^{(k+1)}$, each local PDC j then constructs $\hat{H}_j^{(k+1)}$ and $\hat{\mathbf{c}}_j^{(k+1)}$ from (3.10) by setting $m^{(k+1)}$ to be the sample index that is closest to the time instant $T^{(k+1)}$. Note that $\Delta\theta(m^{(k+1)})$ may not be the most recent measurement sample while constructing $\hat{H}_j^{k+1}$ and $\hat{\mathbf{c}}_j^{k+1}$ matrices. However, in order to ensure that all PDCs use the same time-window of the measurements to form $\hat{H}_j^{(k+1)}$ and $\hat{\mathbf{c}}_j^{(k+1)}$, they all use the same value of $m^{(k+1)}$ as decided globally by the central PDC at every iteration $k+1$.

The A-ADMM algorithm adapted for (3.8) can then be written as follows:

Distributed Prony using asynchronous ADMM

1. The central PDC initializes $T^{(0)}$ and sends it to all local PDCs.
2. PDC j initializes $\boldsymbol{a}_j^{(0)}$, $\mathbf{z}^{(0)}$, and $\mathbf{w}_j^{(0)}$, for $j = 1, \ldots, N$.
3. At iteration k:
   a. Given $T^{(k)}$, PDC j constructs $\hat{H}_j^{(k)}$ and $\hat{\mathbf{c}}_j^{(k)}$ from (3.10) using $m^{(k)}$ decided from $T^{(k)}$.
   b. PDC j updates $\boldsymbol{a}_j$ as:

$$\boldsymbol{a}_j^{(k+1)} = ((\hat{H}_j^{(k)})^T \hat{H}_j^{(k)} + \rho I_{2n})^{-1}((\hat{H}_j^{(k)})^T \hat{\mathbf{c}}_j^{(k)} - \mathbf{w}_j^{(k)} + \rho \mathbf{z}^{(k)}) \tag{3.15}$$

   c. PDC j transmits $\boldsymbol{a}_j^{(k+1)}$, $\mathbf{w}_j^{(k)}$, and k to the central PDC.
   d. The central PDC receives the values of $\boldsymbol{a}_j^{(k+1)}$, $\mathbf{w}_j^{(k)}$, and k only from the active PDCs $j \in \mathscr{S}^{(k)}$.
   e. The central PDC updates $\mathbf{z}$ as:

$$\mathbf{z}^{(k+1)} = \frac{1}{N}\sum_{j=1}^{N}(\boldsymbol{a}_j^{(k+1)} + (1/\rho)\mathbf{w}_j^{(k)}) \tag{3.16}$$

   where $\boldsymbol{a}_j^{(k+1)} = \boldsymbol{a}_j^{(k)}$ and $\mathbf{w}_j^{(k)} = \mathbf{w}_j^{(k-1)}$ for $j \notin \mathscr{S}^{(k)}$.
   f. The central PDC broadcasts $\mathbf{z}^{(k+1)}$, $(k+1)$, and $T^{(k+1)}$ to all local PDCs.
   g. PDC j updates $\mathbf{w}_j$ as:

$$\begin{aligned} \mathbf{w}_j^{(k+1)} &= \mathbf{w}_j^{(k)} + \rho(\boldsymbol{a}_j^{(k+1)} - \mathbf{z}^{(k+1)}), \; j \in \mathscr{S}^{(k)}, \\ \mathbf{w}_j^{(k+1)} &= \mathbf{w}_j^{(k)}, \; j \notin \mathscr{S}^{(k)} \end{aligned} \tag{3.17}$$

Note that the iteration numbers k and $k+1$ are communicated between the PDCs in Steps 3c and 3f of the above algorithm to keep track of the order of the receiving data.

This architecture is similar to Architecture 1 shown in Figure 3.2. The only difference is that more information is exchanged between the central and local PDCs. The aforementioned A-ADMM algorithm converges to the minimizer of (3.8) with the rate of $O(1/k)$ if none of the local PDCs is allowed to be dormant all throughout. In other words, for all $\mathscr{S}^k \in \mathscr{P}$, where $\mathscr{P} \triangleq 2^{\{1,\ldots,N\}}$ is the set of all subsets of N PDCs, $\mathscr{S}^k$ must be active *infinitely often* with probability 1 [22].

By virtue of this algorithm, the real-time nature for solving (3.8) can still be maintained despite asynchronous delays in the SDN.

3.4.3 Architecture 2: distributed Prony using distributed ADMM (D-ADMM)

Consider again the problem (3.8). The S-ADMM and the A-ADMM algorithms discussed for Architecture 1 need a central PDC to update $\mathbf{z}$ at each iteration, and to

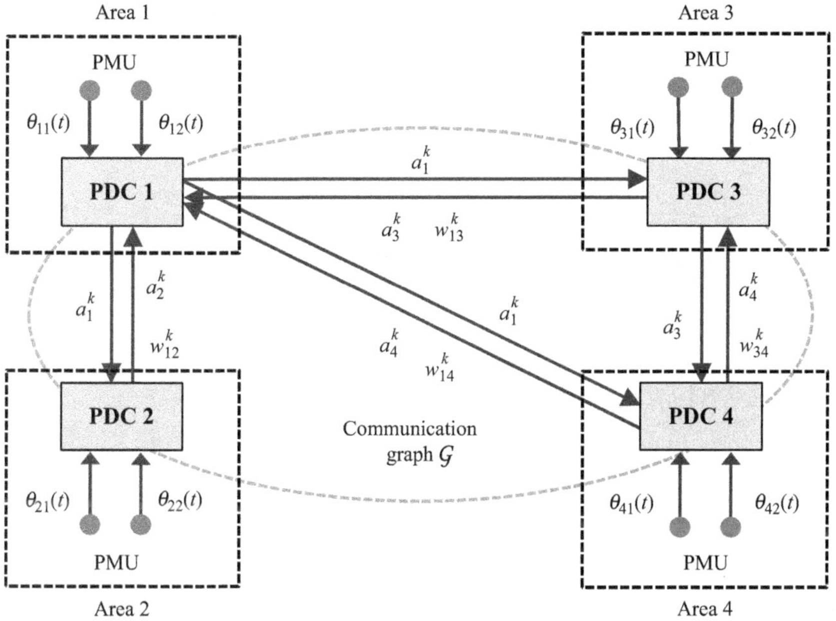

Figure 3.3 Architecture 2 using D-ADMM for a given communication graph $\mathcal{G}$ for a four-area network

broadcast it back to the local active PDCs. Although this architecture preserves the data privacy between the N PDCs, it is not very resilient as the central PDC is directly amenable to failure under extraneous attacks. This problem can be resolved by resorting to a completely distributed version of Architecture 1 as shown in Figure 3.3. The resulting distributed algorithm for solving (3.8) is referred to as distributed ADMM (D-ADMM). In this formulation, each active PDC at each iteration communicates directly with a subset of other active PDCs determined by a communication graph $\mathcal{G}$. Therefore, the need for the central PDC no longer exists. The set of nodes of the communication graph $\mathcal{G}$, denoted by $\mathcal{V}(\mathcal{G})$, are the indexed PDCs of the network, i.e., $\mathcal{V}(\mathcal{G}) = \{1, 2, \ldots, N\}$. The edge set of $\mathcal{G}$ is denoted by $\mathcal{E}(\mathcal{G})$, where $e_{jv} \in \mathcal{E}(\mathcal{G})$ determines the existence of a communication link between PDCs j and v. We consider the following assumptions for the graph $\mathcal{G}$: (1) $\mathcal{G}$ is known a-priori and is considered to be fixed during the iterations. (2) $\mathcal{G}$ is a *simple*[1] and *connected*[2] graph. Before stating the main algorithm, we define an alternative representation of (3.8) using the communication graph $\mathcal{G}$ as follows:

$$\begin{aligned}&\underset{\boldsymbol{a}_1,\ldots,\boldsymbol{a}_N}{\text{minimize}} \ \sum_{j=1}^{N} \frac{1}{2}\|\hat{H}_j\boldsymbol{a}_j - \hat{\mathbf{c}}_j\|^2 \\ &\text{subject to } \boldsymbol{a}_j - \boldsymbol{a}_v = 0, \ \text{for } \ e_{jv} \in \mathcal{E}(\mathcal{G})\end{aligned} \tag{3.18}$$

[1]A graph with no self-loop or multiple edges between two nodes.
[2]A graph that has at least one path between any two arbitrary nodes.

Since $\mathscr{G}$ is connected, then the reformulated problem (3.18) is equivalent to (3.8) [23]. Now, let us define two sets of predecessors and successors of PDC j, denoted respectively by P_j and S_j as

$$P_j = \{\text{PDC } v \mid e_{jv} \in \mathscr{E}(\mathscr{G}),\ v < j\} \tag{3.19a}$$

$$S_j = \{\text{PDC } v \mid e_{jv} \in \mathscr{E}(\mathscr{G}),\ v > j\} \tag{3.19b}$$

Let $n(P_j)$ and $n(S_j)$ denote the number of elements of the sets P_j and S_j, respectively. Also, let us define the Lagrangian associated with (3.18) at iteration k as

$$L_\rho'^{(k)} = \frac{1}{2}\sum_{j=1}^{N}\left(\|\hat{H}_j^{(k)}\boldsymbol{a}_j - \hat{\mathbf{c}}_j^{(k)}\|^2 + \rho\left(\sum_{v\in P_j}\|\boldsymbol{a}_v^{(k+1)} - \boldsymbol{a}_j - \frac{1}{\rho}\mathbf{w}_{vj}^{(k)}\|^2 + \sum_{v\in S_j}\|\boldsymbol{a}_j - \boldsymbol{a}_v^{(k)} - \frac{1}{\rho}\mathbf{w}_{jv}^{(k)}\|^2\right) - \frac{1}{\rho}\left(\sum_{v\in P_j}\|\mathbf{w}_{vj}^{(k)}\|^2 + \sum_{v\in S_j}\|\mathbf{w}_{jv}^{(k)}\|^2\right)\right) \tag{3.20}$$

where $\mathbf{w}_{jv}$ is the dual variable associated with the edge $e_{jv} \in \mathscr{E}(\mathscr{G})$, and $\rho > 0$ is the penalty factor. Using these definitions, we next present the D-ADMM algorithm for solving (3.18).

Distributed Prony using distributed ADMM

1. Each PDC j initializes $\boldsymbol{a}_j^{(0)}$ and $\mathbf{w}_{vj}^{(0)}$, $j \in \mathscr{V}(\mathscr{G})$, $e_{vj} \in \mathscr{E}(\mathscr{G})$.
2. At iteration k, every PDC $j, j = 1, \ldots, N$:
   a. Receives the update of $\boldsymbol{a}_v^{(k+1)}$ for all $v \in P_j$.
   b. Updates $\hat{H}_j$ and $\hat{\mathbf{c}}_j$ using (3.10).
   c. Updates $\boldsymbol{a}_j$ as:

$$\begin{aligned}\boldsymbol{a}_j^{(k+1)} &= \arg\min_{\boldsymbol{a}_j} L_\rho'^{(k)} \\ &= ((\hat{H}_j^{(k)})^T\hat{H}_j^{(k)} + \rho(n(P_j) + n(S_j))I_{2n})^{-1}\boldsymbol{\kappa}_j^{(k)}\end{aligned} \tag{3.21}$$

   where $\boldsymbol{\kappa}_j^{(k)} = (\hat{H}_j^{(k)})^T\hat{\mathbf{c}}_j^{(k)} + \sum_{v\in S_j}\mathbf{w}_{jv}^{(k)} - \sum_{v\in P_j}\mathbf{w}_{vj}^{(k)} + \rho\left(\sum_{v\in P_j}\boldsymbol{a}_v^{(k+1)} + \sum_{v\in S_j}\boldsymbol{a}_v^{(k)}\right)$.

   d. Updates all $\mathbf{w}_{vj}$ for $l \in P_j$ as:

$$\mathbf{w}_{vj}^{(k+1)} = \mathbf{w}_{vj}^{(k)} - \rho(\boldsymbol{a}_v^{(k+1)} - \boldsymbol{a}_j^{(k+1)})$$

   e. Sends $\boldsymbol{a}_j^{(k+1)}$ to all PDCs in $P_j \cup S_j$.
   f. Sends $\mathbf{w}_{vj}^{(k+1)}$ to $v \in P_j$.
   g. Receives $\boldsymbol{a}_v^{(k+1)}$ and $\mathbf{w}_{jv}^{(k+1)}$ from all $v \in S_j$.

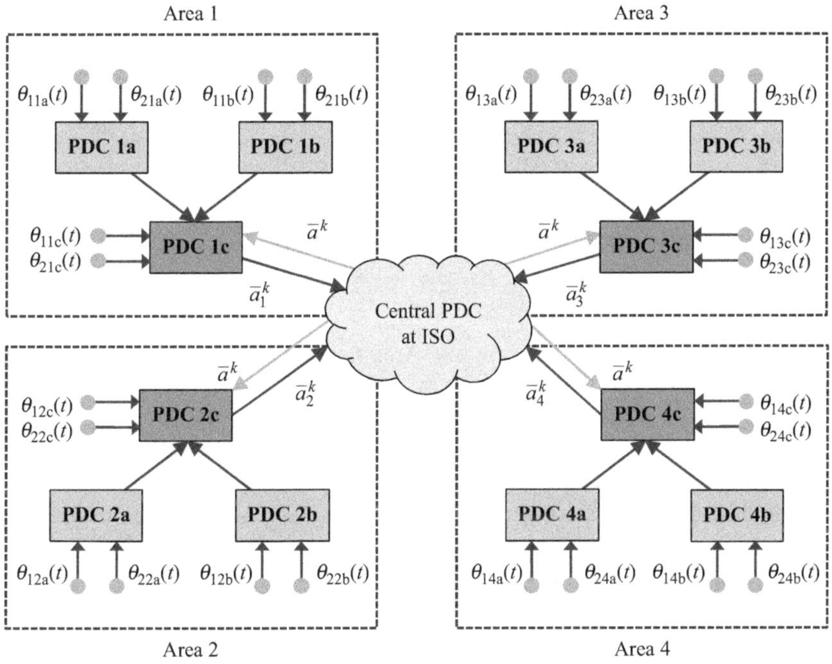

Figure 3.4 Architecture 3 using H-ADMM for a network consisting of four areas

At every iteration k, the primal variables $\boldsymbol{a}_j^{(k)}$ are updated sequentially starting from PDC 1 to PDC N using the most recent available values of $\boldsymbol{a}_v$ for v belonging to its predecessors and successors. PDC j also updates the dual variables $\mathbf{w}_{vj}$ for v belonging to P_j. This algorithm also has a convergence rate of $O(1/k)$, and can be realized using asynchronous distributed ADMM [23].

3.4.4 Architecture 3: distributed Prony using hierarchical ADMM (H-ADMM)

Let us consider again the Prony formulation in Architecture 1 where every area is assumed to contain only one aggregated PDC. However, if p, the number of PMUs, is large, a better strategy will be to create multiple hierarchical layers of PDCs so that the computational load of the global estimation gets divided, as shown in Figure 3.4.

Let us divide the entire network into r computational areas, where each area contains multiple layers of PDCs. Each PDC receives measurements from a subset of the total number of PMUs in that area. We assume these subsets to be disjoint. For each PDC j in layer ℓ of Area l, let Q_j denote the set of the PDCs in layer $\ell - 1$ from which it receives information, and let U_j denote the set containing a single PDC in layer $\ell + 1$ to which it sends information. Also, without loss of generality, let us assume that the final layer in every area consists of exactly one PDC. This PDC

Table 3.1 Comparison between three different architectures

Architecture	Resilience	Data privacy	Convergence rate	Sensitivity to delay	Data volume per PDC
S-ADMM	Low, single point failure	High	$O(1/k)$	Low	High
D-ADMM	High, collaborative	Low	$O(1/k)$	High	High
H-ADMM	Low, single point failure	Low	$O(1/k)$	High	Low

is referred to as the *leader PDC* for any Area l, $l = 1, \ldots, r$. Every leader PDC is connected to the central PDC at the ISO through an inter-area communication link. For example, in the system shown in Figure 3.4, we have $r = 4$ areas with 2 layers of PDCs in each. PDCs 1a and 1b belong to layer 1, while PDC 1c is the leader PDC of Area 1. Similar notations have been used for the other areas. Using these definitions, distributed Prony using the Hierarchical ADMM (H-ADMM) can be described as follows:

Distributed Prony using the hierarchical ADMM

1. Every PDC j initializes $\boldsymbol{a}_j^{(0)}$, $\mathbf{z}^{(0)}$, and $\mathbf{w}_j^{(0)}, j = 1, \ldots, N$.
2. At each iteration k:
   a. PDC j belonging to layer ℓ of Area l constructs $\hat{H}_j^{(k)}$ and $\hat{\mathbf{c}}_j^{(k)}$ from (3.10).
   b. PDC j receives the values of $\mathbf{z}_v^{(k+1)}$ from all PDCs v belonging to Q_j.
   c. PDC j computes $\boldsymbol{a}_j^{(k+1)}$ using (3.12).
   d. PDC j calculates $\mathbf{z}_j^{(k+1)} = \boldsymbol{a}_j^{(k+1)} + \sum_{v \in Q_j} \mathbf{z}_v^{(k+1)}$.
   e. If PDC j is not the leader PDC of Area l, it transmits $\mathbf{z}_j^{(k+1)}$ to the PDC v belonging to U_j. If PDC j is the leader PDC l, then it transmits $\mathbf{z}_l^{(k+1)} \triangleq \mathbf{z}_j^{(k+1)}$ to the central PDC.
   f. The central PDC receives $\mathbf{z}_l^{(k+1)}$ from the leader PDC l, $l = 1, \ldots, r$.
   g. The central PDC calculates $\mathbf{z}^{(k+1)} = \frac{1}{N} \sum_{l=1}^{r} \mathbf{z}_l^{(k+1)}$.
   h. The central PDC broadcasts $\mathbf{z}^{(k+1)}$ to the r leader PDCs through an inter-area communication network.
   i. The leader PDC l broadcasts $\mathbf{z}^{(k+1)}$ to all PDCs in Area l through an intra-area communication network.
   j. Every PDC j in Area l computes $\mathbf{w}_j^{(k+1)}$ using (3.14).

Note that the solution of H-ADMM is equal to that of an equivalent S-ADMM problem with N areas, each containing exactly one PDC. Table 3.1 compares the three proposed architectures in terms of their various properties.

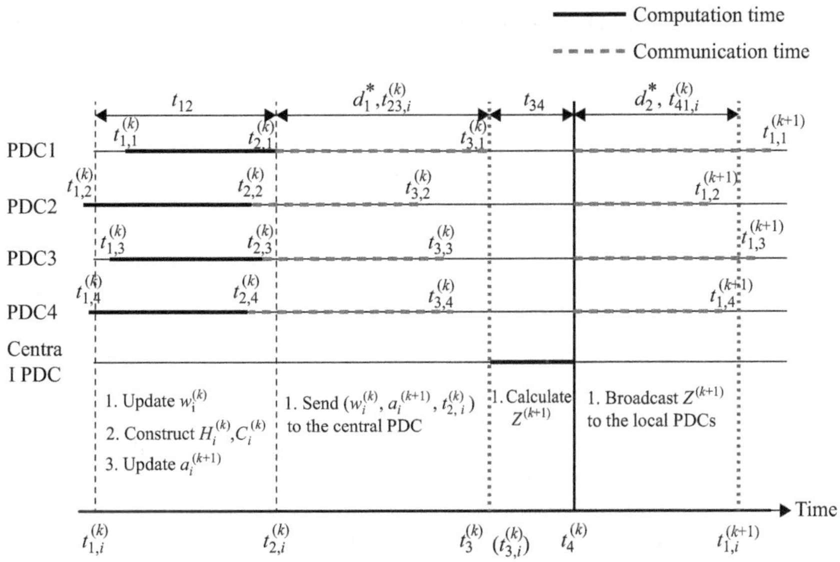

Figure 3.5 Timing diagram for asynchronous ADMM

3.5 Update strategies for asynchronous communication

Communication delay and, therefore, asynchrony in information exchange between estimators is an unavoidable problem in wide-area communication networks. The architectures proposed so far are somewhat agnostic of these delays. The architecture 1 with asynchronous communication, as proposed in Section 3.4.2, partially addresses this problem, but does not provide any insight on how convergence of the ADMM algorithm can be correlated to the stochastic characteristics of the delays. In this section, we address this problem and propose three new variants of the A-ADMM algorithm stated in Section 3.4.2 that can mitigate asynchrony by careful manipulation of the update rules of the algorithm itself. We first consider a probabilistic traffic model for modeling delays in such wide-area networks, and study how these delays enter the process of information exchange between the PDCs. We next propose four different update rules for the proposed Architecture 1 in Section 3.4, by which the convergence rate of ADMM and the accuracy of estimation can be made immune to the asynchrony resulting from the network traffic as much as possible. In practice, the communication between the local PDCs and the central PDC will always involve communication delays, thereby leading to asynchrony in message arrivals at all PDCs. The timing diagram under that condition, as shown in Figure 3.5, will consist of three main delay-sensitive components:

1. Local PMU measurements stream in real-time to the local PDCs. Since this communication happens over a private network, we ignore this communication delay throughout this chapter;

2. The dashed lines between time $t_{2,i}^{(k)}$ and time $t_{3,i}^{(k)}$ at any iteration k show that the local estimates $(\mathbf{w}_i^{(k)}, \mathbf{a}_i^{(k+1)})$ arrive at the central PDC at different instants $t_{3,i}^{(k)}, i = 1, \ldots, N$ due to the variable uplink delays;
3. The green dashed lines between time $t_4^{(k)}$ and time $t_{1,i}^{(k+1)}$ show that the consensus variable $\mathbf{z}^{(k+1)}$ arrives at different instants $t_{1,i}^{(k+1)}, i = 1, \ldots, N$ at different local PDCs due to variable downlink delays.

One trivial way to counteract the asynchrony would be to force all PDCs to wait till they receive every scheduled message at every iteration. This, however, can lead to unacceptably slow convergence times, and, depending on network congestion, can even turn out to be risky in case any message gets lost or delayed for an uncertain period of time. Instead, we wish to counteract asynchrony by defining a set of *flexible deadlines* for message arrival in every PDC, and accordingly modify the update rules based on only those messages that respect these deadlines. In order to understand how these deadlines should be constructed in accordance to the network traffic, we first develop a probability distribution model for the network delays.

3.5.1 Delay model for wide-area communication

Following Reference 24, we model the stochastic end-to-end delay experienced by a message between the central PDC and local PDCs in terms of three components: the minimum deterministic delay, denoted by m; the Internet traffic delay with Probability Density Function (PDF), denoted by ϕ_1; and the router processing delay with PDF, denoted by ϕ_2. Then, the PDF of the total delay at any time t is given as

$$\phi(t) = p\phi_2(t) + (1-p)\phi_1(t) * \phi_2(t), \quad t \geq 0 \tag{3.22}$$

with $\phi_1(t) * \phi_2(t) = \int_0^t \phi_2(u)\phi_1(t-u)du$. Here p is the probability of open period of the path with no Internet traffic, and the router processing delay can be well approximated by a Gaussian density function $\phi_2(t) = \frac{1}{\sigma\sqrt{2\pi}}e^{-\frac{(t-\mu)^2}{2\sigma^2}}$, where $\mu > m$. The Internet traffic delay is modeled by an alternating renewal process with exponential closure period when the Internet traffic is on, with the PDF $\phi_1(t) = \lambda e^{-\lambda t}$, where λ^{-1} models the mean length of the closure period. The benchmark value of all parameters of this model is set as: $p = 0.58, \lambda = 1.39, \mu = 5.3, \sigma = 0.078$, following Reference 24.

We next derive the Cumulative Distribution Function (CDF) of the delay model. First, (3.22) can be rewritten as

$$\phi(t) = \frac{p}{\sigma\sqrt{2\pi}}e^{-\frac{(t-\mu)^2}{2\sigma^2}} + \frac{\lambda(1-p)}{\sigma\sqrt{2\pi}}e^{-\lambda t}\int_0^t e^{\lambda s-\frac{(s-\mu)^2}{2\sigma^2}}ds \tag{3.23}$$

We rewrite the integral part of (3.23) by using the error function $\mathrm{erf}(x) = \frac{2}{\sqrt{\pi}} \int_0^x e^{-t^2} dt$, to get

$$\begin{aligned}\phi(t) &= \frac{p}{\sigma\sqrt{2\pi}} e^{-\frac{(t-\mu)^2}{2\sigma^2}} \\ &+ \frac{\lambda(1-p)}{2} e^{(\frac{1}{2}\lambda^2\sigma^2+\mu\lambda)} e^{-\lambda t} \mathrm{erf}\left(\frac{t-\lambda\sigma^2-\mu}{\sqrt{2}\sigma}\right) \\ &+ \frac{\lambda(1-p)}{2} e^{(\frac{1}{2}\lambda^2\sigma^2+\mu\lambda)} \mathrm{erf}\left(\frac{\lambda\sigma^2+\mu}{\sqrt{2}\sigma}\right) e^{-\lambda t} \end{aligned} \tag{3.24}$$

By using the partial integral method and the first derivative of the error function, $\frac{d}{ds}\mathrm{erf}(s) = \frac{2}{\sqrt{\pi}} e^{-s^2}$, we derive the CDF of the delay model as

$$\begin{aligned}P(t) &= \int_{-\infty}^{t} \phi(s)ds = \frac{1}{2}\left[\mathrm{erf}\left(\frac{\mu}{\sqrt{2}\sigma}\right) + \mathrm{erf}\left(\frac{t-\mu}{\sqrt{2}\sigma}\right)\right] \\ &+ \frac{(p-1)}{2} e^{(\frac{1}{2}\lambda^2\sigma^2+\mu\lambda)} e^{-\lambda t} \left[\mathrm{erf}\left(\frac{\lambda\sigma^2+\mu}{\sqrt{2}\sigma}\right) + \mathrm{erf}\left(\frac{t-\lambda\sigma^2-\mu}{\sqrt{2}\sigma}\right)\right] \end{aligned} \tag{3.25}$$

Random delays from this CDF will next be imposed on the communication links to emulate A-ADMM in Architecture 1.

3.5.2 *Proposed A-ADMM strategies*

3.5.2.1 Strategy I

In this strategy PDCs *skip* messages that do not arrive within a chosen deadline. We define two deadlines or *delay thresholds*, namely $\mathbf{d}_1^* > 0$ and $\mathbf{d}_2^* > 0$ in milliseconds, for the uplink and downlink delays, respectively. Without any loss of generality, assume that the counting of these deadlines start from the instant at which any PDC *sends out* any message at any iteration. For simplicity of notations, we also assume that every local PDC is assigned the same threshold $\mathbf{d}_2^*$ although same exact analyses will hold for different threshold values at different PDCs. Following the timing diagram in Figure 3.5, if any local update $(\mathbf{w}_i^{(k)}, \mathbf{a}_i^{(k+1)})$ does not arrive at the central PDC within time $\mathbf{d}_1^*$, the central PDC skips them, and computes the consensus variable $\mathbf{z}^{(k+1)}$ in (3.16) simply as,

$$\mathbf{z}^{(k+1)} = \frac{1}{|S_1^{(k)}|} \sum_{i \in S_1^{(k)}} (\mathbf{a}_i^{(k+1)} + (1/\rho)\mathbf{w}_i^{(k)}) \tag{3.26}$$

where $S_1^{(k)}$ is the index set of local PDCs whose messages arrive on time, and $S_2^{(k)}$ is the index set for PDCs that can receive $\mathbf{z}^{(k+1)}$ within time, at the kth iteration.

On the other hand, if any local PDC i does not receive $\mathbf{z}^{(k+1)}$ from the central PDC within the delay threshold $\mathbf{d}_2^*$, it updates (3.12)–(3.14) altogether by using the latest local measurements and the previous $\mathbf{z}$. For a practical Internet model, the skipping strategy is most suitable for scenarios that involve exceptionally long delays or possible packet loss. The algorithm for this strategy is listed in Algorithm 1.

Algorithm 1 A-ADMM with Strategy I

procedure CENTRALPDC(ε)
  initialize: $k = 1$, $\mathbf{d}_1^*$
  repeat
    repeat
      wait
      receive updates $(\mathbf{w}_i^{(k)}, \mathbf{a}_i^{(k+1)})$
    until timer $\leq \mathbf{d}_1^*$ or all updates received
    update $\mathbf{z}^{(k+1)}$ by (3.26)
    broadcast $\mathbf{z}^{(k+1)}$ to all local PDCs
    $k \leftarrow k + 1$
  until $\Delta \mathbf{z} = ||\mathbf{z}^{(k+1)} - \mathbf{z}^{(k)}|| \leq \varepsilon$
end procedure
procedure LOCALPDC i
  initialize: $k = 1, \mathbf{w}_i^{(0)} = 0, \mathbf{a}_i^{(0)} = 1, \mathbf{d}_2^*$;
  repeat
    repeat
      wait
    until timer $\leq \mathbf{d}_2^*$, or receive $\mathbf{z}^{(k)}$
    update $y_i(k), \hat{\mathbf{H}}_i^{(k)}, \hat{\mathbf{c}}_i^{(k)}$ by (3.5)
    if global update received **then**
      update $(\mathbf{w}_i^{(k)}, \mathbf{a}_i^{(k+1)})$ by (3.12), (3.14)
    else
      $\mathbf{w}_i^{(k)} = \mathbf{w}_i^{(k-1)}; \mathbf{a}_i^{(k+1)} = \mathbf{a}_i^{(k)}$
    end if
    send $(\mathbf{w}_i^{(k)}, \mathbf{a}_i^{(k+1)})$ to the central PDC
    $k \leftarrow k + 1$
  until termination
end procedure

3.5.2.2 Strategy II

In this strategy PDCs use internal memory to replace messages that do not arrive within the respective deadlines with their values from previous iteration. Memorized data from PDCs, e.g., can be stored and retrieved via a distributed storage service space [25]. Following the timing diagram in Figure 3.5, if any local update $(\mathbf{w}_i^{(k)}, \mathbf{a}_i^{(k+1)})$ does not arrive at the central PDC within time $\mathbf{d}_1^*$, the central PDC computes the consensus variable $\mathbf{z}^{(k+1)}$ in (3.16) as

$$\mathbf{z}^{(k+1)} = \frac{1}{N}\left(\sum_{i \in S_1^{(k)}}\left(\mathbf{a}_i^{(k+1)} + \frac{\mathbf{w}_i^{(k)}}{\rho}\right) + \sum_{i \notin S_1^{(k)}}\left(\mathbf{a}_i^{(l+1)} + \frac{\mathbf{w}_i^{(l)}}{\rho}\right)\right) \tag{3.27}$$

where $l \in \{k-1, k-2, k-3, \cdots\}$ denotes the index of the *latest* message that arrived successfully at the central PDC. Similarly, if any local PDC $i, i \notin S_2^{(k)}$ does not receive $\mathbf{z}^{(k)}$ within its deadline, then it updates (3.12)–(3.14) as

$$\mathbf{w}_i^{(k)} = \mathbf{w}_i^{(k-1)} + \rho(\mathbf{a}_i^{(k)} - \mathbf{z}_i^{(l_i)}) \tag{3.28a}$$

$$\mathbf{a}_i^{(k+1)} = ((\hat{\mathbf{H}}_i^{(k)})^T \hat{\mathbf{H}}_i^{(k)} + \rho I)^{-1}((\hat{\mathbf{H}}_i^{(k)})^T \hat{\mathbf{c}}_i^{(k)} - \mathbf{w}_i^{(k)} + \rho \mathbf{z}_i^{(l_i)}) \tag{3.28b}$$

where $l_i \in \{k-1, k-2, k-3, \cdots\}$ denotes the index of the *latest* message that arrived successfully at the ith local PDC. Note that $l_i \neq l_j$, in general. However, since every PMU data is time-stamped by GPS, the PDCs will have the ability to stamp or decipher the iteration number corresponding to any message they send or receive.

Adjustment in stopping criterion: The usual practice in S-ADMM is to keep track of $\Delta\mathbf{z} = ||\mathbf{z}^{(k+1)} - \mathbf{z}^{(k)}||$, and terminate the algorithm once $\Delta\mathbf{z}$ falls below a chosen tolerance. This step needs to be modified for A-ADMM with Strategy II. The reason is as follows. In case every local update $(\mathbf{w}_i^{(k)}, \mathbf{a}_i^{(k+1)})$, $i = 1, \ldots, N$ arrive after time $\mathbf{d}_1^*$ at the local PDC, then the newest update of $\mathbf{z}^{(k+1)}$ will be computed from its stored latest update, which is exactly same as $\mathbf{z}^{(k)}$ causing $\Delta\mathbf{z}$ to be zero. This may force the algorithm to terminate unexpectedly. To alleviate impact of delayed updates, we first modify the stopping criteria to

$$\Delta\mathbf{z} = \frac{1}{|S_1^{(k)}|} \sum_{i \in S_1^{(k)}} ||(\mathbf{a}_i^{(k+1)} - \mathbf{a}_i^{(l)}) + (1/\rho)(\mathbf{w}_i^{(k)} - \mathbf{w}_i^{(l-1)})|| \tag{3.29}$$

where $(\mathbf{w}_i^{(l-1)}, \mathbf{a}_i^{(l)})$ is the latest local update received from the ith local PDC at the $(k-1)^{th}$ iteration at the central PDC. The tolerance factor, denoted by ε, is set to be a constant value for every iteration (10^{-6} in our simulations). This adjustment is referred to as Adj. I. Another way to avoid unnecessary termination would be to adjust ε adaptively according to the number of non-delayed estimates. This adjustment is referred to as Adj. II. It chooses smaller values of ε for smaller values of l. That is, we define a vector $\varepsilon = [\varepsilon_1, \ldots, \varepsilon_N]$ such that for $|S_1^{(k)}| = 1$ we use ε_1, for $|S_1^{(k)}| = 2$ we use ε_2, and so on, where $\varepsilon_1 < \varepsilon_2 < \cdots < \varepsilon_N$, [20]. The different steps of A-ADMM with Strategy II are shown in Algorithm 2.

3.5.2.3 Strategy II with gradient update

It is well-known that ADMM algorithms typically show steep initial convergence to a ball around the optimal point, and slow convergence toward the optimal thereafter [22]. This fact motivates us to add a gradient update term to (3.28) to improve convergence as:

$$\mathbf{w}_i^{(k)} = \mathbf{w}_i^{(k-1)} + \rho(\mathbf{a}_i^{(k)} - (\mathbf{z}^{(l)} + \gamma(\mathbf{z}^{(l)} - \mathbf{z}^{(l-1)}))) \tag{3.30a}$$

$$\begin{aligned}\mathbf{a}_i^{(k+1)} &= ((\hat{\mathbf{H}}_i^{(k)})^T \hat{\mathbf{H}}_i^{(k)} + \rho I)^{-1}((\hat{\mathbf{H}}_i^{(k)})^T \hat{\mathbf{c}}_i^{(k)} - \mathbf{w}_i^{(k)} \\ &\quad + \rho(\mathbf{z}^{(l)} + \gamma(\mathbf{z}^{(l)} - \mathbf{z}^{(l-1)}))\end{aligned} \tag{3.30b}$$

where γ is the step size whose value can be flexibly adjusted to expedite convergence. That is, now the local PDC $i, i \notin S_2^{(k)}$ uses $\mathbf{z}^{(l)} + \gamma(\mathbf{z}^{(l)} - \mathbf{z}^{(l-1)})$ instead of $\mathbf{z}^{(l)}$ in its

Algorithm 2 A-ADMM with Strategy II

procedure CENTRALPDC(ε)
  initialize: $k = 1$, $\mathbf{d}_1^*$
  repeat
    repeat
      wait
      receive updates $(\mathbf{w}_i^{(k)}, \mathbf{a}_i^{(k+1)})$
    until timer $\leq \mathbf{d}_1^*$ or all updates received
    update $\mathbf{z}^{(k+1)}$ by (3.27), update $\Delta\mathbf{z}$ by (3.29)
    broadcast $\mathbf{z}^{(k+1)}$ to all local PDCs
    $k \leftarrow k + 1$
  until $\Delta\mathbf{z} \leq \varepsilon$
end procedure
procedure LOCALPDC i
  initialize: $k = 1, \mathbf{w}_i^{(0)} = 0$, $\mathbf{a}_i^{(0)} = 1$, $\mathbf{d}_2^*$;
  repeat
    repeat
      wait
    until timer $\leq \mathbf{d}_2^*$, or receive $\mathbf{z}^{(k)}$
    update $y_i(k), \hat{\mathbf{H}}_i^{(k)}, \hat{\mathbf{c}}_i^{(k)}$ by (3.5)
    if global update received **then**
      update $(\mathbf{w}_i^{(k)}, \mathbf{a}_i^{(k+1)})$ by (3.12), (3.14)
    else
      update $(\mathbf{w}_i^{(k)}, \mathbf{a}_i^{(k+1)})$ by (3.28a), (3.28b)
    end if
    send $(\mathbf{w}_i^{(k)}, \mathbf{a}_i^{(k+1)})$ to the central PDC
    $k \leftarrow k + 1$
  until termination
end procedure

update equations. Since the algorithm is solving consensus, this gradient information can improve the proximity of $\mathbf{z}^{(k)}$ to $\mathbf{z}^{(k+1)}$ compared to $\mathbf{z}^{(l)}$. The update at the central PDC remains same as in (3.27).

Table 3.2 compares the three A-ADMM strategies in terms of their sensitivity to different communication bottlenecks.

3.5.2.4 Other strategies

Other heuristic strategies can also be used to mitigate asynchrony by taking advantage of fact that we are solving a consensus problem. One such strategy is to use the spatial correlation between the estimates generated by the local PDCs at every iteration. Local estimates $\mathbf{a}_i^{(k)}$ and $\mathbf{a}_j^{(k)}$ for any two PDCs i and j, located at two different spatial locations

Table 3.2 Comparison of A-ADMM strategies with respect to sensitivity to network characteristics

Strategy type	Packet loss	Packet delay	Delay threshold
Strategy I – skipping	High	High	High
Strategy II – previous update	High	Low	Medium
Strategy II with gradient method	Medium	Low	Low

in the grid, are likely to be close in their magnitudes over iteration k, and exactly equal as $k \to \infty$. The closeness at any given k, especially for smaller values of k, depends on the difference between the initial conditions $\mathbf{a}_i^{(0)}$ and $\mathbf{a}_j^{(0)}$, and individual convergence rates of the two PDCs depending on their network traffic. The idea, therefore, is to keep track of the correlation factors between every pair of local estimates at the central PDC. The correlation coefficient $\rho_{(\mathbf{a}_i^{(k)},\mathbf{a}_j^{(k)})}$ between $\mathbf{a}_i^{(k)}$ and $\mathbf{a}_j^{(k)}$ with expected values $\mu_{\mathbf{a}_i^{(k)}}$ and $\mu_{\mathbf{a}_j^{(k)}}$ and standard deviations $\sigma_{\mathbf{a}_i^{(k)}}$ and $\sigma_{\mathbf{a}_j^{(k)}}$ is defined as

$$\rho_{(\mathbf{a}_i^{(k)},\mathbf{a}_j^{(k)})} = \frac{cov(\mathbf{a}_i^{(k)}, \mathbf{a}_j^{(k)})}{\sigma_{\mathbf{a}_i^{(k)}}\sigma_{\mathbf{a}_j^{(k)}}} = \frac{E[(\mathbf{a}_i^{(k)} - \mu_{\mathbf{a}_i^{(k)}})]E[(\mathbf{a}_j^{(k)} - \mu_{\mathbf{a}_j^{(k)}})]}{\sigma_{\mathbf{a}_i^{(k)}}\sigma_{\mathbf{a}_j^{(k)}}} \tag{3.31}$$

The central PDC computes the correlation matrix $\mathscr{C}^k$ as

$$\mathscr{C}^k = \begin{bmatrix} \rho_{(\mathbf{a}_1^{(k)},\mathbf{a}_1^{(k)})} & \rho_{(\mathbf{a}_1^{(k)},\mathbf{a}_2^{(k)})} & \cdots & \rho_{(\mathbf{a}_1^{(k)},\mathbf{a}_N^{(k)})} \\ \rho_{(\mathbf{a}_2^{(k)},\mathbf{a}_1^{(k)})} & \rho_{(\mathbf{a}_2^{(k)},\mathbf{a}_2^{(k)})} & \cdots & \rho_{(\mathbf{a}_2^{(k)},\mathbf{a}_N^{(k)})} \\ \vdots & \vdots & \cdots & \vdots \\ \rho_{(\mathbf{a}_N^{(k)},\mathbf{a}_1^{(k)})} & \rho_{(\mathbf{a}_N^{(k)},\mathbf{a}_2^{(k)})} & \cdots & \rho_{(\mathbf{a}_N^{(k)},\mathbf{a}_N^{(k)})} \end{bmatrix} \tag{3.32}$$

If at iteration $(k+1)$ any estimate $\mathbf{a}_i^{(k+1)}$ does not arrive on time, then the central PDC scans $\mathscr{C}^k$, and locates the index j such that $\rho_{(\mathbf{a}_i^{(k)},\mathbf{a}_j^{(k)})}$ has the highest magnitude among all entries in the ith row of $\mathscr{C}^k$. If $\mathbf{a}_j^{(k+1)}$ has arrived on time, then it substitutes the missing value $\mathbf{a}_i^{(k+1)}$ by $\mathbf{a}_j^{(k+1)}$ instead of $\mathbf{a}_i^{(l_i+1)}$ as in (3.27). In other words, (3.27) now takes the form

$$\mathbf{z}^{(k+1)} = \frac{1}{N}\left(\sum_{i\in S_1^{(k)}}\left(\mathbf{a}_i^{(k+1)} + \frac{\mathbf{w}_i^{(k)}}{\rho}\right) + \sum_{j\notin S_1^{(k)}}\left(\mathbf{a}_{J_j}^{(k+1)} + \frac{\mathbf{w}_{J_j}^{(k)}}{\rho}\right)\right) \tag{3.33}$$

where $J_j = \arg\max_{i\in S_1^{(k)}} \mathscr{C}^k(j,i)$.

3.6 Case studies of proposed architectures

To verify the distributed Prony algorithms described in Section 3.4, we first consider the IEEE 68-bus system shown in Figure 3.6. The system is divided into five areas, each with one local PDC and three PMUs as shown in Figure 3.6. The simulated measurements are obtained using the Power Systems Toolbox (PST) nonlinear dynamics simulation routine `s_simu` and the data file `data16m.m` [26]. The synchronous generators in this model are assumed to be sixth order for the sake of practicality. A three-phase fault is considered occurring at the line connecting buses 1 and 2. The fault starts at $t = 0.1$ s, clears at bus 1 at $t = 0.15$ s and at bus 2 at $t = 0.20$ s. The measurements are downsampled and the sampling period T is increased up to 0.2 s. Our objective is to estimate the post-fault inter-area oscillation modes of the system. Since there are 16 generators, our proposed algorithms should ideally solve a ninety-sixth order polynomial. However, several of these 96 eigenvalues have negligible residues in (3.3), as a result of which, the practical order can be chosen to be a much smaller number. In fact, in our simulations we show that choosing $2n = 40$ yields a satisfactory estimates of the inter-area modes. The initial 10 samples (2 s) of the measurements are gathered before starting the optimization iterations. We set the penalty factor of the augmented Lagrangian, $\rho = 10^{-9}$.

1. Results of distributed Prony using S-ADMM: We first deploy the Prony algorithm using S-ADMM. Figure 3.7a shows how the estimates of σ and Ω per iteration converge to their global values for four selected slow modes after 50 iterations. The dashed lines show the actual values of σ and Ω for these four modes obtained from

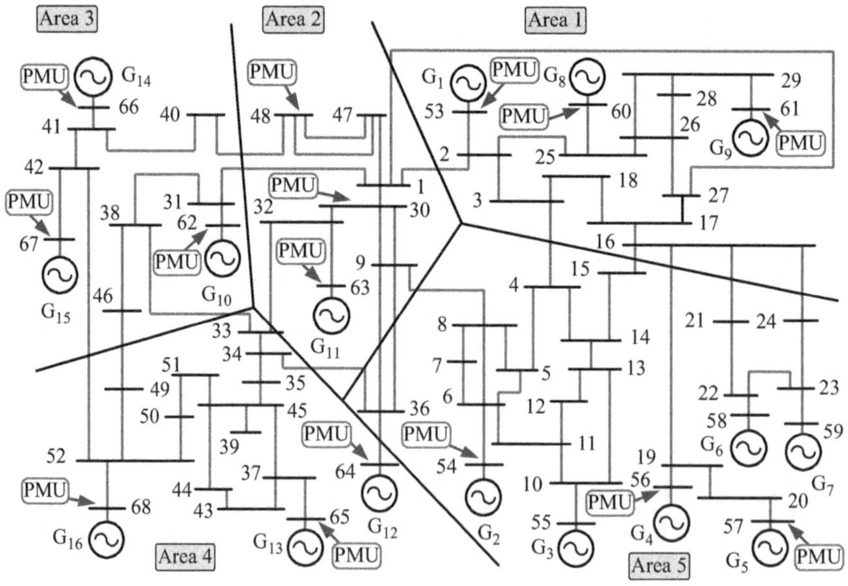

Figure 3.6 IEEE 68-bus model

PST. Figure 3.8 also compares the errors per iteration of S-ADMM with the other three algorithms. As the curves show, the error in all algorithms converge to zero asymptotically.

2. Results of distributed Prony using A-ADMM: For the asynchronous case, the active PDCs in each iteration are chosen randomly with equal probability of 0.5 for a PDC to be either active or dormant. Figure 3.7b shows the estimates of σ and Ω per iteration for each of the four selected modes. Compared to S-ADMM, the convergence of this method is slower due to the stochastic nature of the algorithm.

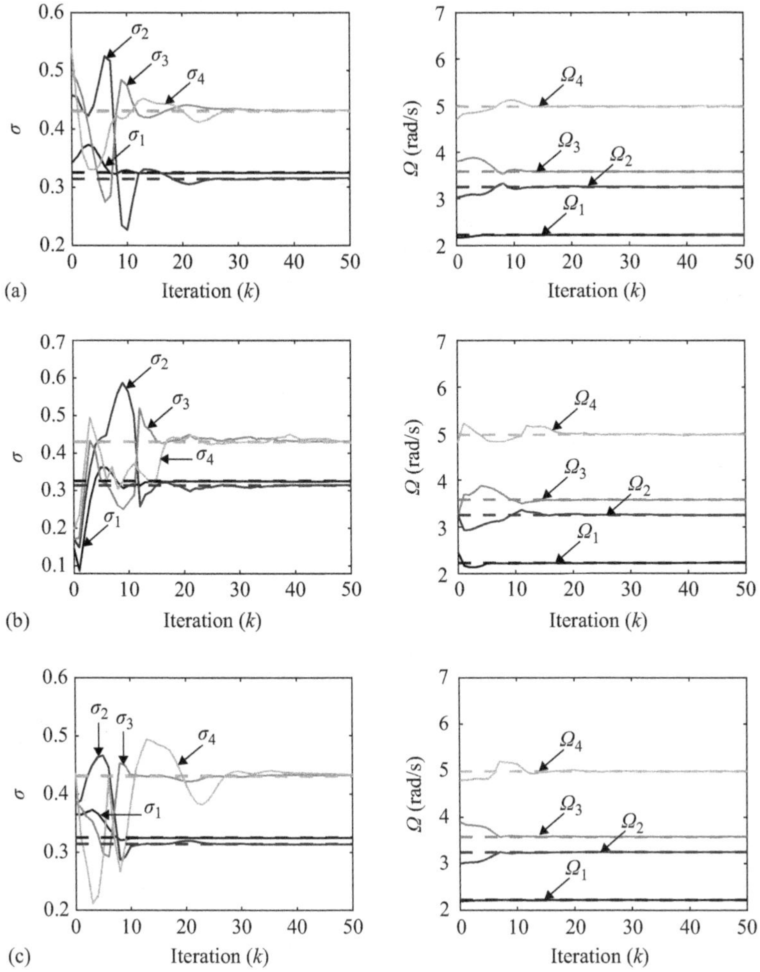

Figure 3.7 Estimated σ and Ω per iteration k (solid lines) versus their actual values (dashed lines). (a) S-ADMM algorithm. The values are calculated at the ISO from $\bar{\mathbf{a}}^k$ (Step 2 of the Prony). (b) A-ADMM. (c) and (d) D-ADMM with $\mathscr{G}_1$ and $\mathscr{G}_1$, respectively. (The values are calculated using $\mathbf{a}_1^k$.) (e) H-ADMM

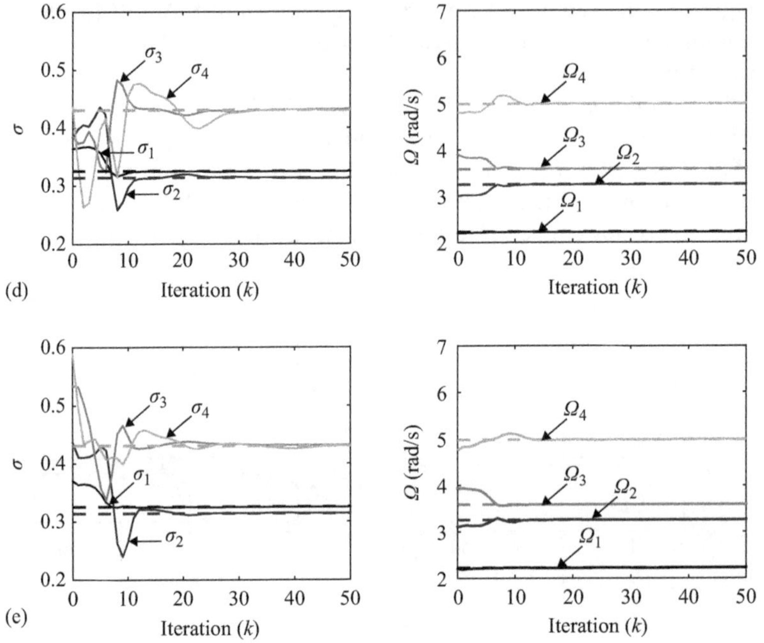

Figure 3.7 (Continued)

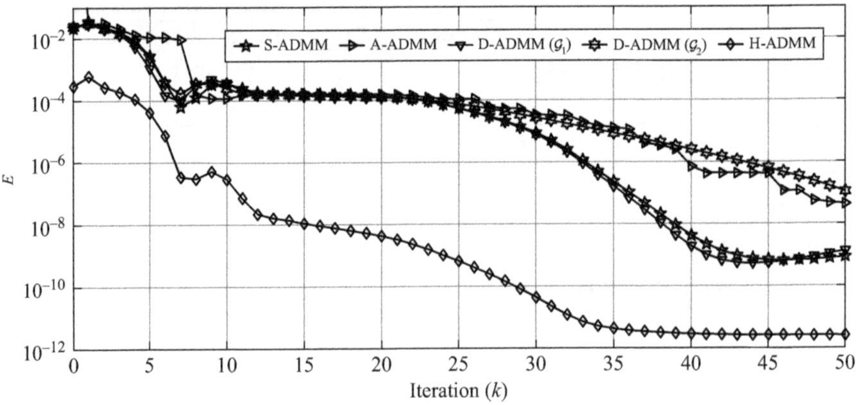

Figure 3.8 Computation of the error term $E(k) = \sum_{j=1}^{N} \frac{1}{2} \| \hat{H}_j^k \boldsymbol{a}_j^k - \hat{\mathbf{c}}_j^k \|_2^2$ *for the four proposed algorithms per iteration*

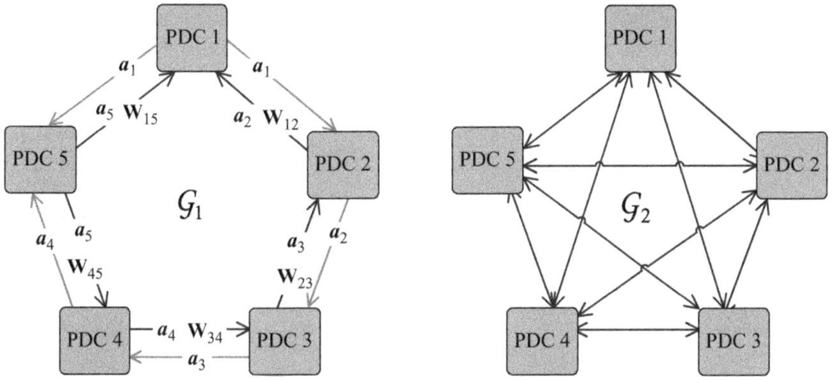

(a) (b)

Figure 3.9 The communication graphs used in Architecture 2. (a) D-ADMM ($\mathscr{G}_1$) and (b) D-ADMM ($\mathscr{G}_2$)

3. Results of distributed Prony using D-ADMM: For the distributed case, we consider the two communication graphs $\mathscr{G}_1$ and $\mathscr{G}_2$ shown in Figure 3.9. Figure 3.7c and d shows the mode estimation per iteration for the four selected eigenvalues using $\mathscr{G}_1$ and $\mathscr{G}_2$, respectively. These figures show that the estimates of σ and Ω per iteration converges to their actual values asymptotically using both $\mathscr{G}_1$ and $\mathscr{G}_2$. However, $\mathscr{G}_2$ has more number of communication links compared to $\mathscr{G}_1$, which makes it a less-favorable choice.

4. Results of distributed Prony using H-ADMM: We now consider each of the three PMUs in every area to be equipped with its own PDC. Two of these PMU-PDC pairs are considered to be in layer 1, while the third pair is considered to be in layer 2. It is worth noting that there is no delay considered in this architecture. Figure 3.7e shows the four selected eigenvalues converge to their global values. Figure 3.8 also shows the estimation error per iteration using this algorithm. It is worth to note that there is no communication delay considered in this architecture.

It should be noted that Figure 3.8 only compares the estimation results per iteration for the proposed algorithms, and not the time needed to execute an iteration. It should be noted that every iteration of H-ADMM involves sequential communication between the hierarchies of PDCs, and hence, the actual time for completing one iteration of H-ADMM will be larger than that of S-ADMM. Similarly, every iteration of D-ADMM may be slower than a corresponding iteration of S-ADMM since the former involves sequential communication, while the latter involves parallel communication between PDCs.

Table 3.3 shows the estimates of the four inter-area eigenvalues obtained from the four algorithms, and compares them with their actual values and the results of the centralized Prony. It can be seen that all these algorithms yield accurate estimates of the slow eigenvalues with relative error less than 1%.

Table 3.3 The estimated slow eigenvalues of the IEEE 68-bus model using centralized Prony (Steps 1 and 2), and the distributed Prony algorithms at $k = 50$

Actual	Centralized	Distributed Prony				
$(-\sigma_i \pm j\Omega_i)$	Prony	S-ADMM	A-ADMM	D-ADMM ($\mathcal{G}_1$)	D-ADMM ($\mathcal{G}_2$)	H-ADMM
$-0.326 \pm j2.226$	$-0.325 \pm j2.223$	$-0.325 \pm j2.223$	$-0.325 \pm j2.223$	$-0.325 \pm j2.223$	$-0.325 \pm j2.223$	$-0.325 \pm j2.223$
$-0.314 \pm j3.251$	$-0.315 \pm j3.253$	$-0.316 \pm j3.254$	$-0.315 \pm j3.253$	$-0.314 \pm j3.253$	$-0.314 \pm j3.253$	$-0.315 \pm j3.253$
$-0.431 \pm j3.581$	$-0.432 \pm j3.585$	$-0.432 \pm j3.584$	$-0.432 \pm j3.585$	$-0.432 \pm j3.587$	$-0.432 \pm j3.586$	$-0.432 \pm j3.584$
$-0.430 \pm j4.984$	$-0.431 \pm j4.987$	$-0.432 \pm j4.987$	$-0.428 \pm j4.980$	$-0.433 \pm j4.986$	$-0.430 \pm j4.983$	$-0.431 \pm j4.988$

The drawbacks of these architectures compared to the centralized case are obvious. S-ADMM, unlike centralized Prony, needs both uplink and downlink communications. D-ADMM suffers from loss of data privacy. H-ADMM involves higher communication delays at the cost of lesser computations.

3.7 Simulation results for asynchronous ADMM strategies

3.7.1 S-ADMM vs A-ADMM

To verify the A-ADMM algorithms described in Section 3.5, we consider the same IEEE 68-bus system, which is described in Section 3.6. We first consider a fundamental comparison of S-ADMM and A-ADMM with Strategy I. From our delay model, we note that $P(X \leq 5.67) = 0.8$, meaning that $\mathbf{d}_1^* = \mathbf{d}_2^* = 5.67$ ms will lead to 20% of the messages be delayed. Using these values, we simulate 1000 runs of the A-ADMM experiments with Strategy I, and plot the expected values of the convergence error in Figure 3.10. The strategy is subdivided into three different cases, namely: (1) Downlink case, where we only consider those runs where downlink deadlines are missed, but uplink deadlines are always met; (2) Uplink case, where the reverse of case (1) happens; and (3) Bi-link case, which is the usual A-ADMM with Strategy I. The left panel of Figure 3.10 shows that compared to the smooth convergence of S-ADMM, the convergence of A-ADMM becomes jittery as more asynchrony is added. Setting the accuracy of estimation to be fixed at 99.5% or more, convergence rate clearly slows down from 36 iterations in S-ADMM to 45 for the Downlink case, 105 for the Uplink case, and 113 for general A-ADMM with bi-directional delay. The right panel of Figure 3.10 shows the convergence and accuracy of the estimated values of the frequency of the three dominant modes. Figure 3.11 shows the correlation plots of every local estimate and the consensus variable over different iterations, illustrating the relative convergence rates of the estimates at the five different PDCs. There are two interesting observations for both S-ADMM and A-ADMM: (1) The correlation coefficients are low to 0.5, as colorful blocks indicate at the first iteration, while all of them converge to the same with the correlation coefficient of 1 at the end of algorithm. (2) Either local estimates or global estimate converges quickly at the first

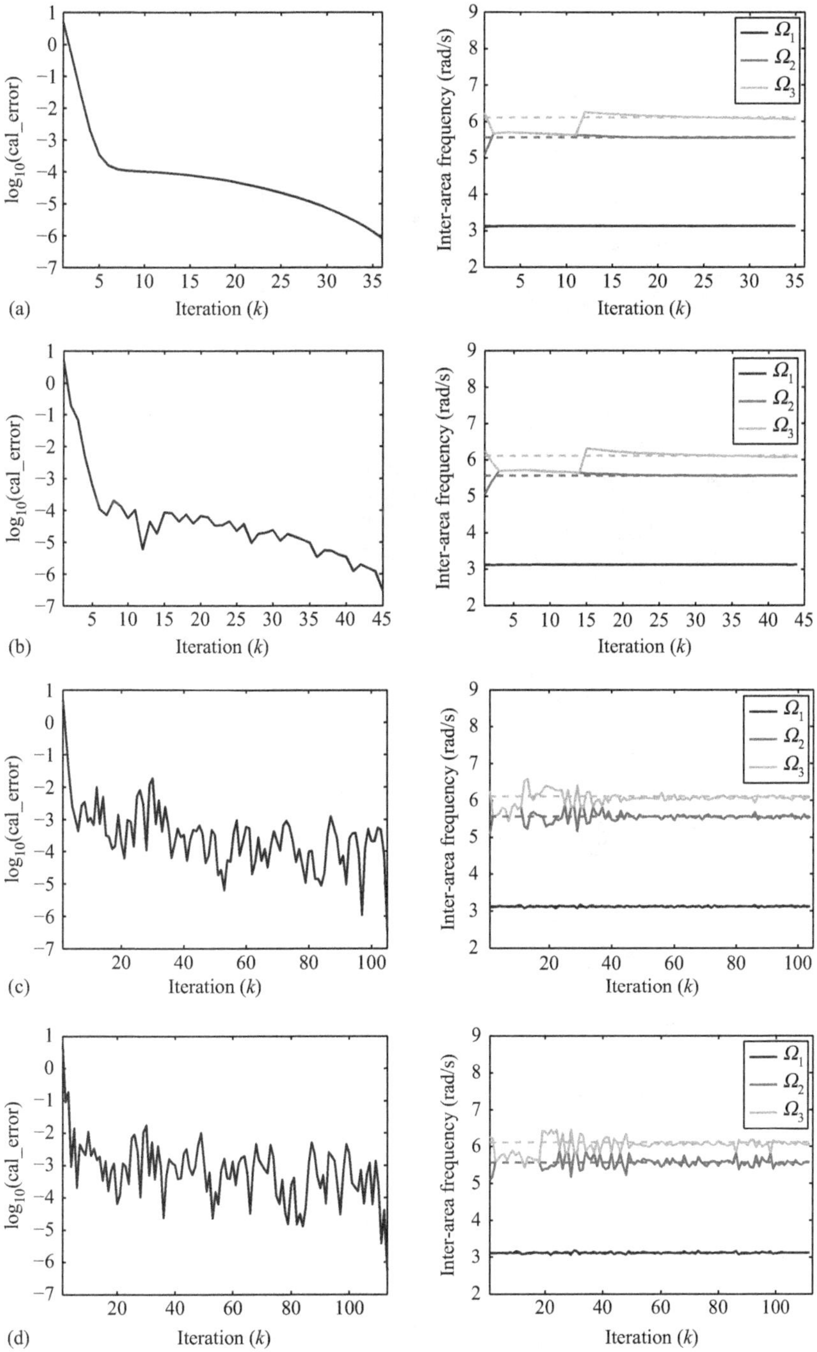

Figure 3.10 Comparison of convergence and accuracy of Strategy I and S-ADMM. (a) S-ADMM; (b) Downlink case of A-ADMM with Strategy I; (c) Uplink case of A-ADMM with Strategy I; and (d) Bi-link case of A-ADMM with Strategy I

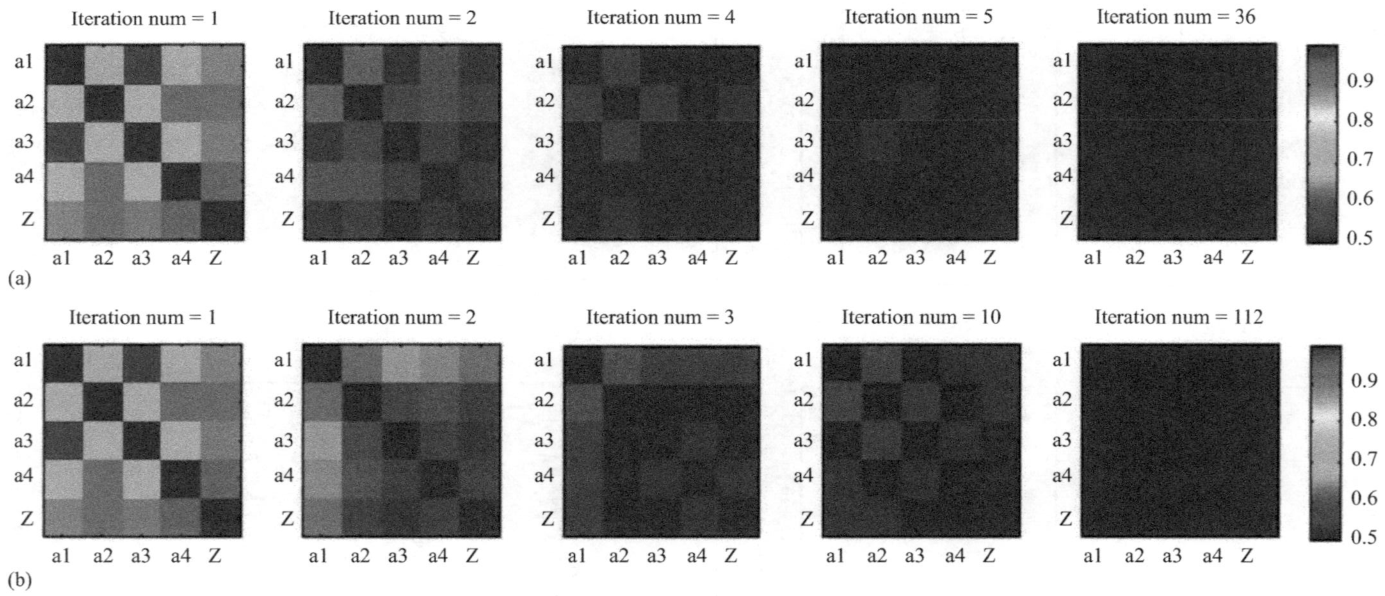

Figure 3.11 Correlation analyses for S-ADMM vs A-ADMM with Strategy I. (a) S-ADMM and (b) Bi-link case of A-ADMM with Strategy I

several iterations, turning from the colorful blocks to the uniform blocks over different iterations.

3.7.2 Sensitivity of A-ADMM to delay thresholds

We next test the impact of $\mathbf{d}_1^*$ and $\mathbf{d}_2^*$ on the convergence and accuracy of the A-ADMM strategies. For each strategy, we run 200 runs for each different choice of the delay thresholds by using 200 pre-generated sets of delays using model (3.25). Figures 3.12 and 3.13 show the relationships between different delay thresholds and the average convergence rate and accuracy.

1. *Strategy I*: From the left subfigure in Figure 3.12a, we see that in the Downlink case, the number of iterations increases slowly when the CDF τ of the delay threshold $\mathbf{d}_2^*$ decreases from 1 to 0.2 as indicated by Down-L. Even when $\tau = 0.1$, meaning that $\mathbf{z}^k$ only has 10% possibility to arrive by the deadline, the algorithm still converges in 100 iterations. However, in Uplink case, once $\tau = 0.1$, the algorithm faces numerical instability, and therefore diverges as indicated by Up-L. Also, as τ decreases from 1 to 0.5, k^* increase dramatically up to 500, while it remains around the value of 400 for $\tau \in [0.2, 0.4]$. As expected, the Bi-link case has the shortest stable range of $\tau \in [0.8, 1]$, and largest value of k^*, compared to the other two cases. If $\tau \leq 0.7$, the algorithm diverges irrespective of the choice of the deadlines, indicating that Strategy I is very sensitive to bi-directional delays. Estimation accuracy of 99.1% or more is maintained by adjusting ε of the stop criterion from 10^{-6} to 10^{-8} or 10^{-7}, as shown in Figure 3.12a. After the adjustment, the Downlink case still has much better convergence than the Uplink case, indicating that message-skipping impacts the averaging step far more than the local least-square update step.

2. *Strategy II – lost data*: Strategy II is further divided into two subcases – namely, for (1) lost data, and for (2) delayed data. Lost data essentially implies that if any message does not arrive on time at any PDC it is considered to be unusable for future iterations. Figure 3.12b shows that similar to Strategy I, the accuracy of this strategy drops dramatically from 99.08 to 96.27% when τ decreases from 0.8 to 0.2 in Downlink case. Even after adjustment of ε, it has worse convergence than Strategy I. This observation implies that if the local PDC i does not receive any message on time it is more advisable to skip the update rather than using out-of-date information about $\mathbf{z}^{(k)}$. For the Uplink case, however, this strategy has significantly better convergence and accuracy guarantees than Strategy I, especially for $\tau \in [0.2, 1]$, implying that when the probability of data loss is high, it is more advisable for the central PDC to use stored values of local updates rather than skipping them.

3. *Strategy II – delayed data*: This is the usual A-ADMM with Strategy II, where delayed data are used in future iterations. It has the best convergence and accuracy guarantees among the three representative cases described so far, partly because of zero packet loss rate, and partly due to limited maximum communication delay. In Downlink case, the blue line in the left subfigure of Figure 3.12c, indicates that the iteration number k^* has the lowest value for each given τ, compared to the blue lines of both Strategy I and Strategy II – lost data before adjustment. In Uplink case, the

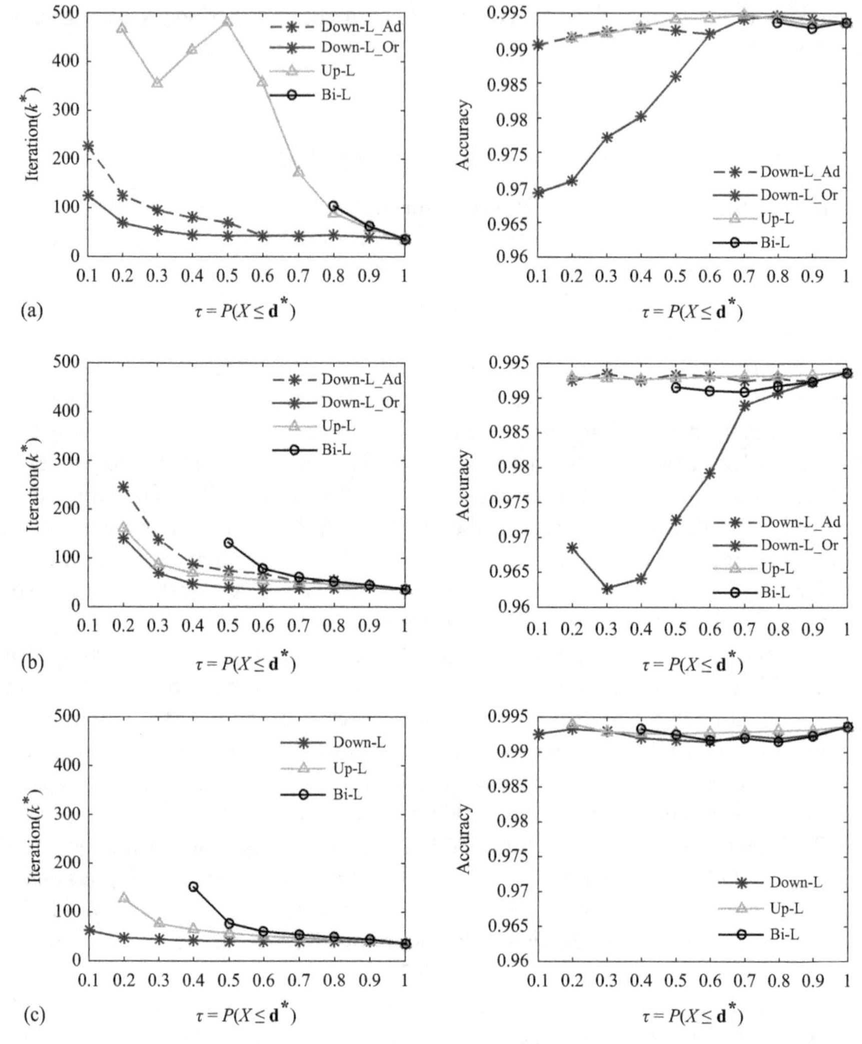

Figure 3.12 Comparison of convergence and accuracy of Strategy I and II with respect to delay threshold. (a) Strategy I – skipping; (b) Strategy II – lost data; and (c) Strategy II – delayed data

magenta line shows smallest k^* in the same effective range of $\tau \in [0.2, 1]$. For the Bi-link case, this not only converges in the smallest number of iterations k^* for a given τ, but also has the longest effective range of $\tau \in [0.4, 1]$ guaranteeing numerical stability.

4. *Hybrid strategies*: So far we have considered cases where Strategies I and II are employed independently for the estimation. We next consider the cases when

Table 3.4 Hybrid control strategies

Strategy Type	Uplink	Downlink
Hybrid-1	Skipping	Lost data
Hybrid-2	Skipping	Delayed data
Hybrid-3	Lost data	Skipping
Hybrid-4	Delayed data	Skipping

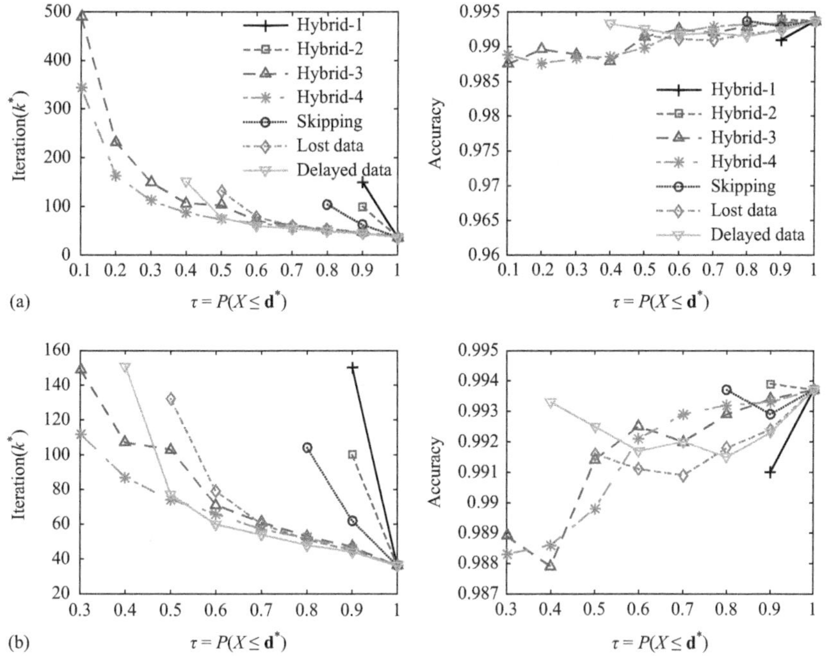

Figure 3.13 Comparison of convergence and accuracy of hybrid strategies with respect to delay threshold. (a) Hybrid strategies and (b) Hybrid vs standard

these two strategies are used in combinations. Four possible such hybrid strategies, shown in Table 3.4, are considered. Figure 3.13 shows the relationship between the delay threshold and the convergence and accuracy guarantees of these hybrid cases. Three main conclusions can be drawn from Figure 3.13: (1) Hybrid-1 and Hybrid-2 have worse convergence because Strategy I is employed in the uplink communication, while Strategy II is applied to the downlink communication. When $\tau \leq 0.8$, they force the algorithm to start to diverge, as shown in the left subfigure of Figure 3.13a; (2) A magnified view of Figure 3.13b shows that

Table 3.5 Sensitivity of gradient update to delay threshold

$\tau = P(X \leq d^*)$	0.1	0.2	0.3	0.4	0.5	0.6	0.7	0.9
Lost data – Downlink (k^*/α)	–	355/0.9914	218/0.9914	160/0.9916	132/0.9917	118/0.9916	108/0.9916	83/0.9913
Gradient method (k^*/α)	1117/0.9897	317/0.9913	218/0.9914	162/0.9916	130/0.9916	113/0.9917	102/0.9918	77/0.9914
Step size γ	−0.4	−0.2	0.001	0.015	0.01	0.9	0.8	1.9
Effective γ range	–	[−0.70.1]	[−0.90.5]	[−1.20.9]	[−1.61]	[−2.21.5]	[−2.52]	[−65]
Lost data – Bi-link (k^*/α)	–	–	–	–	239/0.9917	170/0.9912	148/0.9918	100/0.9917
Gradient method (k^*/α)	–	–	–	628/0.9911	230/0.9918	168/0.9912	133/0.9918	94/0.9917
Step size γ	–	–	–	−0.3	−0.3	−0.3	0.5	1
Effective γ range	–	–	–	–	[−10.3]	[−1.70.6]	[−1.81]	[−4.63.5]
Delayed data – Downlink (k^*/α)	113/0.9915	116/0.9914	114/0.9914	118/0.9915	118/0.9914	111/0.9915	102/0.9912	85/0.9913
Gradient method (k^*/α)	112/0.9916	107/0.9914	100/0.9915	89/0.9915	84/0.9914	77/0.9915	75/0.9913	78/0.9917
Step size γ	0.01	0.2	0.35	0.6	0.8	0.9	1	1.8
Effective γ range	[−0.30.12]	[−0.60.25]	[−10.45]	[−1.40.7]	[−1.80.8]	[−21.2]	[−2.51.7]	[−55.2]
Delayed data – Bi-link (k^*/α)	–	–	–	–	130/0.9913	128/0.9913	119/0.9914	93/0.9913
Gradient method (k^*/α)	–	–	–	196/0.9913	123/0.9914	105/0.9913	101/0.9914	85/0.9914
Step size γ	–	–	–	−0.02	0.3	0.6	0.8	2.3
Effective γ range	–	–	–	[−0.30.2]	[−1.20.5]	[−21]	[−2.51.5]	[−5.53.5]

when the delay thresholds $(\mathbf{d}_1^*, \mathbf{d}_2^*)$ are larger than 5.35 ms, namely $\tau \geq 0.5$, the standard Strategy II-Delayed Data has best convergence with accuracy above 99.13%; (3) Finally, when $\tau \in [0.1, 0.4]$, Hybrid-3 and Hybrid-4 still guarantee convergence, while both Strategy II – lost data and Strategy II – delayed data start to diverge. Therefore, if we need to set the delay thresholds to small values, Hybrid-3 is the best choice.

5. *Strategy II with gradient update*: This strategy is most effective when the initial conditions for the local updates are noticeably different from each other. To demonstrate this method we set $\mathbf{a}_1^{(0)} = 0.01\mathbf{1}, \mathbf{a}_2^{(0)} = 0.1\mathbf{1}, \mathbf{a}_3^{(0)} = \mathbf{1}, \mathbf{a}_4^{(0)} = 10\mathbf{1}$, where $\mathbf{1}$ is a vector of ones with 40 rows (the number of unknown coefficients of the characteristic polynomial is considered to be 40). The convergence guaranteed by S-ADMM in this case is $k^* = 117$, and accuracy (on a scale of 1) is $\alpha = 0.9938$. Table 3.5 shows that the gradient method improves convergence k^* for relatively the same accuracy factor α by tuning the step size γ, considering $\gamma_i = \gamma, \forall i = 1, ..., N$. The symbol "–" in the table means that the algorithm diverges. The stable range of $\tau \in [0.2, 0.9]$ for Downlink and $\tau \in [0.4, 0.9]$ for Bi-link is increased to $\tau \in [0.1, 0.9]$, and $\tau \in [0.5, 0.9]$, respectively, after the gradient update. Table 3.5 also shows the optimal value of γ for a given τ. It should be noted that when τ is large, i.e., when the delay threshold is large, then γ must be chosen to be large as well for optimal convergence. The range of stable γ, however, is increased in that condition.

The heuristic strategy involving spatial correlation analysis for this example shows comparable results to the above, especially when the initial conditions are scattered. However, we defer the details of those results to future for the sake of brevity and space limitation.

3.8 Conclusions

In this chapter, we presented four cyber-physical estimation algorithms for wide-area oscillation monitoring in large transmission-level power grids using synchrophasors. Our algorithms demonstrate how multitudes of geographically dispersed PMUs and PDCs can communicate with each other, and how the various binding factors in the network protocols can pose bottlenecks for their communication. The results provide valuable insights and guidance in deploying future PMU and PDC infrastructures, not only for power systems but also for any generic cyber-physical sensor network where monitoring and control decisions need to be made under critical time-constraints. Our future work will be to evaluate the reliability of the proposed architectures under different cyber-attack scenarios such as denial-of-service (DoS) attacks that can lead to malicious delays, similar to the ones considered in our proposed asynchronous estimation strategies. Besides DoS attacks, analyzing the impact of eavesdropping and data corruption attacks on the reliability and resilience of these CPS architectures will constitute yet another topic of our future research.

References

[1] A. Phadke and J. Thorp, *Synchronized Phasor Measurements and Their Applications*, Springer, NY, 2008.

[2] A. Bose, "Smart Transmission Grid Applications and Their Supporting Infrastructure," *IEEE Transactions on Smart Grid*, 2010, vol. 1(1), pp. 11–19.

[3] P. Myrda, P. Taft, and P. Donner, "Recommended Approach to a NASPInet Architecture," *45th Hawaii International Conference on System Science (HICSS)*, 2012.

[4] D. Bertsekas and J. Tsitsiklis, *Parallel and Distributed Computation: Numerical Methods*, Prentice-Hall, Inc., 1989.

[5] T. Chang, M. Hong, and X. Wang, "Multi-Agent Distributed Optimization with Inexact Consensus ADMM," *IEEE Transactions on Signal Processing*, 2015, vol. 63(2), pp. 482–497.

[6] C. Taylor, D. Erickson, K. Martin, R. Wilson, and V. Venkatasubramanian, "WACS Wide-Area Stability and Voltage Control System: R & D and Online Demonstration," *Proceedings of the IEEE*, 2005, vol. 93(5), pp. 892–906.

[7] R. Hasan, R. Bobba, and H. Khurana, "Analyzing NASPInet Data Flows," *IEEE Power Systems Conference and Exposition (PSCE)*, WA, 2009.

[8] J. Sanchez-Gasca and J. Chow, "Performance Comparison of Three Identification Methods for the Analysis of Electromechanical Oscillations," *IEEE Transactions on Power Systems*, 1999, vol. 14(3), pp. 995–1002.

[9] J. Hauer, C. Demeure, and L. Scharf, "Initial Results in Prony Analysis of Power System Response Signals," *IEEE Transactions on Power Systems*, 1990, vol. 5(1), pp. 80–89.

[10] N. Zhou, D. Trudnowski, J. Pierre, and W. Mittelstadt, "Electromechanical Mode Online Estimation Using Regularized Robust RLS Methods," *IEEE Transactions on Power Systems*, 2008, vol. 23(4), pp. 1670–1680.

[11] A. Messina and V. Vittal, "Nonlinear, Non-stationary Analysis of Interarea Oscillations via Hilbert Spectral Analysis," *IEEE Transactions on Power Systems*, 2006, vol. 21(3), pp. 1234–1241.

[12] S. Boyd, N. Parikh, E. Chu, B. Peleato, and J. Eckstein, "Distributed Optimization and Statistical Learning via the Alternating Direction Method of Multipliers," *Foundations and Trends in Machine Learning*, 2011, vol. 3(1), pp. 1–122.

[13] S. Kar and G. Hug, "Distributed Robust Economic Dispatch in Power Systems: A Consensus + Innovations Approach," *IEEE PES General Meeting*, 2012.

[14] Z. Zhang and M. Chow, "Convergence Analysis of the Incremental Cost Consensus Algorithm Under Different Communication Network Topologies in a Smart Grid," *IEEE Transactions on Power Systems*, 2012, vol. 27(4), pp. 1761–1768.

[15] E. Dall'Anese, H. Zhu, and G. Giannakis, "Distributed Optimal Power Flow for Smart Microgrids," *IEEE Transactions on Smart Grid*, 2013, vol. 4(3), pp. 1464–1475.

[16] A. Dominguez-Garcia and C. Hadjicostis, "Coordination of Distributed Energy Resources for Provision of Ancillary Services: Architectures and Algorithms," in *Encyclopedia of Systems and Control*, J. Baillieul and T. Samad (Eds.), Springer-Verlag, London, 2014, pp. 241–246.
[17] T. Erseghe, "Distributed Optimal Power Flow Using ADMM," *IEEE Transactions on Power Systems*, 2014, vol. 29(5), pp. 2370–2380.
[18] S. Nabavi and A. Chakrabortty, "A Real-Time Distributed Prony-Based Algorithm for Modal Estimation of Power System Oscillations," *American Control Conference (ACC)*, Portland, OR, 2014.
[19] S. Nabavi and A. Chakrabortty, "Distributed Estimation of Inter-area Oscillation Modes in Large Power Systems Using Alternating Direction Multiplier Method," *IEEE PES General Meeting*, National Harbor, MD, 2014.
[20] J. Zhang, S. Nabavi, A. Chakrabortty, and Y. Xin, "Convergence Analysis of ADMM-Based Power System Mode Estimation Under Asynchronous Wide-Area Communication Delays," in *Proceedings of IEEE PES General Meeting*, Denver, CO, 2015.
[21] A. Fouad and P. Anderson, *Power System Control and Stability*, IEEE Press, Piscataway, NJ, 2003.
[22] E. Wei and A. Ozdaglar, "On the O(1/k) Convergence of Asynchronous Distributed Alternating Direction Method of Multipliers," ArXiv preprint, 2013 [online] Available: *http://arxiv.org/abs/1307.8254*
[23] E. Wei and A. Ozdaglar, "Distributed Alternating Direction Method of Multipliers," *IEEE Conference on Decision and Control*, 2012, pp. 5445–5450.
[24] G. Hooghiemstra and P. Van Mieghem, "Delay Distributions on Fixed Internet Paths," *Technical Report*, 20-011-020, Delft University, 2001.
[25] T. Qian, A. Chakrabortty, F. Mueller, and Y. Xin, "A Distributed Storage System for Multi-Resolution Virtual Synchrophasors," *IEEE PES General Meeting*, Washington DC, 2014.
[26] J. Chow and K. W. Cheung, "A Toolbox for Power System Dynamics and Control Engineering Education and Research," *IEEE Transactions on Power Systems*, 1992, vol. 7(4), pp. 1559–1564.

Chapter 4

Cyber-physical strategies for generator coherency in the face of malicious attack

Pirathayini Srikantha[1], *Jin (Wei) Kocsis*[2] *and Deepa Kundur*[1]

Abstract

The modern electric grid is rapidly evolving into a diverse system composed of cyber-physical entities such as distributed energy sources (DERs), data concentrators, and phasor measurement units (PMUs) that allow for intelligent actuation and active monitoring. Cyber-physical integration not only enables enhanced grid security and adaptive resilience but also promotes sustainable grid operations. Security is paramount in a system that is as tightly coupled as the grid. Without appropriate security mechanisms in place, failures in the grid can rapidly propagate vertically and result a catastrophic system-wide collapse which will be very expensive to recover from. Although cyber-physical entities in the grid are used primarily to mitigate threats, if existing communication vulnerabilities in the cyber components are not appropriately administered, these can be leveraged by adversaries to gain unauthorized access to highly sensitive physical grid components which can then be maneuvered to inflict serious damage to the grid. As such, we present two attack-mitigation approaches inspired by biology and differential games that actively engage uncompromised cyber-physical entities to restore system coherency in the face of malicious cyber attacks.

4.1 Introduction

Information and communication infrastructure will be extensively deployed to monitor and control electric power delivery components of today's power grid. While these cyber elements enhance a utility's ability to maintain physical stability, if a subset is compromised by adversaries, disruption of reliable service may occur. Recent research has introduced a variety of adversarial actions on power system cyber infrastructure including false data injection attacks, switching attacks, and attacks on the feedback

[1]Department of Electrical and Computer Engineering, University of Toronto, Toronto, Canada
[2]Department of Electrical and Computer Engineering, University of Akron, Akron, OH, USA

loops of control systems [1–3]. One significant impact of such cyber attacks on power system dynamics is on generator coherency [4]. Phase angle and frequency deviations caused by disturbances will affect the coherency between generators which can lead to adverse effects such as transient grid instability. Once an attack is applied and detected, active approaches, often beyond protection devices, are needed to resynchronize generators.

In this chapter we investigate the application of control-based approaches that leverage distributed energy resources (DERs) to facilitate generator coherency in the face of cyber attacks on smart grid systems. We argue that the coordinated and often stealthy nature of cyber attacks in the context of power systems makes their impact particularly challenging to mitigate. The ability to harness DERs, facilitated through the use of information technology that we explore in this chapter, is a critical advantage to adding system resilience, whereby the grid can protect and recover from insidious attacks. We focus on two cyber-physical approaches for improved resilience.

In the first approach, we leverage biologically inspired nonlinear approaches to model generator coherency [5,6]. Analogies of synchronous generator behavior to bird-like flocks are employed to develop a control paradigm that rapidly addresses security attacks. We demonstrate via simulation studies how information and physical couplings can be synergistically harnessed for restabilizing a power grid under severe fault and cyber attack. Through analysis we assess how hierarchy and the selective use of cyber information can benefit scalability and robustness to information delay. In addition, through a flocking-based paradigm we develop distributed control methodologies that leverage cooperation between DERs and traditional synchronous machines to maintain transient stability in the face of severe disturbances.

Next, we consider the situation in which an adversary controls a subset of DERs for the purpose of generator desynchronization. In lieu of protection devices such as breakers that reduce the system topology, the electric power system employs a complementary subset of uncorrupted DERs that have rapid response capability to aid in generator synchronization. The adversary attempts to construct attack control vectors that maximize state deviations from stable set points to instigate generator clustering. We assert that, in essence, the adversary behaves like an evader attempting to deviate the system from stability [7]. Upon detecting the onset of an attack, the electric power utility (EPU) will design a countermeasure control vector that minimizes state deviations for a fixed time by which time the incident can be isolated. If the EPU takes no counter action, then there is potential for the system to move in the direction favorable to the adversary. The EPU functions like a pursuer. Within this context, we present a nonlinear differential game theoretic formulation that demonstrates how opponents can harness a subset of DERs to steer normal grid operation away from stability while mitigating control can resynchronize generators using complementary system resources.

4.2 System settings

In the traditional power system, passive mechanisms such as circuit breakers are typically utilized to isolate faults. The resulting power system topology reduction can cause increased strain on the system. Moreover, stealthy sophisticated coordinated attacks can bypass these passive protection devices until the system reaches a point where critical power grid components are damaged beyond repair. It is necessary to engage active measurement and actuation components such as phasor measurement units (PMUs) and fast acting DERs with communication capabilities to achieve effective real-time situational awareness, monitoring, and protection [8]. This cyber-physical integration allows the system to engage in mechanisms that enable recovery from faults in a rapid, distributed, and automated manner. In order to understand how cyber components affect physical changes on power components, we first present a unified dynamical model that effectively captures these cyber-physical interactions. Then, we introduce smart grid stability metrics that we use in subsequent sections to design distributed control schemes that maintain robust system operation in the event of faults or attacks.

4.2.1 Cyber-physical dynamical smart grid model

In our power system model, we define a *cyber-physical agent* to consist of a critical physical component (i.e., a synchronous generator), a PMU, a fast acting power source (i.e., DERs), and a local cyber controller which can actuate the corresponding DER [9]. Every synchronous generator is therefore associated with a cyber-physical agent. As it suffices to consider only critical nodes (i.e., buses at which current is actively injected), the power system topology can be Kron-reduced to consist of only the cyber-physical agents. Without the DER, the physical state of agent i is completely determined by the phase angle θ_i and frequency ω_i of the synchronous generator associated with agent i and the dynamical evolution of these states can be effectively modelled via the well-known swing equations [10,11]. In the event of an attack, DER i will begin to absorb or inject power as dictated by the local cyber controller and this cyber actuation will also contribute to changes in the physical state of agent i. In order to capture cyber-physical interactions on system states, Wei *et al.* [5] first proposed the modification of swing equations to the following:

$$M_i\dot{\omega}_i = -D_i\omega_i + P_{m,i} - |E_i|^2 G_{ii} - \sum_{j=1}^{N} P_{ij}\sin(\theta_i - \theta_j + \varphi_{ij}) + u_i \tag{4.1}$$

where N is the total number of agents in the system, θ_i is the rotor phase angle measurement made with respect to a frame of reference rotating at 60 Hz, and ω_i is the relative normalized frequency (i.e., $\omega_i = \dot{\theta}_i$). M_i, D_i, E_i, $P_{m,i,}$ and G_{ii} represent the inertia, damping parameters, internal voltage, mechanical power input, and shunt conductance of generator in agent i, respectively. $P_{ij} = |E_i||E_j||Y_{ij}|$ and $\varphi_{ij} = \tan^{-1}(G_{ij}/B_{ij})$ where Y_{ij}, G_{ij}, and B_{ij} are equivalent Kron-reduced admittance, conductance, and susceptance values. u_i is the power injected (positive) or absorbed (negative) by the DER associated with agent i. u_i is a function of measurements generated by PMUs and is computed by the local cyber controller. This signal is sent to DER i which then actuates this power into the system. Hence, u_i essentially captures the cyber-physical interaction between the PMU, local controller, and DER associated with agent i.

4.2.2 Smart grid stability

System states are subject to certain restrictions for grid stability. In this chapter, we focus on a specific type of stability called the *transient stability* that describes the ability of the power system to return to stable equilibrium within a short time period in the event of a sudden disturbance in the grid [12]. The main objective of the control laws u_i we present later in this chapter is to preserve transient stability when the grid is subjected to malicious attack. Transient stability holds if the following two constraints are satisfied by the system states:

1. Phase angle cohesiveness :

   $|\theta_i(t) - \theta_j(t)| \leq \gamma, \ \forall \, i,j \in N$

2. Exponential frequency synchronization

   $\omega_i(t) \to 0, \ t \to \infty, \ \forall \, i \in N$

The *phase angle cohesiveness* constraint requires the absolute difference between phase angles of all synchronous generators in the system to be within γ where γ is typically $\frac{5\pi}{9}$ [13,14]. When this threshold is not heeded, greater loss in power results which further increases the separation in phase angles. *Exponential frequency synchronization* requires frequencies of all generators to asymptotically reach the nominal value in the event of a fault. Following a severe disturbance, the states of all synchronous generators must return to these stable thresholds within 1–3 s [15].

4.3 Flocking-based control for smart grid resilience

Equipped with the necessary modelling tools and stability metrics, we first present a biologically inspired distributed control mechanism, based on earlier works [6,9,15,16], that is able to effectively restore transient stability in the event of disturbances stemming from attacks or faults. This control scheme is based on flocking behavior commonly observable in nature (i.e., birds). Flocking preliminaries are introduced next. Then, a two-tier control hierarchy architecture consisting of cyber-physical agents divided into multiple clusters based on physical coherency is presented. Within each cluster, only one agent is nominated to perform actual actuation. Since not all agents are not actively participating in the control scheme, scalability of the flocking-based control algorithm is greatly enhanced.

4.3.1 Flocking analogy

There are three main heuristic properties associated with flocking theory and formation control as follows [17]:

- **Flock centering:** Agents remain close to nearby flockmates.
- **Velocity matching:** Agents attempt to match velocity of nearby flockmates.
- **Goal seeking:** Agents aim to obtain a desired velocity toward a specific position in the global space.

Given that all agents in a flock adhere to the above three properties, the state dynamics of these agents are governed by:

$$\begin{aligned} \dot{q} &= p \\ \dot{p} &= \tilde{u} \end{aligned} \tag{4.2}$$

where $q \in \mathbb{R}^N$ and $p \in \mathbb{R}^N$ represent the position and velocity vectors of all N agents in the flock. $\tilde{u} \in \mathbb{R}^N$ is a vector of control signals used by agents to alter their respective physical states to achieve the three aforementioned flocking goals. $\tilde{u}$ has the following general form:

$$\tilde{u} = -\nabla V(q) - L.p + F(p, q, p_r, q_r) \tag{4.3}$$

where the first term is the gradient of the potential energy function $V(q)$ that characterizes the system objectives and constraints, the second term is a velocity consensus protocol where L represents the communication graph of agents, and the third term captures navigational feedback that enables an agent to track its current state with respect to the reference p_r and q_r.

Maintaining transient stability in the power grid also requires critical power system components such as synchronous generators to work toward transient stability goals that are analogous to those of agents in a flock. For instance, requirements on phase angle cohesiveness and frequency stabilization for transient stability directly correspond to the flock centering and velocity matching goals, respectively, in the flock formation control problem.

4.3.2 Two-tier hierarchy for scalable control

To enhance control scalability especially in a large power system, a two-tier hierarchy based on Reference 15 is presented next that divides cyber-physical agents into clusters according to their physical coherency characteristics. Identification of generator coherency is presented in detail in Section 4.3.4. The lead agent selected to represent a cluster consists of the synchronous generator with the greatest inertia in comparison to all other generators in the cluster. It is assumed that significant communication security measures are added to lead agents so that these cannot be easily compromised. This is feasible as the lead agents represent a small percentage of all agents in the system. All other agents in that cluster are secondary agents. Communication occurs only among lead agents via phasor data concentrators (PDCs). PDCs synchronize state measurement information of all lead agents in the system. Actuation is performed by only lead agents not secondary agents. Hence, inter-cluster communications take place on the cyber plane among lead agents and intra-cluster interactions are purely physical as illustrated in Figure 4.1.

In order to reflect this hierarchy, α_i is introduced to the system dynamics from (4.1):

$$M_i\dot{\omega}_i = -D_i\omega_i + P_{m,i} - |E_i|^2 G_{ii} - \sum_{j=1}^{N} P_{ij}\sin(\theta_i - \theta_j + \varphi_{ij}) + \alpha_i u_i \tag{4.4}$$

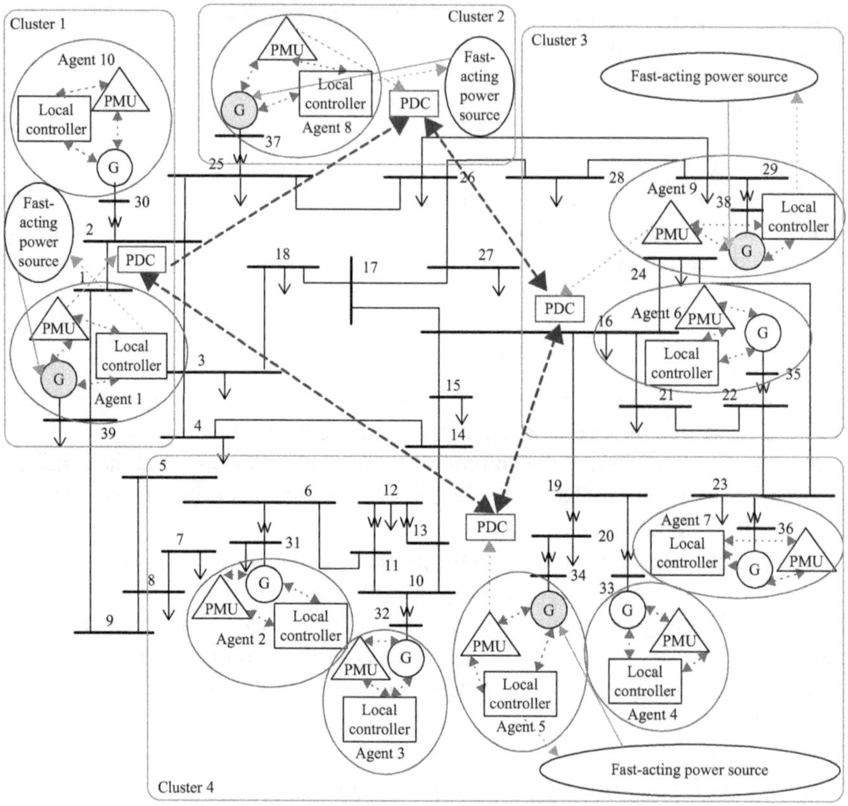

Figure 4.1 *Hierarchical control of a 10-generator 39-bus system. ©2014 IEEE. Reprinted with permission from Reference 15*

where $\alpha_i = 1$ if agent i is the lead agent or 0 otherwise. Phase angle and normalized frequency are both analogous to position and velocity states in the flocking control problem. The swing dynamics of the power system can be expressed in a manner similar to (4.2) by taking the first derivative of (4.4):

$$\begin{aligned} \dot{\theta}_i &= \omega_i \\ D_i\dot{\omega}_i &= -M_i\ddot{\omega}_i - \sum_{j=1}^{N} P_{ij}\cos(\theta_i - \theta_j + \varphi_{ij})(\omega_i - \omega_j) + \alpha_i\dot{u}_i \end{aligned} \tag{4.5}$$

In order to arrive at the above dynamics, it is assumed that the DERs act much faster than the mechanical power input $P_{m,i}$ and the voltage regulation problem is decoupled from the frequency regulation problem. Hence, $P_{m,i}$ and $|E_i|$ are constant terms. Let L be an $N \times N$ matrix and the (i,j) component of L is defined as:

$$l_{ij} = \begin{cases} \sum_{k=1,k\neq i}^{N} P_{ik}\cos(\theta_i - \theta_k + \varphi_{ik}), & \text{if } i = j; \\ -P_{ij}\cos\left(\theta_i - \theta_j + \varphi_{ij}\right), & \text{if } i \neq j, \end{cases} \tag{4.6}$$

L is referred to as the *physical coupling matrix*. Equation (4.5) can be expressed in vector form as follows:

$$\begin{aligned}\dot{\theta} &= \omega \\ D\dot{\omega} &= -M\ddot{\omega} - L\omega + \alpha\dot{u}\end{aligned} \tag{4.7}$$

where $\theta = [\theta_1 \dots \theta_N]^T$ and $\omega = [\omega_1 \dots \omega_N]^T, u = [u_1 \dots u_N]^T$. D, M, and α are diagonal $N \times N$ matrices consisting of D_i, M_i, and α_i in the corresponding diagonal components of the matrices. The $\ddot{\omega}_i$ term must be eliminated to complete the process of transforming the transient stability control problem into the flocking formation control problem. For this, the system must be over-damped (i.e., the coefficient of the double derivative term must be $\mathcal{O}(0.1)$) so that singular perturbation technique can be applied to remove this term and currently, $M_i/D_i \in \mathcal{O}(10)$ [15]. Selecting $\dot{u}$ to have the form:

$$\alpha\dot{u} = \alpha\tilde{u} + L\omega - \alpha B\dot{\omega} \tag{4.8}$$

where B has a form similar to L. Combining (4.7) and (4.8), the following state dynamics result:

$$\begin{aligned}\dot{\theta}_l &= \omega_l \\ \mathcal{M}_l\dot{\omega}_l &= \tilde{u}_l\end{aligned} \tag{4.9}$$

where l is the index of the lead agent and $\mathcal{M}_l = D_{ll} + B_{ll}$. B_{ll} which is the lth diagonal element of matrix B is selected so that $M_l/(D_{ll} + B_{ll}) \in \mathcal{O}(0.1)$. Hence, (4.9) represents the transformation of the transient stability problem into the flocking formation control problem.

The purpose of introducing a two-tier hierarchy is to reduce information flow in the cyber system so that control is made tractable in a large system. Since only lead agents perform actuation and the control law these use must satisfy the general form listed in (4.8), which depends on the states of all agents in the system, in the following, we describe how states of secondary agents can be approximated from states of lead agents. Agents belonging to the same cluster have similar physical coherency. Hence, the state of a secondary agent i can be assumed to be the state of its lead agent k plus a noise term as follows:

$$\begin{aligned}\omega_i &= \omega_k + \Delta_i \\ \theta_i &= \theta_k + \varepsilon_i^0 + \varsigma_i\end{aligned}$$

where ε_i^0 is the difference in phase angles of the lead and secondary agents prior to the attack. Δ_i and ς_i are noise terms with uniform distributions $\mathcal{U}[-a, a]$ and $\mathcal{U}[-b, b]$, respectively, where $|a|, |b| \ll 1$. The physical coupling matrix L can also be approximated in a similar manner. For notational convenience, all terms in the matrices are ordered so that the first C terms correspond to the lead agents and the remaining $N - C$ terms map to the secondary agents. $\mathbf{L}$ can be divided into four sub-matrices as follows:

$$\mathbf{L} = \begin{bmatrix} \mathbf{R}_{C\times C} & \mathbf{S}_{C\times(N-C)} \\ \mathbf{T}_{(N-C)\times C} & \mathbf{U}_{C\times C} \end{bmatrix}$$

S can be approximated as:

$$\hat{\mathbf{S}}(i,j) = -P_{ij}\cos(\theta_{A_i} - \theta_{A_j} - \varepsilon_i^0 + \varepsilon_j^0 + \varphi_{ij}) \tag{4.10}$$

where θ_{A_i} and θ_{A_j} are phase angles of lead agents A_i and A_j of clusters that secondary agents i and j belong to. **R** can be approximated using **S** as follows:

$$\widehat{\mathbf{R}}(i,j) = \begin{cases} \mathbf{R}(i,j), & \text{if } i \neq j; \\ -\sum_{j=1,j\neq i}^{C} \mathbf{R}(i,j) - \sum_{j=C+1}^{N} \widehat{\mathbf{S}}(i,j), & \text{otherwise.} \end{cases} \tag{4.11}$$

Combining (4.8) and (4.9), along with approximations of **R**, the dynamics of the lead and secondary agents can be expressed as follows:

1. *The lead agents (tier-1)*

$$\begin{cases} \dot{\boldsymbol{\theta}}_l = \boldsymbol{\omega}_l, \\ \mathcal{M}_l\dot{\boldsymbol{\omega}}_l = \tilde{\mathbf{u}}_l - \widehat{\mathbf{S}}\boldsymbol{\Delta}, \end{cases} \tag{4.12}$$

   where $\boldsymbol{\Delta} = [\Delta_{C+1}\ldots\Delta_N]^T, \boldsymbol{\theta}_l = [\theta_1\ldots\theta_C]^T$, and $\boldsymbol{\omega}_l = [\omega_1\ldots\omega_C]^T$. The second term $\widehat{\mathbf{S}}$ is a noise term that takes values close to 0.
2. *The secondary agents (tier-2)*

$$\begin{cases} \dot{\boldsymbol{\theta}}_s = \boldsymbol{\omega}_s, \\ \mathbf{D}_s\dot{\boldsymbol{\omega}}_s = -\mathbf{L}_s\boldsymbol{\omega}_s - \mathbf{M}_s\ddot{\boldsymbol{\omega}}_s, \end{cases} \tag{4.13}$$

   where $\boldsymbol{\theta}_s = [\theta_{C+1}\ldots\theta_N]^T$ and $\boldsymbol{\omega}_s = [\omega_{C+1}\ldots\omega_N]^T$.

4.3.3 Control laws for lead agents

Flocking formation control requires $\tilde{u}$ to have the general form as listed in (4.3). $\tilde{u}$ in the dynamics of the lead agents is designed to take the same form as follows:

$$\tilde{\mathbf{u}}_l = -\Phi - \widetilde{\mathbf{L}}\boldsymbol{\omega}_l - c_1(\boldsymbol{\omega}_l - \boldsymbol{\omega}^*) \tag{4.14}$$

where Φ is a potential function gradient, $\widetilde{\mathbf{L}}$ is the frequency consensus cyber coupling matrix, and the last term is the tracking function that compares current frequency $\boldsymbol{\omega}_l$ with the targeted frequency $\boldsymbol{\omega}^*$. Φ is selected as follows:

$$\Phi(i) = \sum_{j=1,j\neq i}^{C} \phi\left(\theta_i - \theta_j\right), \tag{4.15}$$

$$\text{where } \phi(z) = \begin{cases} 0, & \text{if } |z| \leq \frac{5\pi}{9}; \\ 4c_2 z\left(z^2 - \frac{25\pi^2}{81}\right), & \text{otherwise.} \end{cases}$$

and components of $\widetilde{\mathbf{L}}$ take values:

$$\tilde{l}_{ij} = \begin{cases} c_3, & \text{if } i = j; \\ -\frac{c_3}{C-1}, & \text{otherwise.} \end{cases} \tag{4.16}$$

c_1, c_2, and c_3 are positive constants. The control law in (4.14) guarantees transient stability and this is rigorously proven in Reference 15. In order to compute the actual power u_l that lead agent l requires to inject or absorb, (4.8) and (4.14) are combined and integrated. Given that an attack occurs at $t = t_0$, the control law for lead agent l is:

$$\mathbf{u}_l = \begin{cases} \mathbf{0}, & \text{for } t \leq t_0; \\ \int_{t_0}^{t} \dot{\mathbf{u}}_l d\tau = -\mathbf{\Gamma} + \boldsymbol{\eta} - \boldsymbol{\varrho}, & \text{for } t > t_0, \end{cases} \tag{4.17}$$

where $\mathbf{\Gamma} = \int_{t_0}^{t} \Phi d\tau$, $\eta = \int_{t_0}^{t} \left(\widehat{\mathbf{R}} + \widehat{\mathbf{S}}\Psi\right) \omega_l d\tau$, and $\varrho = \int_{t_0}^{t} \left(\widetilde{\mathbf{L}}\omega_l + c_1\omega_l + \mathbf{B}_l\dot{\omega}_l\right) d\tau$. Ψ is defined to be:

$$\Psi_{ij} = \begin{cases} 1, & \text{if the } (C+i)\text{th agent is in the } j\text{th cluster;} \\ 0, & \text{otherwise.} \end{cases}$$

Hence, it is clear that the lead agent can compute its control law by only using phase angle and frequency measurements of other lead agents which is assumed to be readily available via the PDCs. Given that the agents have been divided into clusters that accurately capture physical coherency, controls can be computed to guarantee convergence to states that result in transient stability. This agent clustering method based on Reference 16 is presented next.

4.3.4 Generator coherency identification

Grouping agents with similar physical coherency features is analogous to the clustering of bird-oid (boids) into multiple flocks with distinguishing features. Boids are associated with informational states that govern the dynamics of these entities. In order to make analogous inferences on the clustering of cyber-physical agents, first, the states of the agents are translated from the observational to an informational space. In this case, the information associated with each boid is three-dimensional and the mapping is as follows:

$$\begin{cases} \mathscr{I}_i^1(k) = \theta_i(k) \\ \mathscr{I}_i^2(k) = \omega_i(k) \\ \mathscr{I}_i^3(k) = \delta_i(k) \end{cases} \tag{4.18}$$

where $\theta_i(k)$ is the phase angle, $\omega_i(k)$ is the normalized frequency color red, and $\delta_i(k)$ is the acceleration of agent i at time k. Each boid i is associated with a two-dimensional state $S_i(k) = [p_i(k)\ v_i(k)]^T$ consisting of its position and velocity at time k. The neighbors N_i of boid i are other boids whose distance from boid i is within d_c:

$$\mathscr{N}_i(k) = \{\forall j \,|\, \|\mathbf{p}_i(k) - \mathbf{p}_j(k)\| < d_c\} \tag{4.19}$$

The state dynamics of the boid is modelled as:

$$\begin{cases} \mathbf{v}_i(k+1) = \mathbf{v}_i(k) + \Delta t \sum_{l=1}^{3} w_l \mathbf{g}_{i,l}(k), \\ \mathbf{p}_i(k+1) = \mathbf{p}_i(k) + \Delta t \mathbf{v}_i(k), \end{cases} \tag{4.20}$$

where Δt is the length of a time step and the $w_l \mathbf{g}_{i,l}$ term captures the acceleration induced by the three flocking rules of flock centering, velocity matching, and obstacle avoidance. Neighboring boids do not necessarily belong to the same cluster. It is possible to plot the trajectories of the boids to visually identify multiple distinct flocks

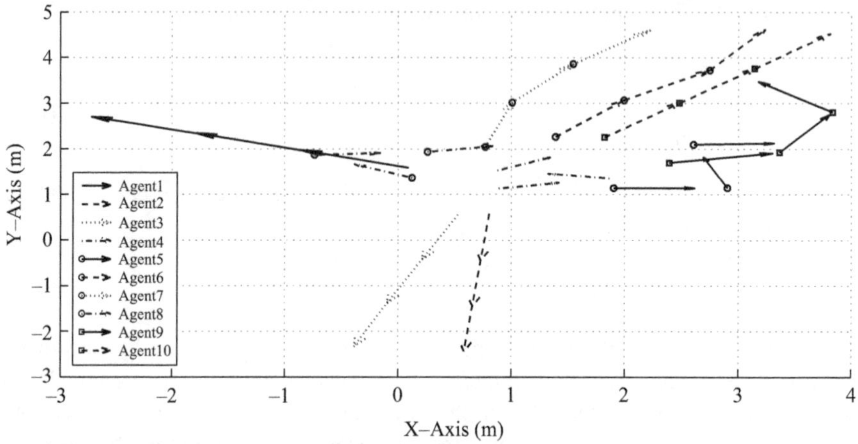

Figure 4.2 Trajectories of boids. ©2014 IEEE. Reprinted with permission from Reference 15

which are then grouped into clusters. A more precise method of clustering based on physical coherency is computing the feature similarity $\zeta_{ij}(k)$ of boids i and j:

$$\zeta_{ij}(k) = \left| \sum_{n=1}^{3} \alpha_n \left(\mathscr{I}_i^n(k) - \mathscr{I}_j^n(k) \right) \right| \tag{4.21}$$

where α_n is the weight associated with the impact of information component n on the interactions of the boid. If $\zeta_{ij}(k)$ is within the threshold $\zeta_{th}(k)$ then boids i and j belong to the same cluster.

4.3.5 Case study

In order to assess the effectiveness of the hierarchical flocking-based transient stability control scheme, the 10-machine 39-bus New England system depicted in Figure 4.1 is considered [15]. The cyber control parameters are configured to be $c_1 = 3$, $c_2 = 5$, and $c_3 = 1/10$. All simulations are conducted via MATLAB/Simulink. At time $t = 0$ s, a short circuit, that may have been caused by an attack, occurs at Line 14–15 which is then removed at $t = 0.1$ s. After the attack has commenced, the physical behavior of the agents is observed for 0.05 s and based on this observation the trajectories of the boids corresponding to the agents are plotted in Figure 4.2. It is clear from these trajectories that three distinct clusters are present and these are $\{Agent_1\}$, $\{Agent_2, Agent_3\}$, and $\{Agent_4, \ldots, Agent_{10}\}$. *Agent* 1, *Agent* 3, and *Agent* 4 are selected to be lead agents of the clusters as these have the highest inertia within their corresponding clusters. Then, hierarchical control laws derived earlier are applied by these lead agents.

As illustrated in Figure 4.3, when there is no active control in place, the system goes into transient instability. Results in Figure 4.4 depict system states when hierarchical control is applied to the lead agents. It is clear that the states of all agents in

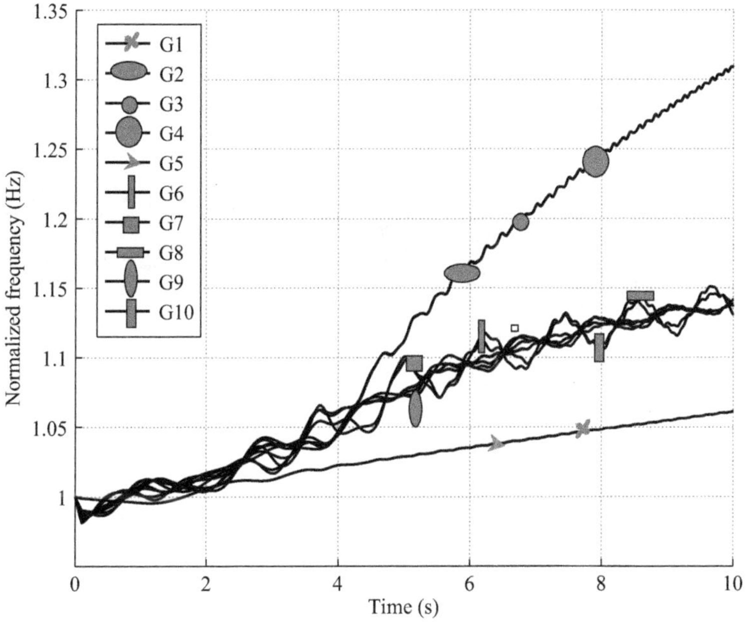

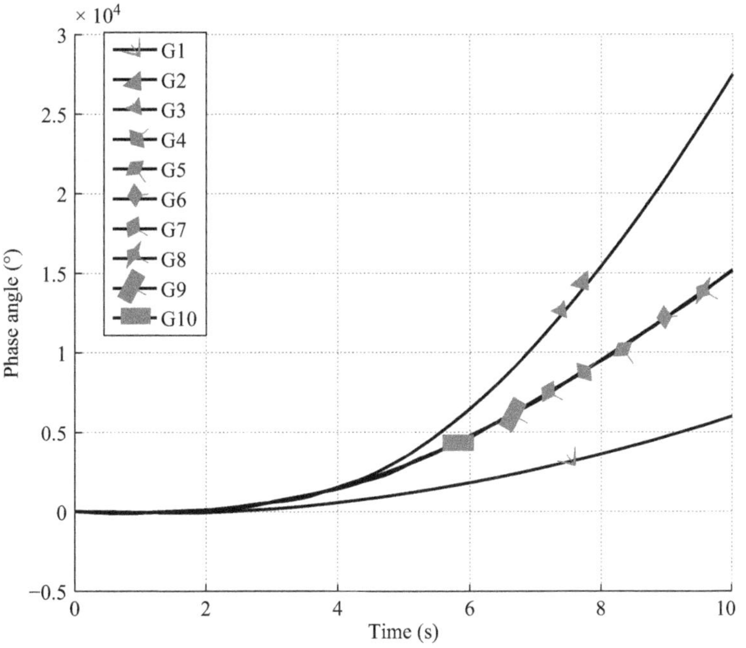

Figure 4.3 System states without active control. ©2014 IEEE. Reprinted with permission from Reference 15

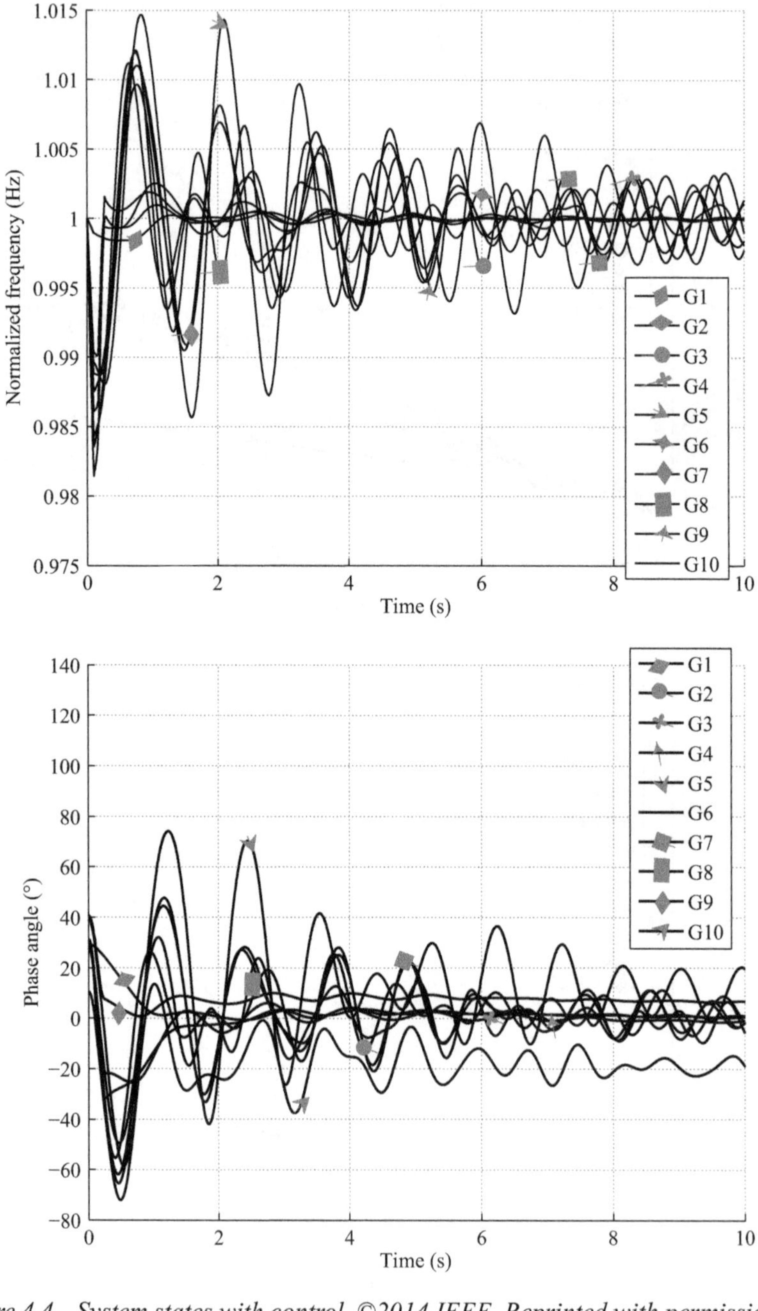

Figure 4.4 System states with control. ©2014 IEEE. Reprinted with permission from Reference 15

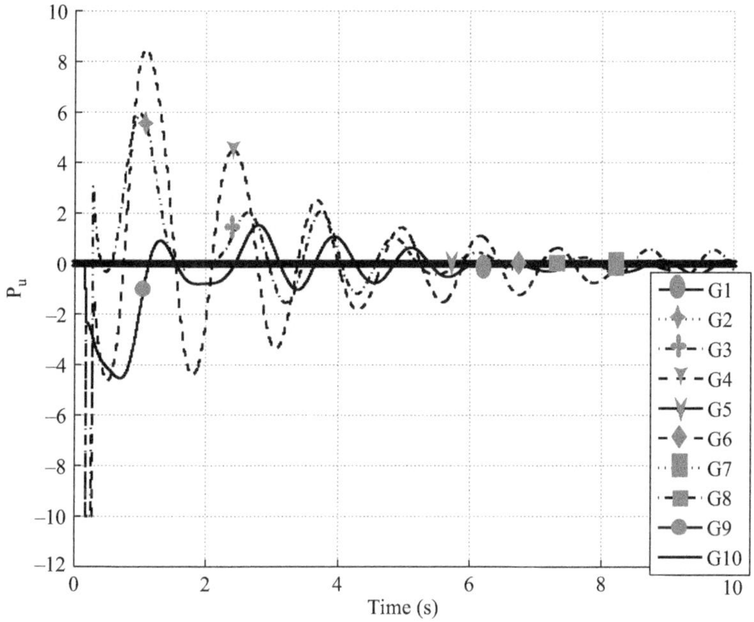

Figure 4.5 Power actuated by DERs in lead agents. ©2014 IEEE. Reprinted with permission from Reference 15

the system remain within the transient stability margins. Finally, the control inputs actually used by lead agents to maintain transient stability in Figure 4.4 is depicted in Figure 4.5. Hence, the selected lead agents based on the trajectories of Figure 4.2 are able to effectively reduce instability effects. From these results, it is clear that the hierarchical flocking-based control scheme effectively utilizes cyber-physical means in a scalable manner to actively mitigate the adverse effects of physical attacks on the grid.

4.4 DER attack-mitigation framework

The hierarchical flocking-based control scheme presented in the previous section effectively utilizes cyber-physical components to maintain grid stability in the advent of a physical attack. However, these cyber-physical components intended to add resilience to the grid can also be exploited via well-documented vulnerabilities associated with standard cyber communication protocols [18]. These vulnerabilities can be leveraged by adversaries to remotely actuate physical devices such as DERs to cause instability in the grid in a stealthy manner. These types of attacks are not only difficult to identify but the ensuing system instability can succeed in severely disrupting grid components. If the attack is identified in a timely manner, we show that the utility will be able to effectively mitigate adverse effects by strategically actuating the remaining unattacked active grid components. We formulate interactions between the attacker and utility as a nonlinear differential game which is then iteratively solved to obtain attack and countermeasure control vectors.

4.4.1 A pursuer-evader analogy for grid destabilization

The attacker in this framework will design control inputs targeting transient stability for a subset of active devices in the grid such as cyber-actuated DERs. As listed in Section 4.2.2, two constraints that dictate transient stability are phase cohesiveness and exponential frequency synchronization. If the phase angles and frequencies deviate significantly from nominal values, transient instability will ensue. Hence, the attacker will aim to maximize state deviations. On the other hand, the utility will attempt to utilize the remaining active devices to minimize these deviations. These opposing goals of the attacker and utility are analogous to those of an evader and a pursuer in the homicidal chauffeur problem [7]. The evader attempts to maximize the gap from the pursuer, while the pursuer attempts to minimize this gap. This analogy translates into a game formulation as detailed in the following section.

4.4.2 Nonlinear attack-mitigation differential game

At time t, the power system is described by state variables $\omega_i(t)$ and $\theta_i(t)$ which are frequency and phase angles corresponding to the generator associated with agent i. The vector x denotes the state variables in a system containing g agents and is defined as follows:

$$x(t) = \begin{pmatrix} \omega_1(t) \\ \cdots \\ \omega_g(t) \\ \theta_1(t) \\ \cdots \\ \theta_g(t) \end{pmatrix}$$

$u_i(t)$ is the power cyber-actuated at time t by an active element such as a DER associated with agent i. Each DER is subjected to an upper and lower actuation limits of P_u^i and P_l^i. If the pursuer (utility) has control over a subset $P \in G$ of cyber-physical agents in the system, then $u^p(t)$, the control vector representing the actuation of all unattacked DERs, is:

$$u^p(t) = \sum_{i \in P} u_i(t) e^i$$

where $e^i \in \mathbb{R}^{G \times 1}$ is a unit vector whose ith component is set to 1 and the remaining components are 0. If the evader (attacker) has access to the remaining controllers represented by the set $E = G \backslash P$, then the attack vector $u^e(t)$ is:

$$u^e(t) = \sum_{i \in E} u_i(t) e^i$$

$L(u^p(t), u^e(t), x(t))$ is the cost incurred by the utility over the attack period $[0\ T]$ which is defined as follows:

$$L(u^p(t), u^e(t), x(t)) = \int_0^T g(t, x(t), u^p(t), u^e(t))dt + q(T, X(T))$$

$g(.)$ is a function of system states and cyber actuation resulting from attack and countermeasure vectors $u^e(t)$ and $u^p(t)$ during the time interval $t = [0\ T]$. $q(T, X(T))$ is the cost of the final state $X(T)$ at the end of the control period. The evader will attempt to maximize utility's cost and the pursuer on the other hand will attempt to minimize these. The duration the evader has to cause an impact on the grid will be T. It is assumed that the utility will be able to isolate the compromised cyber elements and stabilize the grid within this clearing time. Since both players have opposing goals, it is clear that their actions will be governed by a zero-sum differential game subject to the following system dynamics:

$$\dot{x}(t) = F(x(t), u(t)) \tag{4.22}$$

where $F(.)$ is precisely the modified swing equation listed in (4.1) and $u(t) = u^p(t) + u^e(t)$. Assuming that the players are rational but selfish, the best response $u^p(t)$ for the pursuer is to select a strategy that minimizes maximum cost $L(.)$ over all possible strategies $u^e(t)$ that can be selected by the evader. The best response $u^e(t)$ for the evader is to select a strategy that maximizes the minimum cost $L(.)$ over all possible strategies $u^p(t)$ selected by the pursuer. These goals of the pursuer and evader are formally posed in $\mathscr{P}_p$ and $\mathscr{P}_e$, respectively.

$$\begin{aligned}\mathscr{P}_p : \quad & \min_{u^p} \max_{u^e} \quad L(u^p(t), u^e(t), x(t)) \\ & \text{subject to} \\ & \qquad \dot{x} = F(x(t), u(t)) \\ & \qquad p_l^i \leq u_i(t) \leq p_u^i \qquad \forall\, i = 1 \ldots g\end{aligned}$$

$$\begin{aligned}\mathscr{P}_e : \quad & \max_{u^e} \min_{u^p} \quad L(u^p(t), u^e(t), x(t)) \\ & \text{subject to} \\ & \qquad \dot{x} = F(x(t), u(t)) \\ & \qquad p_l^i \leq u_i(t) \leq p_u^i \qquad \forall\, i = 1 \ldots g\end{aligned}$$

Nash Equilibrium (NE) occurs when the strategies of both players computed by solving their respective problems $\mathscr{P}_p$ and $\mathscr{P}_e$ are such that any deviations from these will incur increased penalties for the corresponding player. If the control strategies selected by the pursuer and evader do result in NE, then the following condition also known as the *Isaacs condition* is satisfied [19]:

$$\min_{u^p} \max_{u^e} L(u^p(t), u^e(t), x(t)) = \max_{u^e} \min_{u^p} L(u^p(t), u^e(t), x(t))$$

If the utility does not detect the onset of the attack and performs no actuation, then the attacker will be successful in instigating transient instability using only $u_e(t)$ and this is illustrated for a specific case in Section 4.4.4.

4.4.3 Construction of attack-mitigation control strategies

There are several key challenges in solving $\mathscr{P}_p$ and $\mathscr{P}_e$. Firstly, the derivation of control strategies for the exact system dynamics is not possible due to nonlinearities in $F(.)$. Secondly, there is no guarantee of achieving a unique NE in this system.

In order to overcome computational intractability due to nonlinearities, the continuous-time nonlinear dynamics of the grid is discretized and linearized at every time step. Δt is the length of a time step. Hence, there are K such time steps in the control horizon if $T = \Delta t * K$. This transforms $\mathscr{P}_p$ and $\mathscr{P}_e$ into a form similar to a Linear Quadratic (LQ) two player zero-sum game [20]. The LQ game is a well-known formulation that consists of a quadratic cost function and affine constraints [21]. Optimal control strategies of the LQ game can be obtained analytically and these result in NE if the cost function is convex [22]. However, unlike the typical LQ game, our discretized formulation consists of system dynamic gradients that vary with time due to the application of control signals by the pursuer and the evader at each time step. For this reason, a different approach from that used to solve a typical LQ game is considered instead in this work. This approach iteratively takes into account the deviations δx_k of system dynamics x_k^{new} from the original trajectory x_k^{orig} due to incremental changes δu_k^e and δu_k^p in control signals $u^e(t)$ and $u^p(t)$. In the first iteration, u_k^e and u_k^p are initialized to zero vectors $\forall\, k = 1 \ldots K$. Incremental change in state dynamics, δx_{k+1}, is approximated as follows:

$$\delta x_{k+1} = x_{k+1}^{new} - x_{k+1}^{orig} \approx A_k \delta x_k + B_k^p \delta u_k^p + B_k^e \delta u_k^e \tag{4.23}$$

A_k is the gradient of the original system dynamics equation with respect to x_k (i.e., $\nabla_{x_k} F(x_k, u_k)$). B_k^e and B_k^p are the gradients of the system dynamics with respect to u_k^e and u_k^p (i.e., $\nabla_{u_k^p} F(x_k, u_k)$ and $\nabla_{u_k^e} F(x_k, u_k)$). These gradients are all evaluated in the open loop trajectory x_k^{orig} obtained by setting $u_k^e = 0$ and $u_k^p = 0\ \forall\, k = 1 \ldots K$. Next, the cost function is discretized as follows:

$$\begin{aligned} L(u^p, u^e) = {} & \frac{1}{2}(x_K^{orig} + \delta x_K)' Q (x_K^{orig} + \delta x_K) + \frac{1}{2}\sum_{k=1}^{K-1} [(x_k^{orig} + \delta x_k)' Q (x_k^{orig} + \delta x_k) \\ & + (u_k^{p*} + \delta u_k^p)' Q (u_k^{p*} + \delta u_k^p) - (u_k^{e*} + \delta u_k^e)' Q (u_k^{e*} + \delta u_k^e)] \end{aligned} \tag{4.24}$$

As Q is assumed to be a positive definite matrix, the cost function is convex in δu_k^p and concave in δu_k^e. According to Reference 19, a necessary condition for NE on δu_k^{e*}, δu_k^{p*}, and δx_k^* is the satisfaction of the Hamiltonian and co-state equations listed in (4.25) and (4.26).

$$\begin{aligned} H_k = {} & [(x_k^* + \delta x_k^*)' Q (x_k^* + \delta x_k^*) + (u_k^{p*} + \delta u_k^{p*})' Q (u_k^{p*} + \delta u_k^{p*}) \\ & - (u_k^{e*} + \delta u_k^{e*})' Q (u_k^{e*} + \delta u_k^{e*})] + p_{k+1}' [A_k \delta x_k^* + B^e \delta u_k^{e*} + B_k^p \delta u_k^{p*}] \end{aligned} \tag{4.25}$$

p_{k+1} is the co-state variable at time step $k+1$. Equation (4.25) is derived from the state equation listed in (4.23) and the cost function listed in (4.24).

$$p_k = A'_k[p_{k+1} + [Q(x^*_{k+1} + \delta x^*_{k+1})]'] \quad p_K = 0 \tag{4.26}$$

The co-state equations are obtained by taking a partial derivative of the hamiltonian in (4.25) with respect to δx_k.

From these relations, the control strategies for the evader and pursuer can be derived as listed in (4.27) and (4.28) by obtaining the partial derivatives of the expression in (4.25) with respect to δu^p_k and δu^e_k and setting it to 0. These expressions are then solved for δu^{*p}_k and δu^{*e}_k.

$$\delta u^{*p}_k = -[B^{p'}_k Q(x^*_{k+1} + \delta x^*_{k+1}) + B^{*p'}_k (S_k \delta x_{k+1} + v_{k+1})] \tag{4.27}$$

$$\delta u^{*e}_k = [B^{e'}_k Q(x^*_{k+1} + \delta x^*_{k+1}) + B^{*e'}_k (S_k \delta x_{k+1} + v_{k+1})] \tag{4.28}$$

The matrix S_k and vector v_k are unknown entities but have the general relation $p_k = S_k \delta x_k + v_k$. Equations (4.25–4.28) are manipulated to obtain δx_{k+1}, S_k, and v_k. Solving for these unknown entities will allow computation of the optimal change in control strategies δu^{*p}_k and δu^{*e}_k across a time horizon $k = 1$ to K. The solutions to S_k and v_k are as follows:

$$\begin{aligned}
S_k &= A^{k'}_x [S_{k+1} C A_k + Q C A_k] \\
S_K &= Q \\
v_k &= A^{k'}_x [S_{k+1} C[\delta x_{k+1} + D + E] + v_{k+1} + Q(x_{k+1} + C(\delta x_{k+1} + D + E))] \\
v_K &= Qw
\end{aligned}$$

where

$$\begin{aligned}
\delta x^*_{k+1} &= C * [(A_k \delta x_k + D + E)] \\
C &= [I - B^p_k(-B^{p'}_k Q) - B^p_k(-B^{p'}_k S_{k+1}) - B^e_k B^{e'}_k Q - B^{e'} B^{e'} S_{k+1}]^{-1} \\
D &= B^p_k[-B^{p'}_k Q x_{k+1} - u^{p*}_k - B^{p'}_k v_{k+1}] \\
E &= B^e_k[-B^{e'}_k Q x_{k+1} - u^{e*}_k - B^{e'}_k v_{k+1}]
\end{aligned}$$

S_K is set to be Q and v_K is set to be Q_w where w is the nominal final state of the system. Backward recursion can be used to solve for S_k and v_k. At each iteration, u^{*p}_k and u^{*e}_k are updated to $u^{*p}_k + \delta u^{*p}_k$ and $u^{*e}_k + \delta u^{*e}_k$. This is repeated until no more significant improvements to the control signals are possible.

4.4.4 Case study

The WECC three-machine nine-bus system is implemented in MATLAB/Simulink to which the pursuer and evader controls outlined earlier are applied. There are three generators in this system which are described by six state variables. All control strategies are computed over a time horizon of $t = 0$–10 s. The duration of a time step is $\Delta t = \frac{1}{50} = 0.02$ s which corresponds to the frequency at which PMUs communicate (i.e., 50 Hz). The cost matrix Q is set to be $I \in \mathbb{R}^{2g\times 2g}$ where I is an identity matrix. It is assumed that the cyber-physical active DER element residing on *Agent* 3 is compromised and the utility will use the remaining DERs residing at Generators 1 and 2 to mitigate the attack.

According to Reference 19, the initial conditions of the system will dictate whether the control strategies applied by a pursuer and evader will result in terminating the game in finite time. The game terminates when the evader moves the system away from the capturability threshold. This logic in the homicidal chauffeur problem, where the chauffeur is attempting to maximize distance from the police (pursuer) and police is attempting to minimize distance from the chauffeur, can be applied to our work as well. For the results presented in Figures 4.6 and 4.7, the phase angles and frequencies are considered for the case where the initial system state has deviated from nominal state w by 0.0025 p.u. We further consider two scenarios. First, only the adversary applies its control strategy and the utility applies no counter-control. In the second case, both the adversary and utility apply their respective control strategies.

It is clear from these results that the adversary can inflict significant state deviations when the attack is undetected. However, the impact on the system is reduced when the utility joins the game.

4.5 Final remarks

Two cyber-physical mechanisms that engage in active controls via fast acting storage devices for maintaining generator coherency in the advent of attacks on transient stability in the power grid have been presented in this chapter. The dynamics of the system are highly nonlinear and pose the greatest challenge in tractably computing the control inputs to mitigate the effects of these attacks.

We first capitalized on the underlying physics of the power system to draw analogies of our problem to the well-studied flocking-based formation control problem. The resulting control strategies are demonstrated to be scalable and effective in mitigating adverse effects on system states. This demonstrates the effective use of active cyber-physical components in increasing resilience in the system. On the other hand, cyber components with actuating capabilities can also be exploited by adversaries via well-known communication vulnerabilities. Given that the attacker has access to a subset of these active cyber-physical elements and the utility has access to the remaining elements, we have shown that the goals of the attacker and the utility are similar to that of an evader and a pursuer. This is a classic game theoretic formulation. The resulting nonlinear differential game is solved iteratively to compute the attack and mitigation control vectors resulting in NE. When the attack vector is applied with no active countermeasures in place, the resulting state deviations can trigger transient instability. On the other hand, when both the attacker and the utility apply the attack and countermeasure vectors, at the same time, the system states remain within the nominal thresholds, thereby, increasing the clearing time that can be used by the grid operators to isolate the problem region and activate ancillary services.

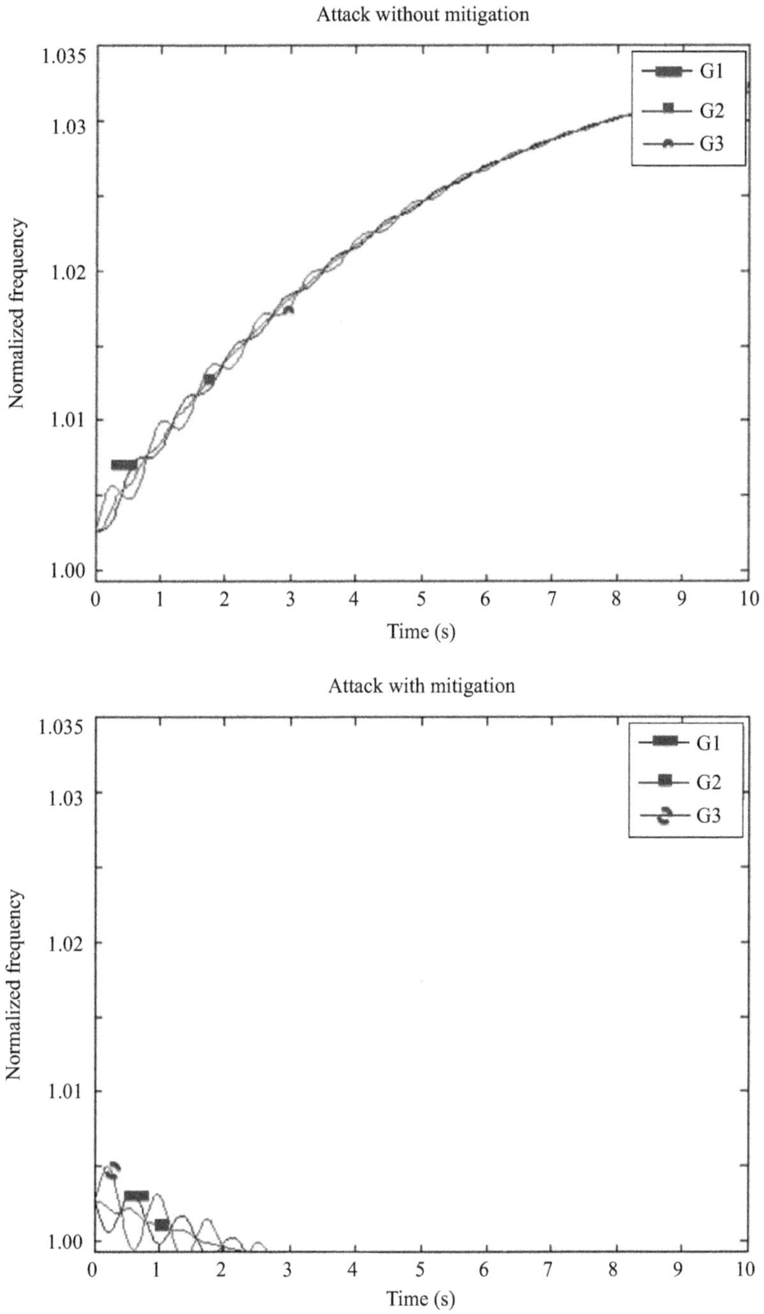

Figure 4.6 Frequencies at each generator

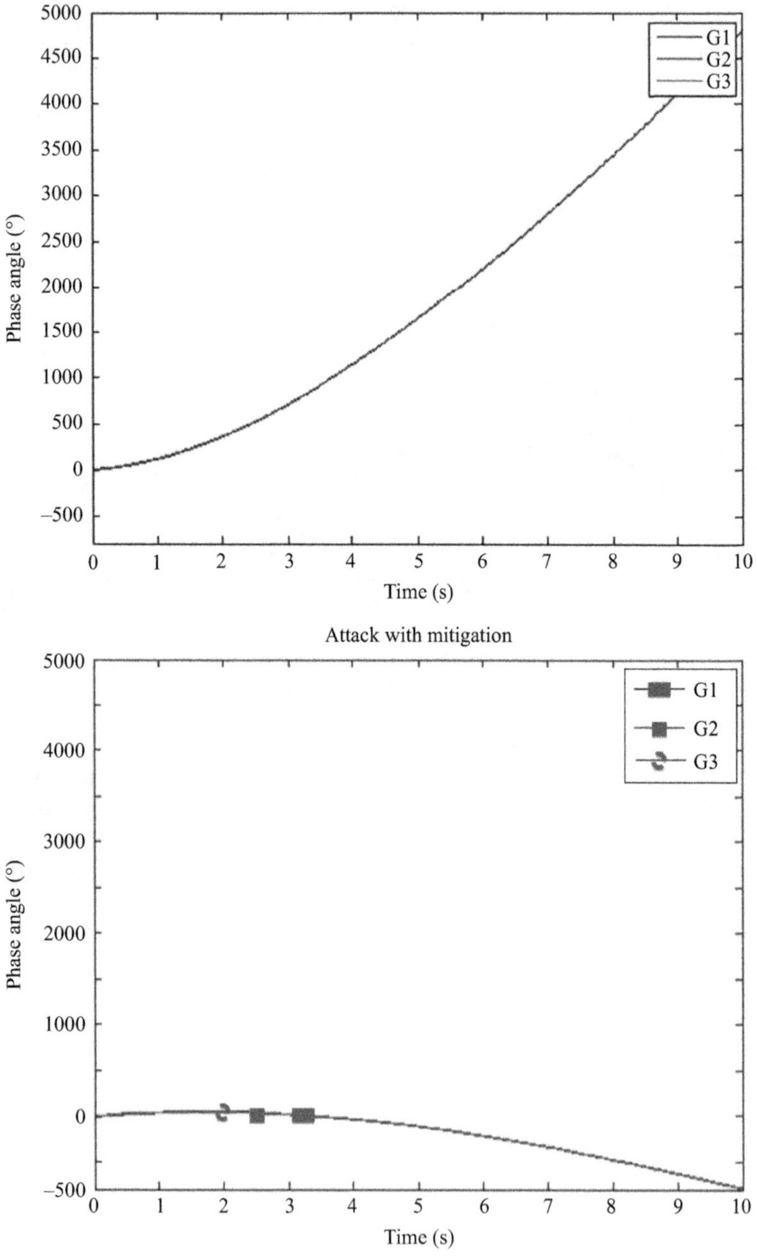

Figure 4.7 Phase angles at each generator

Thus, we have provided two cyber-physical strategies in this chapter when faced with malicious attacks on generator coherency.

References

[1] Y. Liu, P. Ning, and M. K. Reiter, "False data injection attacks against state estimation in electric power grids," *ACM Transactions on Information and System Security*, 2011, vol. 14, no. 13, pp. 21–32.

[2] S. Liu, X. Feng, D. Kundur, T. Zourntos, and K. Butler-Purry, "A class of cyber-physical switching attacks for power system disruption," in *Proc. 7th Cyber Security and Information Intelligence Research Workshop (CSIIRW)*, Oak Ridge National Laboratory, Tennessee, October 2011, pp. 1–4.

[3] C.-W. Ten, C.-C. Liu, and G. Manimaran, "Cyber-Vulnerability of Power Grid Monitoring and Control systems," in *Proc. 4th Cyber Security and Information Intelligence Research Workshop (CSIIRW)*, Oak Ridge National Laboratory, Tennessee, May 2008, pp. 1–3.

[4] M. Jonsson, M. Begovic, and J. Daalder, "A new method suitable for real-time generator coherency determination," *IEEE Transactions on Power Systems*, August 2004, vol. 19, no. 3, pp. 1473–1482.

[5] J. Wei, D. Kundur, T. Zourntos, and K. Butler-Purry, "A flocking-based dynamical systems paradigm for smart power system analysis," in *Proc. IEEE Power & Energy Society General Meeting*, San Diego, CA, July 2012.

[6] J. Wei and D. Kundur, "A multi-flock approach to rapid dynamic generator coherency identification," in *Proc. IEEE Power & Energy Society General Meeting*, July 2013.

[7] B. M. A. Faiya and H. M. Schwartz, "Learning fuzzy controller for the homicidal chauffeur differential game," *in Proc. 20th Mediterranean Conference on Control and Automation (MED)*, July 2012, pp. 247–252.

[8] J. Wei, D. Kundur, and T. Zourntos, "On the use of cyber-physical hierarchy for smart grid security and efficient control," in *Proc. IEEE Canadian Conference on Electrical and Computer Engineering (CCECE)*, Montreal, Canada, April–May 2012.

[9] J. Wei and D. Kundur, "Goalie: goal-seeking obstacle and collision evasion for resilient multicast routing in smart grid," *IEEE Transactions on Smart Grid*, 2015, vol. 7, no. 2, pp. 1–13.

[10] F. Dörfler and F. Bullo, "Synchronization and transient stability in power networks and non-uniform kuramoto oscillators," in *Proc. American Control Conference*, June–July 2010, pp. 930–937.

[11] F. Dörfler, M. Chertkov, and F. Bullo, "Synchronization in complex oscillator networks and smart grids," in *Proc. National Academy of Sciences*, February 2013, pp. 2005–2010.

[12] P. Kundur, *Power System Stability and Control*. McGraw-Hill, NY, 1994.

[13] P. Kundur, J. Paserba, V. Ajjarapu, *et al.*, "Definition and classification of power system stability: IEEE/CIGRE joint task force on stability terms and

definitions," *IEEE Transactions on Power Systems*, August 2004, vol. 19, pp. 1387–1401.

[14] P. Sauer and M. Pai, *Power System Dynamics and Stability*. Prentice Hall, Upper Saddle River, NJ, 1997.

[15] J. Wei, D. Kundur, T. Zourntos, and K. Butler-Purry, "A flocking-based paradigm for hierarchical cyber-physical smart grid modeling and control," *IEEE Transactions on Smart Grid*, 2014, vol. 5, no. 6, pp. 2687–2700.

[16] J. Wei, D. Kundur, and K. Butler-Purry, "A novel bio-inspired technique for rapid real-time generator coherency identification," *IEEE Transactions on Smart Grid*, 2015, vol. 6, no. 1, pp. 178–188.

[17] C. Reynolds, "Flocks, herds, and schools: a distributed behavioral model," *Computer Graphics*, July 1987, vol. 21, no. 4, pp. 25–34.

[18] Y. Yan, Y. Qian, H. Sharif, and D. Tipper, "A survey on cyber security for smart grid communications," *IEEE Communications Surveys and Tutorials*, 2012, vol. 14, no. 4, pp. 998–1010.

[19] T. Basar and G. J. Olsder, *Dynamic noncooperative game theory*. SIAM Series Classics in Applied Mathematics, Academic Press, London, 1982; 2nd edition, 1999.

[20] R. Kwong, Linear quadratic optimal control, ECE1639. [Online]. Available at: http://www.control.utoronto.ca/people/profs/kwong/1639-2010.html. [Accessed 2015].

[21] W. Li and E. Todorov, "Iterative linear quadratic regulator design for nonlinear biological movement systems," *1st International Conference on Informatics in Control, Automation and Robotics*, 2004, pp. 1–8.

[22] P. Abbeel, Optimal Control for Linear Dynamical Systems and Quadratic Cost, CS287. [Online]. Available at: http://www.cs.berkeley.edu/p̃abbeel/cs287-fa12/slides/lqr.pdf. [Accessed 2015].

Chapter 5

Dynamic contingency analysis and remedial action tools for secure electric cyber-physical systems

Joydeep Mitra[1], Mohammed Benidris[2] and Nga Nguyen[1]

Abstract

In recent years, the drive to bring about technological and regulatory changes that concern energy, natural resources, and climate change has gathered significant momentum. Of the numerous changes that the power grid is undergoing, perhaps the most transformative is the increased use of communication and computing technologies. The deployment of new communication, computing, and control technologies has significantly augmented the capabilities of traditional Supervisory Control and Data Acquisition systems. These technologies characterize the "smart grid" and have transformed them into the largest and most complex cyber-physical systems ever built; they also hold the potential to revolutionize power system operation and control paradigms.

The system can often be protected from widespread consequences of failures of system components and other types of faults through timely detection and remedial action. By evaluating the stability of power systems and executing a sequence of remedial actions in real time or faster than real time, power systems can be hardened against cascading failures and unfolding events which can be either initiated by failures of system components, faults, or malicious attacks. This chapter discusses the application of real-time transient stability assessment and remedial action tools in enhancing power systems against potential cascading failures utilizing the developments in communication and computing technologies. Remedial actions are classified into preventive and corrective actions based on the time available to an operator to respond. Direct methods in the form of the energy function are utilized in transient stability screening and generating remedial actions. Different methods of calculating the controlling unstable equilibrium points (an equilibrium point that the trajectory of the system will go to after a destabilizing contingency) such as a Boundary of stability region-based Controlling Unstable equilibrium point method and a

[1]Department of Electrical and Computer Engineering, Michigan State University, Michigan, MI, USA
[2]Department of Electrical and Biomedical Engineering, University of Nevada, Reno, NV, USA

homotopy-based method are presented. Explicit derivations of expressions that are needed for transient stability analysis and remedial actions as well as an illustrative example are also provided.

5.1 Introduction

Modern power systems increasingly operate at lower security margins, due to reduced regulation capability resulting from increased variable generation and market operations. Consequently, they have grown more vulnerable to cascading failures than in any other period in history. Transient instability is a major factor that can lead to cascading failures and is therefore a major concern in power system security. The need for better situational awareness has also become evident, and triggered several ongoing efforts to develop improved methods of determining remedial action alternatives (including special protection schemes, emergency action, and on-line contingency analyses) and assist grid operators in making decisions. However, traditional contingency analyses have only evaluated static contingencies; dynamic contingencies were deemed too computationally intensive to solve in real time. Stability scenarios were evaluated off-line, safe operating zones were determined and documented, and operators were required to be familiar with these scenarios and remedial actions. Recent technological advances have enabled these actions to be incorporated into Supervisory Control and Data Acquisition (SCADA) systems and Energy Management System (EMS), and these are now displayed in modern control rooms. Over the last few years, utilities and independent system operators (ISOs) have been working on integrating expert systems and screening tools into their control center practices.

Following a potentially destabilizing event, the grid can evolve along any of numerous possible trajectories, which may be exacerbated by subsequent events or ameliorated by operator action. Using the tools available today, it is impossible to evaluate or screen every trajectory the system could assume; the computational challenge of performing time-domain simulation of every possible trajectory, within the time available to an operator, is simply unassailable. More recently, there have been theoretical advances that can potentially enable determination of security margins and perform remedial action screening in real time. These methods are predominantly based on "direct" methods, such as Lyapunov or energy function methods. They utilize the advantages of the conservativeness of the direct methods to recursively classify the designated set of contingencies into "stable," "potentially unstable," and "undetermined" subsets. After using direct methods to screen out a large number of stable contingencies, time-domain simulation is performed to check the potentially unstable contingencies. Direct methods have proved to be very effective in reducing the number of contingencies that need further evaluation using time-domain simulations, and thereby made it possible to screen and compute dynamic contingencies in real-time.

Upon identification of critical contingencies, remedial actions also require to be executed that can check the propagation of the disturbance and mitigate or even eliminate system disruption. These remedial actions are divided into preventive and corrective actions depending on system status and they include one or a combination of the following: generation rescheduling/tripping, load shedding, line tripping, voltage control, and utilization of Flexible Alternating Current Transmission Systems (FACTS) devices. This chapter describes emergent methods in real-time transient stability and remedial action screening (based on direct methods) and discusses how they can be combined with developments in telemetry and communication technologies to improve the security and resilience of modern power systems.

5.2 On-line transient stability assessment

As a result of market forces, increased renewable generation, and recent advances in power flow control technologies, power systems are increasingly being operated closer to their stability limits. On-line transient stability assessment (TSA) has become an important feature for systems that operate in such stressed environments. The main challenge in on-line TSA lies in the complexity of the computations and the high computational burden that becomes necessary in order to accurately evaluate contingencies for stability.

Several methods and strategies have been proposed in the area of the on-line TSA. These methods can be classified into artificial intelligence (AI) approaches [1–16] and direct methods [17–24]. Due to the open-access operation of power system networks, the off-line analyses may not be fully correlated to the on-line data and therefore results of the AI approaches may become invalid [19]. On the other hand, direct methods for transient stability screening tools suffer from conservativeness of the results. However, these tools identify and exclude a large number of non-severe contingencies and apply detailed simulations on the severe or undetermined contingencies to reach the requirements of on-line applications. An on-line screening tool based on three levels of filtering using direct methods has been proposed in References 17 and 18. Each level has two time frames: inertial transient filters and post-inertial transient filters. The filters screen the contingencies along the solution trajectory toward the controlling unstable equilibrium point (UEP). Another on-line dynamic contingency screening tool based on the *B*oundary of stability region based *C*ontrolling *U*nstable equilibrium point (BCU) method is proposed in Reference 19. This tool uses six classifiers to screen out a small number of critical contingencies for detailed simulations. Improvements to the screening tool of Reference 19 have been proposed in Reference 20. In the improved screening tool, the authors added another level of classification which is the detection of the islanding mode and improvements in the six classifiers of Reference 19 by tuning the screening thresholds. This chapter presents a screening tool that classifies a designated set of contingencies (C) to stable (C_s) and unstable (C_u) subsets using BCU and homotopy-based (continuation) methods. These methods take advantage of the conservativeness of the Lyapunov methods to divide the classification process into stages. These stages are based on the size of the energy margin along system trajectory.

Existing on-line transient stability analysis tools based on direct methods have introduced several system-dependent thresholds to classify a set of contingencies to stable, unstable, or undetermined subsets. The undetermined contingencies are the contingencies that have numerical convergence problems such as failure to calculate the controlling UEP. In order for these thresholds not to fail in classifying the contingencies, off-line transient stability analyses are required. Also, for the unstable or undetermined contingencies, detailed time-domain simulations are performed. Therefore, setting up the thresholds needs off-line TSAs and may require an excessive use of time-domain simulations.

Several criteria and requirements are suggested in the literature for any on-line transient stability screening candidate. In Reference 19, five requirements were suggested for any on-line transient stability screening tool. These requirements are reliability measure, efficiency measure, on-line computation, speed measure, and performance measure. The reliability measure is defined as the absolute capture of the unstable contingencies. Efficiency measure indicates the ratio between the detected stable contingencies and the actual number of stable contingencies. On-line computation measure can be expressed in terms of the amount of use of off-line computations. Speed measure is used to evaluate computation speed. Performance measure tests robustness of the screening tools against the changes in operating conditions. The reliability, efficiency, and performance measures are related to the screening tool itself. On the other hand, on-line computation and speed measure are comparable indices that can be used to compare between alternative methods.

5.3 System model and transient stability – direct methods

Direct methods have been successfully applied in evaluating the stability of power systems and deriving operating limits [25–31]. Given an n-generator system and assuming uniform damping, the classical model of the equations of motion of the generators with respect to the center of inertia (COI) can be denoted by Reference 29:

$$\dot{\tilde{\delta}}_i = \tilde{\omega}_i, \tag{5.1}$$

$$\dot{\tilde{\omega}}_i = \frac{1}{M_i}(P_{mi} - P_{ei}) - \frac{1}{M_T}P_{COI} - \lambda\tilde{\omega}_i, \tag{5.2}$$

$$\tilde{\delta}_i = \delta_i - \delta_o, \tag{5.3}$$

$$\tilde{\omega}_i = \omega_i - \omega_o, \tag{5.4}$$

$$\delta_o = \frac{1}{M_T}\sum_{i=1}^{n} M_i\delta_i, \tag{5.5}$$

$$\omega_o = \frac{1}{M_T}\sum_{i=1}^{n} M_i\omega_i, \tag{5.6}$$

$$M_T = \sum_{i=1}^{n} M_i, \tag{5.7}$$

where P_{mi} is the mechanical input of machine i, P_{ei} is the electrical power output of machine i, M_i is the inertia constant of machine i, δ_i is the power angle of machine i, ω_i is the angular frequency of machine i, and λ is a uniform damping constant. The electrical power of machine i is given as follows.

$$P_{ei} = \sum_{j=1}^{n} E_i E_j [G_{ij} \cos(\tilde{\delta}_i - \tilde{\delta}_j) + B_{ij} \sin(\tilde{\delta}_i - \tilde{\delta}_j)], \tag{5.8}$$

where n is the number of machines in the system, E_i is the constant voltage magnitude, and G_{ij} and B_{ij} are the conductance and susceptance of the admittance matrix of the network-reduction model. The P_{COI} is computed as follows:

$$P_{COI} = \sum_{i=1}^{n} P_{mi} - \sum_{i=1}^{n} \sum_{j=1}^{n} E_i E_j [G_{ij} \cos(\tilde{\delta}_i - \tilde{\delta}_j) + B_{ij} \sin(\tilde{\delta}_i - \tilde{\delta}_j)], \tag{5.9}$$

The most commonly used energy function in transient stability analysis of power systems can be expressed as [25,26]:

$$V = \frac{1}{2} \sum_{i=1}^{n} M_i \tilde{\omega}_i^2 - \sum_{i=1}^{n} P_i (\tilde{\delta}_i - \tilde{\delta}_i^s) - \sum_{i=1}^{n-1} \sum_{j=i+1}^{n} [C_{ij} (\cos \tilde{\delta}_{ij} - \cos \tilde{\delta}_{ij}^s) - I_{ij}], \tag{5.10}$$

where $P_i = P_{mi} - E_i^2 G_{ii}$, $C_{ij} = E_i E_j B_{ij}$, $\tilde{\delta}_{ij} = \tilde{\delta}_i - \tilde{\delta}_j$, the superscript s denotes the pre-fault stable equilibrium point (SEP), and I_{ij} is the energy dissipated in the network transfer conductances which can be expressed as follows.

$$I_{ij} = \int_{\tilde{\delta}_i^s + \tilde{\delta}_j^s}^{\tilde{\delta}_i + \tilde{\delta}_j} D_{ij} \cos \tilde{\delta}_{ij} \, d(\tilde{\delta}_i + \tilde{\delta}_j), \tag{5.11}$$

where $D_{ij} = E_i E_j G_{ij}$. This term is path dependent and can be calculated only if the system trajectory is known. Several methods are suggested in the literature to approximate this term. In this chapter, the method suggested in Reference 25 is used which can be given as follows.

$$I_{ij} = D_{ij} \frac{\tilde{\delta}_i + \tilde{\delta}_j - \tilde{\delta}_i^s - \tilde{\delta}_j^s}{\tilde{\delta}_i - \tilde{\delta}_j - \tilde{\delta}_i^s + \tilde{\delta}_j^s} [\sin \tilde{\delta}_{ij} - \sin \tilde{\delta}_{ij}^s]. \tag{5.12}$$

Due to the approximation of the path dependent term in the energy function which accounts for transfer conductances, the accurate Exit Point (EP) may not be detected. The EP (see Figure 5.1) is the point at which the projected fault-on trajectory exits the stability boundary of the post-fault SEP. In the method presented in this chapter, an accurate EP is not necessary to find the controlling UEP.

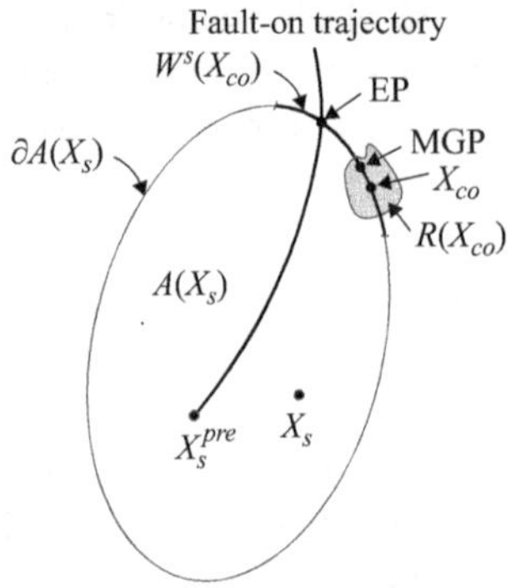

EP	is the exit point,
MGP	is the minimum gradient point,
X_{co}	is the controlling UEP,
$A(X_s)$	is the region of attraction (stability) of the post-fault stable equilibrium point (X_s),
$\partial A(X_s)$	is the boundary of the region of attraction,
$R(X_{co})$	is the region of convergence of the controlling UEP under Newton-Raphson method,
$W^s(X_{co})$	is the stable manifold of the controlling UEP,
X_s^{pre}	is the pre-fault stable equilibrium point.

Figure 5.1 Region of convergence of the controlling UEPs under the use of Newton-Raphson method. See Reference 31 for more details

5.4 Calculation of the controlling UEP

The Controlling UEP, using fault-on trajectory, is a UEP whose stable manifold, $W^s(X_{co})$, contains the EP of the fault-on trajectory as shown in Figure 5.1.

Computing the controlling UEP is crucial because the energy at a controlling UEP is used in computing the critical energy to assess the stability of the system. The dynamics of a power system, as given in (5.1) and (5.2), can be described in a compact form as a set of dynamic equations as follows.

$$\dot{x} = F(x). \tag{5.13}$$

A state vector x is called an equilibrium point x^* of the dynamic system represented in (5.13) if $F(x^*) = 0$. The controlling UEP is one of the UEPs, but it is not an easy task to determine and distinguish it from the other UEPs. A popular method to compute a controlling UEP is to use time-domain simulation of the projected fault-on trajectory to obtain the EP and minimum gradient point (MGP), and then use the MGP as an initial guess to find the controlling UEP by Newton-Raphson method. However, an inaccuracy in calculating the EP could cause a failure in calculating the MGP and then the controlling UEP. Most of the methods in the literature integrate the fault-on trajectory with large time step until the algorithm locates the EP between two time steps. The algorithm then bounds this point and starts searching for the EP using some tools, such as the golden section, linear and quadratic interpolation. Once a more accurate EP is discovered, the algorithm uses it as an initial point to compute the MGP and then the controlling UEP. Therefore, computing the controlling UEP is numerically involved and it requires significant effort to compute an accurate EP and MGP. In this chapter, two methods are used to calculate controlling UEPs, the BCU method and the homotopy-based method. In general, homotopy-based approaches are used to eliminate the problems associated with the choice of appropriate initial starting points. This chapter shows that a controlling UEP can be obtained by using homotopy-based approaches with an approximate EP as an initial point.

5.4.1 Characterization of the region of convergence of the controlling UEPs

The accuracy of direct methods strongly relies on the determination of the controlling UEP. The region of convergence of an equilibrium point can be defined as follows: starting from an initial guess inside the region of convergence, a numerical method succeeds in finding the solution, or it fails if the initial guess lies outside this region. The size and shape of the region of convergence of a controlling UEP can be fractal and different for different numerical methods [31]. Therefore, the presence of fractal shapes of the convergence region of the controlling UEP is the main reason of failure of many numerical methods in finding the correct controlling UEP [31–33]. To find the correct controlling UEP, the initial guess, in using traditional iterative methods, has to be within a certain range of the desired solution. This requirement makes it difficult to find the best guess and renders the method computationally expensive.

In finding the correct controlling UEP, using other iterative methods, the EP and the MGP have to be calculated first [34–40]. The controlling UEP, EP, and the MGP are shown in Figure 5.1. Computationally, the EP is characterized by the first local maximum of the potential energy of the post-fault network along the projected fault-on trajectory. Another method to detect the EP is through detecting the change in the sign of the dot product of the post-fault power mismatch vector and the fault-on speed vector [31]. The MGP is numerically characterized by the first local minimum value of the norm of the vector field of the post-fault trajectory [31]. Most of the reported methods use the MGP as an initial point to generate a sequence of steps to find the controlling UEP. The robustness of finding the controlling UEP depends strongly on the accuracy of the calculation of MGP [35,36,38,40,41]. An inaccuracy in detecting the EP may cause difficulty in computing the MGP. However, detecting an accurate EP is computationally involved and sometimes it requires the use of interpolation methods after bounding the EP in a certain range. Therefore, a numerical inaccuracy in computing the EP might cause failure of numerical methods to calculate the controlling UEP.

Several methods have been proposed to compute the controlling UEP including the BCU method [34]. However, due to the problems associated with the detection of the EP and consequently the MGP, the BCU method occasionally encounters difficulty in finding a solution [38]. A continuation-based method that approximates the stability boundary locally and does not use the energy function to compute the controlling UEP has been proposed in Reference [36]. Despite the long history of applying direct methods in transient stability analysis, the problem of precise determination of the controlling UEP and the speed of computation are still of concern to many researchers. For this reason, some researchers have favored the homotopy method.

5.4.2 BCU-based approach

The BCU method is a systematic method that uses the concept of reduced-state model for computing the controlling UEP [19,31]. This method defines an artificial model that can be solved to find all the equilibrium points of the original model.

For each power system stability model there is a corresponding reduced-state model such that static and dynamic properties of the original model are captured to compute the controlling UEP on the stability boundary of the original model. Computing the controlling UEP using the BCU method on the reduced-state model is easier than computing the controlling UEP of the original model [31]. For example, consider the following generic model [31]:

$$T\dot{x} = -\frac{\partial U}{\partial x}(x,y) + g_1(x,y), \tag{5.14}$$

$$\dot{y} = z, \tag{5.15}$$

$$M\dot{z} = -Dz - \frac{\partial U}{\partial y}(x,y) + g_2(x,y), \tag{5.16}$$

where x, y, and z are state variables, T is a positive definite matrix, M and D are diagonal positive definite matrices, U is the function of system dynamics, and $g_1\,(x,y)$ and $g_2\,(x,y)$ represent the transfer conductances.

The associated artificial, reduced-state model is given as follows [31]:

$$T\dot{x} = -\frac{\partial U}{\partial x}(x,y) + g_1(x,y), \tag{5.17}$$

$$\dot{y} = -\frac{\partial U}{\partial y}(x,y) + g_2(x,y). \tag{5.18}$$

The general procedure of finding the controlling UEP using the BCU method is summarized as follows [34]. The definitions are referred to Figure 5.1.

1. From the fault-on trajectory, determine the EP.
2. Use the EP as an initial point and integrate the post-fault reduced-state model to find the MGP.
3. Use the MGP as an initial point to calculate the controlling UEP of the reduced-state model.
4. Determine the controlling UEP of the original system.

In implementing the above procedure, it is important to be aware of several potential numerical issues [31]. For example, an inaccuracy in calculating the EP may cause an inaccuracy in computing the MGP. If the computed MGP is not sufficiently close to the controlling UEP, the iterative method may diverge or converge to another equilibrium point. Two situations where the EP may not be accurate (either inside the stability region or outside the stability region) are shown in Figure 5.2.

From Figure 5.2, it can be seen that an inaccurate EP can lead to failure of calculating the controlling UEP [31]. A procedure that can be used to solve this problem is called stability-boundary-following procedure [19]. The procedure can be summarized as follows [19]: starting from the given EP, integrate the post-fault model for a few iterations (e.g., three to five iterations) and check if the MGP has been detected. If it has been detected, check if the controlling UEP can be computed using this MGP. If the MGP has not been detected or the controlling UEP cannot be computed (either converges to the post-fault SEP or the distance between the computed controlling UEP and the MGP is larger than a threshold), draw a ray connecting the

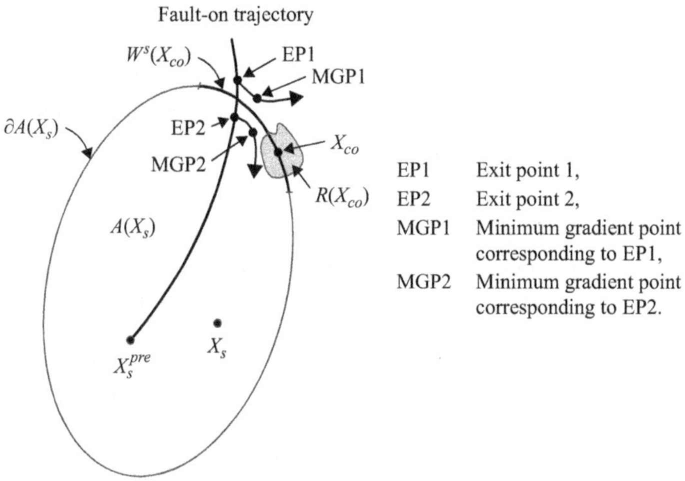

Figure 5.2 The effects of an inaccurate EP on the calculation of the MGP

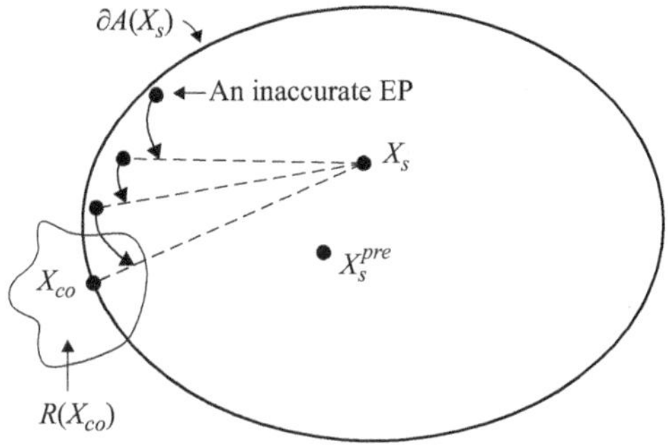

Figure 5.3 A stability-boundary-following procedure. See Reference 19 for details

current point on the post-fault trajectory with the post-fault SEP and search along this trajectory for a new MGP. This procedure is depicted in Figure 5.3 for the case where the EP lies inside the stability region. The complete description of this procedure is given in Reference 19.

5.4.3 Homotopy-based methods

Homotopy-based approaches are known to be reliable in finding a solution. However, homotopy-based approaches are intrinsically slow because these methods map the trajectory of the solution from an easy and known solution to the desired solution. This chapter presents an algorithm that uses homotopy-based approaches with EPs as the initial points to find controlling UEPs. Using approximate EPs rather than computing accurate EPs, as is common practice in finding the controlling UEP, the intrinsic slow speed of computation of homotopy-based approaches is compensated. Further, homotopy-based approaches eliminate the necessity of computing the MGP, which makes the homotopy-based approach comparable with iterative methods in terms of the speed of computation.

Homotopy-based approaches have been successfully applied in computing the closest UEP but not the controlling UEP [42,43]. The closest UEP can be distinguished by having the lowest Lyapunov function value. The approaches reported in these references utilize the strategy of singular fixed-point and the concept of bifurcation to locate the closest UEP. The algorithm proceeds by choosing a set of initial points that, by using homotopy-based approaches, converge to a set of type-1 UEPs (a UEP that has one unstable eigenvalue in the Jacobian matrix). By choosing a set of proper initial points, the resulting set of the type-1 UEPs will include the closest UEP. Chen *et al.* in Reference 36 have used the continuation-based methods without the use of the energy function and approximating the stability boundary locally.

Homotopy is a numerical method to solve and find the roots of nonlinear systems expressed as $F(x) = 0$. It is also called continuous Newton-Raphson method. Homotopy is known to be a reliable method to find solutions and it is not sensitive to the initial conditions as compared with the other iterative methods. However, when multiple solutions exist, a better initial condition helps obtain the solution of interest first. In applying homotopy-based methods, the pre-fault SEP is used as an initial and known solution to calculate the post-fault SEP and the EP is used as an initial and known solution to calculate the controlling UEP.

Computing an accurate EP is not an easy task and computing the MGP depends on the accuracy of the EP. In applying homotopy-based approaches to calculate the controlling UEP, a less accurate EP as a known solution can be used. The justification is that with an EP as the initial guess, homotopy-based approaches can find the controlling UEP first since its location is the closest to the EP in terms of the energy value.

The basic idea of the homotopy method is to find the solutions by path continuation, starting at a known solution x^0 that satisfies $G(x^0) = 0$. The homotopy method traces the solution trajectory by a predictor-corrector algorithm to get a solution of the original equation. The most widely used homotopy function is expressed as follows.

$$H(x,t) = tF(x) + (1-t)G(x) = 0, \tag{5.19}$$

where t changes from 0 to 1 with an incremental step-size through the mapping process, i.e., $H(x,0) = G(x)$ and $H(x,1) = F(x)$.

The function $G(x)$ can be chosen arbitrarily as long as it has a known solution. The Newton-homotopy can be expressed as follows.

$$G(x) = F(x) - F(x^0), \tag{5.20}$$

where x^0 can be any starting point. Therefore, the homotopy function becomes

$$H(x,t) = F(x) - (1-t)F(x^0) = 0. \tag{5.21}$$

It should be noted that in applying homotopy methods, the direction of the search (forward or backward) for the solution is an important factor in determining the correct controlling UEP. The direction of the search depends on the location of the EP with respect to the controlling UEP on the stability boundary. If this direction is unknown and if the forward direction (for example) is performed, the homotopy may converge to a UEP that may not be the controlling UEP. From the extensive studies that were performed, the number of the homotopy iterations to arrive at the controlling UEP starting from the EP ranges from one to three iterations where each homotopy iteration is equivalent to a full Newton-Raphson (or other method) solution. A procedure is assumed so that the solution algorithm proceeds by assuming forward mapping for one homotopy iteration; if the solution diverges or converges to a point far from the EP, the algorithm uses backward mapping.

5.5 Approaches for remedial actions

The current applications of transient stability screening tools and remedial actions are based on off-line studies and the use of AI approaches. These methods may fail if no correlation exists between the off-line studies and the available on-line data. This chapter provides explanations on the application of on-line calculations and real-time control actions. The term "real-time" implies that the process of calculation and the produced remedial actions are ready with a time delay of a few seconds. On the other hand, on-line implies that the data are updated every few seconds but the time delay of the output of the calculation can be a few seconds to several minutes. The off-line studies are used only to omit contingencies that are deemed stable to increase the speed of calculations.

Operators already have information about static security assessment but most of the available information on dynamic security is based on off-line studies and the use of AI methods. What the operators do not have now is the knowledge about the transient stability corrective and preventive actions. For example, in the case of an event, they need to know what the best actions are to be prepared to take an action. The purpose of the remedial action tool is to perform transient stability analyses and provide information to the operators about some suggested actions.

There are two types of remedial actions, namely, corrective actions and preventive actions. Corrective actions can be performed locally or globally depending on the available measurements and means. On the other hand, preventive actions are performed globally. Global corrective actions could be generation tripping, load shedding, and line tripping, and local corrective actions could be control of excitation and output power. The preventive actions could be generation rescheduling and load shedding. The stability assessment and the produced remedial actions can be triggered either periodically on a set of designated contingencies, by occurrence of events or upon operators request. The ability to distinguish between these actions depends on the initial conditions of the system and the triggering events. For example, in a case where there is no disturbance, the remedial action scheme examines the designated contingencies for possible transient stability violations. In case of a possible violation, the scheme lists possible generation rescheduling alternatives based on the sensitivity of every control parameter on the energy margin. However, in case of corrective action mode, the scheme is designed to prevent the possibility of cascading failures. The corrective action scheme is less costly than the preventive action; however, it is the last line of defense and if the correction actions suggested by this scheme cannot be applied, the system is more likely to undergo cascading failures. Therefore, a careful selection of the actions needs to be implemented in this scheme.

After screening, contingencies are divided into two main groups: stable and unstable. With the stable group, no control actions are needed to intervene in system operation. However, with unstable contingencies, control measures are required to prevent unacceptable impact on system stability (preventive actions) or to immediately respond to the occurring disturbance in order to maintain system operation in balance (corrective action).

SCADA/EMS systems have been utilized in utilities for several decades. Therefore, adding the transient stability screening and remedial action tool requires modifications to the existing systems. Essentially, the SCADA/EMS systems assume the system is in the steady condition and do not model the transients in power systems. However, recent implementation of these systems can potentially contribute to better understanding of dynamics.

The philosophy of distinguishing between the corrective and preventive remedial actions is explained as follows. The preventive screening and the suggested remedial actions can be triggered either periodically or upon operator request. The preventive action tool starts with screening a full list of possible contingencies assuming the system resides in a healthy state. On the other hand, the corrective screening and the suggested remedial actions are triggered immediately after an occurrence of an event. Here, instead of screening the entire set of possible contingencies, only the "related" contingencies and the possible contingencies on some critical lines are screened. When a fault occurs at some location in the grid, the neighboring lines will be exposed to overload and/or false tripping due to hidden failures in the protection equipments [44]. Also, some critical lines that carry a large amount of power, transient stability-limited lines, and the inter-ties between areas may be affected by faults in non-neighboring lines. The process of performing remedial action control is shown in Figure 5.4.

5.5.1 Energy margin and sensitivity analysis-based remedial action control

The energy margin (ΔV) is defined as the difference between the value of the energy function V at the instant of clearing, V^{cl}, and at the controlling UEP to determine the critical value V^{u} (the critical value of energy margin is calculated at the controlling UEP). This condition is mathematically expressed by the following equation [45]:

$$\Delta V = V^{u} - V^{cl}. \tag{5.22}$$

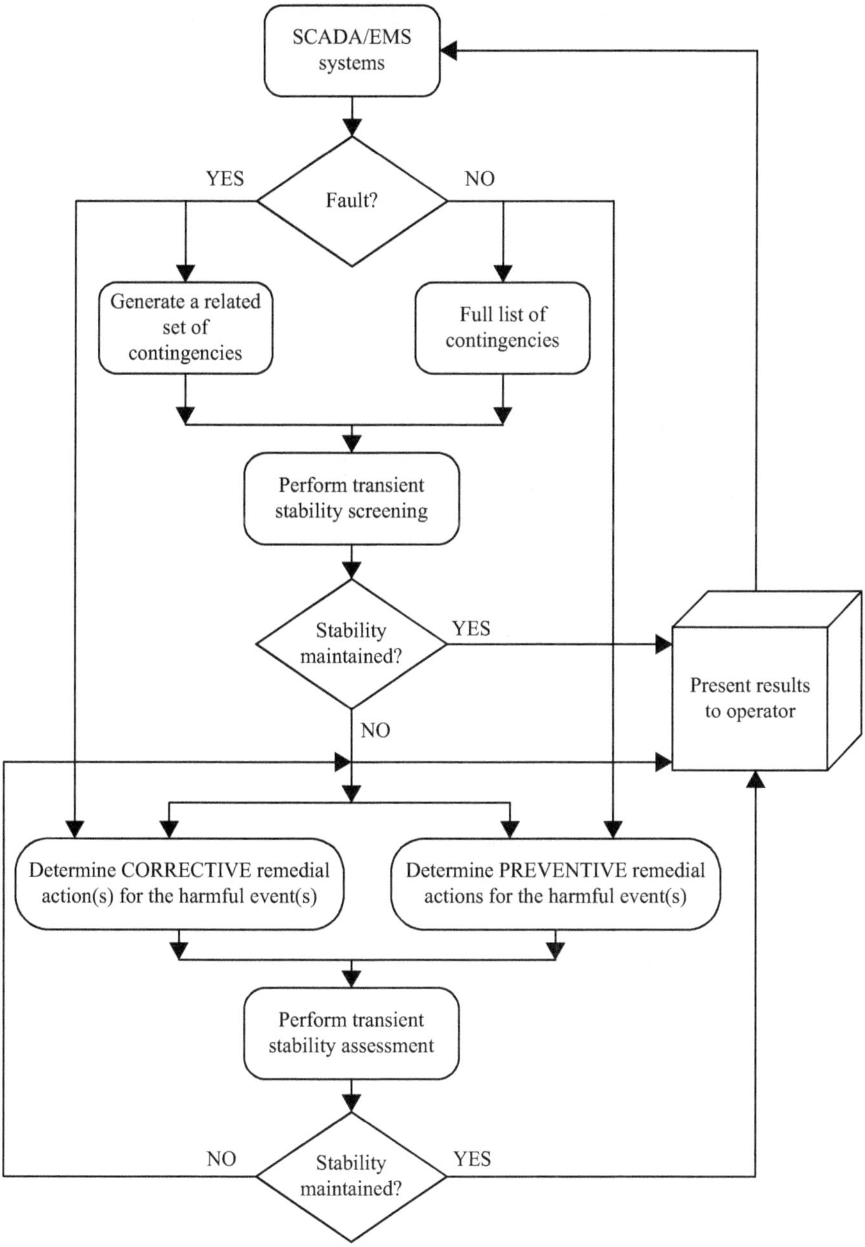

Figure 5.4 Flowchart of the corrective and preventive remedial actions scheme

If the difference is larger than zero, the system is deemed stable; otherwise, it is deemed unstable. The energy margin is calculated as follows [45].

$$\Delta V = -\frac{1}{2}M_{eq}(\tilde{\omega}_{eq}^{cl})^2 - \sum_{i=1}^{n} P_i(\tilde{\delta}_i^u - \tilde{\delta}_i^{cl}) - \sum_{i=1}^{n-1}\sum_{j=i+1}^{n}\left[C_{ij}(\cos\tilde{\delta}_{ij}^u - \cos\tilde{\delta}_{ij}^{cl}) - D_{ij}\frac{\tilde{\delta}_i^u + \tilde{\delta}_j^u - \tilde{\delta}_i^{cl} - \tilde{\delta}_j^{cl}}{\tilde{\delta}_{ij}^u - \tilde{\delta}_{ij}^{cl}}(\sin\tilde{\delta}_{ij}^u - \sin\tilde{\delta}_{ij}^{cl})\right] \quad (5.23)$$

$$= \Delta V_{KE} + \Delta V_{PE}, \quad (5.24)$$

where $M_{eq} = \frac{M_{cr}M_{sys}}{M_{cr}+M_{sys}}$, $\tilde{\omega}_{eq}^{cl} = \tilde{\omega}_{cr}^{cl} - \tilde{\omega}_{sys}^{cl}$, M_{cr} and M_{sys} are the total inertia of all machines with advanced rotor angle and total inertia of the rest of machines, respectively, at the calculated controlling UEP, $\tilde{\omega}_{cr}^{cl}$ and $\tilde{\omega}_{sys}^{cl}$ are the velocity of machines with advanced rotor angle and the rest of machines, respectively, referred to the COI reference frame at the clearing of the disturbance, and ΔV_{KE} and ΔP_{PE} are the change in the kinetic energy and the potential energy, respectively.

5.5.1.1 Sensitivity concept

As can be seen from (5.23), the energy margin is a multi-variable function and can be expressed as:

$$\Delta V = f(P_m, \tilde{\delta}^u, \tilde{\delta}^{cl}, \dot{\tilde{\delta}}^{cl}, E, G_{ij}, B_{ij}). \quad (5.25)$$

When shifting the scheduled power among generators, the changes in clearing speeds, clearing angles, and controlling UEP, and the voltages behind E_i are considerable. The changes in G_{ij} and B_{ij} are neglected because of their small values. Due to the small size of generation changes, it is assumed that the high order terms in the sensitivity equation are neglected and the mode of disturbance is unchanged by generation shifting. The sensitivity equation of the energy margin, ΔV, caused by generation shifting can be approximated as follows [45].

$$\Delta(\Delta V) \approx \sum_{k=1}^{n} \frac{\partial(\Delta V)}{\partial P_{mk}} \Delta P_{mk}, \quad (5.26)$$

where ΔP_{mk} is the change of mechanical power input at machine k and n is the number of machines at which generator outputs are adjusted.

Once the output power of machine k is shifted, the sensitivity of the energy margin is calculated by the partial derivative of ΔV with respect to P_{mk} [46]:

$$\frac{\partial(\Delta V)}{\partial P_{mk}} = -M_{eq}\tilde{\omega}_{eq}^{cl}\dot{u}_{eq,k}^{cl} - (\tilde{\delta}_k^u - \tilde{\delta}_k^{cl}) - \sum_{i=1}^{n}(P_{mi} - G_{ii}|E_i|^2)(u_{ik}^u - u_{ik}^{cl}) + \sum_{i=1}^{n-1}\sum_{j=i+1}^{n} C_{ij}[\sin\tilde{\delta}_{ij}^u(u_{ik}^u - u_{jk}^u) - \sin\tilde{\delta}_{ij}^{cl}(u_{ik}^{cl} - u_{jk}^{cl})]$$

$$+\sum_{i=1}^{n-1}\sum_{j=i+1}^{n} D_{ij}(\sin\tilde{\delta}_{ij}^{u} - \sin\tilde{\delta}_{ij}^{cl})\left[\frac{(u_{ik}^{u} + u_{jk}^{u} - u_{ik}^{cl} - u_{jk}^{cl})}{\tilde{\delta}_{ij}^{u} - \tilde{\delta}_{ij}^{cl}}\right.$$

$$\left.-\frac{(u_{ik}^{u} - u_{jk}^{u} - u_{ik}^{cl} + u_{jk}^{cl})(\tilde{\delta}_{i}^{u} + \tilde{\delta}_{j}^{u} - \tilde{\delta}_{i}^{cl} - \tilde{\delta}_{j}^{cl})}{(\tilde{\delta}_{ij}^{u} - \tilde{\delta}_{ij}^{cl})^{2}}\right]$$

$$+\sum_{i=1}^{n-1}\sum_{j=i+1}^{n}\frac{(\tilde{\delta}_{i}^{u} + \tilde{\delta}_{j}^{u} - \tilde{\delta}_{i}^{cl} - \tilde{\delta}_{j}^{cl})}{(\tilde{\delta}_{ij}^{u} - \tilde{\delta}_{ij}^{cl})}D_{ij}$$

$$\times[\cos\tilde{\delta}_{ij}^{u}(u_{jk}^{u} - u_{jk}^{u}) - \cos\tilde{\delta}_{ij}^{cl}(u_{ik}^{cl} - u_{jk}^{cl})]$$

$$+2\sum_{i=1}^{n}|E_i|\frac{\partial|E_i|}{\partial P_{mk}}G_{ii}(\tilde{\delta}_{i}^{u} - \tilde{\delta}_{i}^{cl})$$

$$-\sum_{i=1}^{n-1}\sum_{j=i+1}^{n}\left(\frac{\partial|E_i|}{\partial P_{mk}}|E_j| + \frac{\partial|E_j|}{\partial P_{mk}}|E_i|\right)B_{ij}(\cos\tilde{\delta}_{ij}^{u} - \cos\tilde{\delta}_{ij}^{cl})$$

$$+\sum_{i=1}^{n-1}\sum_{j=i+1}^{n}[G_{ij}\left(\frac{\partial|E_i|}{\partial P_{mk}}|E_j| + \frac{\partial|E_j|}{\partial P_{mk}}|E_i|\right)\frac{\tilde{\delta}_{i}^{u} + \tilde{\delta}_{j}^{u} - \tilde{\delta}_{i}^{cl} - \tilde{\delta}_{j}^{cl}}{\tilde{\delta}_{ij}^{u} - \tilde{\delta}_{ij}^{cl}}$$

$$\times(\sin\tilde{\delta}_{ij}^{u} - \cos\tilde{\delta}_{ij}^{cl})], \tag{5.27}$$

where

$\dot{u}_{eq,k}^{cl} = \dot{u}_{cr,k}^{cl} - \dot{u}_{sys,k}^{cl}$,

$\dot{u}_{cr,k}^{cl} = \frac{1}{M_{cr}}\sum_{i\in cr} M_i\dot{u}_{ik}^{cl}$,

$\dot{u}_{sys,k}^{cl} = \frac{-1}{M_T - M_{cr}}\sum_{i\in cr} M_i\dot{u}_{ik}^{cl}$,

$u_{ik}^{u} = \frac{\partial\tilde{\delta}_i^u}{\partial P_{mk}}$ is UEP sensitivity coefficient,

$u_{ik}^{cl} = \frac{\partial\tilde{\delta}_i^{cl}}{\partial P_{mk}}$ is clearing angle sensitivity coefficient,

$\dot{u}_{ik}^{cl} = \frac{\partial\tilde{\omega}_i^{cl}}{\partial P_{mk}}$ is clearing speed sensitivity coefficient.

The process to derive the above-mentioned variables is as follows [46].

Sensitivity analysis of the speed and clearing angle
Starting from swing equations during the faulted period,

$$M_i\dot{\tilde{\omega}}_i = P_i - P_{ei} - \frac{M_i}{M_T}P_{COI}, \quad \text{for } i = 1, n-1. \tag{5.28}$$

By differentiating (5.28) with respect to P_{mk}, the set of differential equations are given by [46]:

$$M_i\frac{d^2}{dt^2}u_{ik} = -Q_{ik}^{f} + \sum_{j=1}^{n} A_{ij}^{f}u_{jk}, \quad \text{for } i = 1, n-1, \tag{5.29}$$

where

$$A_{ii}^f = \left(1 - \frac{2M_i}{M_T}\right)\sum_{\substack{j=1\\j\neq i}}^{n} D_{ij}^f \sin\tilde{\delta}_{ij} - \sum_{\substack{j=1\\j\neq i}}^{n} C_{ij}^f \cos\tilde{\delta}_{ij},$$

$$A_{ij}^f = C_{ij}^f \cos\tilde{\delta}_{ij} - D_{ij}^f \sin\tilde{\delta}_{ij} + \frac{2M_i}{M_T}\sum_{\substack{l=1\\l\neq j}}^{n} D_{lj}^f \sin\tilde{\delta}_{lj},$$

$$Q_{ik}^f = \frac{M_i}{M_T} - \theta_{ik} + \sum_{j=1}^{n}\left(\frac{\partial|E_i|}{\partial P_{mk}}|E_j| + \frac{\partial|E_j|}{\partial P_{mk}}|E_i|\right)(B_{ij}^f \sin\tilde{\delta}_{ij} + G_{ij}^f \cos\tilde{\delta}_{ij})$$

$$- \frac{M_i}{M_T}\sum_{j=1}^{n}\sum_{l=1}^{n}\left(\frac{\partial|E_l|}{\partial P_{mk}}|E_j| + \frac{\partial|E_j|}{\partial P_{mk}}|E_l|\right) G_{lj}^f \cos\tilde{\delta}_{lj},$$

$$\theta_{ik} = 1 \text{ for } i = k,$$

$$\theta_{ik} = 0 \text{ for } i \neq k.$$

The solution to the dynamic sensitivity equation can be obtained if the initial conditions are given. The initial conditions are defined based on the assumption that the angular positions and machine speeds do not change instantly. As a result, the initial conditions for u_{ik} and $\dot{u}_{ik}$ with $i = 1, n$ and $k = 1, n-1$ are given by:

$$u_{ik}(0) = u_{ik}^s, \tag{5.30}$$

$$\dot{u}_{ik}(0) = 0, \tag{5.31}$$

where the superscript s denotes the pre-fault SEP. The initial values for u_{nn} and $\dot{u}_{nn}$ are calculated using the following equations:

$$\sum_{i=1}^{n} M_i \frac{\partial\tilde{\delta}_i}{\partial P_{mk}} = 0, \tag{5.32}$$

$$\sum_{i=1}^{n} M_i \frac{\partial\tilde{\omega}_i}{\partial P_{mk}} = 0. \tag{5.33}$$

After getting the initial conditions, the set of second order differential equations is solved to obtain clearing angles sensitivity u_{ik}^{cl} and speed sensitivity $\dot{u}_{ik}^{cl}$.

Sensitivity analysis of SEP and controlling UEP

From the dynamic sensitivity equation, SEP sensitivity $\frac{\partial\tilde{\delta}_i^s}{\partial P_{mk}}$ can be calculated by eliminating the term $M_i \frac{d^2}{dt^2} u_{ik}$. As the system is in pre-fault condition, the pre-fault network parameters are utilized. The set of $(n-1)$ linear equations with n variables is given as:

$$\sum_{j=1}^{n} A_{ij}^{pr} u_{jk}^s = Q_{ik}^{pr}, \quad \text{for } i = 1, n-1, \tag{5.34}$$

where A and Q matrices are calculated using the same concept of (5.29) except that all the parameters are in pre-fault conditions (the superscript pr denotes the pre-fault condition). Equation (5.32) is the nth equation of the set to solve for n variables.

Controlling UEP sensitivity equations can be obtained by modifying SEP sensitivity equations by replacing the superscript s to u and changing the pre-fault conditions to post-fault conditions. The set of equations for controlling UEP sensitivity calculation is given as follows.

$$\sum_{j=1}^{n} A^{f}_{ij} u^{u}_{jk} = Q^{f}_{ik}, \quad \text{for } i = 1, n-1. \tag{5.35}$$

Similar to the SEP sensitivity equations, the nth equation of the controlling UEP sensitivity equations also can be obtained from (5.32).

5.5.1.2 Generation shifting and line distribution factors

To calculate distribution factors of transmission lines due to generation shifting, the AC load flow is linearized to a DC load flow. With the assumption that line reactances are much larger than line resistances, the following approximations are developed [47]:

$$|E_i| = |E_j| = 1, \tag{5.36}$$

$$G_{ij} = \frac{r_{ij}}{r_{ij}^2 + x_{ij}^2} \approx 0, \tag{5.37}$$

$$B_{ij} = \frac{-x_{ij}}{r_{ij}^2 + x_{ij}^2} = \frac{-1}{x_{ij}}, \tag{5.38}$$

$$\cos(\delta_i - \delta_j) \approx 1, \tag{5.39}$$

$$\sin(\delta_i - \delta_j) \approx \delta_i - \delta_j. \tag{5.40}$$

These conditions are applied to calculate all bus phase angles in a network.

$$P_i = \sum_{j=1}^{m} P_{ij} = \sum_{j=1}^{m} \frac{1}{x_{ij}}(\delta_i - \delta_j). \tag{5.41}$$

Then,

$$P = B_x \times \delta, \tag{5.42}$$

where

$$\begin{aligned}
P &= [P_1, P_2, \ldots, P_m]^T, \\
\delta &= [\delta_1, \delta_2, \ldots, \delta_m]^T, \\
B_{xii} &= \sum_{j=1}^{m} \frac{1}{x_{ij}} \text{ for } i \neq \text{reference bus}, \\
B_{xii} &= 0 \text{ for } i = \text{reference bus}, \\
B_{xij} &= \frac{-1}{x_{ij}} \text{ for } i \text{ and } j \neq \text{reference bus}, \\
B_{xij} &= 0, \text{ for } i = \text{reference bus and } j = \text{reference bus}.
\end{aligned}$$

Using (5.42), bus phase angles are calculated as follows [47].

$$\delta = X \times P. \tag{5.43}$$

These steps help in generating matrix X. In calculating the generation shifting sensitivity factors for the generator on bus i, the perturbation on bus i is set to 1 and on the other buses are set to 0. The change in bus phase angles are as follows [47]:

$$\Delta\delta = X \begin{bmatrix} 0 \\ +1 \\ 0 \\ -1 \\ 0 \\ 0 \end{bmatrix}, \tag{5.44}$$

where $+1$ is in row i, -1 is in the reference row, and other rows are zeros.

From (5.44), the distribution factor of line ℓ due to generation shift at generator i is given as follows [47].

$$\begin{aligned}
S_{\ell i} &= \frac{df_\ell}{dP_i} = \frac{d}{dP_i}\left(\frac{1}{x_\ell}(\delta_n - \delta_m)\right) \\
&= \frac{1}{x_\ell}\left(\frac{d\delta_n}{dP_i} - \frac{d\delta_m}{dP_i}\right) \\
&= \frac{1}{x_\ell}\left(X_{ni} - X_{mi}\right),
\end{aligned} \tag{5.45}$$

where X_{ni} and X_{mi} are the nth and m^{th} elements of the $\Delta\delta$ matrix, respectively, and x_ℓ is the reactance of the line ℓ.

The linear relationship between line power flow changes and generation shifting is given by [47],

$$\Delta I_w = \sum_{k=1}^{n-1} S_{wk} \Delta P_{mk} \quad \text{for } k = 1, n-1, \tag{5.46}$$

where w is the monitored line.

5.5.2 Preventive actions based on generation shifting

Operators usually have on their display screen a list of risk indices which are based on the steady state contingency analysis. These indices are updated according to the change in the system configuration. Since these indices are calculated based on the static security assessment, some of the alternatives may not be safe from the transient stability perspective. For example, if the system is operating at an equilibrium point (pre-fault) and a fault occurs that leads to the tripping of a heavily loaded line, the system may still have a post-fault equilibrium point (or, in other words, there is a feasible power flow solution); however, from the transient stability perspective, the system may not be able to move from the pre-fault equilibrium point to the post-fault equilibrium point (the post-fault equilibrium point is determined from the power flow solution under the post-fault conditions). The preventive action scheme is intended to help operators to discard the unsafe alternatives. Also, the suggested actions then can be combined with the other tools and control parameters such as equipment actuation time, cost, and available means to solve an optimization problem.

In the preventive action mode, the scheme identifies possible changes in the initial conditions that can stabilize the designated unstable contingencies. Therefore, the criterion of identifying such actions is based on the changes that can be implemented to make the energy margin ΔV zero or positive. The sensitivity of the energy margin with respect to the control variables (generation rescheduling in this case) can be used in finding the "best" action. However, in finding such actions, the power carrying capabilities of transmission lines, generation limits, and other constraints have to be satisfied and the problem of finding the right or best solution is complicated. For example, in rescheduling the critical generators, the generation shift may cause some critical lines to be heavily loaded. In such cases, other rescheduling options need to be tested and therefore this process is computationally expensive. Another approach is to solve an optimization problem with multiple constraints.

The preventive action scheme works in two stages. It starts with scanning all possible scenarios and classifies them into stable and unstable scenarios based on the energy margin. The second stage performs optimization for unstable contingencies to choose the most effective rescheduling.

If a predicted contingency makes the system unstable, control actions must be implemented to change the initial condition to return the system to a SEP. Several remedial action methods can be implemented to increase the stability margin of the system, which include generation rescheduling and reactive power control. In this chapter, the focus is to use generation redispatch as a preventive control action. Three questions are usually brought up for the cases where generation redispatch is needed [29]: (1) Which generator will be rescheduled? (2) How to determine changing the setting of each generator? and (3) Do these changes introduce any violation in generator and transmission line constraints? These questions can be formulated as an optimization problem as follows [48].

$$\min \{Z = (F_1, F_2, ..., F_{N+L}\}, \tag{5.47}$$

subject to:

$$g(F_i) \geq 0,$$
$$h(F_i) = 0.$$

Equality constraints in the optimization problem are usually from the power flow equations. Inequality constraints relate to branch flow limits, bus voltage magnitude limits, and generation shifting limits. In particular, the objective of the preventive action is to minimize the generation shifting and the changes of line power flow while satisfying the constraints of generation and transmission line limits to increase system energy margin:

$$F_{\min} \leq F \leq F_{\max}.$$

The sensitivity and distribution factor equations can be included in a matrix form to form a constraint of the optimization problem:

$$A \times F = B,$$

where Z is a vector of $N + L$ objective function, $F_{\min}$ and $F_{\max}$ are the limits of F.

$$A = \begin{bmatrix} \frac{\partial(\Delta V)}{\partial P_{m1}} & \cdots & \frac{\partial(\Delta V)}{\partial P_{mn}} & 0 & \cdots & 0 \\ S_{11} & \cdots & S_{1n} & -1 & \cdots & 0 \\ \cdot & & \cdot & & & \\ \cdot & & \cdot & & & \\ \cdot & & \cdot & & & \\ S_{L1} & \cdots & S_{Ln} & 0 & \cdots & -1 \end{bmatrix},$$

$$B = [\Delta(\Delta V)0 \ldots 0]^T,$$
$$F = [\Delta P_{m1} \ldots \Delta P_{mn}\ \Delta I_1 \ldots \Delta I_L]^T,$$
$$\Delta(\Delta V) = \Delta V_{new} - \Delta V_{old},$$

As recommended in Reference 48, it is more reasonable to set the lower bound of all variables to zero. This will reduce the total shifting of all generations.

Solution to the optimization problem defined by (5.47) yields the necessary generation shifting that increases the energy margin to zero or a positive value.

5.5.3 Preventive actions with FACTS devices

Besides generation shifting, installing and operating FACTS is another class of preventive actions to increase system energy margin [49]. Locations of FACTS devices are determined based on the sensitivity of the energy margin, which is calculated by utilizing the structure preserving energy function.

5.5.3.1 Series FACTS controller

Series FACTS controllers can be used for damping in power systems [49]. The control strategy and the locations of the controllers are obtained based on a simplified model

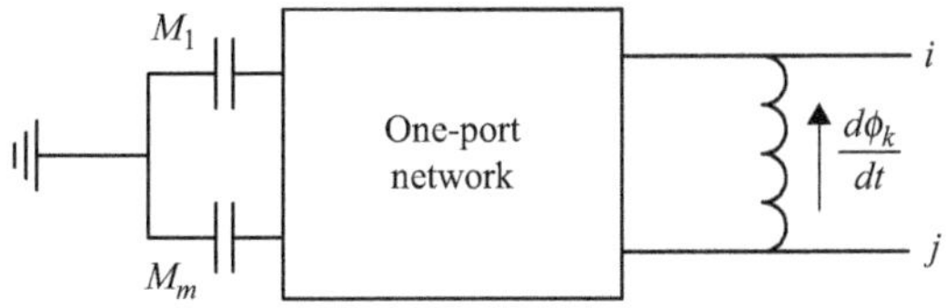

Figure 5.5 One-port network model to analyze FACTS controller response. See Reference 49 for details

of the system. Assuming that the system is lossless, it can be regarded as a one-port network across nodes i and j as shown in Figure 5.5 [49]. The linearized system equations are given by [49]:

$$\begin{bmatrix} M\frac{d^2(\Delta\delta)}{dt^2} \\ 0 \end{bmatrix} = -\begin{bmatrix} K_{gg} & K_{g\ell} \\ K_{\ell g} & K_{\ell\ell} \end{bmatrix}\begin{bmatrix} \Delta\delta \\ \Delta\phi \end{bmatrix}, \tag{5.48}$$

where
$M = Diag[M_1, M_2, \ldots, M_m]$, $K_{gg} = Diag[K_{g1}, K_{g2}, \ldots, K_{gm}]$, $[K_{g\ell}]_{m\times n} = [-K_{gg}\ 0]$

$$K_{gi} = \frac{E_{gi}V_i\cos(\delta_i - \phi_i)}{x'_{gi}},$$

where $K_{\ell\ell}$ is an $n \times n$ symmetric matrix with $K_{\ell\ell}(i,j) = -K_{ij}$, $K_{\ell\ell}(i,i) = \sum_{j=1}^{n} K_{ij}$, $K_{ij} = \frac{V_iV_j}{x_k}\cos\phi_{k0}$.

Eliminating $\Delta\delta$ and converting (5.48) to the state space we get the following [49],

$$\begin{bmatrix} \Delta\dot{\delta} \\ \Delta\dot{\omega} \end{bmatrix} = \begin{bmatrix} 0 & I_{n\times n} \\ A & 0 \end{bmatrix}\begin{bmatrix} \Delta\delta \\ \Delta\omega \end{bmatrix}. \tag{5.49}$$

The eigenvalues of matrix A are: $\pm j\omega_i^m$ for $i = 0, m-1$, where $i = 0$ and $\omega_i^m = 0$ correspond to the frequency of the COI. The eigenvector that corresponds to the eigenvalue $j\omega_i^m$ is $V_i = [V_{i\delta}^t\ V_{i\omega}^t]^t$, where $V_{i\omega} = j\omega_i^m V_{i\delta}^t$. By using the compensation theorem, the capacitor M_j can be replaced by a voltage source $V_{i\omega}(j)$. Replacing the network by its Thévenin equivalent circuit (a voltage source V_{Ck}, which is sinusoidal of frequency ω_i^m, and an inductor L_k^{th}), the presence of a series FACTS controller in line k causes the change in power flow of line k as follows [49].

$$\Delta P_k = \frac{\partial P_k}{\partial\phi_k}\Delta\phi_k + \frac{\partial P_k}{\partial u_k}\Delta u_k \tag{5.50}$$

$$= \Delta P_{k0} + \Delta P_{kc} \tag{5.51}$$

$$= \frac{1}{L_k} + B_k\Delta u_k, \tag{5.52}$$

where $B_k = P_{k0}/x_k$ for a Thyristor-Controlled Series Compensation (TCSC) and $B_k \approx 1/x_k$ for a Static Synchronous Series Compensator (SSSC).

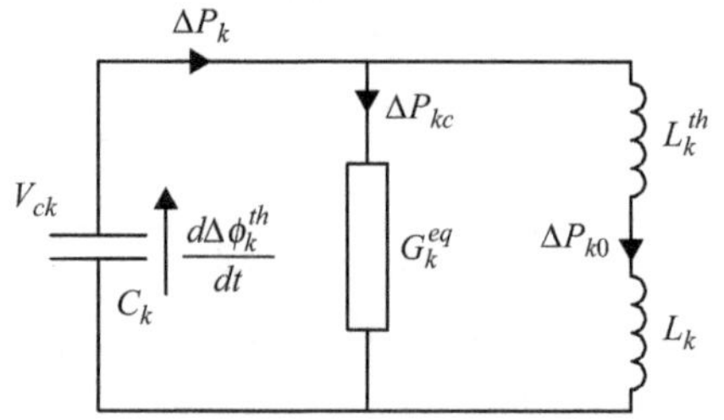

Figure 5.6 Equivalent circuit with the series FACTS controller. See Reference 49 for details

The voltage source V_{Ck} can be replaced by an equivalent capacitor C_k as follows [49].

$$C_k = \frac{1}{\omega_i^m V_{Ck}} \Delta P_k. \tag{5.53}$$

The FACTS controller can be treated as a current source, $\Delta P_{kc} = B_k \Delta u_k$, with a control strategy given as follows [49].

$$\Delta u_k = \frac{G_k^{eq}}{B_k} \frac{d\Delta\phi_K^{th}}{dt}, \tag{5.54}$$

where G_k^{eq} is the equivalent conductance as shown in Figure 5.6 [49].

The optimal location of the series FACTS controller is the one for which $|V_{Ck}|$ is maximum.

5.5.3.2 Shunt FACTS controller

Similar to the series FACTS controller, the shunt FACTS controller also derives a damping control strategy to dissipate energy so that the system reaches a new stable state [49]. Considering the network that includes *nonlinear* inductors and linear capacitors. The total energy in all inductors in the system is given by [49]:

$$W_L = \sum_{k=1}^{n_b} W_k = \sum_{k=1}^{n_b} \frac{V_k^2}{2x_k} - \sum_{j=1}^{N} \int I_{R_j} dV_j, \tag{5.55}$$

and the total energy stored in the capacitors is as follows [49]:

$$W_C = \frac{1}{2} \sum_{i=1}^{m} M_i \frac{d\delta_i^2}{dt}, \tag{5.56}$$

where $V_k^2 = V_i^2 + V_j^2 - 2V_iV_j \cos\phi_k$, $N = n + m$ is the number of bus in the system, n_b is the number of series branches in the network. I_{Rj} is the reactive current injected into bus j and is represented as [49]:

$$I_{Rj} = I_{Rj0} + \frac{\partial I_{Rj}}{\partial V_j} \Delta V_j + \Delta I_{Rj}, \tag{5.57}$$

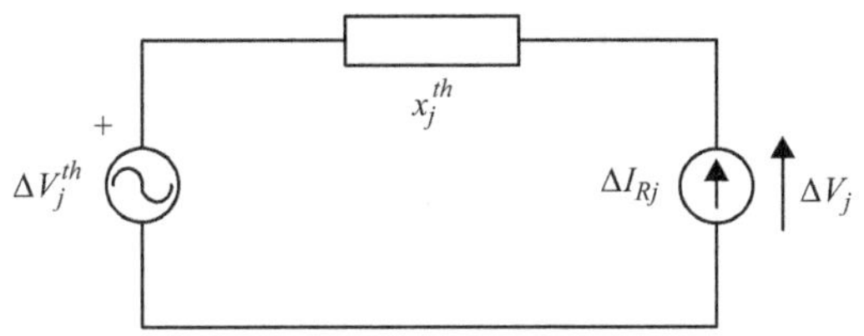

Figure 5.7 Equivalent circuit with the shunt FACTS controller. See Reference 49 for details

where the third term of (5.57) is the supplementary modulation controller which helps in damping rotor swings.

For a conservative system of an L-C network, the total energy is constant and $\frac{d(W_L+W_C)}{dt} = 0$. However, if ΔI_{Rj} is controlled as $\Delta I_{Rj} = -K_{rj}\frac{dV_j}{dt}$, the derivative of the total energy is different from zero due to the change in the total energy and is given as follows [49].

$$\frac{d(W_L + W_C)}{dt} = -\sum K_{rj}(\frac{dV_j}{dt})^2. \tag{5.58}$$

If $K_{rj} > 0$, the derivative of total energy is negative. Therefore, the energy along the trajectory is dissipated. To optimize the control variable, ΔI_{Rj}, the linear network model of reactive current is applied as in Reference 49 and gives the following equation,

$$\Delta I_{Rj} = -K_{rj}\frac{d(\Delta V_j^{th})}{dt}, \tag{5.59}$$

where ΔV_j^{th} is the Thévenin voltage at bus j as shown in Figure 5.7. The optimum location of the FACTS controller, in mode m_i, is at a bus for which $|\Delta V_j^{th}|$ is maximized.

5.5.4 Corrective actions

Although preventive action schemes help in avoiding unstable states, which are results of disturbances, it is possible that the power system may experience an unexpected unstable state. One reason is that preventive actions might be unreasonable to implement considering economic aspects such as the high cost of generation shifting. It is also possible that there is no feasible solution for the preventive action optimization problem. Moreover, many disturbances are the result of natural and random events which usually are not included in the designated list of contingencies. In these cases, the corrective action should be implemented to avoid unfolding events. The most common corrective actions are generation tripping, load shedding, and insertion of braking resistors [29]. Another corrective action that utilizes FACTS controllers has been proposed in Reference 49.

5.5.4.1 Generation tripping

It is assumed that only a small amount of generation is tripped at node k to improve the system stability. As the change in ΔP_{mk} is small compared to the total system generation (which also means a negligible amount of inertia is tripped), the parameters B_{ij}, G_{ij}, $\tilde{\delta}_0$, and $\tilde{\delta}^{cl}$ do not change. Only Y_{kk} (element kk of the admittance matrix) changes due to the change in x'_{dk} (the d-axis transient reactance of generator k) and also $\tilde{\delta}^u$ and M_{eq} [29]. The change in M_{eq} can be expressed as follows [29]:

$$\frac{\Delta M_{eq}}{\Delta P_{mk}} = \frac{M_k}{P_{mk}}. \tag{5.60}$$

The mismatch function of machine i can be expressed as follows [29].

$$f_i = P_{mi} - P_{ei} = P_{mi} - E_i^2 G_{ii} - \sum_{j \neq i} (C_{ij} \sin \tilde{\delta}_{ij} + D_{ij} \cos \tilde{\delta}_{ij}). \tag{5.61}$$

where C_{ij} and D_{ij} are as defined for (5.10).

Therefore, the change in the mismatch function can be expressed as follows:

$$\Delta f_i = -E_i^2 \Delta G_{ii} - \sum_{j \neq i} (\Delta C_{ij} \sin \tilde{\delta}_{ij} + \Delta D_{ij} \cos \tilde{\delta}_{ij}). \tag{5.62}$$

If J is the Jacobian at the equilibrium point without generation tripping, the first estimate of the change in the equilibrium point is given by [29]:

$$\begin{pmatrix} \Delta\tilde{\delta}_1 \\ \vdots \\ \Delta\tilde{\delta}_k \\ \vdots \\ \Delta\tilde{\delta}_n \end{pmatrix} = J^{-1} \begin{pmatrix} 0 \\ \vdots \\ \Delta P_{mk} \\ \vdots \\ 0 \end{pmatrix}. \tag{5.63}$$

In other words, the change in the mismatch functions, Δf_i, equal to zero for $i \neq k$ and equal to ΔP_{mk} for $i = k$. The accuracy of $\Delta\tilde{\delta}$ can be improved if the Jacobian is redefined at the new value of $\Delta\tilde{\delta}$, and then the inverse of this Jacobian is used to recalculate $\Delta\tilde{\delta}$. After determining $\Delta\tilde{\delta}^u$, the change in energy margin is given as [29]:

$$\frac{\Delta(\Delta V)}{\Delta P_{mk}} = \frac{\partial(\Delta V)}{\partial M_{eq}} \frac{\Delta M_{eq}}{\Delta P_{mk}} + \sum_{i=1}^{n} \frac{\partial(\Delta V)}{\partial \tilde{\delta}_i^u} \frac{\Delta \tilde{\delta}_i^u}{\Delta P_{mk}} + \frac{\partial(\Delta V)}{\partial P_{ik}} \frac{\Delta P_{ik}}{\Delta P_{mk}}, \tag{5.64}$$

where:

$$\frac{\partial(\Delta V)}{\partial M_{eq}} = \frac{1}{2}(\tilde{w}_{eq}^{cl})^2, \tag{5.65}$$

$$\frac{\partial(\Delta V)}{\partial P_{ik}} = -(\tilde{\delta}_k^u - \tilde{\delta}_k^{cl}), \tag{5.66}$$

$$\frac{\partial(\Delta V)}{\partial \tilde{\delta}_i^u} = -P_i + \sum_{j=i+1}^{n} \left[C_{ij} \sin \tilde{\delta}_{ij}^u - D_{ij} \left((\sin\tilde{\delta}_{ij}^u - \sin \tilde{\delta}_{ij}^{cl}) \frac{\tilde{\delta}_{ij}^u - \tilde{\delta}_{ij}^{cl} - \tilde{\delta}_{i+j}^u + \tilde{\delta}_{i+j}^{cl}}{(\tilde{\delta}_{ij}^u - \tilde{\delta}_{ij}^{cl})^2} \right.\right.$$
$$\left.\left. + \frac{\tilde{\delta}_{i+j}^u - \tilde{\delta}_{i+j}^{cl}}{\tilde{\delta}_{ij}^u - \tilde{\delta}_{ij}^{cl}} \cos \tilde{\delta}_{ij}^u \right) \right]. \tag{5.67}$$

After generation tripping, the new value of M_{eq} and $\tilde{w}_{eq}$ are $\hat{M}_{eq}$ and $\hat{w}_{eq}$ which can be calculated as follows [29].

$$\hat{M}_{eq} = (1 - K)M_{eq}, \tag{5.68}$$

$$\hat{w}_{eq} \simeq \tilde{w}_{eq} + K(\tilde{w}_{cr} - \tilde{w}_k), \tag{5.69}$$

where $K = \frac{\Delta M_k}{M_{cr}}$. The change of ΔV due to generation tripping is given by [29]:

$$\Delta(\Delta V) = \Delta V_{KE} + \Delta V_{PE} + \Delta V_{ME} + \Delta V_{DE}, \tag{5.70}$$

where

$$\Delta V_{KE} = \frac{1}{2}(\hat{M}_{eq}\hat{w}_{eq}^2 - M_{eq}\tilde{w}_{eq}^2), \tag{5.71}$$

$$\Delta V_{PE} = -P_{ko}\Delta\tilde{\delta}_k^u - \Delta P_k \tilde{\delta}_{ko}^u - \Delta P_k \Delta\tilde{\delta}_k^u - \sum_{i \neq k} P_{io}\Delta\tilde{\delta}_i^u, \tag{5.72}$$

$$\Delta V_{ME} = -\sum_{i=1}^{n-1}\sum_{j=i+1}^{n} C_{ij} \sin \tilde{\delta}_{ijo}^u \Delta\tilde{\delta}_{ij}^u, \tag{5.73}$$

$$\Delta V_{DE} = -\sum_{i=1}^{n-1}\sum_{j=i+1}^{n} D_{ij}\Delta\tilde{\delta}_{ij}^u \left[\frac{\tilde{\delta}_{ijo}^u - \tilde{\delta}_{ij}^{cl} - \tilde{\delta}_{i+jo}^u + \tilde{\delta}_{i+j}^{cl}}{(\tilde{\delta}_{ijo}^u - \tilde{\delta}_{ij}^{cl})^2} (\sin\tilde{\delta}_{ijo}^u - \sin \tilde{\delta}_{ij}^{cl}) \right.$$
$$\left. + \frac{\tilde{\delta}_{i+jo}^u - \tilde{\delta}_{i+j}^{cl}}{\tilde{\delta}_{ijo}^u - \tilde{\delta}_{ij}^{cl}} (\cos\tilde{\delta}_{ijo}^u - \cos \tilde{\delta}_{ij}^u) \right]. \tag{5.74}$$

The subscripts *KE*, *PE*, *ME*, and *DE* stand for kinetic energy, position energy, magnetic energy, and dissipation energy, respectively.

To determine the amount of generation to be tripped, it is assumed that P_u equals the initial power of one generator at plant k. The procedure to estimate the number of units to be tripped is presented as following [29]:

1. For $\Delta P_{mk} = P_u$, calculate $\Delta(\Delta V)$ due to generation tripping,
2. Let $\lambda = \frac{|\Delta(\Delta V)|}{P_u}$, calculate $\frac{-\Delta V}{\lambda}$, and
3. If $\frac{-\Delta V}{\lambda} < P_u$, trip one unit. If $P_u < \frac{-\Delta V}{\lambda} < \gamma P_u$, trip two units ($\gamma$ is a factor slightly less than 2). If $\frac{-\Delta V}{\lambda} > \gamma P_u$, previous steps are repeated for $\Delta P_{mk} = 3P_u$ to see if more units will be tripped and so on.

5.5.4.2 Load rejection

Load rejection is expressed by a change in the impedance of load buses (Y_{bus} matrix will change). Hence, the $\tilde{\delta}^u$ will change too but there will be no change in P_{mi} and M_i. The energy margin will change due to the change in potential energy.

The change in $\Delta\tilde{\delta}^u$ due to load rejection is expressed as [29]:

$$\begin{pmatrix} \Delta\tilde{\delta}_1^u \\ \Delta\tilde{\delta}_2^u \\ \vdots \\ \Delta\tilde{\delta}_n^u \end{pmatrix} = J^{-1} \begin{pmatrix} \Delta f_1 \\ \Delta f_2 \\ \vdots \\ \Delta f_n \end{pmatrix}, \tag{5.75}$$

where

$$\frac{\Delta(\Delta V)}{\Delta G_{ii}} = \frac{\partial(\Delta V)}{\partial P_i}\frac{\Delta P_i}{\Delta G_{ii}} + \sum_{i=1}^{n}\frac{\partial(\Delta V)}{\partial\tilde{\delta}_i^u}\frac{\Delta\tilde{\delta}_{i,G_{ii}}^u}{\Delta G_{ii}} = (\tilde{\delta}_{io}^u - \tilde{\delta}_{io}^{cl})(-E_i^2) + \tilde{\delta}_{io}^u\frac{\Delta\tilde{\delta}_{i,G_{ii}}^u}{\Delta G_{ii}}, \tag{5.76}$$

$$\frac{\partial(\Delta V)}{\partial\tilde{\delta}_i^u} = -P_i + \sum_{j=i+1}^{n}\left[C_{ijo}\sin\tilde{\delta}_{ij}^u - D_{ij}\left((\sin\tilde{\delta}_{ij}^u - \sin\tilde{\delta}_{ij}^{cl})\frac{\tilde{\delta}_{ij}^u - \tilde{\delta}_{ij}^{cl} - \tilde{\delta}_{i+j}^u + \tilde{\delta}_{i+j}^{cl}}{(\tilde{\delta}_{ij}^u - \tilde{\delta}_{ij}^{cl})^2} + \frac{\tilde{\delta}_{i+j}^u - \tilde{\delta}_{i+j}^{cl}}{\tilde{\delta}_{ij}^u - \tilde{\delta}_{ij}^{cl}}\cos\tilde{\delta}_{ij}^u\right)\right], \tag{5.77}$$

$$\frac{\partial(\Delta V)}{\partial C_{ij}} = \frac{\partial(\Delta V)}{\partial\tilde{\delta}_i^u}\frac{\Delta\tilde{\delta}_{i,C_{ij}}^u}{\Delta C_{ij}}, \tag{5.78}$$

$$\frac{\partial(\Delta V)}{\partial D_{ij}} = \frac{\partial(\Delta V)}{\partial\tilde{\delta}_i^u}\frac{\Delta\tilde{\delta}_{i,D_{ij}}^u}{\Delta D_{ij}}, \tag{5.79}$$

$$\Delta(\Delta V) = \frac{\partial(\Delta V)}{\partial G_{ii}}\Delta G_{ii} + \frac{\partial(\Delta V)}{\partial C_{ij}}\Delta C_{ij} + \frac{\partial(\Delta V)}{\partial D_{ij}}\Delta D_{ij}, \tag{5.80}$$

and $\Delta\tilde{\delta}_{i,G_{ii}}^u$, $\Delta\tilde{\delta}_{i,C_{ij}}^u$, and $\Delta\tilde{\delta}_{i,D_{ij}}^u$ are the change in $\Delta\tilde{\delta}_i^u$ due to the change in G_{ii}, C_{ij}, and D_{ij}, respectively.

Again, the accuracy of $\Delta\tilde{\delta}$ can be improved if the Jacobian J at $(\tilde{\delta}^u + \Delta\tilde{\delta}^u)$ is recalculated and then $\Delta\tilde{\delta}^u$ is determined.

5.5.4.3 Application of a braking resistor

The effect of applying a braking resistor is opposite to that of load shedding [29]. Similar to the load shedding, the variation in ΔV is determined by defining the change in the elements of Y_{bus} and then $\Delta\tilde{\delta}_i^u$ and ΔV.

The size of the resistor is calculated using the following procedure [29]:

1. Calculate the change of energy margin $\Delta(\Delta V)$ corresponding to $R_{brake} = 1pu$ based on the procedure presented in load shedding.
2. The size of desired braking resistor equals to $\frac{-\Delta V}{|\Delta(\Delta V)|}$.

5.5.4.4 Application of FACTS controllers

In Reference 49, Padiyar showed that during an unstable contingency, a multi-machine system splits into two coherent groups of generators along a unique cutset. To stabilize the system, the FACTS controllers are installed in this unique cutset. Following a disturbance, the system is separated into two groups: group A and group B. With the assumption that the generators within group A accelerate with respect to the generators in group B, the equations for COI in the two groups with are given as follows [49].

$$\sum_{i \in A} M_i \frac{d^2\delta_{A0}}{dt^2} = \sum_{i \in A} (P_{mi} - P_{ei}) - \sum_{k=1}^{nc} P_{tk}, \tag{5.81}$$

$$\sum_{i \in B} M_i \frac{d^2\delta_{B0}}{dt^2} = \sum_{i \in B} (P_{mi} - P_{ei}) - \sum_{k=1}^{nc} P_{tk}. \tag{5.82}$$

where δ_{A0} and δ_{B0} are center of angles.

From (5.81) and (5.82), the motion equation can be expressed as follows.

$$M_{eq} \frac{d^2\delta_{eq}}{dt^2} = \frac{M_A}{M_T} \sum_{i \in A} (P_{mi} - P_{ei}) - \frac{M_B}{M_T} \sum_{j \in B} (P_{mj} - P_{ej}) - \sum_{k=1}^{nc} P_{tk}, \tag{5.83}$$

where $\delta_{eq} = \delta_{A0} - \delta_{B0}$, P_{tk} is the power flow in line k belonging to the cutset, and nc is the number of series elements in the cutset, $M_A = \sum_{i \in A} M_i$ and $M_B = \sum_{j \in B} M_j$.

Assuming that L_A and L_B are the sets of lines in groups A and B, the total system potential energy can be presented as [49],

$$\begin{aligned} V_{PE} &= \sum_{i \in A} \int_{\tilde{\delta}_{io}}^{\tilde{\delta}_i} (P_{ei} - P_{mi}) d\tilde{\delta}_i + \sum_{j \in B} \int_{\tilde{\delta}_{jo}}^{\tilde{\delta}_j} (P_{ej} - P_{mj}) d\tilde{\delta}_j \\ &= \sum_{k \in L_A} \int_{\phi_{ko}}^{\phi_k} (P_k - P_{ks}) d\phi_k + \sum_{r \in L_B} \int_{\phi_{ro}}^{\phi_r} (P_k - P_{ks}) d\phi_r \\ &\quad + \sum_{k=1}^{nc} \int_{\phi_{ko}}^{\phi_k} (P_{tk} - P_{tks}) d\phi_k \\ &= V_{PE_A} + V_{PE_B} + V_{PE_{cutset}}. \end{aligned} \tag{5.84}$$

If the voltage angles at all buses in each area are identical, the potential energy of all areas equals zero. Hence,

$$V_{PE} = V_{PE_{cutset}} = \sum_{k=1}^{nc} \int_{\phi_{ko}}^{\phi_k} (P_{tk} - P_{tks}) d\phi_k, \tag{5.85}$$

where P_{tks} is the power flow in line k of the cutset in post-fault condition.

To stabilize the system, FACTS controllers are connected to the lines in the cutset thereby maximizing the power flow in these lines. The control algorithm is applied to each line in the cutset as follows [49].

$$\frac{d\phi_k}{dt} = \frac{d\phi_i}{dt} - \frac{d\phi_j}{dt}, \tag{5.86}$$

where ϕ_k is the angle across line k connected between node i in group A and node j in group B. The control algorithm applicable to individual FACTS controller in a line is given below [49]:

- To stabilize the system, the control variables associated with a FACTS controller u_c are chosen so that the power flow in the line P_{tk} is maximized. The controller vector of series FACTS controller can be X_{TCSC} for TCSC or V_r for SSSC. The controller vector of shunt FACTS controller can be B_{SVC} for SVC. If a UPFC (Unified Power Flow Controller) is installed, u_c is a vector of (V_c, β, I_r) with $V_c\angle\beta$ – the injected series voltage.
- The post-fault condition for the control variables is chosen when $\frac{d\phi_k}{dt} = 0$ or $P_k = P_{ks}$ and $\frac{d\phi_k}{dt} < 0$.
- The power flow is minimized by changing values of the control variables when $\frac{d\phi_k}{dt}$ is minimum and smaller than $-\varepsilon$.
- The control variables take their operating values when $\frac{d\phi_k}{dt} = 0$.
- The control variables are selected such that the power flow is maximized when $\frac{d\phi_k}{dt}$ is maximum and larger than ε.
- Repeat step 2 if $|\frac{d\phi_k}{dt}| > \varepsilon$.

These four corrective actions can be applied individually or simultaneously to ensure the system stability.

5.6 Example

This section provides an illustrative example to show the use of the transient stability screening and remedial action procedure. The procedure is applied on the Western Electricity Coordinating Council (WECC) nine-bus test system [31]. The data of this system are given in per unit with 100 MVA and 230 KV base and system frequency is 50 Hz. System configuration is shown in Figure 5.8.

Considering only three phase faults, there will be 12 (6 lines with a fault at each end) $N-1$ possible line fault contingencies. Faults are assumed near to the buses and cleared by opening the associated line. The critical clearing times of the 12 contingencies range from 0.21 s to 0.5 s. In this example, the fault clearing time for all contingencies is assumed 0.3 s; all the classifications are performed according to this clearing time. If the energy margin of a certain contingency is larger than the energy margin at the controlling UEP, the contingency is classified unstable; otherwise the contingency is classified stable. The process of performing screening and suggesting remedial actions is depicted in Figure 5.9.

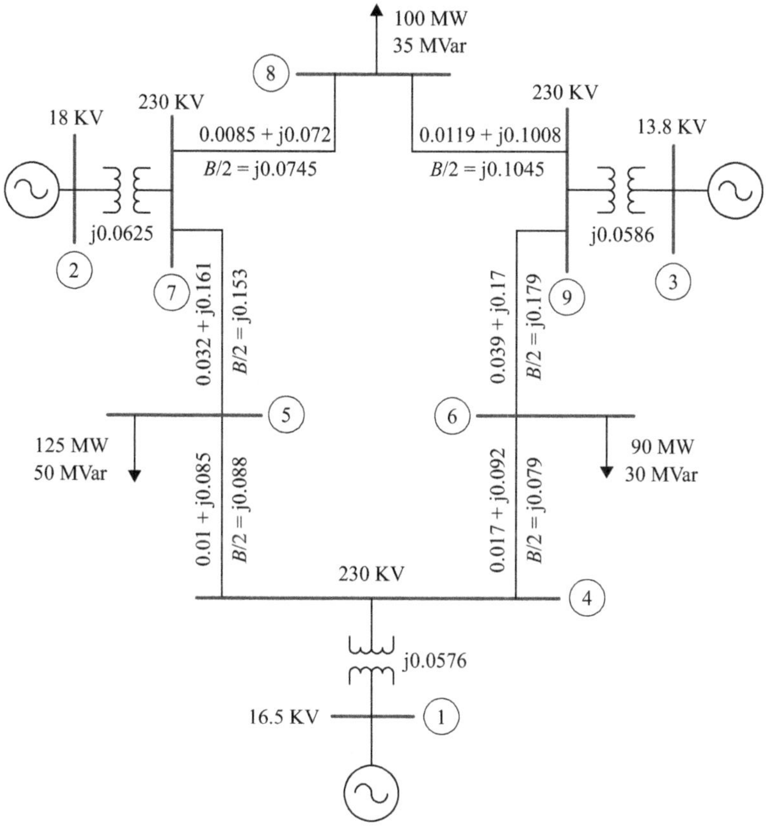

Figure 5.8 Configuration of three-machine, nine-bus system

The pre-fault load flow results are shown in Table 5.1 and the EPs associated with each contingency are shown in Table 5.2. The EPs are calculated using the procedures mentioned earlier.

The EPs of Table 5.2 are used as initial conditions for the corresponding list of contingencies to calculate the controlling UEP of each contingency. The calculated controlling UEPs are given in Table 5.3. The controlling UEPs shown in Table 5.3 can be obtained using the BCU method or the homotopy method described earlier.

For each contingency, the energy margin is calculated to determine contingencies, if any, that cause instability. From this analysis, contingencies 6, 8, 9, and 12 are found unstable. There are several remedial action options; however, the selection of the best option requires a multi-decision tool. This example illustrates the use of generation shift as remedial action. Generation shift results of this system for contingencies 6, 8, and 12 are shown in Table 5.4 where ΔV is the energy margin of the contingency, and ΔP_{mi} the shifting in generation of generator i. Generators in the system are rescheduled to move the system to a new stable state with $\Delta V \geq 0$. For this example, generator 1, which has the highest capacity, is chosen to be the reference generator.

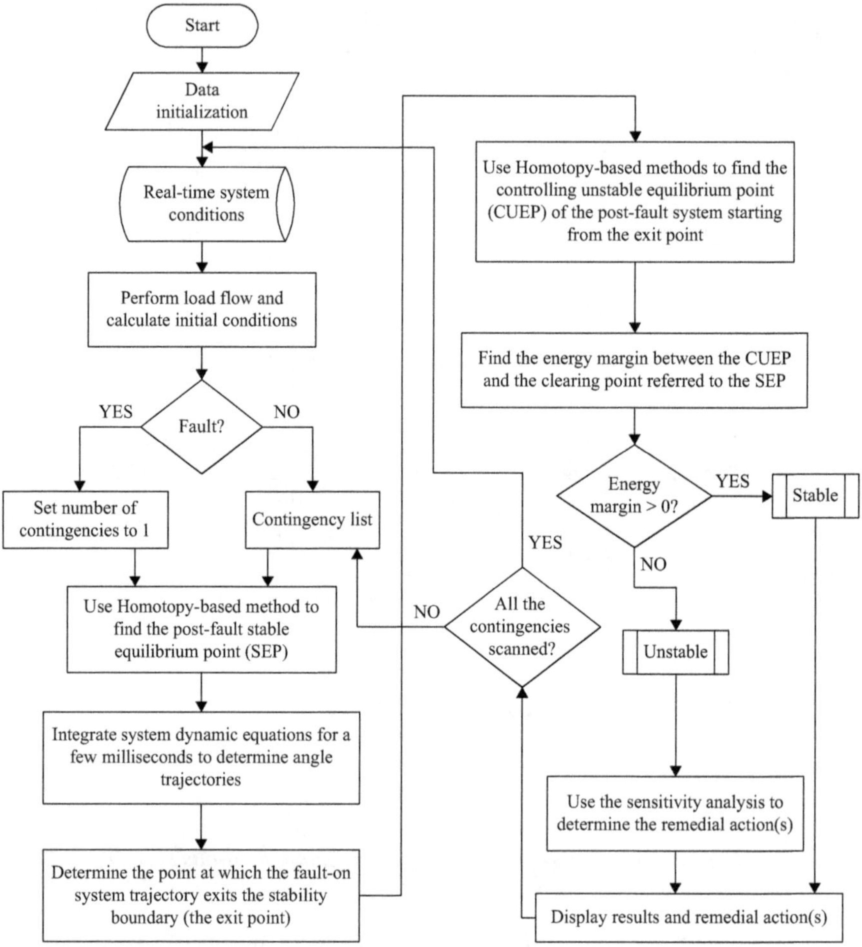

Figure 5.9 Flowchart of the screening and remedial action tool

Table 5.1 Pre-fault load flow results of the WECC test system

	$\|E\|$ (pu)	δ (rad)	$\tilde{\delta}$ (rad)	P_m (pu)
Machine #1	1.107632	0.044823	−0.048248	0.897915
Machine #2	1.106539	0.218307	0.125237	1.343200
Machine #3	1.057698	0.205717	0.112647	0.941900

Table 5.2 Contingency list details and the EP of each contingency

Contingency number	Fault at bus	Line trip		EP (rad)		
		From	To	$\tilde{\delta}_1$	$\tilde{\delta}_2$	$\tilde{\delta}_3$
1	4	4	5	−0.83157	2.05223	2.16748
2	5	4	5	−0.85604	2.24468	1.95048
3	4	4	6	−0.82573	2.00656	2.22575
4	6	4	6	−0.82663	2.00656	2.22575
5	5	5	7	−0.77351	2.07331	1.66662
6	7	5	7	−0.84828	2.68764	0.94769
7	6	6	9	−0.75797	1.83370	2.05405
8	9	6	9	−0.50293	0.51129	2.86278
9	7	7	8	−0.73581	2.28572	0.91888
10	8	7	8	−0.77159	1.74633	2.34681
11	8	8	9	−0.76930	1.74136	2.33940
12	9	8	9	−0.45872	0.48828	2.56452

Table 5.3 The computed controlling UEPs of each contingency

Contingency number	Fault at bus	Line trip		Controlling UEP (rad)			Energy margin
		From	To	$\tilde{\delta}_1$	$\tilde{\delta}_2$	$\tilde{\delta}_3$	
1	4	4	5	−0.8323	2.0742	2.1447	1.8375
2	5	4	5	−0.8323	2.0742	2.1447	2.6324
3	4	4	6	−0.8266	2.0821	2.0540	1.9103
4	6	4	6	−0.8266	2.0821	2.0540	2.5565
5	5	5	7	−0.7598	1.9521	1.8071	1.3057
6	7	5	7	−0.7598	1.9521	1.8071	−0.6056
7	6	6	9	−0.7586	1.8576	1.9979	1.5423
8	9	6	9	−0.7586	1.8576	1.9979	−1.4409
9	7	7	8	−0.5430	2.1797	−0.3764	−1.2008
10	8	7	8	−0.3500	0.0738	2.5861	0.9967
11	8	8	9	−0.2915	−0.1017	2.5004	0.4266
12	9	8	9	−0.2915	−0.1017	2.5004	−2.1234

Table 5.4 Generation shift based on sensitivity calculation

No	Faulted bus	Tripped line		ΔV	ΔP_{m1}	ΔP_{m2}	ΔP_{m3}
		From	To				
6	7	5	7	−0.6056	0.00543	0.00134	−0.00678
8	9	6	9	−1.4409	0.11883	−0.45693	0.33810
9	7	7	8	−1.2008	0.17998	−0.27438	0.09441
12	9	8	9	−2.1234	0.35561	0.04248	−0.39809

5.7 Conclusion

This chapter described recent and emerging methodology that is amenable for integration into emergent cyber-physical energy systems and "Smart Grid" infrastructures. Although some modern control centers no longer use nomograms and remedial action tables, and have incorporated much of this data into expert-system based tools, in most control centers today on-line contingency analysis sweeps still report only static contingencies. At the time of this writing, only Interconnection (PJM) is known to have implemented a form of the BCU method reported here. Further development and testing of the direct methods are still in progress. Chiang [30] has been a pioneer in these efforts, developing and implementing the BCU methods, while Mitra and his team [24] have been developing the homotopy-based method under a grant from the US Department of Energy.

While direct methods make it possible to include dynamic contingencies into online contingency sweeps (the PJM implementation performs dynamic sweeps every 15 min and static sweeps are typically performed every 4 min) they also provide the opportunity to develop more accurate real-time estimates of angle and voltage stability margins with the help of synchro-phasor measurements. Applications are numerous, ranging from improved real-time determination of remedial actions to identification of stressed regions islanding which could prevent system collapse in extreme contingencies.

This chapter has described several different approaches for remedial action to mitigate the impact of stability events. Recent and ongoing developments in power flow control technologies are also likely to provide a significantly increased set of options that will be compatible with improved telemetry, communication, and control technologies that are now being developed for the "Smart Grid."

References

[1] D. Sobajic and Y. Pao, "Artificial Neural-Net Based Dynamic Security Assessment for Electric Power Systems," *IEEE Trans. Power Syst.*, Feb. 1989, vol. 4, no. 1, pp. 220–228.

[2] V. Brandwajn, A. Kumar, A. Ipakchi, A. Bose, and S. Kuo, "Severity Indices for Contingency Screening in Dynamic Security Assessment," in *Proc. of the IEEE Power Eng. Soc. Summer Meeting*, Denver, CO, Jul. 28 – Aug. 2, 1996.

[3] H. Hakimmashhadi and G. Heydt, "Fast Transient Stability Assessment," *IEEE Trans. Power App. Syst.*, Dec. 1983, vol. PAS–102, pp. 3816–3824.

[4] Y. Akimoto, H. Tanaka, J. Yoshizawa, D. Klapper, and K. Price, "Transient Stability Expert Systems," *IEEE Trans. Power Syst.*, Jan. 1989, vol. 4, no. 1, pp. 312–320.

[5] L. Wehenkel, M. Pavella, E. Euxibie, and B. Heilbronn, "Decision Tree Based Transient Stability Method a Case Study," *IEEE Trans. Power Syst.*, Feb. 1994, vol. 9, no. 1, pp. 459–469.

[6] "Power System Dynamic Security Analysis Using Artificial Intelligence Systems: Phase 1–Feasibility Evaluation," Tech. Rep., EPRI TR-103607 Project 3103-02, Final Report, Apr. 1994.
[7] Y. Mansour, E. Vaahedi, A. Chang, B. Corns, J. Tamby, and M. El-Sharkawi, "Large Scale Dynamic Security Screening and Ranking Using Neural Networks," *IEEE Trans. Power Syst.*, May 1997, vol. 12, no. 2, pp. 954–960.
[8] L. Moulin, A. Alves da Silva, M. El-Sharkawi, and R. Marks II, "Support Vector Machines for Transient Stability Analysis of Large-Scale Power Systems," *IEEE Trans. Power Syst.*, May 2004, vol. 19, no. 2, pp. 818–825.
[9] A. Gavoyiannis, D. Vogiatzis, D. Georgiadis, and N. Hatziargyriou, "Combined Support Vector Classifiers Using Fuzzy Clustering for Dynamic Security Assessment," in *Proc. of the IEEE Power Eng. Soc. Summer Meeting*, vol. 2, Vancouver, BC, 2001.
[10] L. Moulin, A. Alves da Silva, M. El-Sharkawi, and R. Marks II, "Support Vector and Multilayer Perceptron Neural Networks Applied to Power Systems Transient Stability Analysis with Input Dimensionality Reduction," in *Proc. of the IEEE Power Eng. Soc. Summer Meeting*, 2002, vol. 3, IEEE, pp. 1308–1313.
[11] I. Kassabalidis, M. El-Sharkawi, R. Marks II, L. Moulin, and A. Alves da Silva, "Dynamic Security Border Identification Using Enhanced Particle Swarm Optimization," *IEEE Trans. Power Syst.*, Aug. 2002, vol. 17, no. 3, pp. 723–729.
[12] Y. Mansour, E. Vaahedi, and M. El-Sharkawi, "Dynamic Security Contingency Screening and Ranking Using Neural Networks," *IEEE Trans. Neural Netw.*, Jul. 1997, vol. 8, no. 4, pp. 942–950.
[13] I. Kamwa, R. Grondin, and L. Loud, "Time-Varying Contingency Screening for Dynamic Security Assessment Using Intelligent-Systems Techniques," *IEEE Trans. Power Syst.*, Aug. 2001, vol. 16, no. 3, pp. 526–536.
[14] Y. Park, G. Kim, H. Cho, and K. Lee, "A New Algorithm for Kohonen Layer Learning with Application to Power System Stability Analysis," *IEEE Trans. Power Syst.*, Dec. 1997, vol. 27, no. 6, pp. 1030–1034.
[15] K. Chan, A. Edwards, R. Dunn, and A. Daniels, "On-Line Dynamic Security Contingency Screening Using Artificial Neural Networks," *IEEE Proc. Generat. Transm. Distrib.* Nov. 2000, vol. 147, no. 6, pp. 367–372.
[16] V. Brandwajn, A. Kumar, A. Ipakchi, A. Bose, and S. Kuo, "Severity Indices for Contingency Screening in Dynamic Security Assessment," *IEEE Trans. Power Syst.*, Aug. 1997, vol. 12, no. 3, pp. 1136–1142.
[17] "Analytical Methods for Contingency Selection and Ranking for Dynamic Security Analysis," Tech. Rep., EPRI TR-104352, Project 3103-03, Final Report, Siemens Energy & Automation Inc., Sep. 1994.
[18] V. Chadalavada, V. Vittal, G. Ejebe, *et al.*, "An On-Line Contingency Filtering Scheme for Dynamic Security Assessment," *IEEE Trans. Power Syst.*, Feb. 1997, vol. 12, no. 1, pp. 153–161.

[19] H. Chiang, C. Wang, and H. Li, "Development of BCU Classifiers for On-Line Dynamic Contingency Screening of Electric Power Systems," *IEEE Trans. Power Syst.*, May 1999, vol. 14, no. 2, pp. 660–666.

[20] H. Chiang, H. Li, J. Tong, and Y. Tada, " On-Line Transient Stability Screening of a Practical 14,500-Bus Power System: Methodology and Evaluations," in *High Performance Computing in Power and Energy Systems*. Springer Heidelberg, New York, Dordrecht, London, 2013, pp. 335–358.

[21] S. Kim, Y. Tada, and S. Hur, "Transient Contingency Screening of Large Power Systems using Tepco-BCU Classifier," in *Proc. of the International Conference on Electrical Engineering, ICEE*, Shenyang, China, 2009.

[22] D. Ernst, D. Ruiz-Vega, M. Pavella, P. Hirsch, and D. Sobajic, "A Unified Approach to Transient Stability Contingency Filtering, Ranking and Assessment," *IEEE Trans. Power Syst.*, Aug. 2001, vol. 16, no. 3, pp. 435–443.

[23] Y. Mansour, E. Vaahedi, A. Chang, *et al.*, "B. C. Hydro's On-line Transient Stability Assessment (TSA): Model Development, Analysis and Post-processing," *IEEE Trans. Power Syst.*, Feb. 1995, vol. 10, no. 1, pp. 241–253.

[24] M. Benidris, N. Cai, and J. Mitra, "A Fast Transient Stability Screening and Ranking Tool," in *Power Systems Computation Conference (PSCC)*, Aug. 2014, pp. 1–7.

[25] T. Athay, R. Podmore, and S. Virmani, "A Practical Method for the Direct Analysis of Transient Stability," *IEEE Trans. Power App. Syst.*, Mar. 1979, vol. PAS-98, no. 2, pp. 573–584.

[26] A. Fouad and S. Stanton, "Transient Stability of a Multi-Machine Power System Part I: Investigation of System Trajectories," *IEEE Trans. Power App. Syst.*, Jul. 1981, vol. PAS-100, no. 7, pp. 3408–3416.

[27] P. Varaiya, F. Wu, and R. Chen, "Direct Methods for Transient Stability Analysis of Power Systems: Recent Results," *Proc. IEEE*, Dec. 1985, vol. 73, no. 12, pp. 1703–1715.

[28] H. Chiang, F. Wu, and P. Varaiya, "Foundations of Direct Methods for Power System Transient Stability Analysis," *IEEE Trans. Circuits Syst.*, Feb. 1987, vol. 34, no. 2, pp. 160–173.

[29] A. Fouad and V. Vital, *Power System Transient Stability Analysis Using The Energy Function Method*. Prentice-Hall, Englewood Cliffs, NJ, 1992.

[30] H. Chiang, C. Chu, and G. Cauley, "Direct Stability Analysis of Electric Power Systems using Energy Functions: Theory, Applications, and Perspective," *Proc. IEEE*, Nov. 1995, vol. 83, no. 11, pp. 1497–1529.

[31] H. Chiang, *Direct Methods for Stability Analysis of Electric Power Systems: Theoretical Foundation, BCU Methodologies, and Applications*. John Wiley & Sons, NJ, 2011.

[32] J. Thorp and S. Naqavi, "Load Flow Fractals," in *Proc. of the 28th IEEE Conference on Decision and Control*, Dec. 1989, vol. 2, pp. 1822–1827.

[33] J. Thorp, S. Naqavi, and N. Chiang, "More Load Flow Fractals," in *Proc. of the 29th IEEE Conference on Decision and Control*, Dec. 1990, vol. 6, pp. 3028–3030.

[34] H. Chiang, F. Wu, and P. Varaiya, "A BCU Method for Direct Analysis of Power System Transient Stability," *IEEE Trans. Power Syst.*, Aug. 1994, vol. 9, no. 3, pp. 1194–1208.

[35] R. Treinen, V. Vittal, and W. Kliemann, "An Improved Technique to Determine the Controlling Unstable Equilibrium Point in a Power System," *IEEE Trans. Circuits Syst. I, Fundam. Theory Appl.*, Apr. 1996, vol. 43, no. 4, pp. 313–323.

[36] L. Chen, Y. Min, F. Xu, and K. Wang, "A Continuation-Based Method to Compute the Relevant Unstable Equilibrium Points for Power System Transient Stability Analysis," *IEEE Trans. Power Syst.*, Feb. 2009, vol. 24, no. 1, pp. 165–172.

[37] J. Lee and H. Chiang, "Convergent Regions of the Newton Homotopy Method for Nonlinear Systems: Theory and Computational Applications," *IEEE Trans. Circuits Syst. I, Fundam. Theory Appl.*, Jan. 2001, vol. 48, no. 1, pp. 51–66.

[38] I. Nazareno, L. Alberto, and N. Bretas, "Problems in the Precise Determination of BCU's Controlling Unstable Equilibrium Points and PEBS's Exit Point Method in Real-Time Transient Stability Analysis," in *Proc. IEEE/PES Trans. and Dist. Conf. Expo.: Latin America*, Nov. 2004, pp. 475–480.

[39] A. Llamas, J. De La Ree Lopez, L. Mili, A. Phadke, and J. Thorp, "Clarifications of the BCU Method for Transient Stability Analysis," *IEEE Trans. Power Syst.*, Feb. 1995, vol. 10, no. 1, pp. 210–219.

[40] A. Xue, S. Mei, and B. Xie, "A Comprehensive Method to Compute the Controlling Unstable Equilibrium Point," in *Third International Conference on Electric Utility Deregulation and Restructuring and Power Technologies*, Apr. 2008, pp. 1115–1120.

[41] J. Lee, "An Optimization-driven Framework for the Computation of the Controlling UEP in Transient Stability Analysis," *IEEE Trans. Autom. Control*, Jan. 2004, vol. 49, no. 1, pp. 115–119.

[42] J. Lee and H. Chiang, "A Singular Fixed-Point Homotopy Method to Locate the Closest Unstable Equilibrium Point for Transient Stability Region Estimate," *IEEE Trans. Circuits Syst. II, Exp. Briefs*, Apr. 2004, vol. 51, no. 4, pp. 185–189.

[43] J. Lee, "A Novel Homotopy-based Algorithm for the Closest Unstable Equilibrium Point Method in Nonlinear Stability Analysis," in *Proc. of the International Symp. on Circuits and Syst.*, May 2003, vol. 3, pp. III-8–III-11.

[44] K. A. Nigim, S. Suryanarayanan, R. Gorur, and R. G. Farmer, "The Application of Analytical Hierarchy Process to Analyze the Impact of Hidden Failures in Special Protection Schemes," *Electric Power System Research*, Dec. 2003, vol. 67, no. 3, pp. 191–196.

[45] A. Fouad and V. Vittal, *Power System Transient Stability Analysis Using the Transient Energy Function Method*. Prentice-Hall, Englewood Cliffs, NJ, 1991.
[46] H. Chiu, "Sensitivity analysis of the transient energy function method," PhD dissertation, Iowa State University, 1989.
[47] A. Wood and B. Wollenberg, *Power Generation Operation and Control*. John Wiley and Sons, New York, NY, 1984.
[48] A. Fouad and T. Jianzhong, "Stability Constrained Optimal Rescheduling of Generation," *IEEE Trans. Power Syst.*, Feb. 1993, vol. 8, no. 1, pp. 105–112.
[49] K. Padiyar, *Structure Preserving Energy Functions in Power Systems: Theory and Applications*. CRC Press, Boca Raton, FL, 2013.

Part 3

New philosophies of control and economics in distribution systems

Chapter 6

Customer modeling and pricing-mechanisms for demand response in smart electric distribution grids

Timothy M. Hansen, Robin Roche**, Siddharth Suryanarayanan†, Anthony A. Maciejewski†, Howard Jay Siegel†,‡, and Edwin K. P. Chong†*

Abstract

We describe and contrast different market mechanisms to incentivize residential electricity customers to perform demand response (DR) via load shifting of *schedulable assets*. A customer-incentive pricing (CIP) mechanism from our past research is discussed, and compared to flat-rate, time-of-use (TOU), and real-time pricing (RTP). The comparison is made using a for-profit aggregator-based residential DR approach to solve the "Smart Grid resource allocation" (SGRA) problem. The aggregator uses a heuristic framework to schedule customer assets and to determine the customer-incentive price to maximize profit. Different customer response models are proposed to emulate customer behavior in the aggregator DR program. A large-scale system consisting of 5,555 residential customer households and 56,588 schedulable assets using real pricing data over a period of 24 h is simulated and controlled using the aggregator. We show that the aggregator enacts a beneficial change on the load profile of the overall power system by reducing peak demand. Additionally, the customers who are more *flexible* with their loads, represented as a parameter in the proposed customer α-model, have a greater reduction on their electricity bill.[1]

*Electrical Engineering and Computer Science Department, South Dakota State University, Brookings, SD, USA
**IRTES, Université de Technologie de Belfort-Montbéliard, Université Bourgogne Franche-Comté, Belfort, France
†Department of Electrical and Computer Engineering, Colorado State University, Fort Collins, CO, USA
‡Department of Computer Science, Colorado State University, Fort Collins, CO, USA
[1]This chapter is an expanded version of a journal article [1]. Project codes and data sets have been made available for use through the open source BSD 3-Clause license at https://github.com/IPEMS

6.1 Customer modeling introduction

The US Energy Information Administration predicts a 24% increase in residential electricity use from a 2013 reference case to the year 2040. Additionally, since 1982 the growth in peak electricity usage has exceeded the growth in transmission capacity by almost 25% each year [2]. Given the combination of these trends, it is expected that peak energy demands will approach, and eventually exceed, the available transmission system capability (the remaining power that could be transferred from generation to consumption). A side consequence of these trends, in addition to the possibility of system outage, is more congestion on transmission lines, leading to increased locational marginal prices (LMP) – different marginal prices (price of providing the next MW of power) at different buses in the transmission network. Studies show that small and targeted reductions in peak demand can have large impacts on wholesale electricity prices [3]. As shown in Reference 2, it is unlikely that additional spending will be allocated for increasing transmission capability, leading to research in the areas of distributed generation (DG) and, in the case of this chapter, shifting or curtailing load during peak hours. Given that residential customers can account for over half of the system peak demand in summertime, such as in markets like the Electric Reliability Council of Texas (ERCOT) [4], residential demand response (DR) programs are attractive solutions for relieving the stress on the system and market. In this work, we define the term DR to be the reduction in peak demand by shifting or shedding loads in response to system or economic conditions to alleviate stress on the electric power system.

As presented in Reference 5, incentives can influence customer behavioral changes. Dynamic pricing programs are one method for accomplishing DR. These utility-offered programs, such as time-of-use (TOU) and real-time pricing (RTP), fluctuate the price of electricity throughout the day in accordance with system load levels to elicit a change in the consumption of electricity [6]. In this chapter, we introduced an additional pricing method, *customer-incentive pricing* (CIP), which provides the residential end-user an additional competitive pricing scheme for participating in a targeted DR program through an aggregator. The aggregator in this chapter is a for-profit entity in a deregulated market structure that interfaces a *DR* exchange market (DRX) and a set of customers. The aggregator uses the combination of many customer schedulable assets of the participating customers to perform large-scale load shifting. In many, if not all energy markets, there is a minimum power rating required to bid into the market (e.g., 0.1 MW in the PJM market [7]). The aggregator entity is able to enact a change on the electric power system load profile by bidding the aggregate load of customer assets into the bulk market through the DRX.

By voluntarily opting into the aggregator DR program, the customer is provided the opportunity to participate in the bulk power market. Residential customers can change electricity use to take advantage of the time-varying rates provided by the utility and aggregator to reduce their electricity bill. The challenges of an effective residential DR program are (a) the uncertainty in the time-varying price of electricity and (b) that as a customer, the benefit received from changing energy usage must exceed the inconvenience caused. To overcome these challenges and to

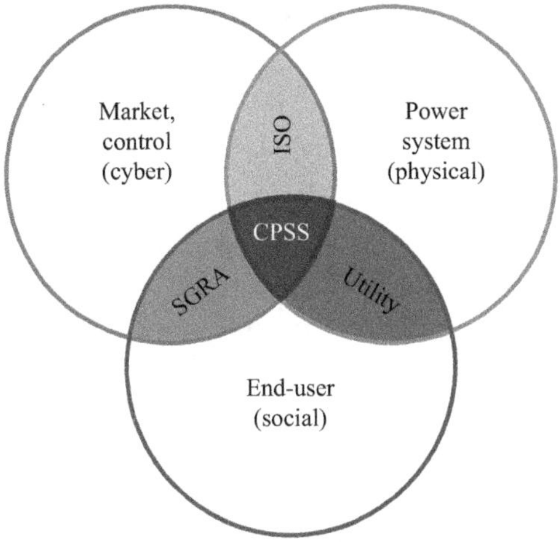

Figure 6.1 The proposed SGRA CPSS lies at the intersection between the power system network, market and controls, and end-user

maximize the benefit of dynamic pricing, we introduce the aggregator-based residential DR program, denoted Smart Grid Resource Allocation (SGRA), where given a set of participating customers with schedulable assets, subject to customer constraints (i.e., availability of customer assets and customer incentive requirements), the aggregator sets the *CIP* and *schedule of assets* to maximize profit of the aggregator while not inconveniencing the participating customer. The SGRA is formulated as a cyber-physical-social system (CPSS), as illustrated in Figure 6.1.

The proposed SGRA CPSS lies at the intersection of the *physical* electric power network, *cyber* market and control layer, and *social* residential end-users. The independent system operator (ISO) interfaces the bulk power market and physical equipment. Local utilities traditionally deliver power to end-user customers through the distribution network. The aggregator entity provides the residential end-user a path to participate in the market. The SGRA problem is solved using resource allocation methods analogous to those used in the computing discipline, where compute tasks must be allocated to machines. In general, such problems are NP-complete [8–10], so heuristic optimization is used to find *near-optimal* solutions. In this chapter, we use heuristic optimization techniques to find near-optimal solutions to the SGRA problem in a reasonable time frame to be used as a day-ahead scheduler of a large number of residential end-user assets.

According to the California Energy Commission (CEC), residential loads are not easily controlled and need to be composed of a large portfolio of assets to provide a strategic DR product [11]. The work in this chapter directly addresses the CEC DR strategies (direct DR participation with the ISO, new market and auction mechanisms, e.g., our proposed DRX, improving customer willingness to participate, and the introduction of time-variant pricing) by offering direct DR participation through

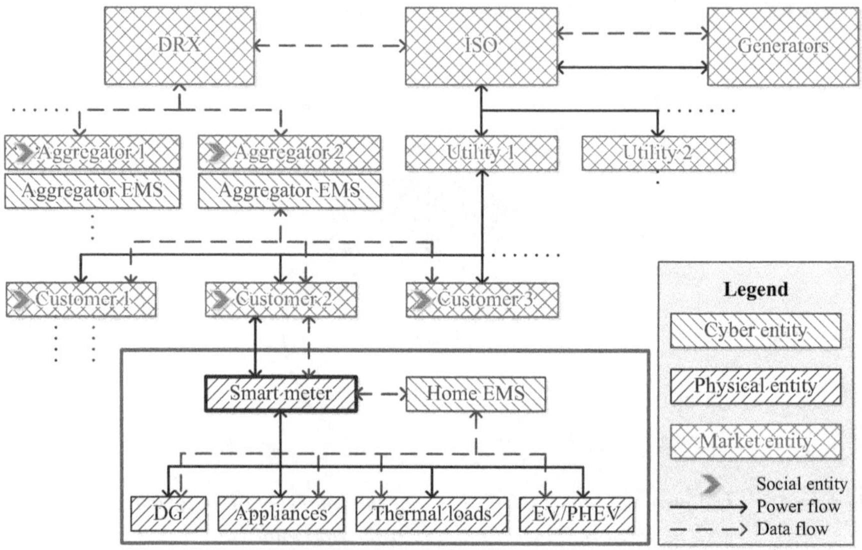

Figure 6.2 The architecture and communication for the proposed aggregator-based residential DR program CPSS, expanded from Reference 1

the customer-aggregator-DRX relationship (see Figure 6.2) and encouraging customer participation with the time-variant CIP mechanism. In this chapter, a heuristic framework is designed and implemented; and customer-behavior is emulated using a proposed α-model. To quantify the impact of the SGRA CPSS, a simulation study is conducted where the aggregator schedules over 56,000 end-user assets from 5,555 customers. Comparisons are made between different utility pricing mechanisms, namely TOU, RTP, and flat-rate.

The rest of the chapter is organized as follows. In Section 6.2.3, the aggregator-based residential DR is formally introduced in the context of CPSS. To mimic customer behavior in large simulation studies, mathematical customer behavior models are defined in Section 6.3. Section 6.4 summarizes commonly used utility pricing mechanisms. The heuristic framework and problem formulation are derived in Section 6.5. A large-scale simulation study of the SGRA problem is conducted in Section 6.6, and Section 6.7 concludes.

6.2 Aggregator-based residential demand response

6.2.1 CPSS

The proposed CPSS for the aggregator-based residential DR program is shown in Figure 6.2. The right side of the CPSS shows the power flow and market structure of the traditional electric power system. The ISO coordinates the generators in the system (supply) and load requirements of the utilities (demand) to ensure power balance reliably and at the cheapest price. The utilities deliver the power to the end-user, in this case residential customers. Industrial and commercial end-users can also be considered in the SGRA by modeling their assets, but this is not presented in this chapter. The left-hand side of Figure 6.2 encapsulates our proposed residential DR program. The DRX is an ancillary market in a fully deregulated market structure that provides DR services to the ISO. The aggregator interfaces the DRX and the residential customer, and provides the positive attributes (e.g., load shifting, distributed storage) of the aggregated customer assets (e.g., DG, electric vehicles) to the ISO. The aggregator coordinates the use of the participating customer assets and brings the result, such as peak shaving, to the DRX for market exchange.

6.2.2 Aggregator

The aggregator is a for-profit market entity engaged in interacting with the customer and the bulk power market in a fully deregulated market structure. As shown in Figure 6.2, the aggregator is situated between the DRX and the customer. The aggregator energy management system interacts with each of the customer home energy management systems (HEMS). In this chapter, we are only considering one aggregator entity, but it is expected that several aggregators may exist within the same distribution area.

The aggregator coordinates a set of participating customers, each with a set of schedulable assets. In this chapter, we are currently only considering schedulable loads in the form of smart appliances (e.g., dishwashers, washer, dryer), but this approach could be extended to other types of assets such as DG, thermal loads (e.g., electric water heaters [12] and heating-ventilation-air-conditioning systems [13]), and electric vehicles in the form of vehicle-to-grid [14] or scheduling vehicle charging cycles [15].

The scheduling problem is proposed as a day-ahead optimization. To make decisions, the aggregator requires information about the customer loads, the forecast utility pricing, and the forecast spot market pricing (SMP) in the bulk electricity market. Using this information, the aggregator must find the *CIP* and *schedule of loads* to maximize its profit. Because it is a day-ahead optimization, there are constraints on the execution time of the optimization technique used. This time constraint and the complexity of the scheduling problem (i.e., the class of problems is, in general, NP-complete) due to the large number of customer assets leads to the use of heuristic optimization. Other objectives could be considered, such as minimizing the peak load, or considering multiple objectives in the form of a multi-objective optimization using Pareto-fronts [16]. In this chapter, we solely optimize for the aggregator profit to demonstrate that a purely economic motivation will affect the desired change of reduced system peak demand.

6.2.3 Aggregator demand response

CIP is a proposed pricing structure that the aggregator would offer all customers to allow the rescheduling of their loads. That is, instead of paying the utility company, the customer pays the aggregator the CIP for electricity. The customer paying the CIP for electricity to the aggregator at the time the asset has been rescheduled *to* is one part of the profit of the aggregator. The sum of these payments over all customers and all rescheduling events is denoted S. The other two components to the aggregator profit are: (a) the aggregator selling a negative load to the spot market where the assets have been rescheduled *from* (denoted N), and (b) the aggregator buying spot market electricity where the assets have been rescheduled *to* (denoted B). The aggregator would, perhaps, need to enter into a leasing agreement with the utility company for the use of the distribution assets, but modeling this and other potential fixed costs are beyond the scope of this chapter.

To reschedule load, the aggregator requires information on the set of schedulable loads. These schedulable loads represent a subset of the system load. For each schedulable load i, the aggregator receives the following information on the schedulable appliances of each participating customer:

- δ_i, the runtime duration (in 15-min intervals);
- p_i, the average power rating (in kW);
- t_{i_start}, the customer scheduled start time; and
- (A_{i_start}, A_{i_dur}), a 2-tuple that represents the customer-defined availability window for load i determined by the availability window start time, A_{i_start}, and the availability window duration, A_{i_dur}.

In this chapter, we assume that the aggregator knows the exact time a load will run (i.e., from t_{i_start} for δ_i time intervals) if it is not rescheduled by the aggregator (i.e., the start time is deterministic). This could be represented as a probability distribution based on historical runtimes (i.e., the start time is stochastic), leading to a stochastic SGRA problem.

Let $\boldsymbol{\lambda}$ be the CIP vector containing 96-elements, where each element λ_t gives the aggregator determined CIP at time interval t. In addition to the information about the schedulable loads, the aggregator possesses information on:

- $\gamma(i, \boldsymbol{\lambda}, t)$, a binary function that represents whether the customer will allow load i to be rescheduled to time t with CIP $\boldsymbol{\lambda}$ ($\gamma = 1$) or not ($\gamma = 0$);
- $s(t)$, the forecast SMP of electricity in the bulk electricity market (in cents/kWh); and
- $r(t)$, the forecast price of electricity from the utility company (in cents/kWh).

Because the customer also has access to the forecast utility price (e.g., RTP and TOU), if the CIP, $\boldsymbol{\lambda}$, does not offer enough of a cost reduction to justify the inconvenience of rescheduling the load, the customer has the opportunity to refrain from participation, as represented by the binary function, γ (described in the customer models in Section 6.3). Therefore, the position of the aggregator is to find the following:

- L, the set of loads the aggregator is rescheduling;
- t_{i_resch}, the rescheduled start time for load i; and
- $\boldsymbol{\lambda}$, the CIP vector

to maximize profit. Let I be the total number of schedulable loads. The cardinality of L is less than or equal to I (i.e., $|L| \leq I$) because the aggregator has information about all I schedulable customer loads, but it does not necessarily have to reschedule all loads.

6.2.4 Aggregator profit function

For the aggregator, let S be the *total income received* for selling electricity to customers, given by (6.1); N be the *total income received* for selling negative load to the spot market given by (6.2); and B be the *total cost paid* to the spot market for buying electricity given by (6.3). The exact payment received from N would depend on policy, such as the outcome of FERC Order 745 [17] and its future iterations; however, we are not addressing energy policy in this chapter. We assume that the aggregator is a well-behaved agent that does not manipulate the market (such as by misrepresenting the sum of the negative load) and is paid the difference from a deterministic baseline load. The calculations for S, N, and B are given as:

$$S = \sum_{i \in L} \left[\gamma(i, \boldsymbol{\lambda}, t_{i\_resch}) \sum_{t=t_{i\_resch}}^{t_{i\_resch}+\delta_i-1} \frac{\lambda_t p_i}{4} \right] \tag{6.1}$$

$$N = \sum_{i \in L} \left[\gamma(i, \boldsymbol{\lambda}, t_{i\_resch}) \sum_{t=t_{i\_start}}^{t_{i\_start}+\delta_i-1} \frac{s(t) p_i}{4} \right] \tag{6.2}$$

$$B = \sum_{i \in L} \left[\gamma(i, \boldsymbol{\lambda}, t_{i\_resch}) \sum_{t=t_{i\_resch}}^{t_{i\_resch}+\delta_i-1} \frac{s(t) p_i}{4} \right] \tag{6.3}$$

The forecast aggregator profit P is given as:

$$P = N + S - B \tag{6.4}$$

The terms are illustrated in Figure 6.3.

6.3 Customer models

6.3.1 Customer overview: Gamma parameter

Each customer under agreement with the aggregator has a set of schedulable loads. In this chapter we are only considering flexible, non-interruptible smart appliances (e.g., clothes dryers to avoid thermal losses) according to the definitions given in Reference 18. Each customer load has an *availability window* associated with it, describing the times during the day that their schedulable load can be rescheduled by the aggregator. In addition to the availability window, each customer has an option to veto the offer from the aggregator through the use of the γ utility variable. If the customer does not feel that the offer from the aggregator is worth the added inconvenience of rescheduling a particular load, the customer can set $\gamma = 0$ for that load on a day-to-day basis.

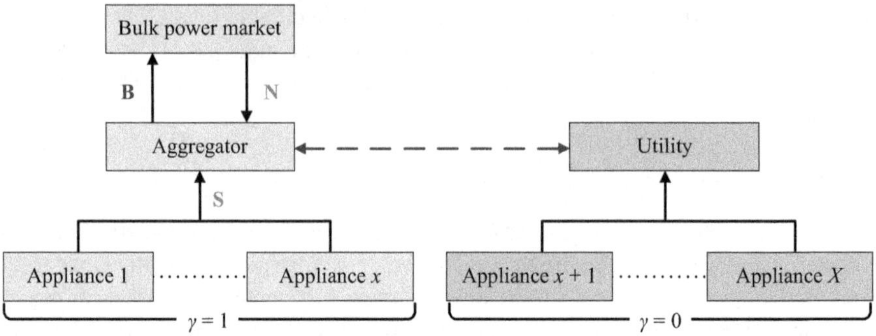

Figure 6.3 The exchange of money between the aggregator, customer, bulk power market (through the ISO), and utility. Each customer has an appliance-by-appliance choice of electricity provider. The customers of appliances {1, . . . , x} pay the customer-incentive pricing to the aggregator for use of these appliances for DR. The CIP offers for appliances {x + 1, . . . , X} were not worth the inconvenience, so those customers instead purchase electricity from the utility company for those appliances. This selection of appliances occurs on a daily basis and could be a manual choice or automated through the use of a HEMS or smart phone app. The solid arrows represent the money flowing in the system. The dashed arrow indicates the possible need for a relationship between the aggregator and utility company for use of the electrical distribution equipment

The customer sets $\gamma = 1$ to accept an offer for a particular asset. This interaction occurs on a day-by-day basis and is illustrated in Figure 6.3. This process could be automated on the customers' side, e.g., through a HEMS or smart phone app.

Only those loads that are agreed on for DR between the customer and aggregator utilize the CIP. The base load and those loads not used for DR will utilize the status quo of the utility company, e.g., RTP and TOU. This choice of supplier is a powerful new tool for the customer and offers the customer an avenue to participate in the bulk power market (through the aggregator entity), which may reduce the customer electricity bill and offers a freedom of choice of electricity supplier. Each customer has a baseline load and a set of schedulable loads, as described in Section 6.3.3. To determine the value of γ for each aggregator customer schedulable appliance, we use the alpha model described in the following section.

6.3.2 Alpha model

6.3.2.1 Alpha model overview

A key assumption in the proposed DR methods is customer participation. We model the behavior of each customer for determining whether or not they will allow the aggregator to reschedule their smart appliances using the proposed alpha-model.

In the alpha-model, each schedulable load i has an associated threshold metric for "customer comfort" in percent, α_i. Let c_{i_0} be the original cost of running load i at the utility price and c_{i_sch} be the rescheduled cost of running load i at the CIP offered by the aggregator. For the owner of load i to allow it to be rescheduled (i.e., $\gamma = 1$), the inequality $c_{i\_sch} \leq \alpha_i c_{i\_0}$ must hold. This new model allows flexibility for the customer on a load-by-load basis. Additionally, the customer is always guaranteed (if its loads are used by the aggregator), to save $1 - \alpha_i$ times the cost of running load i compared to paying the utility price. Customer inconvenience is captured through the γ value and availability window, as opposed to the time-dependent models in References 19 and 20. The customer γ values are private, and the aggregator is assumed to operate without receiving this information explicitly.

To determine α_i for simulation studies, we use three different methods: constant-α, Gaussian, and the coefficient-of-variation-based (CVB) method. The three methods are used to determine the effect of increasing amounts of variance in the α value and the impact on aggregator profit versus customer savings. The constant-α method sets $\alpha_0 = \alpha_1 = \cdots = \alpha_{I-1} = y$, where y is a user-defined constant. In the Gaussian method, the α value for each schedulable appliance is sampled from a random variable $\sim \mathcal{N}(\mu_\alpha, \sigma_\alpha)$. The CVB method is used to introduce ordered heterogeneity into the randomness, and is described in detail in the following section.

6.3.2.2 Coefficient-of-variation-based method

We use the CVB method to generate the α values for each load i, similar to generating task execution times for a heterogeneous suite of machines in a computing environment [21]. We offer an analogous method of generating load α values for a heterogeneous suite of customers. This approach is taken because it is assumed that customers will act similarly when using similar load types (e.g., more flexible with laundry, less flexible with the TV).

Let μ_a be the desired average α value for all loads, σ_a be the desired coefficient-of-variation of the load types, and σ_c be the desired coefficient-of-variation of the customers within a load type. For each load type k, we sample from a Gamma distribution with mean μ_a and standard deviation σ_a to obtain the mean α value for load type k, denoted $\mu_{a,k}$. For each customer that owns load type k, obtain α_i by sampling a Gamma distribution with mean $\mu_{a,k}$ and standard deviation σ_c. This gives similar α values for each type of load, and thus similar customer behavior.

Let $\mathcal{G}(v, \theta)$ be a Gaussian distribution with shape v and scale θ. We can define these values as functions of the mean μ and variance σ^2, given in (6.5) and (6.6), respectively. The CVB method is then provided as pseudocode in Figure 6.4.

$$v(\mu, \sigma^2) = \frac{\mu^2}{\sigma^2} \tag{6.5}$$

$$\theta(\mu, \sigma^2) = \frac{\sigma^2}{\mu} \tag{6.6}$$

A parameter sweep was performed on the input values μ_a, σ_a, and σ_c. A representative result is shown in Section 6.6 using the inputs $\mu_a = 0.75$, $\sigma_a = 0.10$, and

for load type k **do**
  $\mu_{a,k} = \mathcal{G}(v(\mu_a, \sigma_a^2), \theta(\mu_a, \sigma_a^2))$
  for asset i **do**
    if asset i is of type k **then**
      $\alpha_i = \mathcal{G}(v(\mu_{a,k}, \sigma_c^2), \theta(\mu_{a,k}, \sigma_c^2))$
    end if
  end for
end for

Figure 6.4 Pseudocode for the CVB method for determining asset α values

$\sigma_c = 0.05$. In general, the magnitude of the CIP is sensitive and positively correlated to μ_a (i.e., as μ_a increases, the CIP proportionally increases with respect to the RTP). Values of σ_a and σ_c are positively correlated with the noise level of the CIP.

6.3.3 Customer loads

Two types of loads are assumed to be available for each customer in this study: baseline and schedulable (smart) appliances. The baseline load is divided into thermal, modeled as air conditioning [22] and electric water heaters [12], and other non-schedulable loads. The non-schedulable loads are probabilistically generated for each customer based on the data in Reference 23.

A probabilistic model for 18 generic schedulable appliance types is given in Reference 1. The penetration level gives the probability that an appliance is present for a given customer; if it is present, the rated power of the appliance, as well as the start hour, is obtained from a normal distribution. The values probabilistically reflect the actual energy use of an average household. Similar to the non-schedulable loads, a set of schedulable loads corresponding to each customer is generated.

Each probabilistically generated load i has an associated availability window, (A_{i_start}, A_{i_dur}), that describes the time-window that the customer has allocated for load i to be scheduled. Recall that t_{i_start} is the originally scheduled starting time for load i. Let $\mathcal{U}(\delta_i, 96)$ be a uniform random variable in the interval $[\delta_i, 96]$. In our study, to generate the availability window for each load i, an interval of duration $\mathcal{U}(\delta_i, 96)$ is generated around the starting time t_{i_start}. That is, $A_{i\_dur} = \mathcal{U}(\delta_i, 96)$ and $A_{i\_start} = t_{i\_start} - \frac{A_{i\_dur}}{2}$. Other methods can be used to determine the start time and duration of a set of schedulable loads, such as an $M_t/G/\infty$ queue [24].

6.4 Pricing mechanisms

The different utility pricing mechanisms used for this chapter are a flat rate tariff, TOU, and RTP. The RTP information used in the simulation is real data from the ComEd Residential RTP program [25] from July 23, 2015, shown as the solid lines in

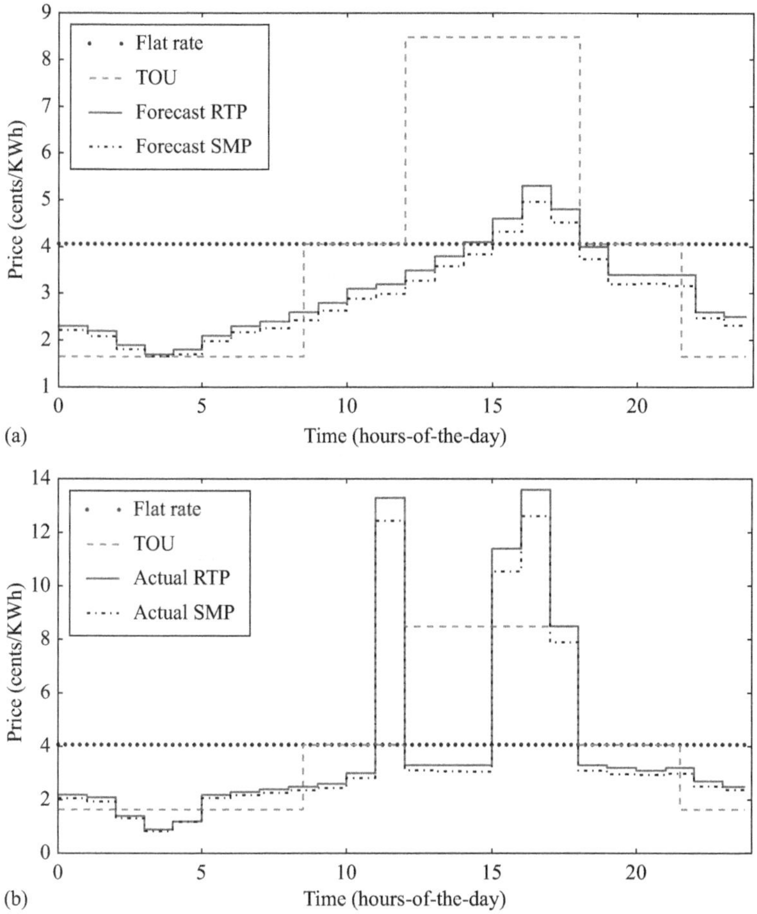

Figure 6.5 Real-time (RTP) [25] and spot market pricing (SMP) [27] from July 23, 2015, compared to flat-rate and time-of-use (TOU). (a) The day-ahead forecast prices and (b) the actual market-clearing prices. The flat-rate and TOU are the same for the day-ahead and real-time periods

Figure 6.5. The RTP market is modeled after the ComEd Residential RTP program [25]. The price of electricity changes every hour in response to the PJM real-time hourly market price. At approximately 4:30 pm, a forecast for the next day's hourly prices are provided to the customer. At the start of each hour, the actual price for that hour is provided. This data is given as 24 1-h intervals.

To make a fair comparison to the RTP, the flat rate used the mean value of the actual RTP, which was 4.06 cents/kWh. The TOU schedule was taken for the summer season from Pacific Gas and Electric (PG&E) [26], summarized in Table 6.1. The day is split into three time periods: off-peak, partial-peak, and peak, where the price

Table 6.1 Time-of-use rate periods

Period	Time	Price (cents/kWh)
Off-peak	21:30–8:30	1.65
Partial-peak	8:30–12:00, 18:00–21:30	4.06
Peak	12:00–18:00	8.48

of electricity in each subsequent period is more expensive. To determine the rates for each period to provide a fair comparison to the RTP and flat rate, the partial-peak was set at the mean price (i.e., 4.06 cents/kWh). The off-peak rate was chosen as the twenty-fifth percentile of the RTP curve, and the peak was set proportionally to keep the total mean of the TOU equivalent to the mean of the RTP and flat rate. These prices were chosen to provide a CIP comparison to the real RTP data for the day in question. The results should not be used to make general conclusions about the performance between RTP and TOU.

The aggregator also uses information about the forecast and real-time SMP in the scheduling decisions (i.e., in (6.2) and (6.3)). The SMP information used in the simulation was also real data from July 23, 2015, obtained from PJM [27]. The day-ahead forecast pricing is given in Figure 6.5(a) and the actual pricing is given in Figure 6.5(b).

6.5 Heuristic framework

6.5.1 Problem formulation

In this section, we formally define the heuristic framework in terms of the optimization function and heuristic implementation. We use heuristic optimization in this chapter because, in general, scheduling problems are NP-complete [8–10], and thus exact optimal methods are non-tractable. The heuristic optimization problem is set up as follows:

$$\max_{t_{i\_resch} \forall i \in L, \boldsymbol{\lambda}=(\lambda_0,\ldots,\lambda_{95})} P \tag{6.7}$$

subject to

$$A_{i\_start} \leq t_{i\_resch} \leq A_{i\_start} + A_{i\_dur} \qquad \forall i \in L \tag{6.8}$$

and

$$t_{i\_resch} \in \mathbb{Z} \qquad \forall i \in L \tag{6.9}$$

$$\lambda_t \in \mathbb{R} \qquad t = 1, \ldots, 96 \tag{6.10}$$

$t=0$	$t=1$	...	$t=95$	$load_0$	$load_1$	...	$load_I$
λ_0	λ_1	...	λ_{95}	$t_{0,sch}$	$t_{1,sch}$	...	$t_{I,sch}$

Figure 6.6 The chromosome structure for the genetic algorithm. The genes $\lambda_0, \ldots, \lambda_{95}$ represent the customer-incentive pricing vector, one element for each 15-min interval in the 24-h period. The genes $t_{1\_sch}, \ldots, t_{I\_sch}$ are a representation of the aggregator schedule for the I customer loads that are schedulable. If the t_{i_sch} represents the original start time t_{i_start}, then the aggregator has not chosen to reschedule that appliance for use in DR

6.5.2 Genetic algorithm implementation

In this research, the steady-state Genitor [28] version of genetic algorithm (GA) is used to implement the heuristic framework. We use a GA as an example global search heuristic as GAs have been shown to work well in many optimization problems, such as resource allocation in computing [29–31], economic dispatch [32], and unit commitment [33], but any heuristic optimization method can be used to implement the described framework. If multiple objectives are used, the GA can easily be extended to generate Pareto fronts, e.g., with NSGA-II [16,34].

The implemented chromosome structure is broken into two parts, each with its own gene type, shown in Figure 6.6. The first portion of the chromosome is dedicated to the CIP vector, $\boldsymbol{\lambda}$, containing 96 genes representing the CIP (in cents/kWh) for each of the corresponding 15-min intervals of the day. The second portion of the chromosome represents the schedule of loads, containing one gene for each of the I schedulable customer loads. Let t_{i_sch} be a real value in the interval [0, 1] representing the scheduled start time of load i. The time interval that each load i is scheduled is calculated as $t_{i\_resch} = A_{i\_start} + t_{i\_sch}A_{i\_dur}$. If $t_{i\_resch} = t_{i\_start}$, then the load has not been used in DR (i.e., $i \notin L$). The [0, 1] representation of t_{i_sch} is used to avoid violating the customer-defined availability constraints of the loads given in (6.8).

No duplicate chromosomes are allowed in the initial population of Genitor to prevent premature convergence. Genitor is a steady-state algorithm that maintains a ranked list of chromosomes (in our study, ranked by (6.4)) to keep the best solutions between generations, i.e., elitism. In each generation, two parent chromosomes are selected using the linear bias function (as defined in Reference 28) to perform the global search. The linear bias selection function uses a linear bias parameter, $[1, 2] \in \mathbb{R}$, to bias selection toward the better solutions. A linear bias parameter of 1.5 means the best-ranked solution has a 50% greater chance of being selected than the median solution.

After two parent chromosomes are selected using the linear bias function, crossover and mutation search operators are performed. The former uses a two-point crossover performed separately on each of the two portions (CIP and schedule) of the chromosome. After crossover, two new children chromosomes are created. Within each child chromosome, each gene has a probability of mutation that randomly

```
initialize starting population
sort population by (6.4)
repeat
    select two parent chromosomes via linear bias
    create two children chromosomes using crossover
    mutate children genes probabilistically
    insert children chromosomes into population
    remove two worst chromosomes from population
until stopping criteria
return best chromosome from population according to (6.4)
```

Figure 6.7 Pseudocode for the steady-state Genitor algorithm

generates a new value for that gene. These new chromosomes are evaluated in terms of the objective function (given in (6.7)), inserted into the sorted population, and the worst two chromosomes in the population are removed, leading to a fixed population size. The complete algorithm is shown as pseudocode in Figure 6.7.

A parameter sweep was used to determine the best parameters to use for the GA in the scope of this problem. The population size was 100, the linear bias parameter was 1.4, and the probability of mutation was 0.01. The stopping criteria were defined as 500,000 total iterations or 10,000 iterations without an increase in the objective function.

To increase the search speed and quality of the Genitor solution, the initial population was seeded. Let ω be a real value in the interval [0, 1]. To seed the CIP vector, $\boldsymbol{\lambda}$, in 50 chromosomes in the initial population, we use the seeding function, denoted $\sigma(t,\omega)$, for each time-window $t=1,\ldots,96$, given by (6.11). The schedule for the customer load was randomly generated for each seed. The 50 seeds were generated using values $\omega=\frac{n}{49}$, $n=0,\ldots,49$.

$$\sigma(t,\omega)=\begin{cases}\omega s(t) & s(t)\geq r(t)\\ \omega r(t) & s(t)<r(t)\end{cases} \tag{6.11}$$

The rest of the chromosomes in the initial population are randomly generated. For each gene in the CIP vector, representing the cost in cents/kWh at time t, a random value is generated in the interval $[0,\max(r(t),s(t))]$. For each gene in the schedule, representing the scheduled time of load i, a random value in the interval [0, 1] is generated. The data in Figure 6.5(a) is used by the GA to determine $\boldsymbol{\lambda}$ and the schedule of loads.

6.6 Simulation study

6.6.1 Simulation setup

A total of 56,588 schedulable loads (i.e., $I=56,588$) from the 5,555 customers were randomly generated using the method from Section 6.3.3. The schedulable customer

Table 6.2 Aggregator forecast profit (in USD)

	RTP	**TOU**	**Flat rate**
Constant-$\alpha = 0.75$	699.06	863.33	954.10
Gaussian$\sim \mathcal{N}(0.75, 0.1)$	649.78	782.79	760.51
CVB	660.48	786.87	810.59
Constant-$\alpha = 1.0$	872.34	1101.38	1272.45

loads correspond to 18.7% of the total energy used by the 5,555 customers, equivalent to 30.6 MWh. To capture the aggregator DR in steady-state, a 2-h window was added to the start and end of the simulation time. Any appliance load that occurs within these windows did not contribute toward the objective function (i.e., only the 24-h window was used for the objective function calculation). Different cases were simulated using the different pricing mechanisms from Section 6.4 and the different alpha-models in Section 6.3.2, summarized in Table 6.2. Each case used the same set of customer appliances, generated α-values (within a table row), and price (within a table column). The stopping criteria for the Genitor algorithm were 500,000 total iterations, or 10,000 iterations without an increase in the objective function of the best solution (whichever occurs first). On average across all cases, the Genitor ran for 423,813 iterations in 188 min on a Dell Server with an Intel Xeon E5-2560 running at 2.4 GHz using a C++ implementation on a Ubuntu virtual machine.[2]

6.6.2 Results

The total system load before and after the aggregator-based DR is shown in Figure 6.8(a). This result is for the CVB method for generating α values (the CVB row in Table 6.2) for each of the three utility pricing mechanisms. In all three shown cases, the aggregator was able to reduce the peak of the system by approximately 2–2.5 MW, equating to a 12.5% peak reduction. All load shapes were similar when using the other methods of generating α values. To obtain better resolution on the effect of the aggregator, the schedulable load of the system is isolated and shown in Figure 6.8(b). Here, the difference in pricing mechanisms becomes apparent. In all cases, over half of the schedulable load at the peak (around 5 pm) is moved off-peak. The reason this value is not higher is due to the customer availability windows described in Section 6.2.3. Because of this constraint, not all of the load can be moved to off-peak hours. The large negative difference in load before and after DR, especially at the peak, directly corresponds to the component of aggregator profit obtained by selling negative load, N, to the spot market. The positive difference in load (i.e., when the "after" load is higher than the "before") is the portion of the load that contributes to the $S - B$ component of the aggregator profit function.

[2]The simulation code and data sets are made available at https://github.com/IPEMS.

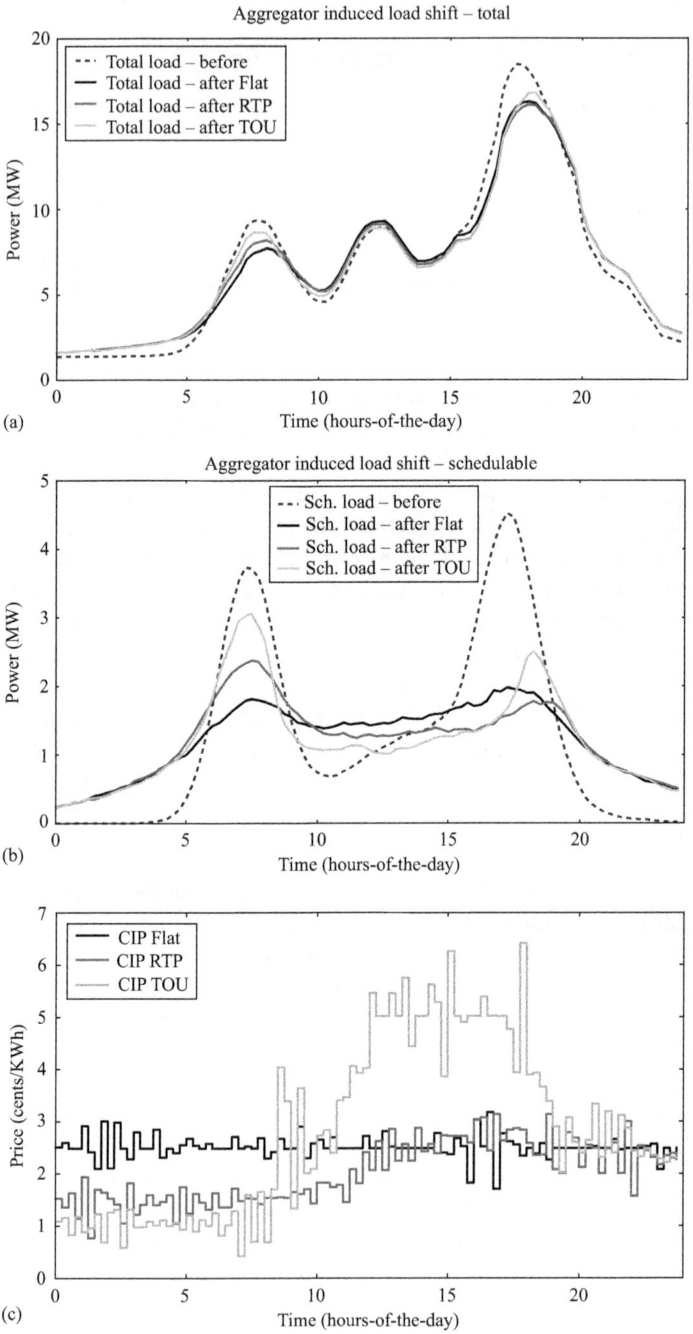

Figure 6.8 *The change in load from before and after the aggregator demand response action compared to the aggregator CIP. This data is for the CVB method for generating the appliance α-values. (a) The overall system load of the 5,555 customers. (b) The schedulable load. (c) The CIP*

Table 6.3 Aggregator profit after the market clears (in USD)

	RTP	TOU	Flat rate
Constant-$\alpha = 0.75$	960.53	1124.11	1135.88
Gaussian$\sim \mathcal{N}(0.75, 0.1)$	904.85	1048.19	944.89
CVB	918.02	1056.20	991.85
Constant-$\alpha = 1.0$	1102.87	1352.47	1452.05

Interesting details between the different utility pricing mechanisms emerge in Figure 6.8(b). In the TOU case, the aggregator pushes as much load as possible off of the "peak" price at the second peak load (18:00 h), and off of the "partial-peak" price at the first peak load (8:30 h). During the morning partial-peak price and peak price (8:30 am to 6:00 pm), the TOU case has the lowest schedulable load. In the flat rate case, the aggregator choice purely depends on the forecast SMP (from Figure 6.5(a)) and the schedulable load curve is flattened. The RTP case depends on the forecast SMP *and* RTP, which have very similar shapes because the ComEd utility passes the SMP to the customer with a slight markup. The forecast peak and valley of the RTP/SMP occurs at 4–5 pm and 3–4 am, respectively. As such, the aggregator tries to move as much load off of the peak RTP time as possible, and tries to move load from the morning load peak toward the price valley. The aggregator-determined CIPs for the CVB cases are shown in Figure 6.8(c). The determined CIP is always lower than the forecast prices in each utility pricing case, and the majority of the time the CIP is lower than the actual RTP. In the flat rate and TOU cases, the CIP is always lower than the utility price, indicating the customer receives a competitive, and reduced, rate of electricity for participating with the aggregator.

The final objective value, i.e., forecast aggregator profit P, for the different cases are given in Table 6.2 (based on Figure 6.5(a)). When evaluated for the *actual* RTP (for the RTP cases) and SMP, the aggregator schedule and CIP determined by the Genitor resulted in the aggregator profit given in Table 6.3 (based on Figure 6.5(b)). This increase in profit from forecast to actual is because the actual SMP at the peak period was much higher than forecast (as shown in Figure 6.5), leading to an increase in profit from the N component (6.2) of the profit function.

In each case, the constant-α values of 0.75 and 1.0 (denoted C75 and C100, respectively) allowed the aggregator to make the most profit. This is because the GA was able to find CIPs close to the optimal solutions in these cases – a CIP of 75% and 100% of the utility rate in the C75 and C100 cases, respectively – compared to the other alpha methods. The CIPs for C100 and C75 are shown in Figure 6.9(a) and (b), respectively. This is most obvious in the CIPs for the flat rate and TOU in the C100 case in Figure 6.9(a). The CIP flat rate curve found by the GA is very close to the utility flat rate of 4.06 cents/kWh, shown in Figure 6.5. Similarly, the partial-peak and peak values for the CIP TOU case are very similar to the utility TOU price. The same trend can be found in Figure 6.9(b) for the C75 case, but the CIP values are near 75%

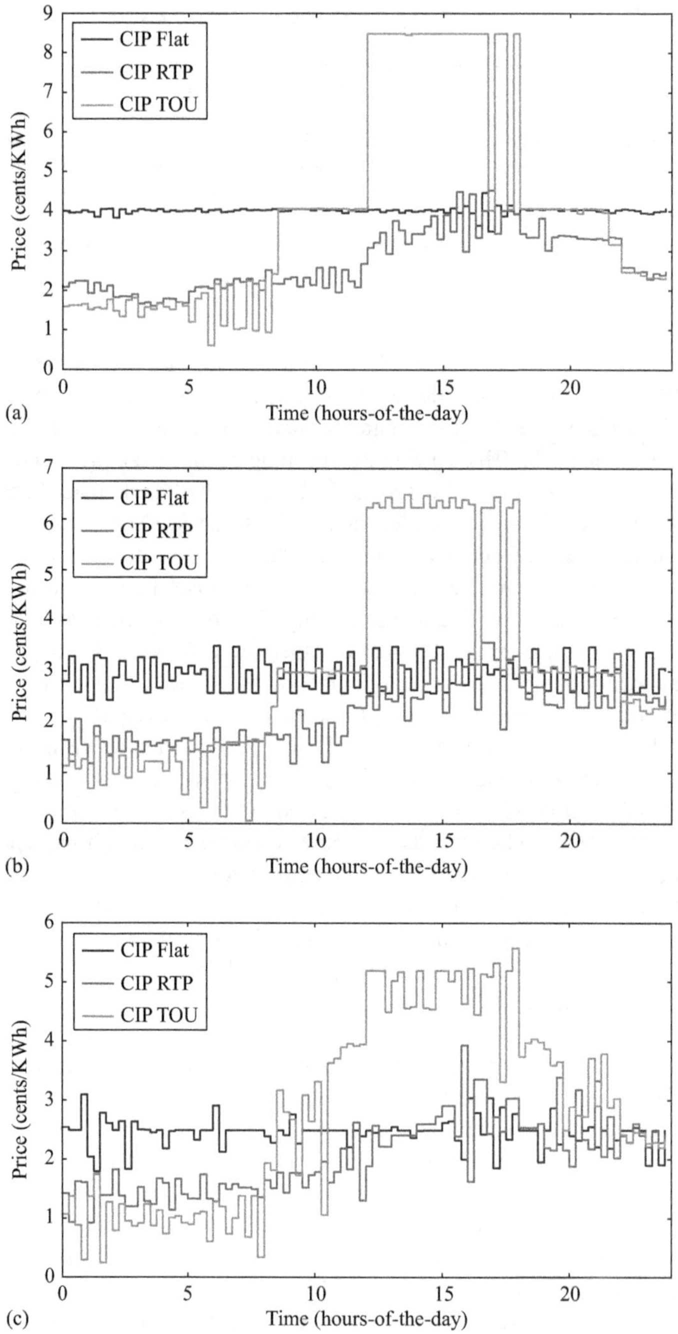

Figure 6.9 A comparison between the utility pricing mechanisms of the aggregator-determined CIPs for the (a) constant-$\alpha = 1.0$, (b) constant-$\alpha = 0.75$, and (c) Gaussian$\sim \mathcal{N}(0.75, 0.1)$ cases. The CVB case is provided in Figure 6.8(c)

Table 6.4 Forecast total customer savings (in USD)

	RTP	TOU	Flat rate
Constant-$\alpha = 0.75$	400.62	606.59	326.15
Gaussian$\sim \mathcal{N}(0.75, 0.1)$	414.12	658.00	429.78
CVB	426.66	665.63	436.93
Constant-$\alpha = 1.0$	244.95	360.28	12.88

of those in the utility pricing cases. The C100 case is most likely unrealistic as the customer is giving up the convenience of their loads for no monetary benefit, as shown in Table 6.4, and is used as a pseudo-upper-bound on the profit of the aggregator.[3]

In all alpha models, the RTP pricing is the worst performing in terms of aggregator forecast profit and actual aggregator profit. This could be due to the difficulty in the search space for finding the correct CIP, but also because loads that run greater than 1 h (or cross the hourly boundary during their duration) will cost two separate prices per kWh. This makes it difficult for the aggregator to schedule many long-running appliances from many customers at a significant profit. In general, for the results from this single day study, the aggregator made the most profit, in descending order of profit, for flat rate, TOU, and RTP; and in terms of customer alpha models for constant-α, CVB, and Gaussian (Tables 6.2 and 6.3). The significant differences in pricing methods and aggregator profit could be due to the method used for determining the values of the flat rate and TOU price in Section 6.4. For example, a weighted average of the price compared to the peak load may have resulted in a different mean value for price (i.e., the flat rate price and the partial-peak price of TOU). Additionally, the choice of 25% for the TOU off-peak price could be changed to any value and would impact the results. As these results are for a single sampling of the α-values for the stochastic Gaussian and CVB methods for a single day, no general conclusion can be made for the relative performance of the two methods without conducting simulations to estimate their statistical performance.

From a customer perspective, the total savings of all 5,555 customers are given in Tables 6.4 and 6.5 when using the forecast and actual data, respectively, for the 24-h period under consideration. An increase in savings similar to the aggregator's increase in profit is found in the RTP case, due to the large increase in peak RTP that the customer choosing CIP no longer has to pay. For the TOU and flat rate cases, there is no difference in forecast and actual savings because there is no uncertainty in the price of electricity for the customer. As a customer, participating with the aggregator in the TOU and flat rate cases provides a guaranteed, known benefit. When participating with the aggregator in the RTP case, the customer is offered stability in the price of electricity as the aggregator CIP is guaranteed the day-ahead.

[3]Monetary benefits may not be the only consideration for early adopters of new technology. Instead, some customers are motivated by altruistic reasons, such as environmental benefits (i.e., "being green") [35].

Table 6.5 Actual total customer savings (in USD)

	RTP	TOU	Flat rate
Constant-$\alpha = 0.75$	972.93	606.59	326.15
Gaussian$\sim \mathcal{N}(0.75, 0.1)$	983.32	658.00	429.78
CVB	1002.27	665.63	436.93
Constant-$\alpha = 1.0$	819.32	360.28	12.88

Table 6.6 Percent of rescheduled loads

	RTP (%)	TOU (%)	Flat rate (%)
Constant-$\alpha = 0.75$	83.4	72.6	94.9
Gaussian$\sim \mathcal{N}(0.75, 0.1)$	80.6	70.3	87.8
CVB	82.7	71.3	92.6
Constant-$\alpha = 1.0$	86.1	73.0	95.4

The CVB method is consistently the highest performing in terms of customer savings, which is a promising result as it is the closest to actual human behavior. By assigning similar appliances similar α-values (through the setting of the $\mu_{a,k}$ for appliance type k), the aggregator-determined CIP and schedule benefits the average customer. In the best case (CVB and RTP), the average customer saved \$0.18. Although this is for a single day, a rough extrapolation of this value would save the customer around \$5.50 per month. In general, the aggregator makes less profit and the customer saves more when the average α-value is decreased, and vice versa.

The maximum customer savings from all cases was \$0.51. This customer had an average $A_{i\_dur} > 15$ h on 12 loads with a total energy of 92.9 kWh. The average alpha value for this customer was $\alpha = 0.76$. Using the same extrapolation estimate, this customer, for being flexible with the loads, could save between \$15 and \$16 per month, which is a large percentage of the monthly bill. This is indicative of the possible monetary benefits from the customer being more flexible with loads (in the availability of the reschedulable load and the customer α-values) and bringing more energy (i.e., an asset set with a large kWh rating) to the aggregator to participate in DR.

In Table 6.6, the percent of reschedulable loads that the aggregator utilized in DR is provided (i.e., $^{|L|}/_{l}$). It is interesting to note that even though the TOU was the second best performing algorithm in terms of aggregator profit in most cases (some cases it performed the best), and had the highest *forecast* customer savings, TOU had the least number of rescheduled loads. This occurs because the TOU has relatively long periods of constant price, and the difference between the price periods is relatively large. This allows the aggregator to shift load in small time displacements and receive large rewards for both the customer and the aggregator. This becomes more apparent in the heat maps in Figure 6.10, specifically in Figure 6.10(b).

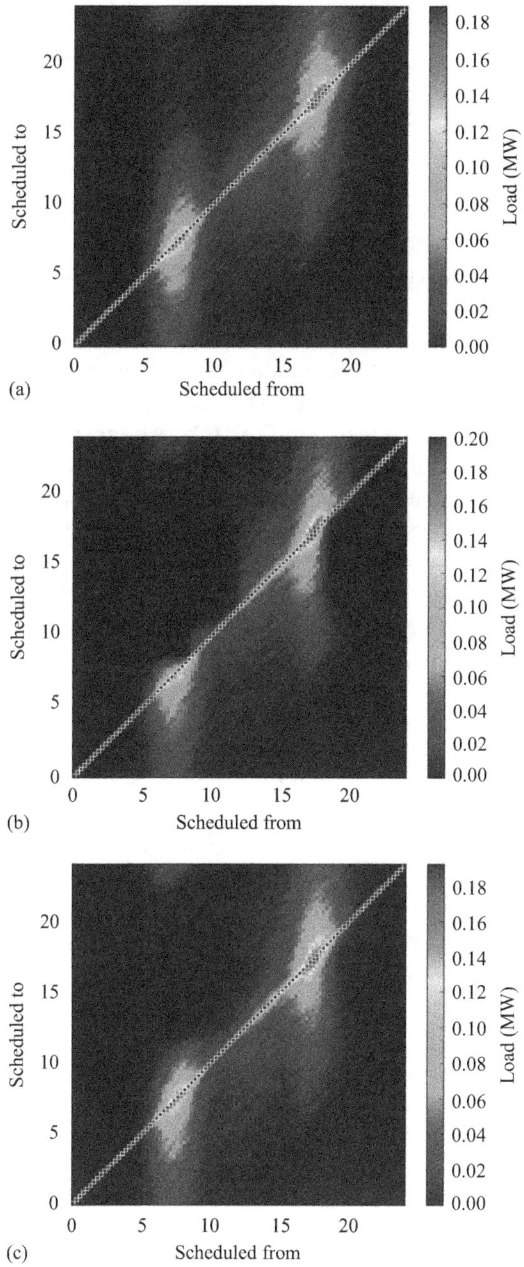

Figure 6.10 Heat maps for the rescheduled load in the CVB alpha model case for the (a) flat rate, (b) TOU, and (c) RTP utility pricing mechanisms. For each (x, y) coordinate, the color of the square in the heat map indicates the magnitude of the load that was rescheduled from time x ($t_{i\_start} = x$) to time y ($t_{i\_resch} = y$)

To further characterize the impact of the aggregator-based residential DR program, for Figure 6.10 we use the heat map visualization methods from Reference 36. For each (x, y) coordinate, the color of the square in the heat map indicates the magnitude of the load that was rescheduled from time x ($t_{i\_start} = x$) to time y ($t_{i\_resch} = y$). For $x > y$ (i.e., below the diagonal), the plot indicates the load was scheduled earlier, and $x < y$ indicates the load was scheduled at a later time. At a $t_{i\_start} = x$, the larger the value of $|y - x|$, the further away the load was scheduled by the aggregator DR. This distance is limited by the customer defined availability window. The heat maps only show the load that was rescheduled. The load that was not rescheduled (i.e., $\notin I$) would appear on the diagonal ($x = y$), but for visualization purposes it is set to 0 MW. We do this because the magnitude of the load that is not scheduled is much larger than any individual (x, y) magnitude where $x \neq y$, reducing the resolution of the interesting portions of the heat map. The flat rate, TOU, and RTP cases for the CVB alpha model are shown in Figure 6.10(a)–(c), respectively. The flat rate and RTP cases are very similar, except the flat rate case schedules the loads from the first peak to a much later time than the RTP. This can be seen in the verticality of the loads near 8:00 am between the two heat maps, i.e., the loads scheduled *from* the times around 8:00 am are rescheduled *to* more times in the y-direction. The magnitude of the two heat maps are very similar, but there is much more load rescheduled in the flat rate case (as was shown in Table 6.6). In the TOU case, it is very interesting to see how the aggregator schedules load around the change in price periods. The aggregator in the TOU case does not schedule load *to* the partial-peak price in the morning, shown as the horizontal flat transition at $y = 8{:}30$ between $5{:}00 < x < 10{:}00$. Similarly, the aggregator does not schedule load *from* the partial-peak price in the evening, shown as the vertical flat transition at $x = 18{:}00$.

6.7 Conclusions

We presented an aggregator-based residential DR approach for scheduling residential customer assets. A CIP structure to compensate the customer for the inconvenience of rescheduling their assets is discussed and compared to three utility rate structures. This new pricing structure gives the customer a near real-time choice of electricity supplier in a fully deregulated market scenario. A heuristic framework was designed to perform an optimization on the profit of the aggregator. To validate the heuristic framework, a system comprised 5,555 customer households and 56,588 schedulable loads was simulated using a GA implementation of the framework. The CIP found by the GA was, in general, lower at all times than the customer would pay via all forms of real utility pricing data (or with reasonable modifications), i.e., RTP, TOU, and flat rate tariff for a given day in summer 2015. Despite this, the aggregator was able to make a profit by selling negative peak load to the spot market. This showed an example of optimizing for purely economical reasons in the form of aggregator profit, and enacting an overall change on the system peak load. This change benefits the customers of the aggregator (in the form of reduced cost of electricity for schedulable loads), the aggregator (in the form of a profit), and also those customers *not* participating with the aggregator (because the overall system peak is lowered as a common good).

Reducing the peak demand of the electric power system provides benefits by reducing the cost of electricity by lowering the deployment of expensive generators during peak hours. By reducing the peak, we can reduce the capacity factor of dirty diesel-fired peaking generators. Moreover, as peak demand increases, the available transmission capacity will also need to increase. By reducing the peak demand, we can defer building new transmission lines; a costly, long-term project. As more asset types with more capabilities, such as electric vehicles and HVAC, are considered in similar studies, the savings can be expected to increase.

References

[1] T. M. Hansen, R. Roche, S. Suryanarayanan, A. A. Maciejewski, and H. J. Siegel, "Heuristic optimization for an aggregator-based resource allocation in the Smart Grid," *IEEE Transactions on Smart Grid*, Jul. 2015, vol. 6, no. 4, pp. 1785–1794.

[2] United States Department of Energy, "The Smart Grid: An introduction," 2009. [online]. Available: http://energy.gov/oe/downloads/smart-grid-introduction-0

[3] P. Siano, "Demand response and smart grids—A survey," *Renewable and Sustainable Energy Reviews*, Feb. 2014, vol. 30, pp. 461–478.

[4] A. Zipperer, P. Aloise-Young, and S. Suryanarayanan, "On the design of a survey for reconciling consumer behaviors with demand response in the smart home," in *North American Power Symposium (NAPS), 2013*, Sep. 2013, p. 6.

[5] P. C. Stern, "Information, incentives, and proenvironmental consumer behavior," *Journal of Consumer Policy*, Dec. 1999, vol. 22, no. 4, pp. 461–478.

[6] M. H. Albadi and E. El-Saadany, "A summary of demand response in electricity markets," *Electric Power Systems Research*, 2008, vol. 78, no. 11, pp. 1989–1996.

[7] PJM Forward Market Operations, "PJM Manual 11: Energy & Ancillary Services Market Operations, Revision 83," July 2016. [online]. Available: http://www.pjm.com/~/media/documents/manuals/m11.ashx. Accessed: Aug. 24, 2016.

[8] E. G. Coffman, *Computer and Job-Shop Scheduling Theory*. John Wiley & Sons, New York, NY, 1976.

[9] D. Fernandez-Baca, "Allocating modules to processors in a distributed system," *IEEE Transactions on Software Engineering*, Nov. 1989, vol. 15, no. 11, pp. 1427–1436.

[10] O. Ibarra and C. E. Kim, "Heuristic algorithms for scheduling independent tasks on nonidentical processors," *Journal of the ACM*, Apr. 1977, vol. 24, no. 2, pp. 280–289.

[11] California Energy Commission. "2013 Integrated Energy Policy Report." Accessed: Mar. 28, 2014. [Online]. Available: http://www.energy.ca.gov/2013_energypolicy/

[12] P. Du and N. Lu, "Appliance commitment for household load scheduling," *IEEE Transactions on Smart Grid*, Jun. 2011, vol. 2, no. 2, pp. 411–419.

[13] K. F. Fong, V. I. Hanby, and T. T. Chow, "HVAC system optimization for energy management by evolutionary programming," *Energy and Buildings*, Mar. 2006, vol. 38, no. 3, pp. 220–231.

[14] E. Sortomme and M. A. El-Sharkawi, "Optimal scheduling of vehicle-to-grid energy and ancillary services," *IEEE Transactions on Smart Grid*, Mar. 2012, vol. 3, no. 1, pp. 351–359.

[15] M. A. A. Pedrasa, T. D. Spooner, and I. F. MacGill, "Coordinated scheduling of residential distributed energy resources to optimize smart home energy services," *IEEE Transactions on Smart Grid*, Sep. 2010, vol. 1, no. 2, pp. 134–143.

[16] R. Friese, T. Brinks, C. Oliver, A. A. Maciejewski, and H. J. Siegel, "Analyzing the trade-offs between minimizing makespan and minimizing energy consumption in a heterogeneous resource allocation problem," in *2nd International Conference on Advanced Communications and Computation*, Oct. 2012, pp. 81–89.

[17] United States Federal Energy Regulatory Commission, "FERC Order 745: demand response compensation in organized wholesale energy markets," Mar. 2011.

[18] Z. Chen, L. Wu, and Y. Fu, "Real-time price-based demand response management for residential appliances via stochastic optimization and robust optimization," *IEEE Transactions on Smart Grid*, Dec. 2012, vol. 3, no. 4, pp. 1822–1831.

[19] L. Gkatzikis, T. Salonidis, N. Hegde, and L. Massoulié, "Electricity markets meet the home through demand response," in *IEEE Conference on Decision and Control*, Dec. 2012, pp. 5846–5851.

[20] T. Bapat, N. Sengupta, S. K. Ghai, V. Arya, Y. B. Shrinivasan, and D. Seetharam, "User-sensitive scheduling of home appliances," in *ACM SIGCOMM Workshop on Green Networking*, Aug. 2011, pp. 43–48.

[21] S. Ali, H. J. Siegel, M. Maheswaran, D. Hengsen, and S. Ali, "Representing task and machine heterogeneities for heterogeneous computing systems," *Tamkang Journal of Science and Engineering,* Special Tamkang University 50th Anniversary Issue, Nov. 2000, vol. 3, no. 3, pp. 195–208. Invited.

[22] N. Lu and Y. Zhang, "Design considerations of a centralized load controller using thermostatically controlled appliances for continuous regulation reserves," *IEEE Transactions on Smart Grid*, Jun. 2013, vol. 4, no. 2, pp. 914–921.

[23] R. Roche, "Agent-based architectures and algorithms for energy management in smart grids." PhD dissertation. Université de Technologie de Belfort-Montbéliard, Belfort, France, Dec. 2012.

[24] T. M. Hansen, E. K. P. Chong, S. Suryanarayanan, A. A. Maciejewski, and H. J. Siegel, "A partially observable Markov decision process approach to customer home energy management systems," http://www.engr.colostate.edu/sgra, in press.

[25] ComEd. "ComEd residential real-time pricing program." Accessed: Aug. 18, 2013. [Online]. Available: https://rrtp.comed.com/live-prices/
[26] Pacific Gas and Electric. "PG&E time-of-use rates." Accessed: Mar. 21, 2016. [Online]. Available: http://www.pge.com/TOU
[27] United States Federal Energy Regulatory Commission. "PJM daily report archives." Accessed: Jan. 14, 2014. [Online]. Available: http://www.ferc.gov/market-oversight/mkt-electric/pjm/pjm-iso-archives.asp
[28] D. Whitley, "The GENITOR algorithm and selective pressure: why rank-based allocation of reproductive trials is best," in *3rd International Conference on Genetic Algorithms*, Jun. 1989, pp. 116–121.
[29] L. D. Briceño, H. J. Siegel, A. A. Maciejewski, *et al.*, "Heuristics for robust resource allocation of satellite weather data processing on a heterogeneous parallel system," *IEEE Transactions on Parallel and Distributed Systems*, Nov. 2011, vol. 22, no. 11, pp. 1780–1787.
[30] T. D. Braun, H. J. Siegel, N. Beck, *et al.*, "A comparison of eleven static heuristics for mapping a class of independent tasks onto heterogeneous distributed computing systems," *Journal of Parallel and Distributed Computing*, Jun. 2001, vol. 61, no. 6, pp. 810–837.
[31] F. D. Croce, R. Tadei, and G. Volta, "A genetic algorithm for the job shop problem," *Computers and Operations Research*, Jan. 1995, vol. 22, no. 1, pp. 15–24.
[32] D. C. Walters and G. B. Sheble, "Genetic algorithm solution of economic dispatch with valve point loading," *IEEE Transactions on Power Systems*, Aug. 1993, vol. 8, no. 3, pp. 1325–1332.
[33] S. A. Kazarlis, A. G. Bakirtzis, and V. Petridis, "A genetic algorithm solution to the unit commitment problem," *IEEE Transactions on Power Systems*, Feb. 1996, vol. 11, no. 1, pp. 83–92.
[34] D. Kalyanmoy, A. Pratap, S. Agarwal, and T. A. M. T. Meyarivan, "A fast and elitist multiobjective genetic algorithm: NSGA-II," *IEEE Transactions on Evolutionary Computation*, Apr. 2002, vol. 6, no. 2, pp. 182–197.
[35] A. Zipperer, P. A. Aloise-Young, S. Suryanarayanan, *et al.*, "Electric energy management in the smart home: Perspectives on enabling technologies and consumer behavior," *Proceedings of the IEEE*, Aug. 2013, vol. 101, no. 11, pp. 2397–2408.
[36] T. Hansen, S. Suryanarayanan, A. A. Maciejewski, H. J. Siegel, and A. V. Modali, "A visualization aid for demand response studies in the Smart Grid," *The Electricity Journal*, Apr. 2015, vol. 28, no. 3, pp. 100–111.

Chapter 7

Application of the agent-based technology to invoke demand flexibility under the SG-BEMS integration framework

Luis A. Hurtado[1], *Phuong H. Nguyen*[1] *and Wil L. Kling*[1]

Abstract

Modern electrical infrastructure is being challenged by increasing uncertainty from the introduction of renewable energy sources (RES) and distributed energy resources (DER). To cope with the challenges such technologies introduce, a considerable amount of attention has recently been given to the concept of "demand flexibility." It is presented as an alternative to complement the current control and operation methodologies of the power grid, and to assist the penetration process of new technologies. Conventionally, flexibility is harnessed from power generation units and used by SOs to maintain the balance between supply and demand of the power system. However, as the penetration of stochastic generation and new forms of demand increase, this capability of conventional generation units might not be sufficient to cope with increasing uncertainty in both supply and demand. In response, throughout the literature, the necessity for a new flexibility source is highlighted. Demand side becomes active component in the control and operation of the power system, with the advances in ICT and computational intelligence in a so called cyber-physical social system (CPSS). The flexibility offered by the end-users, through for instance building energy management systems (BEMS), has the potential to help not only resolve network and system problems, but also accommodate a higher amount of renewables, increase asset utilization, and reduce peak demand. Yet, this is not a straightforward transition toward a smart energy system or smart grid (SG). Advanced energy management systems are required to manage the flexible demand while integrating emerging technologies. This chapter introduces a SG-BEMS interoperation framework and demonstrates the capabilities and benefits of multi-agent systems (MASs) in enabling the correct operation of the emerging SG, while unlocking the flexibility potential of the built environment.

[1]Department of Electrical Engineering, Electrical Energy Systems, Eindhoven University of Technology, Eindhoven, The Netherlands

7.1 Introduction to the concept of demand flexibility

The electric power system today is a dynamic system under constant change, toward varied levels of modernization. The recent changes are a natural consequence of the growing population and economy, new technologies, environmental issues, and green legislation. With the introduction of renewable energy sources (RES) and distributed energy resources (DER), a transition is evident from a centralized to a decentralized energy production system. Together with the new types of loads and the electrification of mobility, these new forms of power generation challenge the operation of the power system and make necessary the use of advanced monitoring and controlling strategies. In this context, the smart grid (SG) has arisen as the concept that encapsulates the different goals and technologies necessary for the stable operation of the power system. It allows the integration of the demand side in the monitoring, operation, and support tasks of the electric system [1] through the use of ICT and computational intelligence, becoming a cyber-physical social system (CPSS). Traditionally, the demand is considered uncontrollable, but well predictable in a certain aggregation level. The imbalance between supply and demand comes from unforeseen demand fluctuations and generation units' failures. To deal with this, system operators (SOs) make use of reserve capacity. This capacity of the system to react and adapt in tolerable time to these unforeseen events is known as flexibility [2]. In this chapter, a general introduction to the concept of demand flexibility is provided.

7.1.1 Demand side management and demand response

For a long time, the demand side has been considered inelastic and variable during time. However, a key feature of the demand is its high diversity in usage of appliances. This means that there is a low probability that the aggregated power demand in a moment of time, is equal to the sum of the demand of all the appliances in the network at the same moment. This is known as the coincidence factor,[1] i.e., not all loads consume at the same time, noncoincidence. This factor is not only useful for system planning, but also means that the customer has the capacity to fill in load valleys, and to shift or to shed demand. Such capacity of the demand is recognized to be a more cost-effective alternative to expanding the system transmission and generation capacity, as well as useful tool for load-leveling, which has been a long recognized issue of the power system [3]. This extension of the customer's role in the power system, allows for new relations and interactions between the system's members. This social aspect is what distinguishes CPSS from traditional power systems.

Following the oil crisis of the 1970s, demand side management (DSM) was considered as a measure for load management, which included: energy efficiency, demand response (DR), and strategic load growth. In general, DSM refers to the long- and short-term measures designed to change the consumption patterns in such a way that they will influence properly the load shape of the utility. With the introduction of

[1] Ratio between the system's capacity required to supply a certain number of loads and the capacity of the system if all the loads were to consume their maximum simultaneously.

intermittent and uncontrollable generation, DSM was soon recognized as a potential tool to deal with the uncertainty intrinsic of these technologies, and to assist maintaining demand-supply balance [4]. Within the measures under the DSM umbrella, DR mechanisms are designed to directly influence the demand of customer in the short term in response to supply conditions, for instance through market prices, or control signals.

Several strategies are proposed for the implementation of DSM and DR across the power systems [5]:

- ***Time of use pricing (TOU):*** Tariffs for different time intervals are defined to reflect the production and investment cost structures. Prices are high during peak periods and lower during off-peak periods.
- ***Dynamic prices:*** Hourly tariffs are established for the demand on a day-ahead or intra-day basis.
- ***Fixed load capping:*** Load volume limits (up and down) are established for time intervals during the day.
- ***Dynamic load capping:*** Hourly load volume limits (up and down) are established for the day.
- ***Direct load control:*** Part of the customer's load is placed under the control of a different party, e.g., the utility. This party can remotely control shutdown/start-up process or operation cycles of the customer's appliances under the scheme.

7.1.2 Operational flexibility

As mentioned, the operational flexibility inherent to traditional generation units is used by the SOs to maintain the system balance. This behavior of generation units can be divided into three main characteristics: (a) ramping rate (MW/min), i.e., how fast the unit reacts; (b) power capacity (MW), i.e., how much power can be delivered for support activities; and (c) energy capacity (MWh), i.e., how long can the response be sustained [6]. Flexibility is used by unit commitment and ancillary services to manage uncertainty in the coordination of supply operations in power systems. These mechanisms aim to schedule and dispatch capacity in response to sudden and unexpected supply-demand imbalances. To do so, SOs make use of automated, i.e., primary and secondary control, or manual, i.e., tertiary control[2] [8], power reserves [2]. Here, uncertainty comes from load and generation forecasting errors, and random generation failures. In the event of a major load fluctuation, the primary reserves are activated, this is done almost immediately. To relieve the primary reserves and if the imbalance exceeds the 30 s mark, secondary reserves are activated. Finally, if the deviation lasts longer than 15 min, the tertiary reserves are used.

Primary control is used to stop the frequency change and stabilize it on a quasi steady-state value. Once the frequency has been stabilized by primary reserves, secondary control is activated to bring the frequency back to its nominal value within

[2]Tertiary control can be activated automatically or manually. In the Netherlands [7], tertiary control is activated manually.

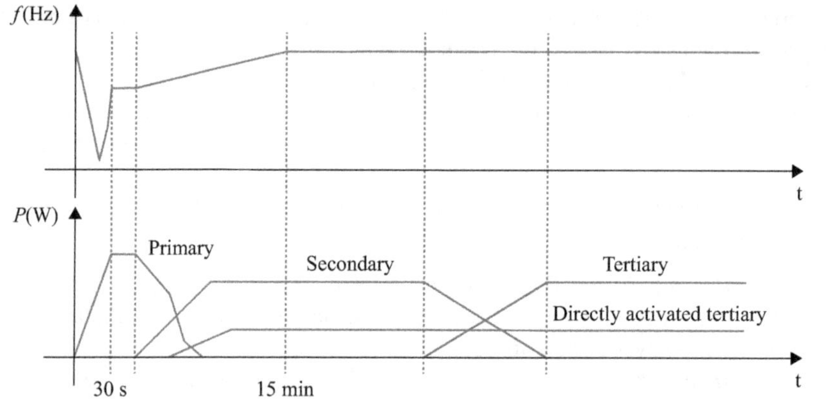

Figure 7.1 Principle frequency deviation and subsequent activation of reserves

15 min. Finally, tertiary control reserves are activated to free up secondary reserves [8]. This process can be seen in Figure 7.1. In case of large disturbances, tertiary reserves can be used to supplement secondary control. However, in most control areas, tertiary reserves deployment is the responsibility of the market parties. As the penetration of renewable technology increases, new forms of uncertainty are introduced in the power system operation. In such systems, SOs need to capture the increasing reserve capacity requirements given by the stochastic behavior of these sources. However, these new technologies do not share inertia with the system and cannot effectively provide reliable balancing services. Hence, several propositions have been made to address this specific issue, such as, cross-border power exchange and market coupling. Nonetheless, it is recognized that the SOs require new flexibility sources to ensure the system stability. With the introduction of the DSM and DR, the concept of demand flexibility started to gain relevance in the power system literature. This concept will be discussed in detail next.

7.1.3 Demand flexibility

As the SG starts emerging, a more decentralized, less dispatchable, and less predictable system is expected. This creates a necessity for having more elastic demand to offer grid support services. With the advances in ICT and computational intelligence, flexibility can be provided not only by traditional generating units or energy storage systems, but also by the demand side. This is known as the demand flexibility, defined as: "*the changes in consumption/injection of electrical power from/to the power system from their current/normal patterns in response to certain signals, either voluntarily or mandatory*" [9].

The high diversity of devices present at the demand side makes flexibility vary among customers. The ***non-curtailable*** systems, inelastic demand/production, make the base load or base production of the customer. This capacity cannot be interrupted, independent of the grid situation, e.g., emergency systems. ***Curtailable*** devices are

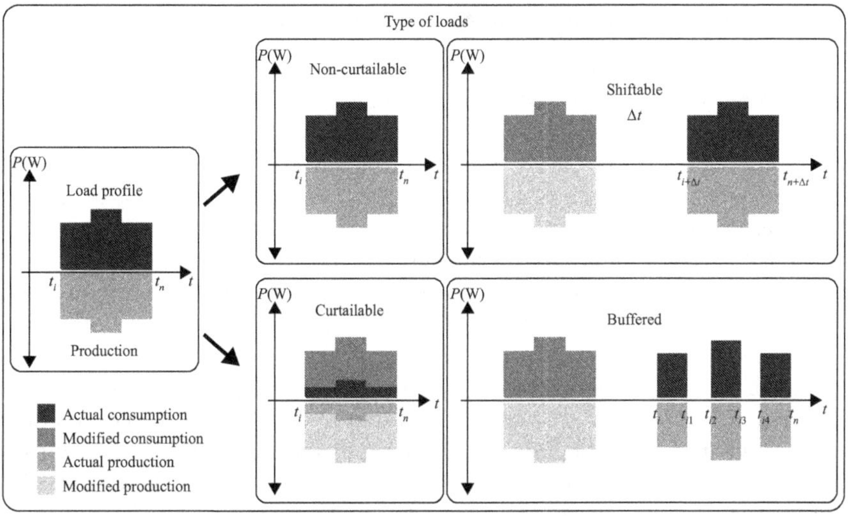

Figure 7.2 Diversity of the demand

those consumption or generation systems that cannot be shifted without affecting the end use of the system. However, they can be switched off, e.g., lighting, TV. ***Shiftable*** demand or production can be moved in time without affecting the operation and the end use of the service, e.g., laundry, dish washer, batteries. ***Buffered*** systems represent the power consumption or generation that is decoupled by some kind of storage, like the thermal inertia of a building. Together, these type of loads define the type of flexibility and the potential each customer can offer. However, it is an intrinsic characteristic of the demand that the load mix, and therefore the flexibility, changes over time. It depends not only on the type of devices but also on lifestyle, user preferences, and the social context. Figure 7.2 illustrates the load diversity of the customer, not only consumption but also production. In this figure, the modified energy profile is depicted in lighter shade.

In order to access and activate the flexibility potential of the demand side several strategies are proposed throughout the literature. In general, these translate into the use of direct, e.g., on-off, or indirect, e.g., price, control signals in response to system or market situations. Furthermore, the use of flexibility can be divided based on the type of use and time horizon. Market and commercial use, i.e., portfolio balancing and market participation, occurs in the day-ahead and intra-day time scales. Portfolio balancing refers to the minimization of the differences between scheduled and the actual operation of the flexible sources. The technical uses occur in the operation, i.e., near real-time time scale. Within the technical uses of flexibility, two sub-groups can be differentiated. The first one is the technical system uses. Here, flexibility is used to ensure the correct operation of the power system as a whole, i.e., system balancing (frequency control), system restoration, and black start. The second is the technical network uses. Here, flexibility is used to optimize local network operations, i.e., network constraint management, voltage/VAr optimization, restoration, power flow stabilization.

7.1.4 Emerging energy systems

As discussed, under the transition toward an SG, the new forms of uncertainty introduced to the system could lead to more frequent supply-demand mismatches and network problems, e.g., voltage variations, network congestion. This has resulted in an extensive research work and demonstration projects, which focus on the concept of SGs and emerging distribution systems. The general objective of these activities is to guarantee efficient and reliable operation of the power system, while safely integrating more RES, EVs, and DER technologies, by enabling the customer to participate actively in the electricity market and facilitate the operation of the system. In Europe there are 459 SG-related projects up until 2014, with a total investment of around €3.15 billion Euro[10].

Under this new system context, new roles and actors need to be specified and identified, while new services and markets start to develop and to operate next to the existing ones. An effort for standardization and to enable the creation of smart energy services and products is the Universal Smart Energy Framework (USEF), which establishes guidelines for a functional SG. Under this standard, a new role is distinguished, i.e., aggregator, and existing ones are modified. With the integration of DER technologies, electric vehicles (EVs), and other new technologies, the customer becomes an active part of the energy system, by feeding and consuming energy to/from the grid. This new role of the customer is known as prosumer. The role of the aggregator is to accumulate flexibility from the active demand and supply, in an economic efficient way to meet the energy demand of its portfolio. Figure 7.3 shows in a graphical way the emerging energy system's value chain.

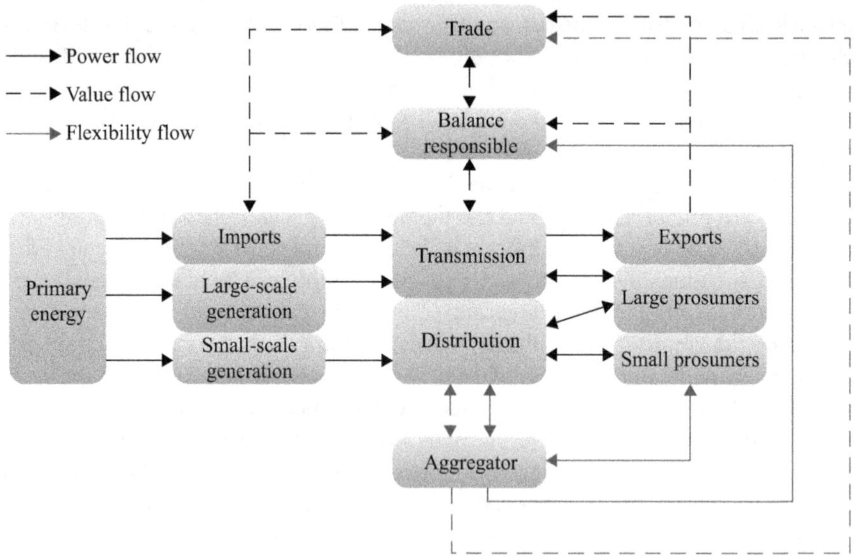

Figure 7.3 The emerging power system value chain: cyber, physical, and social entities

As the system evolves into a SG structure, new control methodologies and frameworks are needed. The SG-BEMS interoperation framework is introduced in the following section.

7.2 The SG-BEMS interoperation framework

It is expected, unless measures are taken, that the electrical infrastructure will reach the margins for stable operation and will be vulnerable due to the increasing uncertainty originated from the stochastic nature of renewables [11]. Advances in IT, control, and forecasting capabilities have made DSM and DR viable options to increase the power system flexibility, while compensating for the increasing uncertainty. However, buildings, i.e., prosumer domain, and the grid have been treated, so far, as independent and unique control systems, operated based on their own local information [12]. Here, a functional framework is presented based on the Smart grid architecture model (SGAM) [13]. The main objective is to fully invoke the flexibility from the built environment, and to achieve the energy efficiency and sustainable goals of the power system, with comfort as a necessary metric for this flexibility. This section presents the framework domains, services, and operation schemes for demand flexibility management.

7.2.1 Framework domains

The SG domains embody the physical areas related to the electricity supply chain, i.e., electricity generation, transmission, distribution, and consumption. Only the transmission, distribution, and prosumer domains are discussed in the following sections. The power generation domain is not discussed as it is not part of the scope of the SG-BEMS framework.

7.2.1.1 The transmission domain

The transmission domain represents the infrastructure that connects the large power generation units to the distribution grids and thereby the end-users. Within this domain, the transmission system operators (TSOs) has the responsibility to ensure reliable and uninterrupted power supply, to maintain the system balance, i.e., frequency, and to maintain and expand the high voltage (HV) grid.

7.2.1.2 The distribution domain

The distribution domain exemplifies the infrastructure that connects the transmission system to the end-users, i.e., distributes power among end-users. Traditionally, the distribution grids are operated by the distribution system operators (DSOs). Their main function is to sustain reliable power supply to the connected systems, while maintaining the distribution assets. This means that the DSOs control must support the prevention of asset overloading, the regulation of voltage magnitude, and the maintenance of power quality and security. Furthermore, as the DER penetration increases, the DSOs will need to handle reverse power flows, local balancing, and capacity issues.

7.2.1.3 The prosumer domain

The two main purposes of the prosumer domain are comfort and energy management. In buildings, the central objective is to provide the occupants with a comfortable environment. This is evident when looking at the energy breakdown of the built environment, about 50% of the total electrical energy consumed is used for comfort management [14]. This strong correlation is crucial to reveal flexibility from the built environment. Furthermore, as the penetration of DER technologies in this domain increases, the energy management task of the customer gets more complex.

Being responsible for about one-third of the energy consumed in cities [15], commercial and industrial buildings have the potential to significantly contribute to the efficient operation of the smart energy system, accommodate a higher amount of RES, increase asset utilization and reduce peak demand [16–18, 6]. Therefore, the SG-BEMS framework focus primarily on the integration of large-scale buildings into the SG.

7.2.2 Roles and responsibility

Within the SG context new roles are required to handle the new requirements and services of the system. These are explained as follows:

7.2.2.1 The aggregator

The role of the aggregator is to procure flexibility from the prosumer, in an economic and efficient way not only to meet the flexibility demand of its portfolio, but also to offer ancillary services to the end-users and operators. It collects information from the portfolio of customers, the system responsible, and balance responsible to create a flexibility request.

7.2.2.2 Building energy management system responsible (BEMS)

The role of the BEMS is to ensure the correct operation of the building by aggregating the information from the different building's zones and systems; thus, acting as a policy enforcer. It serves as the interface to SG and building manager. Its responsibility is twofold: it is in charge of the correct operation of the environment, i.e., comfort and energy management; and it is responsible for determining the flexibility potential for time interval, based on the system current status, operational boundaries, and the information received from the aggregator.

7.2.2.3 The balance responsible party (BRP)

The BRP is responsible for actively balancing energy supply and demand, following economic and technical optimization principles. It uses information from its portfolio to forecast the energy supply and demand, and it tries to find the optimal solution for the energy supply problem.

7.2.2.4 Distribution system operator

The DSO is accountable for the operation of the distribution system. It collects system information to predict network problems and tries to optimize the local system operation. It aggregates the information from its portfolio of prosumers. It acts as a feeder policy enforcer, deciding the appropriate requests/control actions for every system connected to the distribution system. It also receives data from aggregator and balance responsible.

7.2.3 Services

The use of demand flexibility and resources can occur in different time horizons and be used for different purposes, such as:

7.2.3.1 Network services

These services are provided by the prosumer to the DSR in response to a request from the aggregator or local information at the point of connection (POC).

- ***Voltage regulation:*** ensuring the voltage limits are not violated.
- ***Congestion management:*** ensuring the network elements, e.g., transformers, are not overloaded for periods longer than thermal limits of the elements allows.
- ***System restoration:*** enabling the orderly restart of the power system after a blackout.

7.2.3.2 System services

These services are provided by the prosumer to the BRP in response to a request from the aggregator.

- ***System balancing:*** ensuring frequency stability.
- ***Local balancing:*** preventing deviations from the scheduled and measured inter-area power exchange, i.e., balancing of an aggregator's portfolio.
- ***System restoration:*** enabling the orderly restart of the power system after a black out.

7.2.4 Operation scheme

Within the USEF operation structure, i.e., planning, validation, operation, and settlement [19], the SG-BEMS framework distinguishes four operation regimes, as depicted in Figure 7.4. Within these four regimes, the procurement of demand flexibility follows a different purpose and interactions [20].

7.2.4.1 Green regime

In the green regime or nominal operation, i.e., the power system is without problem, the aggregator and its portfolio of prosumers are acting to maintain the balance within the distribution network area, i.e., internal balance. During this operation scheme, market-related services are only provided. This means that little to none interaction exists between the aggregator and the DSO or the BRP.

7.2.4.2 Yellow regime

This regime indicates the operation of the power system is in danger. The network constraints, voltage/VAr regulation, and internal balance are within but close to the established limits. The DSO and BRP start interacting with the aggregator in order to keep the system from going into the orange or red regimes, i.e., procurement of flexibility.

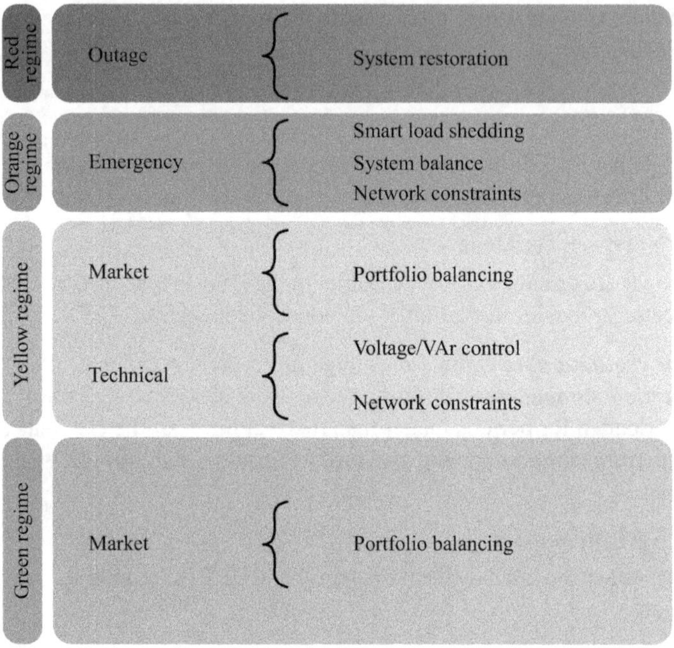

Figure 7.4 SG-BEMS operations scheme

7.2.4.3 Orange regime

If the flexibility is not enough to restore the normal operation, the system goes into the orange regime. Here the system is still within its boundaries. However, to prevent the system from going into the red regime, intelligent load shedding and network reconfiguration need to take place. The DSO and BRP need to take over the control of the area.

7.2.4.4 Red regime

In this regime the system goes into system restoration mode.

To cope with the complexity of this integration framework, a shift is required, from centralized energy management systems and control, to a decentralized structure that integrates both computational and distributed intelligence. The new control philosophy is discussed in the following section.

7.3 New control philosophy

Power system control is a highly diverse topic, covering a wide spectrum that includes power generation control, load flow control, reactive power control, system stabilization, among others. However, the main objective of power system control is to supply customers with constant and uninterrupted electricity. This control philosophy can be summarized in three main points: (a) provide the interconnected system security/reliability, (b) avoid violations of transmission/distribution system limits, and (c) maintain system balance.

In the transition to SG, larger amount of data is required to cope with the new uncertainty sources, which are unprecedented in the power system. This new system brings a new class of control and sensing systems, with higher sampling rate and wider range of applications. New opportunities are opened for dynamic operation and control. However, this translates into a larger control system. A system with control and monitoring devices spreading over large distances, a real-time full-state feedback control scheme that communicates measurements and commands to every controller is impractical. Therefore, the use of decentralized scheme using minimal communication, integrating computational and distributed intelligence is required.

To cope with new challenges in power systems, new strategies have been recently proposed. Among them, agent-based control is one of the most popular solutions. Nowadays, multi-agent systems (MASs) are being applied to a wide range of applications, including but not limited to, condition monitoring, system restoration, market simulation, network control and automation, demand monitoring and control [12,21]. Moreover, agent-based control schemes are now being applied in practice to facilitate the integration of DERs and the control of the emerging power system [10].

7.3.1 Agent-based control

By dividing the large control task into smaller sub-tasks, MAS-based systems aim to tackle complex problems relying on the cooperation, coordination, and negotiation of individual units, called agents. The notion of an agent has been given multiple definitions, having a common basis that an agent is an entity embedded in an environment, and capable of autonomously reacting to changes in that environment [22]. Therefore, an intelligent agent should exhibit: (a) reactivity, i.e., ability to react to changes in its environment in a timely fashion; (b) proactiveness, i.e., goal-directed behavior; and (c) the ability to interact with other intelligent agents, i.e., coordinated or competitive behavior.

A MAS is formed by two or more agents or intelligent agents, each with a local goal. The lack of a global goal is an intrinsic characteristic to MAS. Nonetheless, it is the task of the designer to make sure that by achieving the local goals, the global objective is fulfilled. For this purpose, agents can be organized in a single layer or in a hierarchical structure. In the first, agents are restricted to gather information and produce an action for a part of the system. In such systems, the social nature of agents allows them to communicate, share information, and cooperate or negotiate with other agents, in order to achieve their individual goals. In a hierarchical structure, superior agents have authority over other agents in the control task. Typically, such agents take upon more responsibility, acting as coordinators of inferior agents, providing policies, and information. Furthermore, the superior agent of a MAS-controlled area can also coordinate the operation of its area based on information exchanged with other agent-based systems. This makes such structure scalable when a superior agents are considered subordinates or peers of other agents units.

7.3.2 The SG-BEMS structure

Due to the characteristics of the demand and the SG domains, a hierarchical agent structure is preferred. In such system, two main control areas are identified: (a) the SG area and (b) the built environment area. Within these two areas, multiple control levels can be realized. Within the SG area, the transmission and distribution domains are separated. At the building area, different comfort zones levels can be established, corresponding to the structural division of a building, i.e., floors and rooms. Therefore, the SG-BEMS structure is a dual agent-based control system that addresses the interoperation of both the power grid and the building, as it is illustrated in Figure 7.5.

Although the agents are organized in a hierarchical way, each agent is autonomous to operate in their own environment. For instance, the zone agents in the MAS-BEMs platform will be the autonomous to switch off local energy systems, e.g., lights, based on occupancy or other relevant information. The advantages are that such structure offers ranges from the minimization of the communication effort by sharing only relevant information to interested agents, to the scalability of the structure.

7.3.2.1 The agent-based SG control

As mentioned, the MAS-SG encompasses the transmission and the distribution domains. It is represented by a transmission agent, responsible for monitoring and operation of the transmission system; the distribution agents, which are in charge of

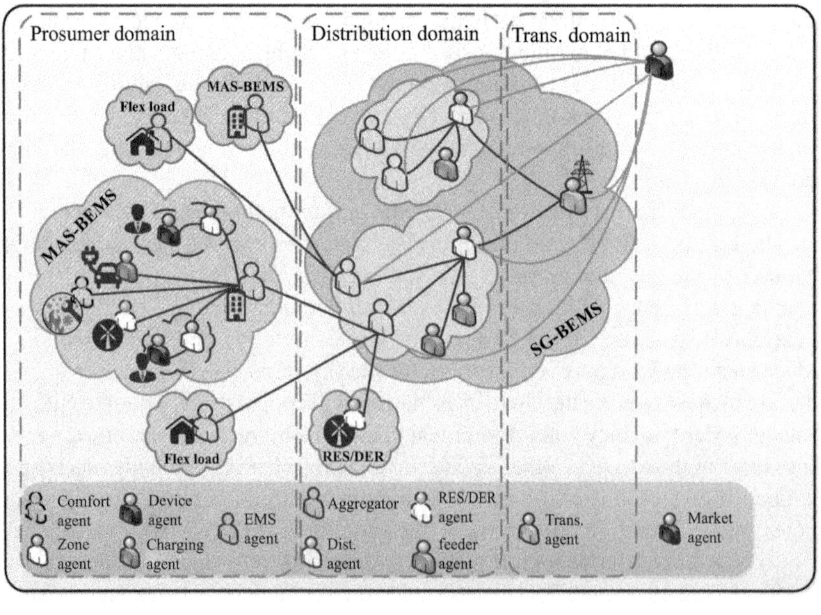

Figure 7.5 SG-BEMS agent architecture diagram

monitoring and controlling the distribution network operation; the feeder agents in charge of the operation of LV feeders and MV sub-rings; RES/DER agents responsible for the operation of the generation units connected to the distribution network; and aggregator agents responsible for procuring the demand flexibility required for a stable system operation.

7.3.2.2 The agent-based building control

The MAS-BEMS is formed by three hierarchical management levels according to the building structure:

- ***BEMS agent:*** It is located at the highest level in the BEMS structure, and takes charge of solving the optimization problem while being the link to the distribution network. This agent is able to accept and prioritize requests made by agents and operators outside the building premises, i.e., the aggregator agent. Based on the information received, it tunes up the building optimization, which defines the control variable values for the comfort systems.
- ***Zone agent:*** It is responsible for the floor operation and the rooms within the floor. The zone agent acts as an aggregator of the room information, monitoring the local zone and informing BEMS agent if the current status is within the parameters or not.
- ***Device agent:*** It is placed at local zone, i.e., room level, and takes charge of the operation of the room within the building. It is responsible for gathering information to assess the comfort levels to determine the building flexibility.
- ***Comfort agent:*** It is located at the same level with the zone agent but aims to monitor and control centralized systems, i.e., HVAC. These systems are usually comfort systems designed to operate for larger parts of the building or even for the whole building.
- ***Charging agent:*** It is located at the same level with the zone agent but aims to control the charging process of EVs.

7.4 Application of the agent-based control architecture

The SG-BEMS framework aims to enable the intertwined operation of the SES domains and the built environment. Through the use of flexible demand, the purpose of agent-based SG-BEMS structure is to ensure the correct operation of the system without significant impacts on the demand operation. This closely aligns with the concepts of DSM and DR, which have been extensively explored throughout the literature. However, a clear correlation between comfort and demand flexibility, in the context of grid support, is still lacking. In this section, an effort toward establishing this relation will be presented.

7.4.1 The models

In a CPSS as the one described in the SG-BEMS framework, the demand side becomes key in the operation, monitoring and control task of a power system, acting as active prosumers. The two main aspects of the prosumer domain are comfort management and energy consumption. However, the central objective is to provide the occupants with a comfortable environment. The strong correlation between energy and comfort is crucial to define the demand flexibility potential of the prosumer domain. The following subsections describe more in detail these two aspects of the building customer domain.

7.4.1.1 Comfort formulation

Comfort is a complex and subjective human perception, which varies according to each person and each particular environmental context. Traditionally, it is controlled by a combination of a centralized management system and human interventions, e.g., lights in local zones. Different standards have been developed to guarantee comfort levels. For instance, $ASHRAE55$ and $ISO7730$, for thermal comfort; $ISO8995-1$, for visual comfort; and $ASHRAE62.1$, for indoor air quality.

As a building consists of a high number of components differing in characteristics and operation times, the building is usually divided into multiple zones, e.g., office rooms, common areas, halls, floors, etc. Nonetheless, each zone has a particular energy demand and control variables. Here, comfort is conceptualized as a function of thermal comfort (i.e., temperature) [23] and extended to air quality (i.e., humidity).[3] Both are modeled through a Gaussian function representing the degree of satisfaction, with the average comfort value as the mean (μ) and a standard deviation (σ). This guarantees the operation of the system in a range instead of a single value, which is closer to the subjective nature of comfort perception, as expressed in the following equation:

$$comf = \underbrace{\omega\, e^{\left[\frac{-(T-\mu_T)^2}{2\sigma_T^2}\right]}}_{\text{Thermal comfort}} + \underbrace{(1-\omega)e^{\left[\frac{-(RH-\mu_{RH})^2}{2\sigma_{RH}^2}\right]}}_{\text{Air quality comfort}}, \tag{7.1}$$

where ω is a weight factor, T is the temperature, μ_T is the mean temperature, σ_T is the standard deviation for the thermal comfort, RH is the relative humidity, μ_{RH} is the mean humidity, and σ_{RH} is the standard deviation for air quality comfort.

In this model, there are two state variables: the zone temperature (T), and the zone relative humidity (Rh) and the zone CO_2 concentration (Φ_{CO_2}) levels. Moreover, the air in the zone is assumed to be fully mixed, i.e., uniform temperature distribution, with constant density, and the pressure losses in the zone and the effect of the building orientation, i.e., solar gains, are neglected. Finally, occupancy (N) and the weather, i.e., temperature (T_{env}), water concentration ($\Phi_{env}^{H_2O}$), and CO_2 concentration ($\Phi_{env}^{CO_2}$), are the uncontrolled inputs.

[3]The carbon dioxide is considered as a constraint of the optimization problem.

7.4.1.2 Building thermal dynamics

With the aforementioned assumptions, the thermal dynamics of the building represented through lumped capacity model described by the energy conservation principle:

$$\frac{dT(t)}{dt} = \frac{1}{M_{air}\ c_{p,a}}(Q_{in} + Q_{heater} - Q_{loss}) \tag{7.2}$$

where, Q_{in} represents the internal gains due to the heat generation of occupants; Q_{heater} represents the heat contribution of the heating system used; Q_{loss} is used to model the heat losses through the envelope of the zone; M_{air} is the mass of the enclosed air; and $c_{p,a}$ is the heat capacity of air.

The energy transferred to the building is proportional to the energy transferred by the heating system, as expressed in (7.3), and by heat contribution of the occupants, as expressed in (7.4).

$$Q_{heater} = \dot{v}_{s,w}\rho_w c_{p,w}(T_{s,w} - T_{r,w}) \tag{7.3}$$

$$Q_{in} = N\ Q_{gen} \tag{7.4}$$

where, $\dot{v}_{s,w}$ is the water supply flow rate; ρ_w is the density of water; $c_{p,w}$ is the heat capacity of water; $T_{s,w}$ is the temperature of the supply water; $T_{r,w}$ the temperature of the return water; N is the number of people; and Q_{gen} is the metabolic heat rate.

The energy removed from the building is the energy lost to the environment through the building envelope. These are represented through the conduction and convective heat transfer mechanisms,

$$Q_{loss} = \sum_{i=1}^{n} U_{i,in}A_i(T_i - T) \tag{7.5}$$

where, n is the number of interface elements, e.g., walls; $U_{i,in}$ is the heat transfer coefficient of wall i internal surface; A_i is the area of the wall i; and their temperature, T_i, is given by:

$$\frac{dT_i(t)}{dt} = \frac{U_{i,in}A_i(T - T_i) + U_{i,out}A_i(T_i - T_{env})}{M_i\ c_{p,i}} \tag{7.6}$$

where, T_{env} is the outdoor air temperature; $U_{i,out}$ is the heat transfer coefficient of wall i outer surface; M_{air} is the mass of the wall i; and $c_{p,a}$ is the heat capacity of the wall material.

7.4.1.3 Building air quality dynamics

Air is a mixture of multiple elements in different concentrations. Indoor air quality is traditionally measured through the water content, i.e., relative humidity, and CO_2 concentration dynamics in the air. These dynamics can be represented through the mass and component balances in the air volume. The concentration change in time of an element k is proportional to the particles of that element added and extracted from the volume, as expressed in the following equation:

$$\frac{d\Phi_k}{dt} = \frac{1}{V_z}\left(\dot{v}_s\Phi_s^k - \dot{v}_r\Phi_k + N\Phi_{gen,k}\right) \tag{7.7}$$

where, V_z is the enclosed air volume; $\dot{v}_s$ is the air supply flow rate; $\dot{v}_r$ is the air removal flow rate; $\Phi_{gen,k}$ is the metabolic generation rate of k; and the concentration of element k in the supplied air, Φ_s^k, is given by:

$$\Phi_s^k = \frac{1}{\dot{v}_s}\left(\dot{v}_r\Phi_k + \dot{v}_{in}\Phi_{env}^k - \dot{v}_{out}\Phi_k + \Phi_{humid}^k\right) \tag{7.8}$$

where, Φ_{env}^k is the concentration of k in the outdoor air; $\dot{v}_{in}$ is the outdoor air supply flow rate; $\dot{v}_{out}$ is the exhaust air flow rate; and Φ_{humid}^k is the humidifier mass removal rate of element k, with $\Phi_{humid}^k = 0, \ \forall\ k = CO_2$.

Finally, using the Ideal gas law, relative humidity and the CO_2 concentration can be rewritten as follows:

$$Rh(t) = 100\frac{\Phi_{H_2O}}{\Phi_{H_2O}^{sat}} \tag{7.9}$$

$$[ppm]\Phi_{CO_2} = 1000\frac{\Phi_{CO_2}R\,T_z}{M_{CO_2}P_z} \tag{7.10}$$

where, Φ_{H_2O} is the concentration of water particles in the air; Φ_{CO_2} is the concentration of CO_2 particles; R is the universal gas constant; T_Z is the air temperature; M_{CO_2} is the molecular weight of CO_2; P_z is the air pressure in the building; and the saturated concentration of water is given by Antoine's equation:

$$Log_{10}(\Phi_{H_2O}^{sat}) = 8.07131 - \frac{1730.63}{T - 39.73} \tag{7.11}$$

7.4.1.4 Building energy demand formulation

Here, energy consumption systems in the building are categorized into comfort and non-comfort systems. The power demand of the first type corresponds to the consumption of the comfort systems including the heat pump for heating purposes, P_{heater}, and the HVAC for air quality and supplementary heat, P_{HVAC}. Whereas power demand of a non-comfort system corresponds to individual systems in local zones, P_i, e.g., lights and computers, and to the demand of the EVs connected to the building's POC.

The total consumption of a building is expressed in the following equation:

$$P_{total} = \underbrace{P_{HVAC} + P_{heater}}_{\text{comfort}} + \underbrace{\sum_{i=1}^{N} P_i + \sum_{j=1}^{M} P_j^{EV}}_{\text{non-comfort}} \tag{7.12}$$

The power consumed by a typical HVAC system is a function of the supplied air temperature and flow rate. In turn, air temperature and flow rates are functions of the individual systems that form the HVAC system as follows:

$$P_{HVAC} = P_{fan,s} + P_{fan,r} + P_{hcoil} + P_{ccoil} \tag{7.13}$$

where, $P_{fan,s}$ and $P_{fan,r}$ are the power consumed by the supply and return fans, which are proportional to $\dot{v}_s$ and $\dot{v}_r$ [24]. The power consumed by the heating coil, i.e., P_{hcoil}, and the power consumed by the cooling coil, i.e., P_{ccoil}, are a function of the air flow rates (supply and return), the difference in the indoor and outdoor temperatures, and their respective efficiencies.

The power demand of the heat pump is a function the required heat power and the coefficient of performance, *COP*, of the machine. The required heat is the energy used to compensate for the thermal losses, and it is proportional to the volumetric flow of water through the system, $\dot{v}_{s,h}$, the temperature of the supplied water, $T_{s,w}$, and the zone temperature, T_z. Finally, the *COP* describes the ratio between the useful heat produced and the work input.

$$P_{heater} = \frac{\dot{v}_{s,h}\rho_w c_{p,w}(T_{s,w} - T_z)}{COP} \tag{7.14}$$

where, ρ_w is the water density and $c_{p,w}$ is the heat capacity of water.

7.4.2 Optimization problem

A CPSS involves multiple objectives which might be conflicting, i.e., comfort maximization and energy minimization, and a consideration to enable grid support services. These objectives are handled in two different levels in the SG-BEMS structure: (a) Network-level optimization, e.g., congestion and voltage management and (b) customer-level optimization, i.e., minimization of comfort loss.

At the first level, the task of the aggregator is to collect enough flexibility from the customers in the network to solve the network issues described earlier. This translates into an optimization problem for the aggregator role, i.e., the minimization between the demand and offer of flexibility.

Minimize

$$\Delta\ Flex = Flex_d - \sum_{j=1}^{B} Flex_{s,j} \tag{7.15}$$

where $Flex_d$ is the flexibility demand, e.g., amount of power to solve a congestion or voltage issue; $Flex_{s,j}$ is the flexibility offer of building j, i.e., amount of power the customer can shift or shed; and B is the number of customers in the network.

At the second level, the problem of the BEMS agent involves a multi-objective optimization problem. The first objective relates to the problem described by (7.15), with $comf \in [0\ ,\ 1]$. Thus, the maximization of comfort can be rewritten as the minimization of discomfort. The second objective is the minimization of the energy consumed in the building. In this work, we limit the energy optimization problem to the minimization of the energy consumed by the comfort systems. These two objective functions are represented by the following equations:

$$f_1(x) = discomf = 1 - comf \tag{7.16}$$

$$f_2(x) = P_{total} \tag{7.17}$$

where x is a solution vector formed by the thermal and air quality comfort control parameters, i.e., $\dot{v}_s$, $\dot{v}_{s,h}$, and $\dot{v}_r$.

In order to support the grid, i.e., flexibility offer, without jeopardizing the comfort levels, the BEMS agent of the CPSS needs to handle both optimization problems in (7.16) and (7.17) simultaneously under optimal conditions for the distribution network. Therefore, the optimization problem can be rewritten as follows:

Minimize

$$f(x) = (f_1(x), f_2(x)) \tag{7.18}$$

Subject to

$$\Phi_{CO_2}(t) \leq \Phi_{max} \tag{7.19}$$

$$comf \geq comf_{min} \tag{7.20}$$

$$\int_{toa}^{tod} SoC_j(t)dt \geq SoC^j_{min}; \ \forall j \tag{7.21}$$

where Φ_{CO_2} is the CO_2 level at instant t, Φ_{max} is the maximum CO_2 concentration level allowable, and $comf_{min}$ is the minimum comfort level acceptable. The first two constraints are used to not allow the comfort satisfaction fall out of the defined comfort ranges at all times. The last constraint ensures that the EVs present in the building are supplied at least with the required energy, SoC^j_{min}. Where the state of charge SoC of EV j is proportional to power demand of the EV's battery $P(t)^{EV}_j$, the voltage of the battery V^{EV}_j, and the battery's capacity Q_0, as described in the following equation:

$$SoC_j(t) = \frac{1}{Q_0} \int_{toa}^{t} \frac{P(t)^{EV}_j}{V^{EV}_j} dt \tag{7.22}$$

As the optimization objectives (7.16) and (7.17) are two different functions, it is a challenge for the BEMS to find a single optimal solution for the optimization problem in (7.18). Pareto-optimal solutions must be found to represent the best trade-off and/or the best compromise by using different approaches [25]. Here, the weighted aggregation method is used to aggregate all the objectives of the problem into a single one through a weighted combination. Thus, (7.18) can be rewritten as:

$$Wf_1(x) + (1 - W)\sigma f_2(x), \tag{7.23}$$

where W is a non-negative weight and σ is a normalization factor that allows the two objectives to be treated equally. The main advantage of this method is that it allows

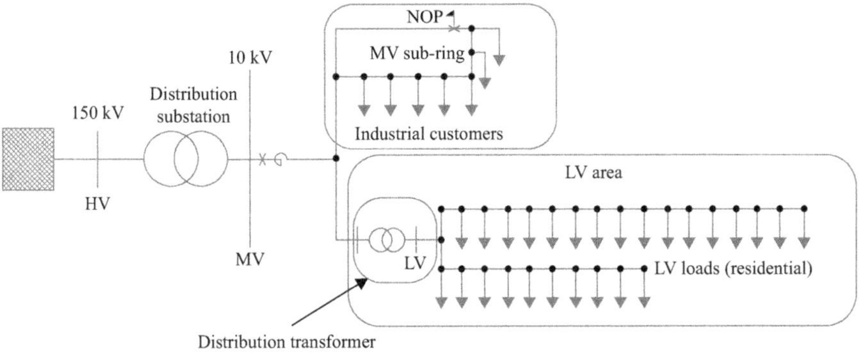

Figure 7.6 Single-phase schematic of the distribution grid

using a single objective algorithm and turning the weight W to adjust dynamically the important role of either $f_1(x)$ or $f_2(x)$. However, it requires the algorithm to be applied repeatedly to find the desirable number of non-dominated solutions.

7.4.3 Simulation-based case study

This section describes a simulation setup, designed and implemented to demonstrate the performance of the SG-BEMS agent platform, and shows the potential of demand flexibility for grid support applications. The test system consists of two parts, a Matlab/Simulink MV-LV distribution grid serving as the physical environment; while the agents are developed in the middleware environment of Java/JADE. Figure 7.6 shows the single-line diagram of the distribution grid. It represents a three-phase balanced system formed by seven industrial loads in a single MV sub-ring, and 27 residential loads in a single LV feeder (two branches). There are two voltage transforming steps, a HV-MV (150–10 kV) step in a distribution substation, and a MV-LV (10–0.4 kV) in a distribution transformer. The MV loads are connected through self-owned step down transformers. PV systems are randomly added to the LV and MV loads, with a penetration of 40% and 50%, and with a capacity of 3 kWp and 15 kWp, respectively.

The LV loads are modeled using a static profile obtained from measured data of typical Dutch LV feeder [16]. At this level, the comfort level is only determined by temperature. To do so, heating systems are placed in each LV load model to represent the relation between weather variations, comfort, and energy. At the MV sub-ring, six loads are modeled as a single comfort zone. The energy demand of these loads is formed by the action of the comfort systems, as described earlier, and a base load which is constant throughout the time the building is occupied. The sizes and base loads of these six buildings are varied to represent different types of buildings. Finally, the remaining load is modeled after a typical Dutch two-floor office building, with 24 comfort zones and a parking load of 10 EVs.

As mentioned, the agent-based control system is implemented in JADE, a platform that provides a java environment for the development of agents. A TCP/IP communication is established between the two software platforms, with

Matlab/Simulink as the server client. The prosumer domain is formed by 34 BEMS agents, 33 of these are single layer structure, whereas the MAS structure of the two-floor building, is a hierarchical one, with 159 agents organized in three levels. The distribution domain is formed by four agents, including the LV and MV aggregators, the distribution agent and the feeder agent. The system response is simulated in seconds, whereas the MAS-based SG-BEMS interaction, i.e., observation and feedback, is done in 15-min intervals.[4] The environment parameters of each load, i.e., comfort levels, active and reactive power, POC loading and voltage, and those of the MV and LV networks, i.e., network loading, losses and power exchange, are exchanged between the Matlab/Simulink model and the JADE agents. Between the JADE agents, multiple ACL (Agent Communication Language) messages are exchanged. These messages contain the necessary information for the agents to take the adequate control decisions, and for the BEMS agents to define flexibility while taking into account the effect on comfort. In total there are 196 agents interacting every 900 s in this application of the SG-BEMS framework. Finally, at MV loads, a particle swarm optimization is ran within the BEMS agent to determine the optimal operation parameters of the comfort systems, i.e., $\dot{v}_s$, $\dot{v}_{s,h}$ and $\dot{v}_r$, in response to the weather and flexibility request variations. The resulting parameters give the optimal relation between energy demand and comfort for the next time step. This is explained in detail in [2,12]. At the LV loads, on-off controllers are used to operate the heating systems in response to weather and flexibility request variations. Moreover, occupancy is a key parameter to define the relevance of comfort in the operation of the MV and LV loads. During the time a building is occupied, the comfort is considered a hard constraint of the system and thus tight boundaries are established. While during the time the building is unoccupied, these boundaries are relaxed, leaving more space for a more flexible operation. Different occupancy profiles were used for each building. Each load has an average number of occupants with a random variation in time, as expressed in the following equation:

$$N = \begin{cases} \mathscr{N}(N_{av}, 1) & \text{if } t_{in} < t < t_{out} \\ 0 & \text{if } t \leq t_{in} \vee t \geq t_{out} \end{cases} \tag{7.24}$$

where, $\mathscr{N}$ is the number of people, $\mathscr{N}_{av}$ is the average number of occupants, *rand* represents a random variation in time, t_{in} the arrival time, and t_{out} the leaving time, these two randomly selected between different ranges of time. Finally, the energy demand is also weather dependent. The models use measured weather data of a typical winter day in the Netherlands. Table 7.1 shows the cable data used for the simulations.

In addition, a graphical user interface is used to display the information and details of the two control domains. These are shown in Figure 7.7. These interfaces are flexible enough to adapt to the structural difference presented in all the domains, with a minimum modification.

[4]This interval corresponds to the program time unit (PTU) period adopted by the Dutch transmission system operator TenneT, for scheduling and settlement of the electricity market participants. However, it can be extended to different time horizons.

Table 7.1 Cable information

Sections	Length (m)	*R* (Ω/km)	*L* (mH/km)	Description
1	1140	0.07	0.246	MV bus cable
7	287 (avg)	0.532	0.311	MV-sub-ring cable
27	10 (avg)	0.69	0.47	LV cable

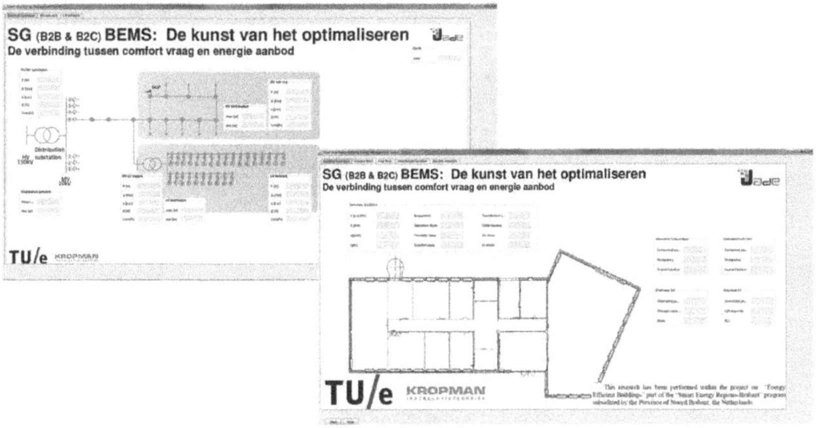

Figure 7.7 SG-BEMS user interface

7.4.4 Simulation results

The proposed dual agent-based platform tries to offer grid support services, i.e., capacity and voltage management, by influencing the active power demand of the prosumers. However, depending on the building comfort status, the MAS-BEMS has the autonomy to choose whether it modifies its behavior or not, i.e., flexibility. If so, the flexibility offer is found based on how far the comfort level is from the established boundaries. In this section some results illustrating the performance of the agent-based SG-BESM platform, and the effect of flexible demand in grid support within the SG-BEMS framework are presented.

7.4.4.1 Congestion management

In this simulation congestion problems, i.e., overloading, are examined at two levels of the system. These correspond to the entire network and the MV sub-ring, i.e., industrial loads. For the sake of computation time, the LV network is not monitored and therefore not used for congestion management. Three strategies are used for the aggregator's flexibility request. Under a discrete strategy, the aggregator sends a request at the moment the network or MV sub-ring is overloaded, after the situation is restored, the aggregator releases the prosumers, and allows them to return to normal operation. As a result, if the overloading lasts longer than a couple of PTUs, the

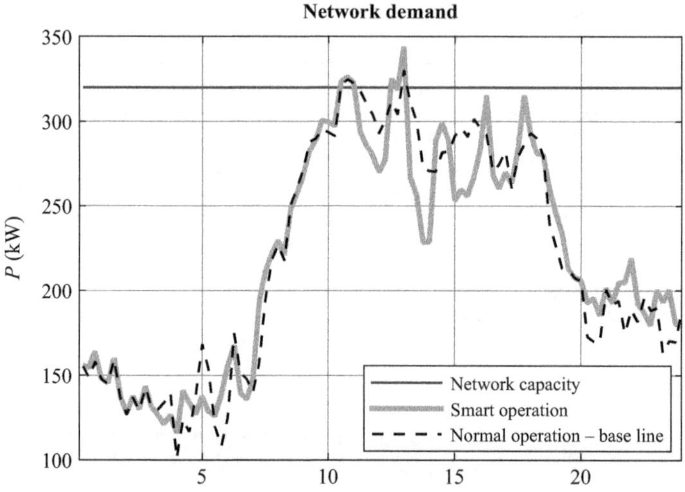

(a)

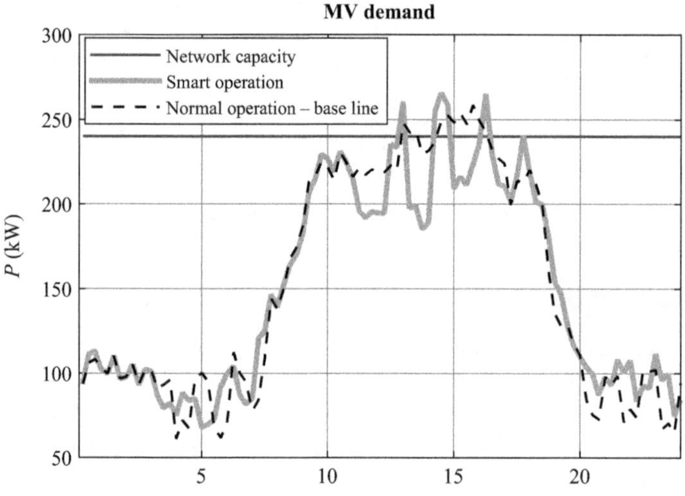

(b)

Figure 7.8 Congestion management under discrete flexibility request. (a) Network active loading. (b) MV sub-ring loading

capacity of the network is violated again, creating a switching like behavior. This is illustrated in Figure 7.8. All the results are compared against a reference, i.e., base line case. This represents the situation in which no request is created, independent of the network situation.

The second strategy follows a decreasing flexibility request. This means that once an overloading issue is detected (at any level), the aggregator sends a request similar to the one of the previous strategy. However, after the situation is resolved,

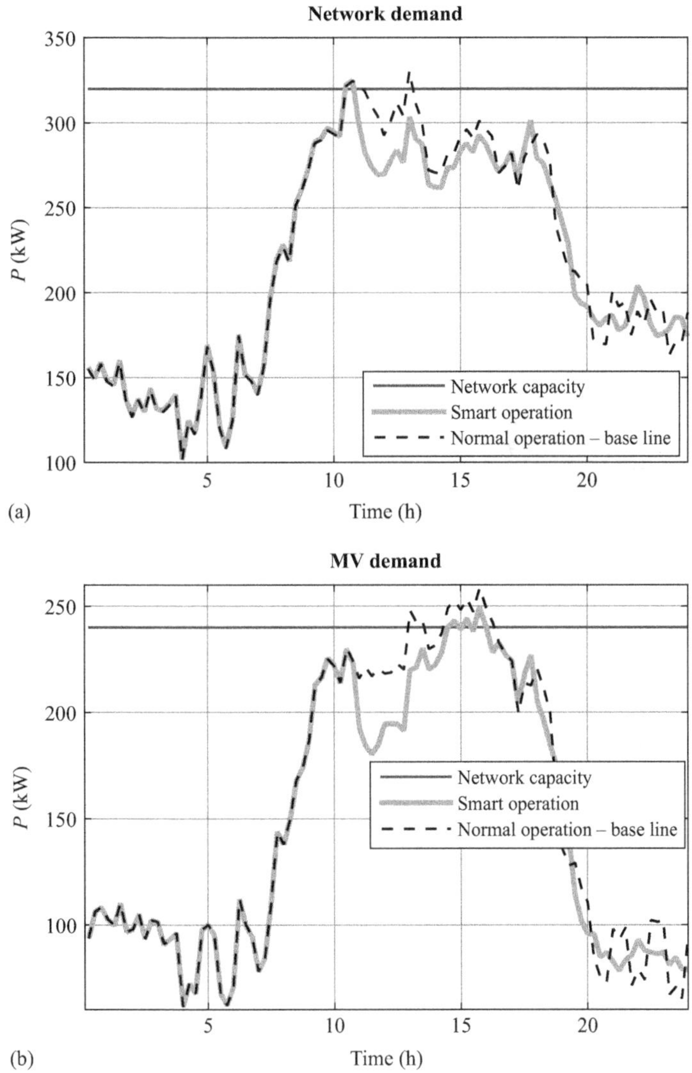

Figure 7.9 Congestion management under decreasing flexibility request. (a) Network active loading. (b) MV sub-ring loading

the aggregator sends a decreased request. This continues until the request reaches the zero crossing. This is illustrated in Figure 7.9.

Finally, the last request strategy is a mixture of the two previous strategies, i.e., a bounded strategy. This means that after a overloading situation is detected, a flexibility request is created for each MV load, and after the normal situation is restored, a decrementing request is sent until it has decreased to 70% of its starting value. This

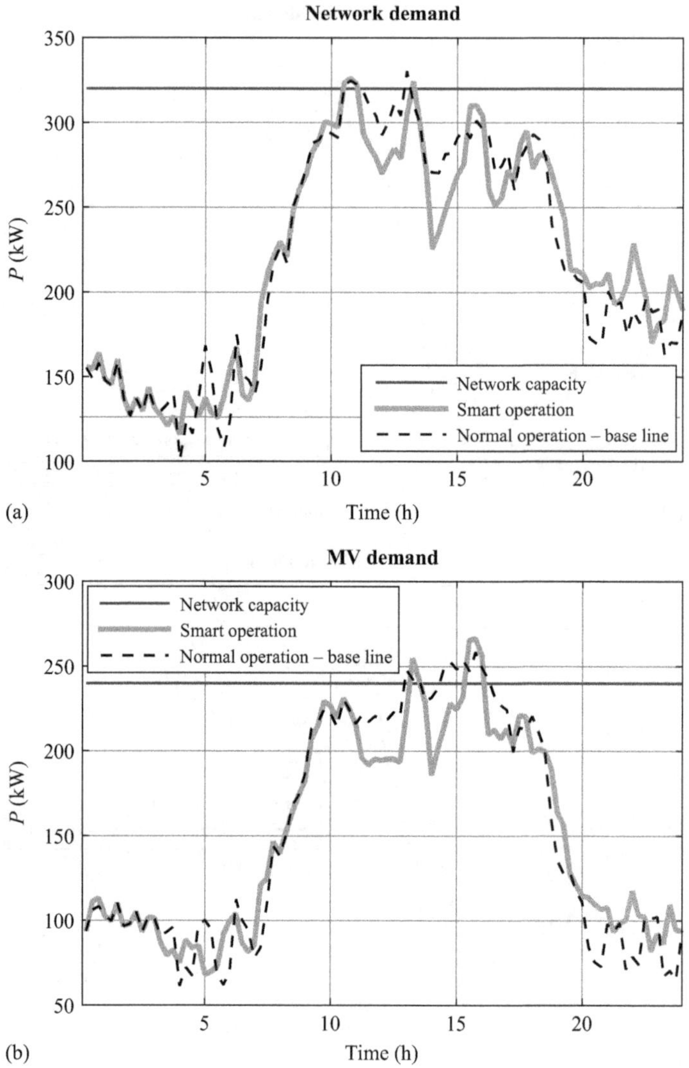

Figure 7.10 Congestion management under bounded flexibility request. (a) Network active loading. (b) MV sub-ring loading

means that a higher request will last longer than a lower one, which tries to compensate the switching behavior showed by the first strategies. This is illustrated in Figure 7.10.

Every flexibility value or offer is determined by the BEMS agent based on the load's share of the total overload, i.e., amount of power above the loading capacity. As the share increases so does the request value, which means that the impact of such load will be greater, with *Flex* $\in \{0 - 40\%\}$. The request is defined by the aggregator based on the ratio between the total demand and the rated capacity of the network,

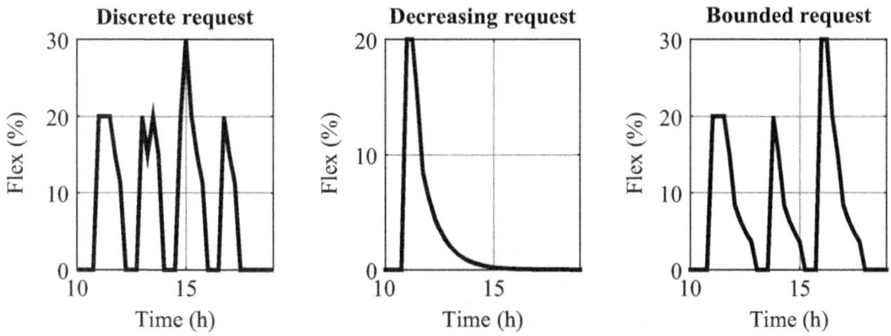

Figure 7.11 Flexibility request for one load

e.g., transformer or cable capacity. A situation is considered as overloading if the rated capacity of the system is surpassed and sustained for a certain amount of time. Figures 7.8–7.10, show, additionally, the rated capacity defined for the distribution network, i.e., 0.32 MW, and the MV sub-ring, i.e., 0.24 MW. In these figures it can be seen that the distribution network and the MV sub-rings are both overloaded at different moments in time. This corresponds to different requests in time, depending on the strategy used. Figure 7.11 shows the request for the three strategies described before. It can be seen how the discrete strategy results in a higher number of requests. As discussed earlier, if the overloading lasts longer than a couple of PTUs, a switching like behavior is observed in the resulted network load profile. This can be also seen in the bounded request strategy, although, with a lesser degree. Such issue can be addressed through the use of forecasting agents to determine the duration of the overloading issue, and create the flexibility request accordingly, i.e., promotiveness.

7.4.4.2 Voltage management

Voltage variations are investigated in the LV network only (Figure 7.6), as the voltage at the MV sub-ring is rather stable due to the HV-MV transforming station. The PV penetration in the LV network is increased to 100%, while the simulated day is one during the summer. Following the SG-BEMS framework, the agent platform, through the BEMS agents, monitors the voltage along the feeder and decides on the appropriate actions based on both, the parameters of the network, i.e., LOC, and those of the loads, e.g., comfort. As voltage violation is deemed to be a local issue, i.e., loads closer to the affected point have a bigger impact on the voltage profile of the affected point, each load reacts to measurements made at the POC. Three voltage areas are defined: (a) green area (normal operation) $\Delta\, u < |3|\% \rightarrow u = 1.0 \pm 3\%$ [*p.u.*]; (b) yellow area (stressed operation) $|3|\% \leq \Delta\, u < |5|\%$; and (c) red area (emergency) $\Delta\, u \geq |5|\%$. At the green area, the loads are operated in response to the local optimization targets and aggregator requests. As the load enters the yellow area, flexibility starts being dispatched. If the situation is not solved, the system might enter the red are, at which curtailment starts, e.g., PV production curtailment. Figure 7.12 shows the voltage profile of the upper branch of the LV feeder, under no control. As it can be seen, under the aforementioned assumptions and conditions, the LV network exhibits

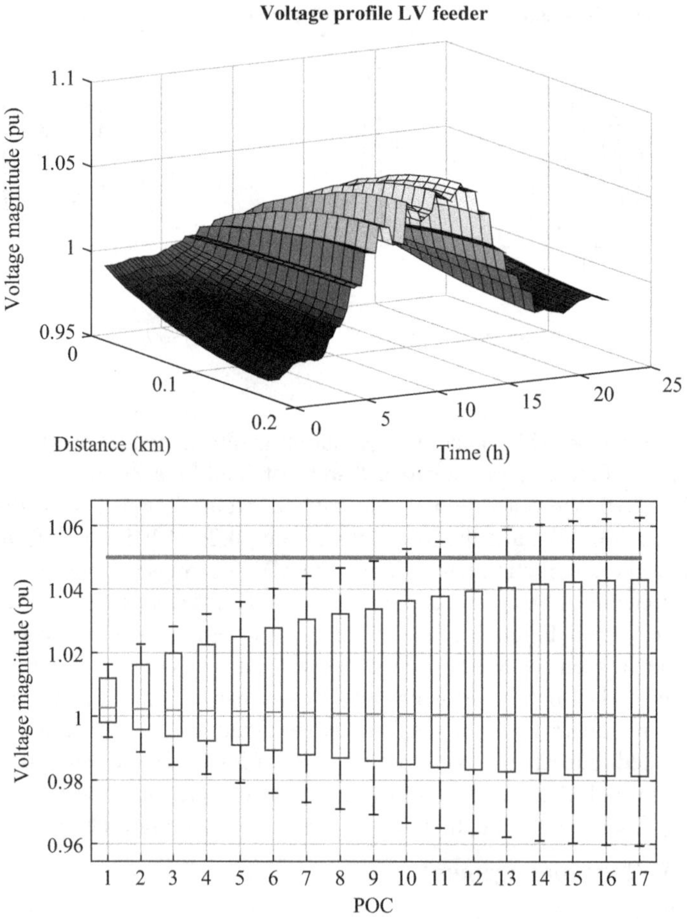

Figure 7.12 LV voltage feeder profile, no control

not only over voltages due to the high energy production from the PV systems during the day time, but also low voltages levels. The figure also shows the voltage distribution for each POC. It can be seen that the connection points at the end of the feeder show a bigger distribution of the voltage, in ranges close to the limits, while violations start occurring after the tenth connection.

Under the SG-BEMS framework, two different strategies are used. The first one involves no cooperation among the BEMS agents. Here cooperation is assumed to mean data exchange between the agents. Figure 7.13 shows the obtained voltage profile under this strategy. It can be observed that under this strategy the profile of the LV branch is considerably improved, reducing the number of violations. However, it can also be observed that under this strategy the voltage profile exhibits a switching behavior during the PV peak production time. This is due to the disconnection of some

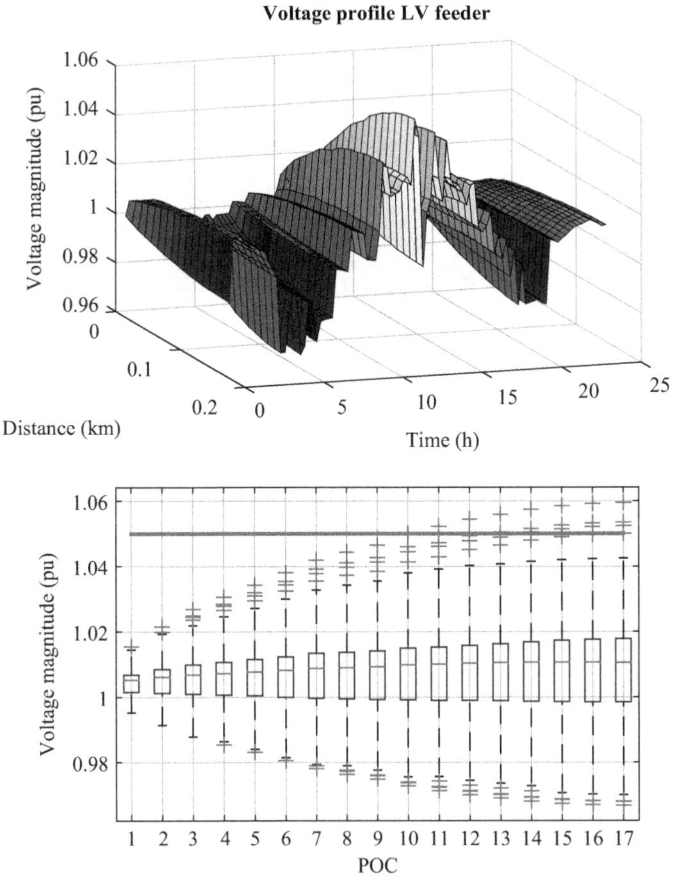

Figure 7.13 Agent control: no cooperation

units to avoid over voltage. The figure also shows the voltage distribution over the connection points. Here it can be seen how the variation range is reduced compared to uncontrolled case. However, the switching behavior results in outliers, i.e., grey crosses on the figure, outside the voltage limit.

The disconnection process of the previous strategy is uncoordinated, and therefore it results in the disconnection of more units than required. Thus, the strategies aims to solve this issue. By the coordinated operation of the BEMS agents during voltage violation moments, i.e., exchange of information, only the required power to avoid the over voltage is disconnected. Figure 7.14 shows the resulting profile under this strategy. It can be seen that the result is a more steady voltage. This can also be observed looking at the voltage distribution also shown in the figure. It can be seen how the number of outliers were reduced to zero. This indicates a more steady voltage behavior. A violation still remains, but this is due to the fact that strategies are reactive rather than proactive.

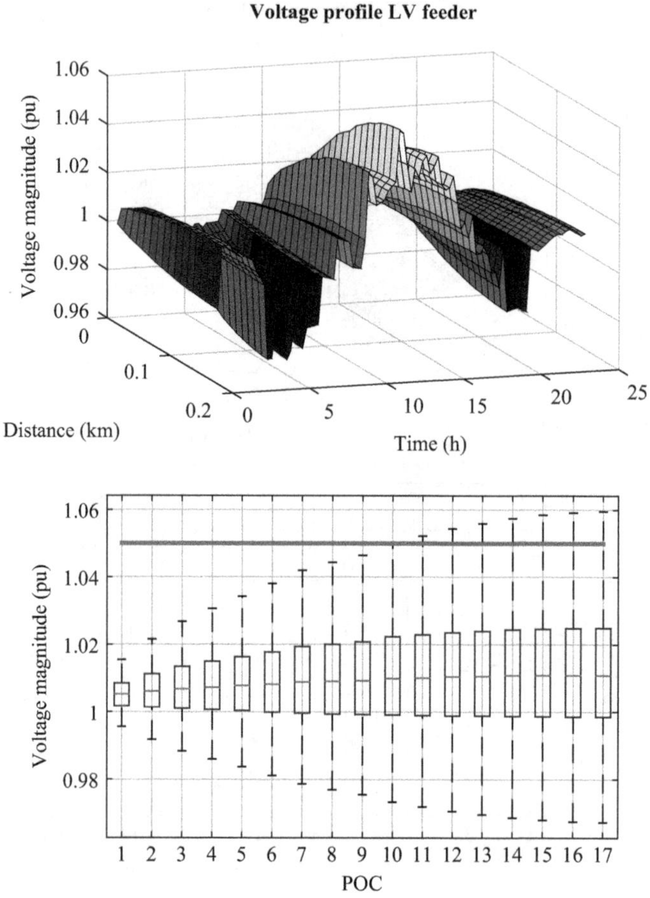

Figure 7.14 Agent control: supervised cooperation

Figure 7.15 shows the PV power and energy production of the LV feeder for the uncontrolled case and the two different operation strategies. This corresponds to the power generation for a day of July 2013 of 27 PV systems of 12 kW. It can be observed that the disconnection process starts after the peak production has been reached and it last for the rest of the afternoon. This is the result of a low demand and full batteries. During the first part of the day, the PV production is used to supply the needs of the customer and charge the batteries. However, as the batteries reach their maximum charge, they cannot longer be used as a buffer for voltage regulation, and disconnection starts (Orange regime). Furthermore, it can be seen that despite the fact that both strategies result in the disconnection of PV units, the supervised cooperation strategies reduces the amount of disconnected power significantly, without violating the voltage limits in the LV branch.

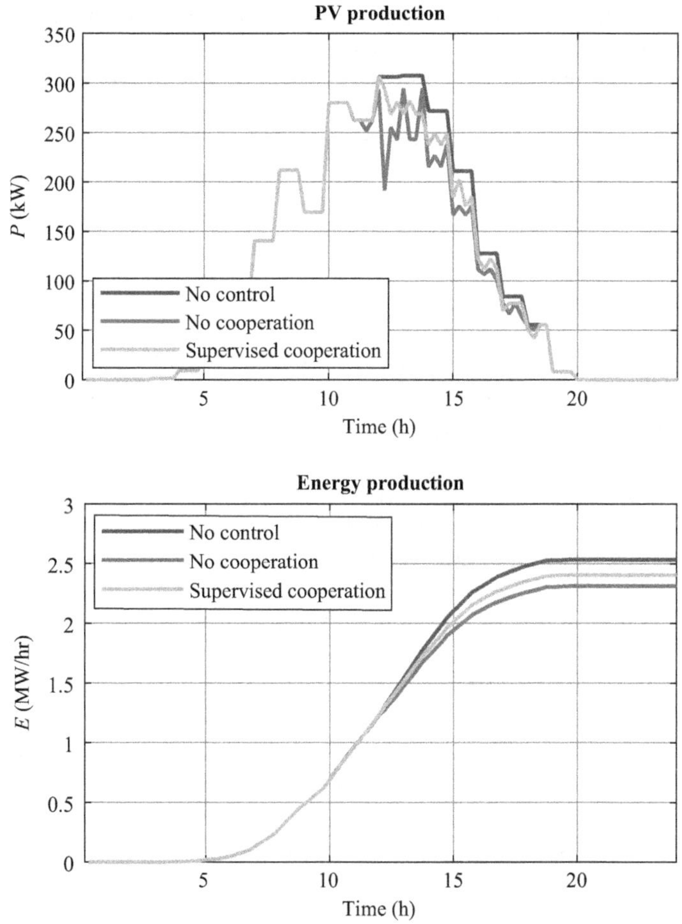

Figure 7.15 LV feeder PV production

Throughout the chapter, demand flexibility has been highlighted as a suitable option to assist the integration process of renewable and distributed generation, and to ensure the reliable and economic operation of the power system. From the presented results, it is clear that an interoperation framework that enables the smart behavior of prosumers in benefit of not only local goals but also the grid operation, requires an adequate flexibility request signal and an appropriate agent decision making strategy. However, it is clear that the SG-BEMS platform and the SG-BEMS framework have the potential to invoke the demand flexibility of the built environment and effectively offer grid support, helping the integration of RES/DER technologies into the power system. Furthermore, in the emerging CPSS multiple and new stakeholders, that were not considered in the traditional operation of the power system, have to cooperate and coordinate their operation in order to ensure the sustainability, reliability, and efficiency of the power system operation. In this emerging system, not only the technical but also the social and economic aspect of the power system and its stakeholders operation needs to be taken into account, new markets and business models need to be developed and implemented. Finally, as prosumers vary in interests and characteristics, both technical and social, demand flexibility should be considered as a part of a more general approach, limited by the socio-technical characteristics of both the power system and the prosumers.

References

[1] J. G. Slootweg, C. E. P. J. Córdova, C. M. Portela, and J. Morren, "Smart Grids – Intelligence for Sustainable Electrical Power Systems," in *Telecommunications Energy Conference (INTELEC), 2011 IEEE 33rd International*, 2011, pp. 1–8.

[2] L. A. Hurtado, E. Mocanu, P. H. Nguyen, M. Gibescu, and W. L. Kling, "Comfort-Constrained Demand Flexibility Management for Building Aggregations Using a Decentralized Approach," in *4th International Conference on Smart Cities and Green ICT Systems SMARTGREENS*, 2015, pp. 157–166.

[3] I. Lampropoulos, W. L. Kling, P. F. Ribeiro, and J. van den Berg, "History of Demand Side Management and Classification of Demand Response Control Schemes," in *PES General Meeting*, 2013, pp. 1–5.

[4] G. Strbac, "Demand Side Management: Benefits and Challenges," *Energy Policy*, December 2008, vol. 36, no. 12, pp. 4419–4426. [Online]. Available: http://linkinghub.elsevier.com/retrieve/pii/S0301421508004606

[5] X. He, I. Azevedo, and L. Meeus, "Shift, Not Drift: Towards Active Demand Response and Beyond," THINK project: Final report, 2013.

[6] A. Ulbig and G. Andersson, "On Operational Flexibility in Power Systems," in *IEEE Power and Energy Society General Meeting*. San Diego, CA: IEEE, 2012, pp. 1–8.

[7] TenneT, "Imbalance Management TenneT Analysis Report," Tech. Rep., 2011.

[8] Entso-E, "Policy 1: Load-Frequency Control and Performance," European Network of Transmission System Operators for Electricity (ENTSO-E), Tech. Rep., 2009.

[9] CENELEC, "SG-CG/M490/L Flexibility Management Overview of the Main Concepts of Flexibility Management," CENELEC-Smart Grid Coordination Group, Tech. Rep., 2014.

[10] A. Catalin, F. Covrig, M. Ardelean, *et al.*, "Smart Grid Projects Outlook 2014", European Commission: Joint Research Centre-Institute for Energy and Transport, Report EUR 26651 EN, 978-92-79-38374-8, 2014, Luxembourg.

[11] A. Strassheim, J. E. S. De Haan, M. Gibescu, and W. L. Kling, "Provision of Frequency Restoration Reserves by Possible Energy Storage Systems in Germany and the Netherlands," in *International Conference on the European Energy Market, EEM*, 2014, pp. 1–5.

[12] L. A. Hurtado, P. H. Nguyen, and W. L. Kling, "Smart grid and smart building inter-operation using agent-based particle swarm optimization," *Sustainable Energy, Grids and Networks*, 2015, vol. 2, pp. 32–40.
[13] CEN-CENELEC-ETSI Smart Grid Coordination Group, "Smart Grid Reference Architecture," Tech. Rep. November 2012.
[14] P. Zhao, S. Suryanarayanan, and M. G. Simões, "An Energy Management System for Building Structures Using a Multi-Agent Decision-Making Control Methodology," *IEEE Transactions on Industry Applications*, 2013, vol. 49, no. 1, pp. 322–330.
[15] K. Park, Y. Kim, S. Kim, K. Kim, W. Lee, and H. Park, "Building Energy Management System based on Smart Grid," in *2011 IEEE 33rd International Telecommunications Energy Conference (INTELEC)*. IEEE, October 2011, pp. 1–4.
[16] L. A. Hurtado, P. H. Nguyen, and W. L. Kling, "Multiple Objective Particle Swarm Optimization Approach to Enable Smart Buildings-Smart Grids," in *18th Power Systems Computation Conference, PSCC*, 2014, pp. 1–8.
[17] P. Morales-Valdés, A. Flores-Tlacuahuac, and V. M. Zavala, "Analyzing the Effects of Comfort Relaxation on Energy Demand Flexibility of Buildings: A Multiobjective Optimization Approach," *Energy and Buildings*, December 2014, vol. 85, pp. 416–426.
[18] D. S. Kirschen, A. Rosso, J. Ma, and L. F. Ochoa, "Flexibility from the Demand Side," in *IEEE Power and Energy Society General Meeting*. San Diego, CA: IEEE, 2012, pp. 1–6.
[19] A. Backers, F. Bliek, M. Broekmans, *et al.*, "An Introduction to the Universal Smart Energy Framework," Tech. Rep., 2014.
[20] CEN-CENELEC-ETSI Smart Grid Coordination Group, "Smart Grid Use Case Management Process," Tech. Rep. November 2012.
[21] S. D. J. Mcarthur, E. M. Davidson, V. M. Catterson, *et al.*, "Multi-Agent Systems for Power Engineering Applications – Part I : Concepts , Approaches, and Technical Challenges," *IEEE Transactions on Power Systems*, 2007, vol. 22, no. 4, pp. 1743–1752.
[22] M. Wooldridge and N. R. Jennings, "Intelligent Agents: Theory and Practice." *The Knowledge Engineering Review*, January 1995, vol. 10, no.2, pp. 115–152.
[23] R. Yang and L. Wang, "Development of Multi-Agent System for Building Energy and Comfort Management Based on Occupant Behaviors," *Energy and Buildings*, October 2013, vol. 56, pp. 1–7.
[24] Y. Yao, Z. Lian, Z. Hou, and X. Zhou, "Optimal Operation of a Large Cooling System Based on an Empirical Model," *Applied Thermal Engineering*, November 2004, vol. 24, no. 16, pp. 2303–2321.
[25] M. Reyes-Sierra and C. a. Coello Coello, "Multi-Objective Particle Swarm Optimizers: A Survey of the State-of-the-Art," *International Journal of Computational Intelligence Research*, 2006, vol. 2, no. 3, pp. 287–308.

Chapter 8

Grid integration of electric and hybrid electric vehicles in cyber-physical-social systems

Wencong Su[1], *Rui Ma*[1], *and Shengyao Xu*[1]

Abstract

In this chapter, we will explore the challenges and opportunities of grid integration of electric and hybrid electric vehicles in cyber-physical-social systems (CPSS). Transportation systems that move people, goods, and services in societies worldwide pose unprecedented environmental, economic, and social challenges, particularly with the growing urgency to conserve energy, cut back on carbon emissions and pollution, avoid crashes, and relieve congestion. Advances in intelligent transportation systems and Smart Grid offer great promise to address these challenges and have the potential to revolutionize future transportation systems. In the last decade, the government around the world has spurred efforts to boost the utilization of transportation electrification technologies because of their low-pollution emissions, energy independence, and high fuel economy. An ever-increasing number of electric and hybrid electric vehicles will radically change the traditional view of the power industry, transportation industry, social environment, and business world. Research on grid integration of electric and hybrid electric vehicles typically addresses topics at the vehicle-grid boundary, such as peak load impacts and agent-based charging control. While researchers around the world are making significant advances in these areas, there is very little work addressing the coupled cyber-physical-social effects of electric and hybrid electric vehicle charging with the mobility-focused, transportation ecosystem to meet the dynamic needs of a changing society. It is important to recognize that today's critical infrastructure is an interdependent network of networks. A single network consists of millions of subnetworks and individual agents. The "tie point" is critical to its reliability, cost-effectiveness, and resiliency. As such "tie points," the emerging deployment of electric and hybrid electric vehicle charging facilities would complicate the understanding and design of interdependent critical infrastructure systems. For example, as a transportation tool and electricity carrier, electric and hybrid electric vehicle can be charged at any charging facility and at any time, which

[1]Department of Electrical and Computer Engineering, University of Michigan, Dearborn, MI, USA

brings more spatial and temporal uncertainty to the power grid's load forecast. The retail electricity price and parking fee may also have an impact on customer behavior, eventually leading to a change in traffic flow. Besides engineering considerations, the placement of electric and hybrid electric vehicle charging stations are constrained by applicable local policy and regulation, financial incentives, and public interests. We will discuss the state of the art of grid integration of electric and hybrid electric vehicles in a CPSS environment. Moreover, we will present a future perspective to enable the dramatic increase of electrified vehicles, and ultimately lead to (1) reduced fossil fuel consumption; (2) reduced carbon emissions and pollution; (3) increased customer satisfaction; (4) increased reliability and efficiency for moving people and goods; (5) improved efficiency of intelligent transportation systems; (6) accelerated adoption of Smart Grid technologies; and (7) increased use of infrastructure capacity.

8.1 Introduction

The concept of Cyber-Physical System (CPS) has drawn the increasing attentions in the last decade. Generally speaking, CPS is a new generation of engineered system that is built from, and depends on, the seamless integration of cyber infrastructure (e.g., networking, controller, software) and physical infrastructure (e.g., generator, energy storage device). CPS covers a wide range of applications such as manufacturing, automotive, transportation, power grid, and healthcare [1]. CPS research aims to integrate knowledge and engineering principles across the computational and engineering disciplines (e.g., networking, control, software, human interaction, learning theory, as well as electrical, mechanical, chemical, biomedical, material science, and other engineering disciplines) to develop new CPS science and supporting technology [2].

It is worth mentioning that tremendous efforts have been made in advancing CPS technology, especially in the energy sector and Smart Grid. The electrical grid is the largest and most complex infrastructure and industrial investment ever built by humankind [3]. The entire electrical grid is indeed an ultra-large-scale network for delivering electricity from suppliers to millions of consumers across tens of thousands of miles of transmission and distribution lines. The concepts and applications of CPS technology stand to have a significant impact on the ultra-large-scale and complex power grid. A great amount of cyber and physical components can greatly affect the power grid spanning over power system planning, operation, management, monitoring, and protection.

CPS innovations bring both new opportunities and challenges to our research and education community. Just to name a few: (1) interdependency; (2) uncertainty; and (3) scale.

1. Interdependency: It is important to recognize that today's CPS-like critical infrastructure is indeed an interdependent network of networks (e.g., power grids, communication, water, transportation, and control) [4]. There is a clear need to address a complex set of interdependencies between the power grid and other critical infrastructures. For example, at the edge of the distribution grid, a

large-scale penetration of electric vehicles (EVs) will create the interdependency of the power grid and transportation. In the following sections, we will use EVs as an application platform to demonstrate the interdependency between different infrastructures. However, the ideas contained here can apply to more generally coupled and large-scale interdependent CPS-like infrastructures.

2. Uncertainty: As the electricity demand in the residential distribution system increases, dynamic pricing becomes possible and necessary to help economy and efficiency of the system. Meanwhile, as the advent of new cyber infrastructure (e.g., advanced metering infrastructure and wide-area monitoring system) in Smart Grid, the so-called "flexible loads" or "dispatchable loads" can adjust their own electric energy usage on the basis of a real-time set point. For example, as a transportation tool and electricity carrier, EVs can be charged at any charging facility and at any time, allowing more spatial and temporal diversity, but also introducing more uncertainty into the power grid's load forecast. On the other hand, the new physical components, such as distributed generators (DGs) and distributed energy storage devices (DESDs), facilitate the grid integration of distributed renewable energy resources. However, the inherent intermittency and variability of a renewable energy resources (e.g., wind and solar) have complicated implications for power system operations. These renewable energy resources tend to fluctuate dramatically depending on the time of day and time of year. Unlike conventional generation sources, wind and solar power output is highly uncertain and unpredictable.
3. Scale: The CPS-like Smart Grid infrastructure consists of hundreds of thousands of distributed components. The vision we promote will require the equivalent concept of the Internet of Things. The same concept might shift the CPS paradigm broadly across multidisciplinary research areas. Despite significant inroads into Smart Grid research in recent years, there are still many open questions to be addressed to fully support system engineering of highly dynamic power systems.

The economic and societal potential of CPS is vastly greater than what has been realized, ultimately resulting a new concept of **cyber-physical-social systems (CPSS)**. CPSS paradigm encompasses the physical assets, computing infrastructure, human cognition, and changes the way people interact with complex interdependent systems. Figure 8.1 illustrates the general concepts of CPSS.

As shown in Figure 8.1, the "social" part begins to play a critical role in such an integrated system. The "social" elements include, but are not limited to, customer behaviors, policy, regulation, and economics. In this chapter, due to the space constraints, we may not be able to cover every aspect of "social" components. Instead, we will use the grid integration of electric and hybrid electric vehicles as an application platform to illustrate some social aspects of CPSS. It is worth to note that the electrified vehicles include: hybrid electric vehicle (HEV), plug-in hybrid electric vehicle (PHEV), extended range electric vehicle (EREVs), pure electric vehicle or battery electric vehicle (BEV), and fuel cell vehicle (FCV). For the sake of simplicity and consistence, we intend to use EVs throughout this chapter.

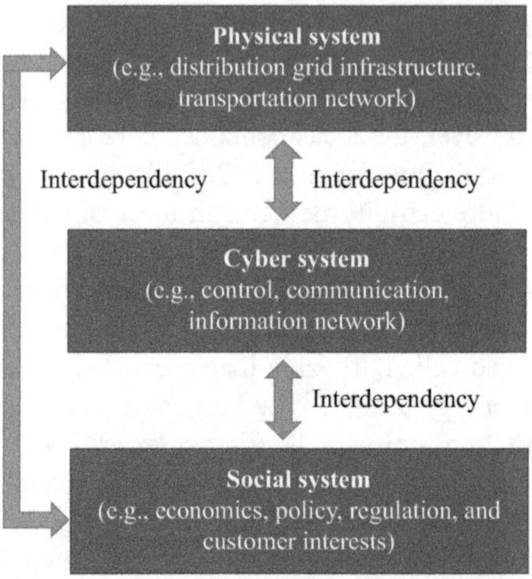

Figure 8.1 *Cyber-physical-social systems*

In short, this chapter is structured as follows: Section 8.2 is an introduction on electrified transportation system with three major CPSS components (i.e., cyber, physical, and social components) and their inter-relationships. Section 8.3 discusses future research trends of electric and hybrid electric in a CPSS environment considering the emerging connected and automated vehicle technologies. Section 8.4 summarizes the major contributions of this chapter.

8.2 Electrified transportation system in a CPSS environment

This section provides a brief overview of electrified transportation systems. Moreover, the cyber, physical, and social components and their inter-relationships in electrified transportation systems will be detailed in this section. The CPSS concept will dramatically transform the way people interact with cyber systems, physical systems, and social systems.

Transportation systems that move people, goods, and services in societies worldwide pose unprecedented environmental, economic, and social challenges, particularly in response to the growing urgency to conserve energy and cut back on carbon emissions and pollution [5]. Advances in transportation electrification not only offer great promise with regard to addressing these stubborn problems, they also have the potential to revolutionize future transportation systems [6–8].

Electric vehicles, which play important roles in the revolution proposed earlier, were first on the road back in 1800s. Ever since, after a series of breakthroughs in the batteries and electric motors, the development of electric vehicle continues to surprise the public. After a time of rises and falls in twentieth century, the soaring gasoline price brought electric vehicle back to people's life in 1970s. The interest in electric vehicle continued because there was an increasing intention on environmental issues in 1990s.

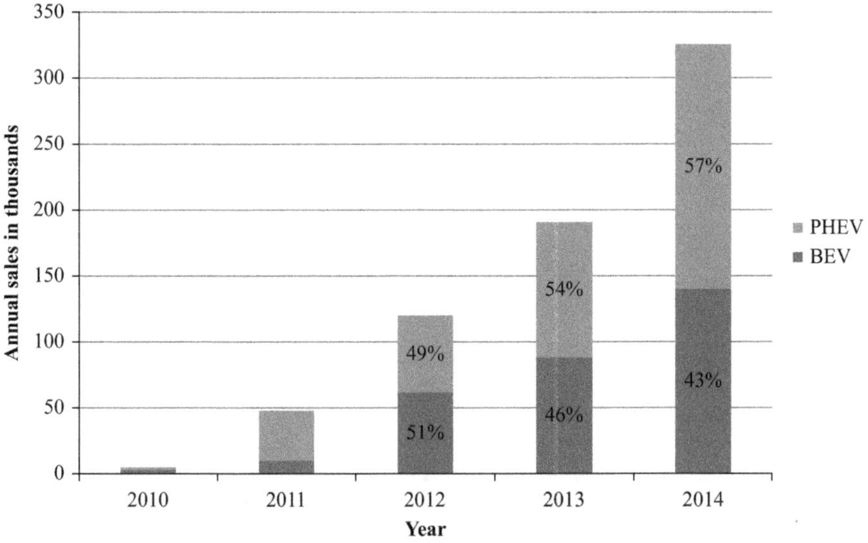

Figure 8.2 Global EV sales [11]

Though problems like high production cost still exist, more and more manufacturing companies started to develop their own model of EVs and invest on related projects. With the support from government policies [9] and the development of electric vehicle manufacturing, battery technology as well as electric motor, companies like Honda, Toyota in Japan and GM, Ford and especially Tesla in the United States all have their own model of electric cars like Prius, Volt, Focus Electric, and Tesla model series [10]. It can be foreseen that as the weighting of electric vehicle becomes more important to these companies, hundreds of models of EVs would be released in the upcoming future.

In the last decade, all around the world [11], efforts to boost the use of EV technologies have accelerated because of their low-pollution emissions, energy independence, low operating cost, high performance, flexible fueling, and high fuel economy. The federal/state agencies and private sectors have put a lot of effort into accelerating the introduction and penetration of advanced electric drive vehicles in the market. If one believes that the early 1900s infrastructure boom was the engine that powered US economic growth 80 years ago, then one could assume that a similar and even more powerful development of transportation electrification could accomplish even more in the next several decades [12]. Figure 8.2 shows a strong growth in EV sales globally. As a result, an ever-increasing number of EVs will radically change the traditional views of the power industry [13], transportation industry [14], social environment [15], and business world [16,17].

The successful rollout of the electrification of transportation is dependent on the affordability, availability, quality, and resilience of the services that our nation's critical infrastructures provide. To realize the full vision of the electrification of transportation, the underlying transportation and electricity distribution infrastructures must change significantly from today's systems. All of these transportation

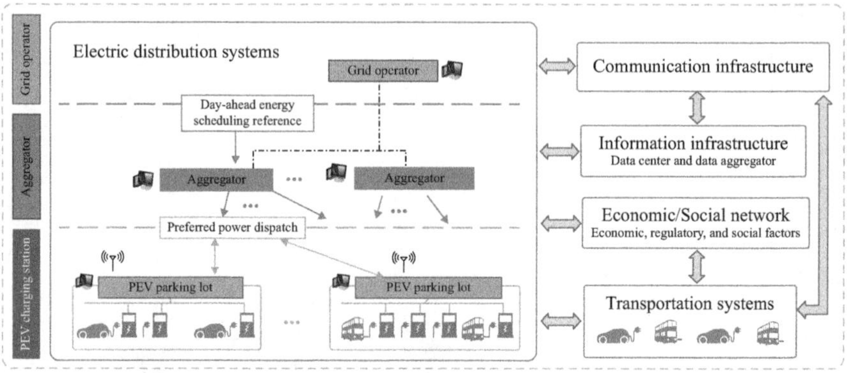

Figure 8.3 The electrified transportation system in a CPSS environment

electrification services are in turn dependent on communication, information, and social networks in a large CPSS infrastructure. However, to date, a well-justified theoretical framework that electrification of transportation depends on is not constructed yet. Figure 8.3 presents the CPSS structure in the application of electrified transportation system. From power system perspective, the proposed hierarchical framework consists of three operating layers: plug-in electric vehicle (PEV) charging stations, aggregators, and grid operators. The coupled electric distribution system and transportation system will complicate the system operations. Figure 8.3 also illustrates the interdependency between multiple critical infrastructure systems, which is primarily created by the EV charging facility. For example, the information sharing is highly depending on the communication infrastructure. The social/economic network (e.g., human factors, regulation, and policy) will have a significant factor on the operations of both electric distribution system and transportation system.

8.2.1 "Physical" infrastructure

The electrification of transportation brings both opportunities and challenges to existing critical infrastructures. The EV charging facility is the major physical infrastructure in supporting the successful rollout of EVs. It determines how, when, and where EV drivers recharge their cars or trucks. Depending on the location, technology, and other factors, the EV charging infrastructure can be categorized into two groups: wired (plug-in) and wireless charging. Figure 8.4 shows a variety of EV charging technology and infrastructure.

For wired (plug-in) charging, there are four different charging levels listed in Figure 8.4. The alternating current (AC) charging can also be seen as slow charging and direct current (DC) charging could be considered fast charging. Normally slow charging would take up to 6–8 h to charge the battery of an electric vehicle, and a DC charging would have done that in 30–40 min. Considering time efficiency, DC fast charging would be a more preferable charging method for the vehicles [18]. But considering non-peak rate of electricity occurs at night, AC slow charging can save money for vehicle owners while charging during night.

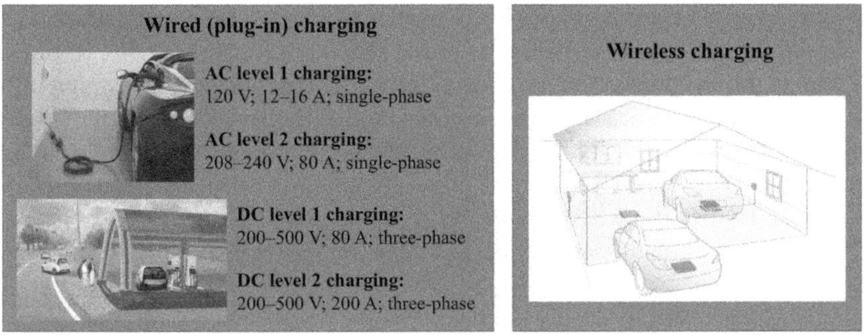

Figure 8.4 EV charging technologies

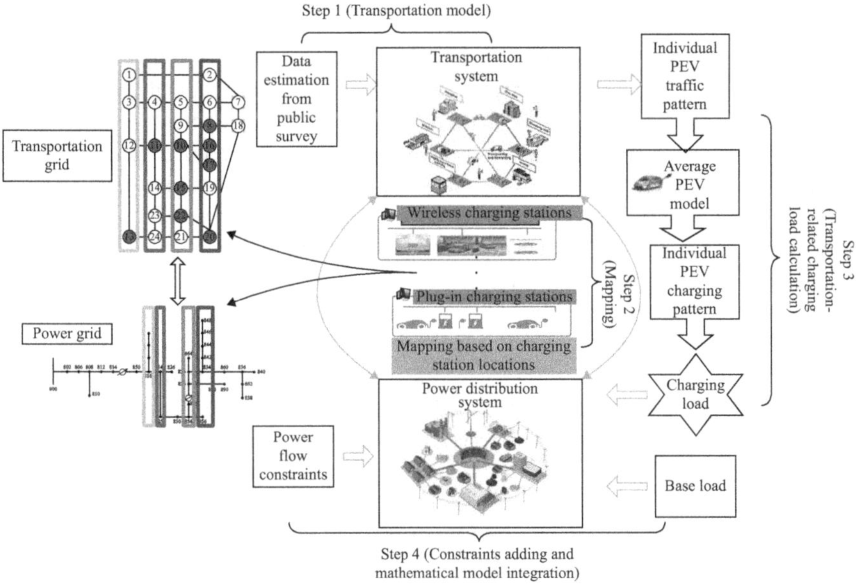

Figure 8.5 System architecture of coupled physical layer for electrified transportation systems [19]

The EV charging facility is an excellent example to illustrate and understand the interdependency across multiple critical infrastructure systems. Because of the nature of transportation tool and electricity carrier, EV charging facility can be considered as "tie points" of multiple critical infrastructure systems (e.g., transportation network and power grid). Figure 8.5 shows the system architecture of coupled physical layers (e.g., power network and transportation network) of electrified transportation systems [19]. A mapping between the transportation system and the power system is defined by the "tie points" – EV charging facilities. EVs can be charged at any charging facility and at any time, which brings more spatial and temporal uncertainty to the power grid's load forecast. The retail electricity price and parking fee may also have an impact on customer behavior, eventually leading to a change in traffic flow. For instance, commercial charging station capacities and prices will be significant factors in causing drivers to choose particular routes and change their behavior. In essence, EV charging infrastructures interconnect a power grid system and a charging infrastructure. Besides engineering considerations, the placement EV charging station is constrained by applicable local policy and regulation, financial incentives, and public interests. In the last decade, there is a growing interest in investigating the planning, operation, and management of EV charging facilities [12]. The optimal decisions on the physical infrastructure are subject to a suite of constraints from cyber and social infrastructure.

8.2.2 "Cyber" infrastructure

The cyber infrastructure, such as communication, data, and control, will play a primary role in supporting the electrification of transportation.

1. Communication: A recent department of energy (DOE) report [20] gives an extensive overview of the Smart Grid benefits and communication needs. One section addresses the specific challenges and opportunities presented by EVs. In order to make rates or programs specifically available to customers with EVs, utilities must offer them special services. The services include being able to (i) enroll, register, and initially set up communications between a vehicle user and the utility or an alternative energy supplier (one-time setup); (ii) repeatedly reestablish communications for each EV charging session (repeat communications and rebinding); (iii) provide EV charging (and other) status information to customer information channels (e.g., web and display devices); and (iv) correctly bill EV customers according to their selected rates or programs. Like in-home charging, public charging will need to match supply and demand; unlike in-home charging, public charging may have to do this very quickly and accurately, since vehicle owners will likely want to avoid a long delay once their vehicle is plugged in [20]. It is estimated that the bandwidth for both load balancing and billing purposes will be between 9.6 and 56 kilobits per second (kbs) [20]. A variety of communication protocols can achieve reliable two-way communication networks. Since EVs can be recharged at various locations (e.g., municipal parking decks and office parking lots), it is critical to maintain the compatibility of communication technologies. The communication infrastructure for the Smart Grid has recently been investigated in References 21 and 22. In addition to certain bandwidth, reliability, and power consumption requirements, security issues need to be considered when different wireless communication technologies are being applied. For example, wireless billing security is viewed as a major concern of EV users. The actual vehicle's location needs to be kept confidential for user privacy. Another security issue is unauthorized transactions by a third party or hacker. In general, security in the communication network at public charging facilities for EVs is of critical importance [23]. Recently, based on different perspectives, a number of papers have investigated security-related issues in the communication networks of the charging infrastructure. Table 8.1 summarizes the vehicle-to-grid communication standard [24].

Table 8.1 Vehicle-to-grid communication standard for North America, 2011 [24]

SAE J	Title	Scope	Status
SAE J2847/1	Communication between PEV and the utility grid	Identifies the communication medium and criteria for the PEV to connect to the utility for Level 1 and Level 2 AC energy transfer	Published November 2013
SAE J2847/2	Communication between PEV and off-board DC chargers	Identifies additional messages for DC energy transfer to the PEV. The specification supports DC energy transfer via forward power flow (FPF) from grid to vehicle.	Published August 2012
SAE J2847/3	Communication between PEV and the utility grid for reverse power flow	Identifies additional messages for DC energy transfer to the PEV. The specification supports DC energy transfer via DC reverse power flow (RPF) from vehicle to grid.	Published December 2013
SAE J2847/4	Diagnostic communication for PEVs	Establishes the communication requirements for diagnostics between PEVs and the EV supply equipment (EVSE) for charge or discharge sessions	Work in process
SAE J2847/5	Communication between plug-in vehicles and their customers	Establishes communication requirements between PEVs and their customers for charge or discharge sessions	Work in process
SAE J2836/1	Use cases for communication between PEVs and the utility grid	Identifies the equipment (system elements) and interactions to support grid-optimized AC or DC energy transfer for plug-in vehicles, as described in SAE J2847/1. Key system elements include the vehicle's rechargeable energy storage system (RESS), power conversion equipment (charger and/or inverter), utility meter, etc.	Published April 2010
SAE J2836/2	Use cases for communication between PEV and EVSE	Establishes use cases for communication between PEVs and the off-board charger, for energy transfer, and for other applications. J2836/2 use cases must be supported by SAE J2847/2.	Published September 2011

(*Continues*)

Table 8.1 (Continued)

SAE J	Title	Scope	Status
SAE J2836/3	Use cases for communication between PEV and the utility grid for reverse power flow	Establishes use cases for communication between PEVs and the public electric power grid, a home branch circuit, or an isolated micro grid, for reverse energy transfer and other applications	Published January 2011
SAE J2836/4	Use cases for diagnostic communication for PEVs	Establishes diagnostic use cases between plug-in electric vehicles and the EVSE	Work in process
SAE J2836/5	Use cases for communication between PEVs and their customers	Establishes the use case for communication between PEV and their customers. The use case scenarios define the information to be communicated related to customer convenience features for charge on/off control, charge power curtailment, customer preference settings, charging status, EVSE availability/access, and electricity usage. Also addresses customer information resulting from conflicts to customer charging preferences.	Published May 2015
SAE J2931	Electric vehicle supply equipment (EVSE) communication model	Establishes the digital communication requirements for the EVSE as it interfaces with a home area network (HAN), an energy management system (EMS), or the utility grid systems. Provides a knowledge base that addresses the communication medium's functional performance and characteristics. Is intended to complement J1772 while addressing the digital communication requirements associated with Smart Grid interoperability.	Published September 2012
SAE J2931/2	Inband signaling communication for PEVs	Establishes the requirements for physical layer communications using in-band signaling between PEVs and the EVSE. Enables onward communications via an EVSE bridging device to the utility smart meter or HAN. This onward communication is known as Frequency Shift Keying (FSK) and is similar to Power Line Carrier (PLC), but uses the J1772 control pilot circuit.	Work in process
SAE J2344	Use cases for communication between PEV and the utility grid	Gives technical guidelines on the safety of EVs during normal operations and charging. To be considered when designing EVs having a gross vehicle weight rating of 4,563 kg (10,000 lb) or less for use on public roads.	Published March 2010

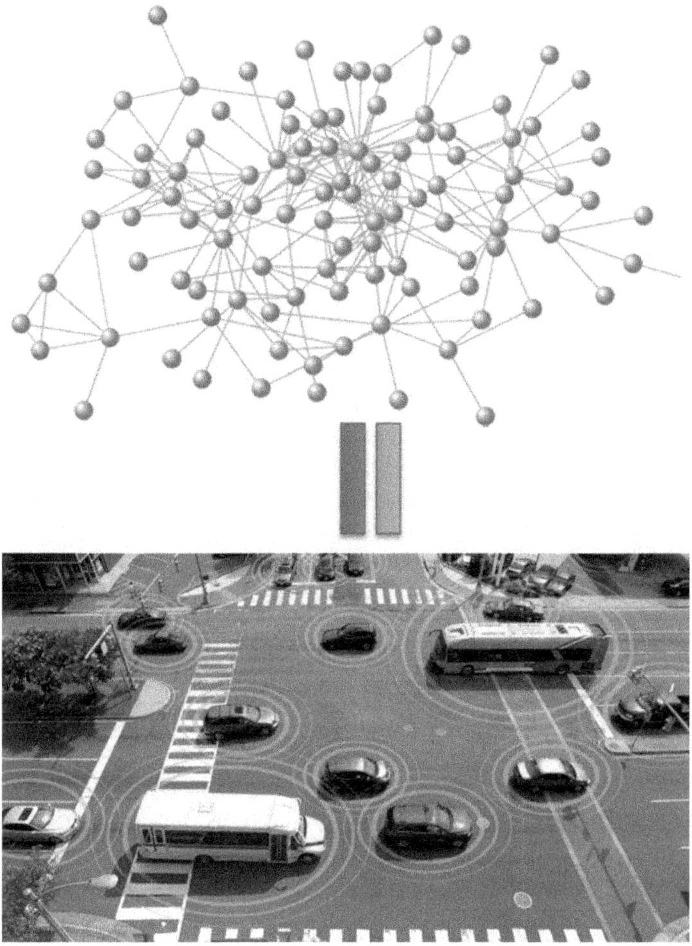

Figure 8.6 An illustrated concept of connected vehicle technologies

2. *Data*: In the last few years, it is also interesting to see that the emerging concepts of connected vehicle will even more complicate the application of EV charging in a cyber environment. Figure 8.6 presents the illustrated concepts of connected vehicle technologies. The development and deployment of a fully connected transportation system provide connectivity (i) among vehicles to enable crash prevention; (ii) between vehicles and the infrastructure to enable safety, mobility, and environmental benefits; and (iii) among vehicles, infrastructure, and wireless devices to provide continuous real-time connectivity to all system users. The additional connectivity can also significantly benefit EV drivers such as finding a best charging spot by sharing and exchanging information with

other, so as to make EVs more affordable, attractive, and sustainable. Vehicle-to-vehicle (V2V) and vehicle-to-infrastructure (V2I) technologies will increase the number of sensors to more than hundreds of thousands. Therefore, it is rather intuitive that the state-of-the-art data processing architecture may no longer be sustainable under such big data explosion. For example, in addition to existing traffic information collection technologies (e.g., remote control camera, speed detectors, variable messages signs), the recent development of V2V and V2I can generate even more massive data. Similarly, from power system perspective, the Advanced Metering Infrastructure (AMI) has been extensively deployed across our nation. Today, a large number of smart meters and sensors capture raw data (e.g., image, video, and text) continuously, and the data is stored in the centralized data centers of different divisions (e.g., electric utility, transportation department). Without being able to handle the high-volume, high-velocity, and high-variety data (big data) efficiently, the system operators of such an integrated transportation system will not be able to provide robust services to a large number of EVs with advanced connectivity. As EVs are equipped with connected vehicle devices, their CPSS-like role will be more emphasized. As shown in Figure 8.6, each dot represents a car with cyber (e.g., V2V and V2I), physical (e.g., dynamic system), and social (e.g., driver behaviors, human factors) features.

3. *Control*: It is important to note that the growing deployment of charging facility may pose potential threats to the aging and overburdened power grid infrastructure. If the control of EV charging facility is not carried out properly, the large numbers of EVs parked there might degrade the power system's stability and reliability. For instance, several hundred EVs might arrive at parking lots in a short period of time (e.g., during the early morning hours when people arrive at work). Under AC Level 2, the hourly charging load of an EV is approximately equivalent to the energy use of an average household. At this level, multiple EV charging loads connected to one feeder at peak time could cause a serious transformer overload during such a short time period. Figure 8.7 illustrates some EV charging control schemes. As shown in Figure 8.7, the sophisticated control can reduce the charging cost based on dynamic pricing (e.g., time-of-use pricing and real-time pricing), shift the peak load, and reduce the peak demand. In addition to satisfying the customer interests, the EV control schemes can facilitate the smooth integration of intermittent renewable sources and provide ancillary services.

The existing EV charging control approaches can be divided into three categories, namely, centralized control, decentralized control, and distributed control. Table 8.2 summarizes the comparison between three control approaches. It is worth to note that the performance of EV charging control highly depends on other types of cyber infrastructure such as communication network. The majority of existing literature focuses on the centralized control for EV charging infrastructure [25–28]. In most cases, the dedicated communication links are required to exchange data between the central controller (e.g., utility, aggregator) and the local agents (e.g., EV charging

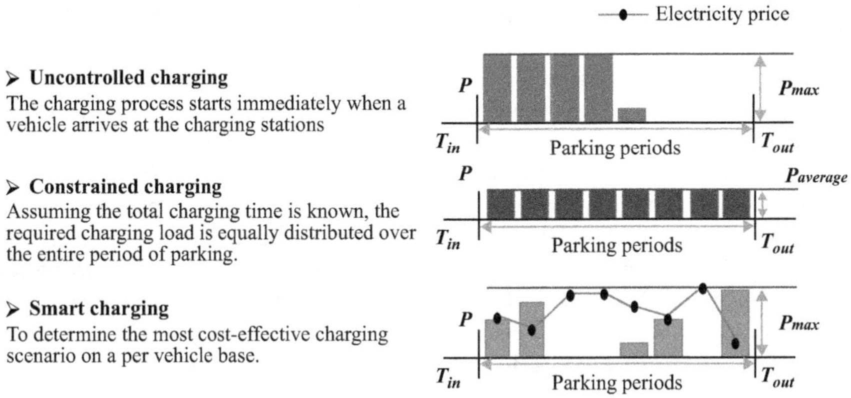

Figure 8.7 Three EV charging control schemes

Table 8.2 Comparison between centralized, decentralized, and distributed control [39]

	Pros	Cons
Centralized control	• Easy to implement • Relatively low capital cost	• Computational burden • Not easy to expand • Single point of failure • Requires high-bandwidth dedicated communication links
Decentralized control	• Local information only • No need for two-way high-speed communication	• Less chance to achieve global optimal solution
Distributed control	• Easy to expand • Low computational cost • Avoid single point of failure • Suitable for large-scale system	• Need synchronization • May be time consuming for local agents to reach consensus • Convergence rates may be affected by the communication network topology • Need two-way communication infrastructure • Upgrading cost on the existing control and communication infrastructure

stations). For example, Luo *et al.* [24] proposed a two-stage optimization model in order to minimize the peak load and the load fluctuation. Hua *et al.* [25] introduced an online adaptive EV charging scheduling framework to optimize EV charging schedules and reduce flow limit, voltage magnitude limit, three-phase voltage imbalance limit, and transformer capacity violations. Jin *et al.* [26] investigated the EV charging with energy storage from an electricity market perspective with joint consideration for the aggregator energy trading in the day-ahead and real-time markets. Sundstrom *et al.* [29] proposed a centralized flexible charging optimization scheme for PEVs considering power grid constraints. Khodayar *et al.* [30] studied the coordinated integration of PEV fleets and renewable energy sources (e.g., wind) in power systems. The centralized control approaches are suitable for relatively small-scale systems without reconstructing the existing communication and control networks. However, as the market penetration of EVs is continuously growing, there are some technical barriers on the centralized control of EV charging such as heavy computation burden and single point of failure. The decentralized control is an intermediate solution to address the above-mentioned challenges [31–33]. The overall objective is to maximize the benefits of local agents without coordinating with others. Gan *et al.* [31] introduced a decentralized algorithm to exploit the elasticity of PEV charging loads to fill the valleys in load profile. Su *et al.* [32] proposed the computational intelligence-based algorithms to locally manage the power allocation at a municipal PEV parking deck. Guo *et al.* [33] proposed a two-stage economic operation framework for a single PEV parking deck to determine the retail electricity price and parking fee. He *et al.* [34] formulated a locally optimal scheduling optimization problem, which aims to minimize the total cost of PEVs in the current ongoing PEV set in the local group. Cao *et al.* [35] proposed an intelligent method to control EV charging loads in response to time-of-use (TOU) price in a regulated market. Since there is no communication link between the different local agents, there is no guarantee that the decisions made by each local agent can contribute to the global optimal decision of the entire system as a whole. It is well known that the participating agents (e.g., EV charging stations) may have partial or complete conflicts such as attracting more EV customer. The strong interactions between different agents may prevent the entire system from achieving the global optimal operation. In general, the overall closed-loop performance of the decentralized control is limited because of the lack of communication link and information exchange. This leads to the emerging interests in distributed control in which local agents can share information through two-way communication links in order to find the global optimal decision. The emerging distributed control heavily relies on the two-way communication link among a cluster of agents that only exchange local information with neighbors [36–38]. For example, Wen *et al.* [36] develop a distributed optimization algorithm to solve the PEV-charging selection problem. Xu [37] used the multi-agent concepts and consensus algorithms to solve the PEV charging power allocation proposed in Reference 31. Rahbari-Asr *et al.* [38] developed a novel cooperative distributed algorithm for charging control of PEVs that solves the constrained nonlinear optimization problem using Karush-Kuhn-Tucker conditions and consensus networks in a distributed fashion. The distributed control network is more vulnerable to cyber-attacks than their centralized counterparts. In short, the interdependency between various sub-networks within the same category of cyber infrastructure needs to be well studied. Figure 8.8 illustrates the economic operation of multiple EV charging stations using a distributed control method. As shown in Figure 8.8, the cyber layer (e.g., control) is also highly related to the social layer (e.g., economics). To date, the interdependency between various cyber and social infrastructure has not been fully investigated.

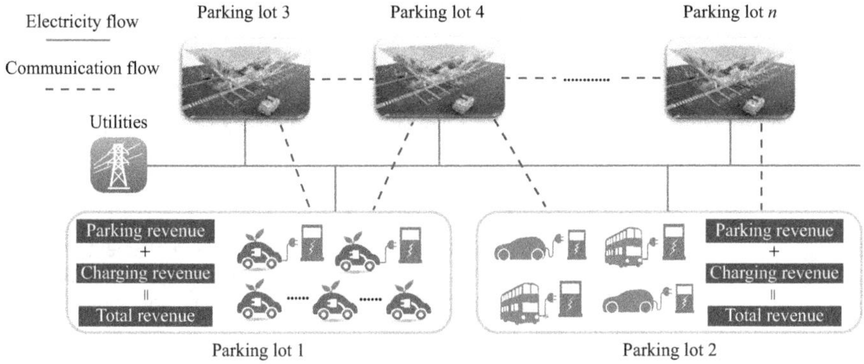

Figure 8.8 An illustrated structure of economic operation of multiple EV charging stations using a distributed control manner

8.2.3 "Social" considerations

Despite the benefits contributed by EVs, there are significant "social" challenges associated with grid integration of large number of EVs. Besides engineering considerations, the "social" considerations include, but are not limited to: policy, regulation, economics, and customer interests.

1. *Policy*: Table 8.3 shows a selected list of current national policy initiatives. As of today, EVs are still substantially more expensive than gasoline and hybrid vehicles. The savings in the cost of fuel (given current gasoline prices) does not justify the higher upfront cost to purchase the EV at the current level. Federal and state government subsidies, based on the grounds that EVs are cleaner and hence offer more social benefits than gasoline vehicles, are being offered to offset their higher purchase cost [40–43]. Similar subsidies were offered for hybrid vehicles when they were initially introduced to the market, and their impacts were studied. The second impediment to consumer adoption of EVs is its small driving range per charge, which leads to range anxiety. A well-developed and well-planned charging infrastructure could help alleviate range anxiety and increase EV sales [44–46].
2. *Regulation*: The local regulation may complicate the deployment and design of EV charging facility. For example, the local regulation determines the optimal sizes and locations of EV parking facility, ultimately impacting the physical infrastructure. The planning process for commercial and residential EV charging stations requires extensive permit approvals and documentation from a variety of agencies, such as utilities, governing authorities, enthusiasts, original equipment manufacturers, and business owners. The local regulation can be even stricter in population centers, considering the building code, zoning requirements,

Table 8.3 A selected list of national policy initiatives as of 2014

China	Purchase subsidies for vehicle of up to RMB 60 000.
Denmark	Exemption from registration and road taxes. DKK 70 million for development of charging infrastructure.
Finland	EUR 5 million reserved for vehicles participating in national EV development program, ending 2013.
France	EUR 450 million is rebates given to consumers buying efficient vehicles, with 90% of that amount from fees on inefficient vehicles. Remaining 10% (EUR 45 million) is a direct subsidy. EUR 50 million to cover 50% of EVSE cost (equipment and installation).
Germany	Exemption from road taxes, four regions nominated as showcase region for BEVs and PHEVs.
India	INR 100 000 or 20% of cost of vehicle, whichever is less. Reduced excise duties on BEVs/PHEVs. The National Mission for Electric Mobility will facilitate installation of charging infrastructures.
Italy	EUR 1.5 million for consumer incentives, ending in 2014.
Japan	Support to pay for 1/2 of the price gap between EV and corresponding ICE vehicles, up to YEN 1 million per vehicle. Support to pay for 1/2 of the price of EVSE (up to 1.5 million per charge).
Netherlands	Tax reduction on vehicles amounting to 10–12% net of the investment. 400 charging points supported through incentives.
Spain	Incentives up to 25% of vehicle purchase price before taxes, up to EUR 6,000. Additional incentives of up to EUR 2 000 per EV/PHEV also possible. Public incentives for a pilot demonstration project. Incentives for charging infrastructure in collaboration between the national government and region administrations.
Sweden	EUR 4 500 for vehicles with emission of less than 50 g of CO_2/km. EUR 20 million for 2012–2014 super car rebate. No general support for charging points besides R&D funding (EUR 1 million in 2015).
United Kingdom	GBP 37 million for thousands of charging points for residential, street, railway, and public sector locations. Available until 2015.
United States	Up to USD 7 500 tax credit for vehicles, based on battery capacity. Phased out after 200 000 vehicles from qualified manufacturers. A tax credit of 30% of the cost, not to exceed USD 30 000 for commercial EVSE installation. A tax credit of up to USD 1 000 for consumers who purchase qualified residential charging. USD 360 million for infrastructure demonstration projects.

community or design guidelines, and electrical source and metering. To address these challenges, some innovative infrastructure projects showcase the new concept of physical infrastructure to relief the constraints. For instance, the existing street lighting systems are used to provide a cost-effective and simple EV charging solution. EV charging stations can be set up at almost any location where suitable parking and street lighting system is available, simply by replacing conventional street lights with EV charging cable.

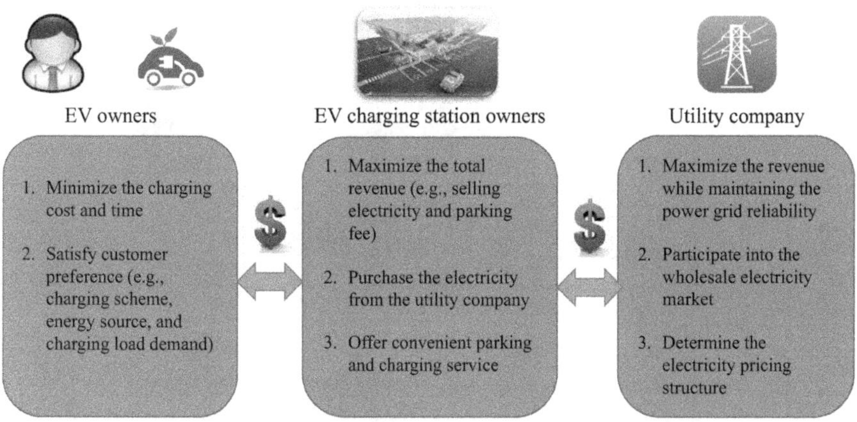

Figure 8.9 The economic operations of an EV ecosystem

3. *Economics*: The majority of current research mainly focuses on the control strategy of EV charging load from utility and end-user perspectives, such as minimizing the system operation cost, promoting demand response, and integrating renewable energy resources. While researchers around the world are making significant advances in these areas, there is little work addressing the economic operations of EV charging facility from both utility, end-user (EV driver), and EV charging station owner's perspective. Figure 8.9 illustrates the economic operations of an EV ecosystem (EV owners, EV charging facility owners, and distribution companies) and their interests. As the figure shows, it is important to introduce the economic inter-relationship among utility, end-user (EV driver), and EV charging station owner. At the early development stage, most of installed charging stations offer free or flat-rate charging [47], with a goal of stimulating public awareness of EVs. The parking deck operators have to consider the price of acquiring electricity from the grid, the initial EV battery state of charge, charging power, parking fee, and parking fee rebate rate. Figure 8.10 illustrates the roles of customer and charging stations in terms of transaction. The well-justified and dynamic business model of an EV ecosystem is needed considering ownership, investment plans, and cost recovery mechanisms. Here is an example. EV charging stations in a public parking facility might have split ownership. For example, one entity might own the parking infrastructure (e.g., a university might own the parking lot) and another entity (perhaps a private investor) might own the charging stations and associated equipment. There might even be a combination of owners. For example, utilities might own and operate the smart metering infrastructure and the power line that connects with the transformers. The general business model is not clearly defined yet.

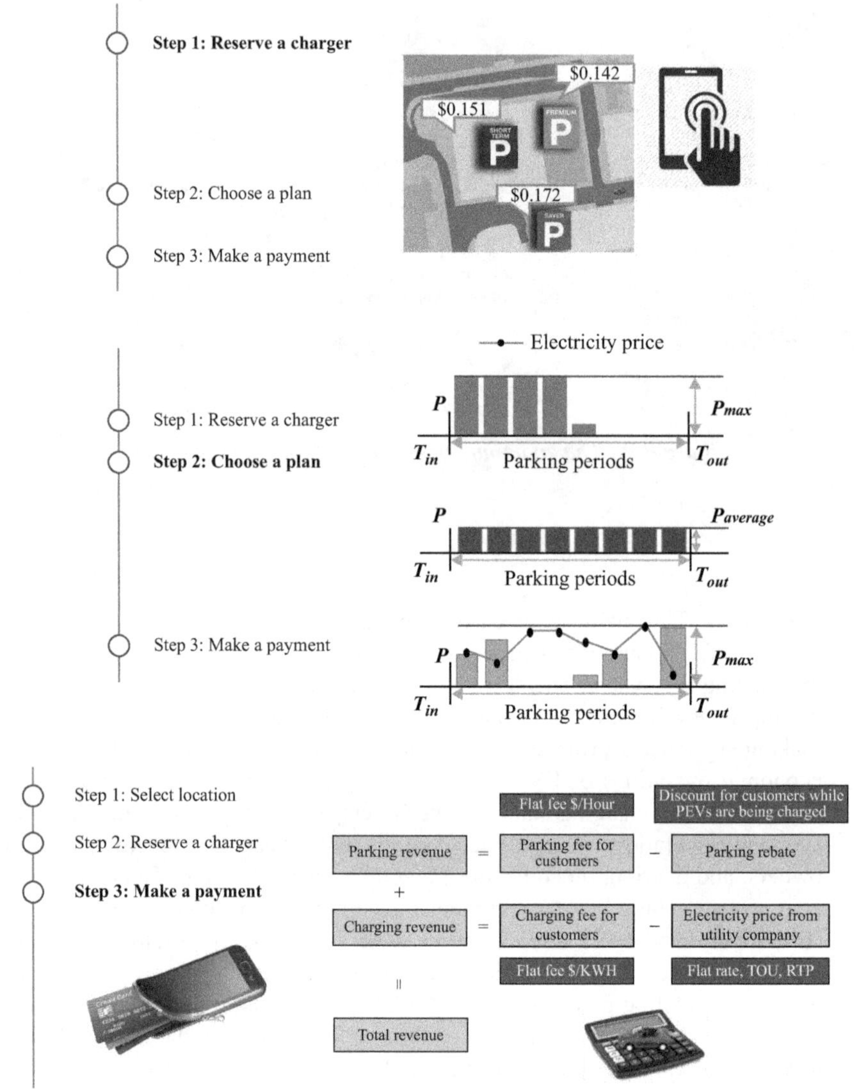

Figure 8.10 The roles of EV customers and charging stations

8.3 Future research trends

To fully implement the electrified transportation system with large penetration of EVs, we must study the highly integrated system as a whole, addressing the synergy and interdependency among different physical, cyber, and social networks/components. We need to establish the scientific foundations and engineering principles needed to realize a CPSS approach to mobility that is capable and dependable beyond what we can achieve today. We need to study the interaction of cyber, physical, and social components of the highly integrated system whose dynamics have historically been modeled separately but need to be modeled jointly in terms of the discrete, continuous, and uncertain behaviors of the subsystems and its environment.

To the best of our knowledge, the future research needs to advance the science, technology, and engineering of CPSS using a systematic approach:

1. A well-justified theoretical framework that these interdependent CPSS infrastructure systems depend on is not constructed yet. Without a clear architecture, it is impossible to develop a technical specification for key functionalities and features against yet-to-be defined system requirements.
2. Novel modeling, control, and simulation of large-scale interdependent CPSS.
3. A better understanding of interaction and interdependency among various networks, subnetworks, and component.

It is important to mention that a single project effort in each area could potentially contribute to the state of the art, but will fail to address the broad aspects of CPSS-like electrified transportation systems. The electrified transportation system consists of millions of subsystems and agents. It is expected to see a new interdisciplinary paradigm shift to support the CPSS interactions broadly across the engineering, social, and economic sciences.

Science of CPSS: CPSS must move beyond the classical fundamental models of computation and physics. A unified science would support composition, bridge the computational versus physical notions of time and space, cope with uncertainty, and enable CPSS to interoperate and evolve.

Technology of CPSS: We need to develop new design, analysis, and verification tools that embody the scientific principles of the CPSS and incorporate measurement, dynamics, and control of such a large-scale CPSS. These tools should offer important perspectives into behaviors and interactions of CPSS. New building blocks are also needed, including hardware computing platforms, operating systems, and middleware.

Engineering of CPSS: CPSS opens a new opportunity to rethink the principles and methods of systems engineering that are built on the foundations of CPSS science and technology. Attention will be given to system architectures, designs, and integrations as well as the exploration of design spaces that will produce certifiably dependable systems. The more complex the CPSS-like electrified transportation, the more realistic tests will need to be performed, and with greater accuracy and better social considerations. It is very challenging to capture critical details in pure mathematical models and software-only simulation. On one hand, pure mathematical modeling inevitably ignores many critical details of complex CPSS to remain analytically tractable. On the other hand, given the limited computational power, software-only simulation can only provide off-line system performance estimates in the context of what-if analysis or sensitivity analysis. The emerging hardware-in-the-loop testing has drawn increased attentions and allow faster, more accurate and more reliable studies under multiple scenarios in near real conditions and without risk, prior to deployment in the field. However, the "social" aspects of complex system are not fully considered in the hardware-in-the-loop testing. The majority of existing work on EVs is mainly focused on cyber and physical aspects. An innovative and universal testing platform will have a significant impact on CPSS's research and education in the near future.

8.4 Conclusions

CPSS are expected to play a major role in the design and development of future engineering systems. In this chapter, we use the grid integration of EVs in electrified transportation systems as an application platform to demonstrate a variety of cyber, physical, and social components and their interdependency. The ideas contained here can also apply to more generally coupled and large-scale CPS infrastructures (healthcare, transportation, manufacturing, and industry). These interdependencies will affect long-term planning, day-ahead scheduling, and real-time operation of EV charging facilities. The successful rollout of the transportation electrification is highly dependent on the affordability, availability, quality, and resilience of the services that our nation's critical infrastructures in a CPSS environment. The ultimate goal is to enhance the smart operation of interdependent critical infrastructure systems that promote economic efficiency for moving people and goods, reduce fossil fuel consumption, encourage customer participation, reduce regulated emissions, increase use of infrastructure capacity, maintain grid reliability, increase customer participation, and support grid integration of sustainable mobility with multidisciplinary complexities. Overall, there is a growing interest on CPSS theory, algorithms, tools, applications, and testbed. However, there is still a significant amount of investigation needed before the full vision of CPSS comes to fruition.

Acknowledgment

This work is in part supported by the US Department of Energy Graduate Automotive Technology Education Grant (Award# DE-EE0005565) and the University of Michigan Office of Research.

References

[1] "Cyber-physical systems: situation analysis of current trends, technologies, and challenges," *National Institute of Standards and Technology (NIST)*, March 9, 2012. [Online] http://events.energetics.com/NIST-CPSWorkshop/pdfs/CPS_Situation_Analysis.pdf

[2] Baheti R. and Gill H., "Cross-Cutting Research Directions." In: T. Samad and A.M. Annaswamy (eds) *Technology overview, success stories, and research challenges*, IEEE Control Systems Society, 2011. [Online] http://www.ieeecss.org/general/impact-control-technology

[3] National Academy of Engineering, "Greatest engineering achievements of the twentieth century," July 2016. [Online] http://www.greatachievements.org/

[4] Buldyrev S.V., Parshani R., Paul G., Stanley H.E., and Havlin S., "Catastrophic cascade of failures in interdependent networks." *Nature*, February 2010, 464, pp. 1025–1028.

[5] Underwood S.E., "Disruptive innovation on the path to sustainable mobility: creating a roadmap for road transportation in the United States." In: Meyer and Beiker (eds) *Lecture notes mobility in road vehicle automation*, Springer, Berlin, Germany, 2014.

[6] Chan C.C., "The state of the art of electric, hybrid, and fuel cell vehicles." *Proceedings of the IEEE*, 2007, 95(4): pp. 704–718.

[7] Su W., Rahimi-Eichi H., Zeng W., and Chow M.Y., "A survey on the electrification of transportation in a smart grid environment." *IEEE Transactions on Industrial Informatics*, 2012, 8(1): pp. 1–10.

[8] Lopes J.A.P., Soares F.J., and Almeida P.M.R., "Integration of electric vehicles in the electric power system." *Proceedings of the IEEE*, 2011, 99(1), pp. 168–183.

[9] Ahman M., "Government policy and the development of electric vehicles in Japan." *Energy Policy*, 2006, 34(4): pp. 433–443.

[10] "Behind the wheel: Ford commits billions to EV development," Chatham, Newstex, 2015. [Online] http://0-search.proquest.com.wizard.umd.umich.edu/docview/1747545047?accountid=14578

[11] "Global EV outlook 2013: understanding the electric vehicle landscape to 2020," April 2013. [Online] http://cleanenergyministerial.org/Portals/2/pdfs/EVI_GEO_2013_FINAL_150dpi.pdf

[12] Su W., Wang J., and Hu Z., "Planning, control and management strategies for parking lots for PEVs." In: Rajakaruna S., Shahnia F., and Ghosh A. (eds) *Plug-in electric vehicles in smart grid: management and control strategies*, Springer, Berlin, Germany, 2015, pp. 61–98.

[13] Kempton W. and Tomic J., "Vehicle-to-grid power implementation: from stabilizing the grid to supporting large-scale renewable energy." *Journal of Power Sources*, 2005, 144(1): pp. 280–294.

[14] Su W., Wang J., Zhang K., and Chow M.Y., "Framework for investigating the impact of PHEV charging on power distribution and transportation networks." *38th Annual Conference of the IEEE Industrial Electronics Society*, Montreal, Canada, October 2012.

[15] Green E.H., Skerlos S.J., and Winebrake J.J., "Increasing electric vehicle policy efficiency and effectiveness by reducing mainstream market bias." *Energy Policy*, 2014, 65: pp. 562–566.

[16] Bessa R.J. and Matos M.A., "Economic and technical management of an aggregation agent for electric vehicles: a literature survey." *European Transactions on Electrical Power*, 2011, 22(3): pp. 334–350.

[17] DeForest N., Funk J., Lorimer A., *et al. Impact of widespread electric vehicle adoption on the electrical utility business-threats and opportunities*. Center for Entrepreneurship and Technology, UC Berkeley, CA, 2009. [Online] https://ikhlaqsidhu.files.wordpress.com/2010/03/nr_utilities_final_8-31-09.pdf

[18] Botsford C. and Szczepanek A., "Fast charging vs. slow charging: pros and cons for the new age of electric vehicles." *International Battery Hybrid Fuel Cell Electric Vehicle Symposium*, May 2009.

[19] Xiong J., Zhang K., Guo Y., and Su W., "Investigate the impacts of PEV charging facilities on integrated electric distribution system and electrified transportation system," *IEEE Transactions on Transportation Electrification*, 2015, 1(2): pp. 178–187.

[20] US Department Energy. *Communication requirements for smart grid technologies*. US Department of Energy Washington, DC, 2010.

[21] Sauter T., and Lobashov M., "End-to-end communication architecture for smart grids." *IEEE Transactions on Industrial Electronics*, 2011, 58(4): pp. 1218–1228.

[22] Gungor V., Sahin D., Kocak T., *et al.*, "Smart grid technologies: communications technologies and standards." *IEEE Transactions on Industrial Information*, 2011, 7(4): pp. 529–539. [Online] http://ieeexplore.ieee.org/stamp/stamp.jsp?tp=&arnumber=6011696

[23] Khurana H., Hadley M., Ning L., and Frincke D.A., "Smart-grid security issues." *IEEE Security Privacy Magazine*, 2010, 8(1): pp. 81–85.

[24] Society of Automotive Engineers International, "SAE Ground Vehicle Standards," 2012. [Online] http://publicaa.ansi.org/sites/apdl/Documents/Meetings%20and%20Events/EDV%20Workshop/Presentations/Pokrzywa-ANSI-EDV-0411.pdf

[25] Luo Z., Hu Z., Song Y., Xu Z., and Lu H., "Optimal coordination of plug-in electric vehicles in power grids with cost-benefit analysis-part I: enabling techniques." *IEEE Transactions on Power Systems*, November 2013, 28(4): pp. 3546–3555.

[26] Hua L., Wang J., and Zhou C., "Adaptive electric vehicle charging coordination on distribution network." *IEEE Transactions on Smart Grid*, November 2014, 5(6): pp. 2666–2675.

[27] Su W., Wang J., Zhang K., and Huang A.Q., "Model predictive control-based power dispatch for distribution systems considering plug-in electric vehicle uncertainty." *Electric Power Systems Research*, January 2014, 106: pp. 29–35.

[28] Jin C., Tang J., and Ghosh P., "Optimizing electric vehicle charging with energy storage in the electricity market." *IEEE Transactions on Smart Grid*, March 2013, 4(1): pp. 311–320.

[29] Sundstrom O. and Binding C., "Flexible charging optimization for electric vehicles considering distribution grid constraints." *IEEE Transactions on Smart Grid*, March 2012, 3(1): pp. 26–37.

[30] Khodayar M.E., Wu L., and Shahidehpour M., "Hourly coordination of electric vehicle operation and volatile wind power generation in SCUC." *IEEE Transactions on Smart Grid*, September 2012, 3(3): pp. 1271–1279.

[31] Gan L., Topcu U., and Low S., "Optimal decentralized protocol for electric vehicle charging." *IEEE Transactions on Power Systems*, May 2012, 28(2): pp. 940–951.

[32] Su W. and Chow M.Y., "Performance evaluation of an EDA-based large-scale plug-in hybrid electric vehicle charging algorithm." *IEEE Transactions on Smart Grid*, March 2012, 3(1): pp. 308–315.

[33] Guo Y., Xiong J., Xu S., and Su W., "Two-stage economic operation of microgrid-like electric vehicle parking deck." *IEEE Transactions on Smart Grid*, May 2016, 7(3): pp. 1703–1712.

[34] He Y., Venkatesh B., and Guan L., "Optimal scheduling for charging and discharging of electric vehicles." *IEEE Transactions on Smart Grid*, September 2012, 3(3): pp. 1095–1105.

[35] Cao Y., Tang S., Li C., Zhang P., *et al.,* An optimized EV charging model considering TOU price and SOC curve, *IEEE Transactions on Smart Grid*, March 2012, 3(1): pp. 388–393.

[36] Wen C.K., Chen J.C., Teng J.H., and Ting P., "Decentralized plug-in electric vehicle charging selection algorithm in power systems," *IEEE Transactions on Smart Grid*, December 2012, 3(4): pp. 1779–1789.

[37] Xu Y., "Optimal distributed charging rate control of plug-in electric vehicles for demand management." *IEEE Transactions on Power Systems*, May 2015, 30(3): pp. 1536–1545.

[38] Rahbari-Asr N. and Chow M.Y., "Cooperative distributed demand management for community charging of PHEV/PEVs based on KKT conditions and consensus networks," *IEEE Transactions on Industrial Informatics*, 2014, 10(3): pp. 1907–1916.

[39] Su W. and Wang J., "Energy management systems in microgrid operations." *The Electricity Journal*, 2012, 25(8): pp. 45–60.

[40] Armaroli N. and Balzani V., "Towards an electricity-powered world." *Energy and Environmental Science*, 2011, 4(9): pp. 3193–3222.

[41] Diamond D., "The impact of government incentives for hybrid-electric vehicles: evidence from US states." *Energy Policy*, 2009, 37(3): pp. 972–983.

[42] Bahn O., Marcy M., Vaillancourt K., and Waaub J., "Electrification of the Canadian road transportation sector: a 2050 outlook with TIMES-Canada." *Energy Policy*, 2013, 62: pp. 593–606.

[43] Beresteanu A. and Li S., "Gasoline prices, government support, and the demand for hybrid vehicles." *International Economic Review*, 2011, 52(1): pp. 161–182.

[44] Gardner L.M., Duell M., and Waller S.T., "A framework for evaluating the role of electric vehicles in transportation network infrastructure under travel demand variability." *Transportation Research Part A*, 2013, 49: pp. 76–90.
[45] Wu D. and Aliprantis D.C., "Modeling light-duty plug-in electric vehicles for national energy and transportation planning." *Energy Policy*, 2013, 63: pp. 419–432.
[46] Tamor M.A., Gearhart C., and Soto C., "A statistical approach to estimating acceptance of electric vehicles and electrification of personal." *Transportation Research Part C: Emerging Technologies*, 2013, 26: pp. 125–134.
[47] U.S. Department of Energy, "Plug-in electric vehicle handbook for public charging station hosts." DOE/GO-102012-3275, 2012. [Online] http://www.afdc.energy.gov/pdfs/51227.pdf

Chapter 9

Increasing local renewable energy use in smart neighborhoods through coordinated trading

Berk Celik[1], *Robin Roche*[1], *David Bouquain*[1] *and Abdellatif Miraoui*[1]

Abstract

As distributed generation (DG) gains in popularity throughout the world, neighborhoods are expected to turn into small microgrids that may be able to operate autonomously from the grid when needed. Such cases may arise to reduce energy costs or when facing an outage of the distribution system.

This feature is only possible if such neighborhoods are constructed with an advanced metering infrastructure, and homes are equipped with smart meters and home energy management systems to make local resources (DG, loads, storage, and electric vehicles) accessible and controllable. This, in turn, enables self-consumption mechanisms, where smart homes consume their own generated energy, when coupled with efficient energy management strategies.

As the possibilities offered by a typical smart home are limited, mainly due to cost and comfort constraints, more possibilities are offered when several homes can coordinate their actions with each other, i.e., by sharing their resources and scheduling their use appropriately.

By communicating with each other, homes can form groups (based on grid topology and economic criterion), estimate the total available energy capacity of the neighborhood, and collaboratively allocate energy generation, consumption, and storage over time. Homes are modeled as rational decision-makers (agents) in the neighborhood, and can cooperate with each other to meet the needs of their occupants.

9.1 Introduction

Due to their low emissions, the interests in distributed generation (DG) from renewable energy sources (RES) and in electric vehicles (EVs) are growing. As a consequence, it

[1]IRTES Laboratory (EA 7274) and FCLAB Research Federation (FR CNRS 3539), UTBM, Université Bourgogne Franche-Comté, Belfort, France

is forcing the current electricity infrastructure to become "smarter" through improved control, monitoring, and planning of the power system [1]. While power used to flow unidirectionally from bulk generation to loads, the introduction of DG on the consumer side is making power flow bidirectional, which implies a significant shift from usual business operations.

Several other trends are also pushing for a modernization of the electric grid, known as the "smart grid," such as electricity consumption increases with population growth, integration of new types of loads (EVs), and insufficient investments in T&D infrastructure. Recent developments in power electronics and in information and communication technologies have made such modernization technically feasible, and often at a reasonable financial cost.

With the smart grid, consumers gain the opportunity to become active participants through so-called smart homes, e.g., by controlling their appliances in response to system conditions [2, 3]. Due to the increasing number of distributed resources and active participants, traditional control strategies for the electricity network are not suitable anymore, as computational and bandwidth needs tend to increase exponentially with the number of resources [4, 5]. Decentralized control is thus becoming more and more popular to accommodate for such a growth in the number resources to coordinate.

At the scale of a smart home, several resources can be used: DG typically in the form of photovoltaic (PV) panels, battery energy storage, and demand-side management in the form of demand response (DR). By enabling some loads to be stopped or shifted in time in order to reduce the energy consumption at a given time, DR programs bring flexibility to the consumer side. For example, the cycle of a dishwasher may be shifted by a few hours, from the evening to the middle of the night, without impacting user comfort. DR may also be supplemented by the use of DG output for self-consumption, instead of selling the generated energy back to the grid, as is commonly done where feed-in-tariffs are used [6].

The concept of DR was introduced several decades ago [7], but has only gained widespread popularity over the last few years due to technological progress in advanced metering infrastructure (AMI) which provides two-way communication between consumers and utilities. This communication channel is used for sharing information on consumption and electricity price [8].

Depending on grid constraints (e.g., peak shaving to remove a congestion) and/or electricity prices, DR programs enable shaping the load curve. Two main types of DR programs are direct load control (DLC) and price or incentive-based programs [9]. In DLC programs, customers allow the utility to control household appliances for a specific duration or offer a specific amount of load reduction, typically during critical conditions (contingencies, blackouts, etc.) in order to help secure grid reliability [10, 11]. Examples include interruptible/curtailable DR programs, in which users receive a financial compensation for reducing their load upon request. In price-based programs, the decision-making process to adjust the load is left to the end-users, who are expected to modify their load profile depending on electricity prices. The basis of price-based programs is to influence end-users by motivating them to shift some of their load from on-peak to off-peak periods, e.g., by scheduling load run times [12]. DR thus provides benefits to the utility (which increase the flexibility of the load profile) and the end-users (which may save on their electricity bill) [13].

Comfort and cost reduction level are crucial issues for active participation: end-users need to find a trade-off between the loss of comfort and the savings [14]. If an end-user thinks that the DR program may not be worth using, he may decide to turn it off, and become a passive consumer again. To solve this issue, DR programs need to take into account user preferences, which are typically determined by their living habits. Depending on their usage and characteristics, household appliances are categorized to ease the control process without disturbing user comfort. Non-deferrable loads (such as lights, computers, and television) are not controlled by the DR program, while deferrable loads are, and can be shifted in time by up to a few hours with minimal impact on user comfort. Deferrable loads are divided into two sub-groups according to their controllability characteristics: deferrable-shiftable loads and deferrable-thermal loads [15, 16]. Deferrable-shiftable appliances, such as washing machines and dish washers, are scheduled later in time as there is no necessity of instant usage. It is, however, necessary to finish the appliance operation cycle before the ending time defined by the consumer. Deferrable-thermal appliances, such as refrigerators and air conditioners, have limited controllability because of the narrow control availability of these appliances. The user may define maximum and minimum temperature limits, between which the appliance is restricted to operate to maintain comfort.

Although individual customers are encouraged to take advantage of participation in DR programs, single household energy management may not achieve the expected goal (e.g., total load reduction) due to uncoordinated programs. The integration of residential DG and storage, such as PV panels, provides additional flexibility that may also benefit both sides, i.e., the utility and the end-users. Moreover, coordinating the use of resources at the neighborhood level (i.e., from several smart houses) may provide more opportunities for optimization toward a defined objective (typically costs), without increasing the impact on end-users' comfort [17–19]. For example, unused surplus energy of PV panels may be shared with other neighbors, instead of being fed back to the grid.

In this chapter, the potential of local, coordinated resources management at the neighborhood level is evaluated through the design of a dedicated energy management system. The approach uses a rolling horizon optimization algorithm, and is simulated with a small neighborhood formed by five smart houses with building-integrated PV and battery systems. The chapter especially focuses on the ability to consume the renewable energy generated within a neighborhood, by coordinating electricity appliances and battery operations. Several control paradigms are compared: centralized (by the utility), distributed (by the end-users), baseline (simple battery charge and discharge), and selfish control (no coordination ability).

The contributions of this chapter include the following:

- An investigation of the effectiveness of coordination between multiple smart homes, with a focus on appliance and storage scheduling.
- A day-ahead, rolling horizon optimization method, to minimize the overall cost of electricity for end-users, as well as increase the local share of renewable energy usage.
- A mechanism for homes to trade energy with each other through an incentive price and by exchanging limited information (e.g., their average load).
- A "proportional source-matching" method, where smart homes can take decisions about purchasing or selling electricity according to the aggregated energy profile of the neighborhood.
- A multi-agent model and simulation to define the interactions between homes and the aggregator.
- A model with several time steps, e.g., a 30-minute time step, for communication and trading, and a 1-minute resolution for physical models.

The remainder of this chapter is organized as follows. In Section 9.2, the system structure is introduced at the smart home level and the neighborhood level. In Section 9.3, the algorithm for the neighborhood area control is presented. In Section 9.4, simulation results are analyzed. In Section 9.5, an overview of the approach is discussed, and in Section 9.6, the chapter is concluded.

9.2 System model

The studied system structure is divided into two levels: the smart home level and the neighborhood level. A bottom-up approach is used to model smart home electricity consumption, with appliances, PV, and battery systems. At the neighborhood level, smart homes are assumed to be connected to the same feeder and to be served by the same utility service. Finally, the deployed electricity pricing scheme is presented.

9.2.1 Modeling approach

An agent-based approach is used to model the neighborhood entities. A multi-agent system (MAS) is a collection of agents whose aims are to achieve a set of predefined objectives. An agent is defined as an intelligent, rational decision-maker that has the ability to communicate with other agents and to interact with its environment. This environment may be defined as everything around the agent except itself. An agent may cooperate or compete with other agents depending on the observed environmental conditions [20].

Although MAS are widely studied in computer science, applications also extend to other science and engineering areas, including power system engineering. With the development of smart grid technologies and the increased need for decentralized topologies, MAS provides a suitable solution for the coordination of multiple local resources. Moreover, agents can have autonomous and intelligent operation characteristics, e.g., when the structure of the system changes or when an unscheduled fault occurs [21].

Table 9.1 Controllability of smart home appliances

House appliances	Rated power (W)	Controllability
Electric space heater	500	–
Electric water heater	800	–
Light (4)	125	–
Microwave	1,250	–
Vacuum cleaner	1,200	–
Television	150	–
Computer	250	–
Iron	1,000	–
Hair dryer	200	–
Toaster	800	–
Coffee maker	350	–
Washing machine	800	x
Clothes dryer	1,300	x
Dish washer	1,200	x
Total rated power	9,925	–
Photovoltaics	Variable	–
Battery	Variable	x

In this chapter, two types of agents are designed for neighborhood modeling: aggregator and home agents. In the system structure described in Section 9.2.3, a two-layer hierarchical architecture is used. All agents within a layer share the same general characteristics. For instance, a single home model is used for all homes, whether they have PV panels or not. Each home is assumed to have a user interface, e.g., in the form of a website or a smartphone app, so the user can enter his preferences for appliance operations, and monitor his electricity consumption, generation, and storage. Home agents thus have the ability to schedule appliance run times based on price information received through the AMI from the aggregator agent. Depending on their habits, end-users can choose different appliance working preferences and consume different amounts of power during the day. Also, while some consumers may define wide appliance control intervals to be more flexible, others may only be willing to allow very limited flexibility. In spite of that, home agents are modeled as highly motivated entities whose objective is to increase renewable energy usage in the neighborhood area and to minimize the total electricity cost (see Section 9.3).

9.2.2 Smart home model

At the home level, the home agent is in charge of controlling the available resources. These resources include appliances, PV panels, and battery storage. However, PV panels and battery storage may not be present at all houses to represent the limited penetration rate of these technologies.

Appliances are classified into two groups depending on their controllability (Table 9.1): must-run appliances (non-deferrable) and controllable appliances

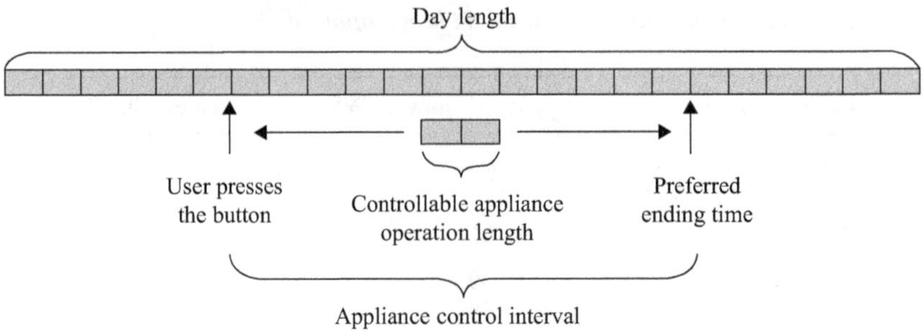

Figure 9.1 Controllable appliance control

(deferrable). Fourteen appliances are modeled and 3 (washing machine, clothes dryer, and dish washer) of 14 are considered as controllable appliances. Thermal loads are not considered as controllable in order to reduce the complexity of the control algorithm. The number of smart houses in the neighborhood is denoted N, with each house denoted by $n = \{1, \ldots, N\}$. Each smart home n has $A(n)$ electricity appliances. The numbers of controllable and must-run appliances are denoted by $Xa(n)$ and $Ya(n)$, respectively. Each of these appliances is denoted $xa(n)$ or $ya(n)$. The power demand profile of a smart home is divided into equal discrete time slots $h = \{1, \ldots, H\}$ over the scheduling horizon H. Each appliance power rating is noted $P_{app}(n, xa)$ or $P_{app}(n, ya)$. In the following, all variables named P with any combination of index or exponent are powers. For controllable appliances, e.g., a control interval is defined by the user, as shown in Figure 9.1. The end-user defines this time interval for the home agent to choose the best time to run these appliances. This time interval must be at least equal to the appliance cycle duration. For each time slot, the total appliance electricity consumption of a smart home $P_c(n, h)$ is given by

$$P(n, xa, h) = \begin{Bmatrix} P_{app}(n, xa), & \phi_s(n, xa) \leq h \leq \phi_e(n, xa) \\ 0 & \text{otherwise} \end{Bmatrix} \tag{9.1}$$
$$\{\phi_s(n, xa), \phi_e(n, xa)\} \in [\alpha(n, xa), \beta(n, xa)]$$

$$P(n, ya, h) = \begin{Bmatrix} P_{app}(n, ya), & \phi_s(n, ya) \leq h \leq \phi_e(n, ya) \\ 0 & \text{otherwise} \end{Bmatrix}, \quad \{\phi_s(n, ya), \phi_e(n, ya)\} \in H \tag{9.2}$$

$$P_c(n, h) = \sum_{xa=1}^{Xa(n)} P(n, xa, h) + \sum_{ya=1}^{Ya(n)} P(n, ya, h) \tag{9.3}$$

where $\phi_s(n, xa)$ and $\phi_e(n, xa)$ are the beginning and the ending operation times of the controllable appliances, respectively, and $\phi_s(n, ya)$ and $\phi_e(n, ya)$ are the beginning and ending operation times of the must-run appliances, respectively. $\alpha(n, xa)$ and

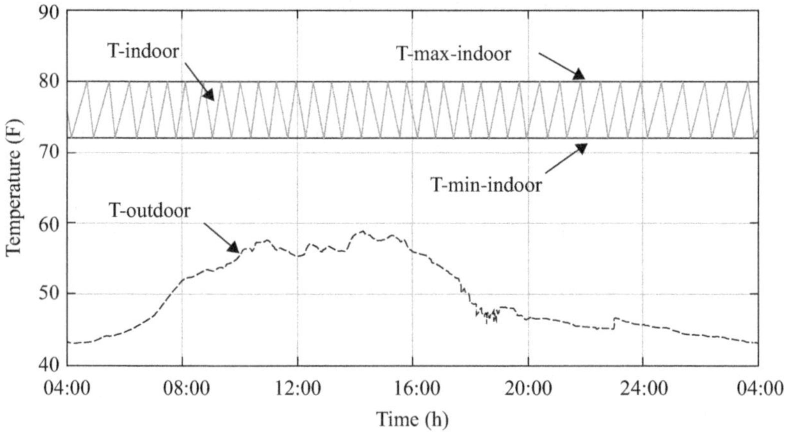

Figure 9.2 Outdoor and indoor temperature with an electric space heater

$\beta(n, xa)$ are, respectively, the beginning and ending times of the authorized run time interval specified by the end-user for the appliance. By this way, the user comfort is guaranteed with the appliance operation interval for appliance scheduling. This interval should be chosen large enough to schedule appliances efficiently for earning some profits. It is also considered that when must-run appliances start to run, they consume the rated power value as long as decided by the end-user [22].

Moreover, for realistic appliance operation, the operation of some appliances depends on other appliances. For example, the clothes dryer should not run before the washing machine finishes its work. Therefore, the proposed constraint is formulated as

$$\phi_s(n, cd) > \phi_s(n, wm) + d_{app}(n, wm) \tag{9.4}$$

where $d_{app}(n, wm)$ is the operation length of the washing machine.

For thermal loads, such as the electric heater and the water heater, the dynamic behavior is given by equations from [23]. The on/off status of the thermal loads is derived to keep the indoor and water temperature between the set points defined by the end-user. The indoor and the outdoor temperature of houses are given in Figure 9.2 with the maximum and the minimum thermal limit.

As smart homes may be equipped with generation, renewable power production from PV panels should also be taken into account. Building-integrated rooftop PV systems are considered and their output is given by

$$P_{pv}(n, h) = N_s(n) \cdot N_p(n) \cdot P_{STC}(n) \cdot \left(G(h)/G_{STC}\right), \tag{9.5}$$

where $N_s(n)$ and $N_p(n)$ are the number of series and parallel-connected PV modules, respectively, and $P_{STC}(n)$ is the maximum power generation value of the PV module at G_{STC}, the irradiance value at the standard test conditions. $G(h)$ is the irradiance value at time h.

However, homes with PV panels do not significantly increase the complexity of the energy management algorithm, as solar energy is uncontrollable due to the natural dependence on solar radiation. The only option to increase the usage of the renewable energy is to schedule electricity appliances during high PV generation hours. However, to increase the local use of renewable generation, the user needs to define wide enough appliance control intervals $[\alpha(n, xa), \beta(n, xa)]$ to contain PV generation hours, which only occur during the day and may not correspond to the intervals typically expected by users. Therefore, to maximize the use of surplus energy generation during the day, smart homes need to be equipped with battery storage systems. The battery storage system formulation is given below:

$$P_b(n,h) = P_{pv}(n,h) - P_c(n,h) \tag{9.6}$$

If $P_{pv}(n,h) > P_c(n,h)$, the battery is charging:

$$P_b(n,h) = \begin{pmatrix} P_{\max\_ch}(n), & P_b(n,h) \geq P_{\max\_ch}(n), SOC_{batt}(n,h) < SOC_{\max}(n) \\ P_b(n,h), & P_b(n,h) < P_{\max\_ch}(n), SOC_{batt}(n,h) < SOC_{\max}(n) \\ 0, & SOC_{batt}(n,h) \geq SOC_{\max}(n) \end{pmatrix}, \tag{9.7}$$

where $P_{\max\_ch}(n)$ is maximum charge power and $SOC_{\max}(n)$ is the maximum state-of-charge (SOC) value of the battery storage. Otherwise, if $P_{pv}(n,h) \leq P_c(n,h)$, the battery is discharging:

$$P_b(n,h) = \begin{pmatrix} P_{\max\_dis}(n), & P_b(n,h) \leq P_{\max\_dis}(n), SOC_{batt}(n,h) > SOC_{\min}(n) \\ P_b(n,h), & P_b(n,h) > P_{\max\_dis}(n), SOC_{batt}(n,h) > SOC_{\min}(n) \\ 0, & SOC_{batt}(n,h) \leq SOC_{\min}(n) \end{pmatrix}, \tag{9.8}$$

where $P_{\max\_dis}(n)$ is maximum discharge power and $SOC_{\min}(n)$ is the minimum SOC value of the battery storage.

When a house is equipped with a battery system, many more possibilities arise for energy management. For example, home agents can decide to use electricity appliances without consuming electricity from the main grid by storing PV energy in the battery for use during hours without generation. Besides, to maximize their profit, home agents also gain the opportunity to sell extra generated renewable power when it is not needed.

For each smart home with PV and/or a battery, the total net power, surplus generation, and self-consumed energy (consumed power satisfied by the PV generation of the same home) for each time slot is given by

$$P_{net}(n,h) = P_c(n,h) + P_b(n,h) - P_{pv}(n,h) \tag{9.9}$$

$$P_{net}(n,h) = \begin{pmatrix} P_{net}(n,h), & P_{net}(n,h) \geq 0 \\ 0, & P_{net}(n,h) < 0 \end{pmatrix} \tag{9.10}$$

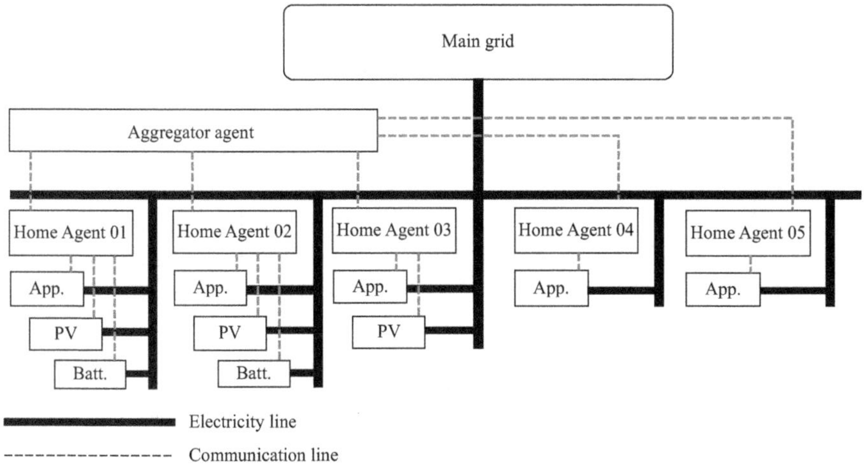

Figure 9.3 Structure of the modeled neighborhood

$$P_s(n,h) = \begin{pmatrix} 0, & P_{net}(n,h) \geq 0 \\ P_{net}(n,h)\cdot(-1), & P_{net}(n,h) < 0 \end{pmatrix} \quad (9.11)$$

$$P_{sc}(n,h) = P_c(n,h) - P_{net}(n,h) \quad (9.12)$$

9.2.3 Neighborhood model

A two-layer system architecture, similar to the one described in [22], is chosen and shown in Figure 9.3. It includes an aggregator and five smart homes. In this work, the coordination of a single neighborhood is studied, thus one aggregator with five smart homes is considered. However, a similar approach could be used in a larger system. All smart homes are assumed to have two-way communication ability with the aggregator through AMI.

At the upper level, the aggregator agent (which can also be a utility) sends price data to home agents and also aggregates the information (see Section 9.3) received from home agents and sends the aggregated information back to each home agent. This enables home agents to have information about the neighborhood electricity profile through communicating with the aggregator, which may be more mindful of privacy than direct communication of home agents with each other. At the bottom level, the energy management of the smart homes is performed by the home agents using price and aggregated power data received from the aggregator agent. Smart homes are equipped differently, i.e., they may be equipped with PV and storage or not.

9.2.4 Electricity price model

A time-varying price signal is chosen to evaluate the scheduling algorithm, in order to enable customers to increase their benefits. In this work, time-of-use (TOU)

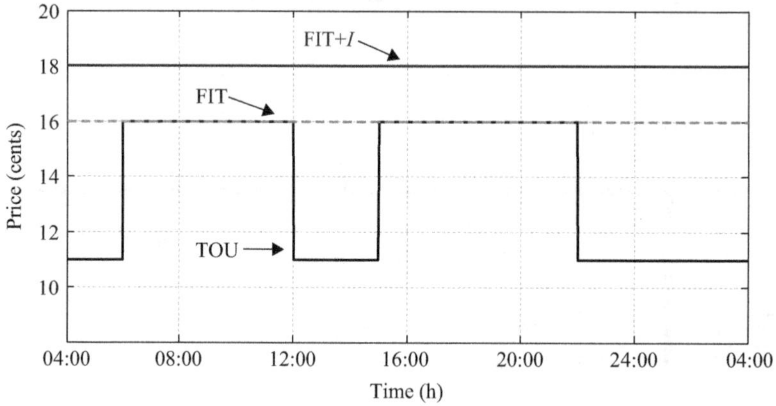

Figure 9.4 Time-of-use, feed-in-tariff, and incentive price

pricing $\lambda_c(h)$ (€0.11/kWh off-peak, €0.16/kWh on-peak) from the French electricity (Figure 9.4) is deployed. This scheme is used throughout the year, and does not change with seasons or week days. TOU is a form of price-based DR, and relies on using different prices for different time blocks: the higher the demand, the higher the price. This scheme is expected to help utilities mitigate demand peaks, while enabling end-users to avoid high price values by scheduling their appliance run times to low-price periods. A feed-in-tariff (FIT) λ_{FIT} (€0.16/kWh) is also used to encourage the development of renewable energy installations (typically PV) by consumers. An FIT is the price used for selling local generation to the grid, and is generally higher than the TOU price. The difference is typically supported by taxes on electricity.

In this study, an incentive price is used to increase the interest in self-consumption, e.g., through batteries and load scheduling to the high generation hours. The incentive price λ_I is set to €0.2/kWh for smart homes. In this study, λ_I is integrated to the current electricity tariffs. With TOU, FIT, and incentive, the daily electricity cost of a smart home (without scheduling) is given by

$$C(n,d) = \sum_{h=1}^{H} (P_{net}(n,h) \cdot \lambda_c(h) + P_s(n,h) \cdot \lambda_{FIT} + P_{sc}(n,h) \cdot (\lambda_{FIT} + \lambda_I - \lambda_c(h))) \cdot \Delta h \quad (9.13)$$

where $C(n,d)$ is the electricity bill of the smart home n for day d and Δh is the time interval of time h. According to (9.13), the integration of λ_I to the total electricity price is given in Figure 9.4. If a smart home uses its own generated energy for its consumption, it earns extra benefits $(\lambda_{FIT} + \lambda_I)$ compared to just selling with λ_{FIT} to the grid. Therefore, when a smart home uses the self-generated electricity, it is receiving $(\lambda_{FIT} + \lambda_I - \lambda_c(h))$ without paying for consumption.

Although Figure 9.3 shows a system structure similar to the ones used in [24, 22], this approach is significantly different in that it does not rely on wholesale markets, but

Table 9.2 Comparison of the control approaches (+: used, −: not used)

	Baseline	Selfish	Distributed	Centralized
Appliance scheduling	−	+	+	+
Communication ability	−	+	+	+
Energy trading	−	−	+	+
Objectives	Basic appliance and generation management	Minimize electricity bills of the smart home	Minimize electricity bills of the smart home with electricity trading	Minimize electricity bill of neighborhood with electricity trading

focuses only on local trading, and assumes a government incentive for encouraging RES use.

9.3 Control Strategies

In this section, four algorithms are presented (Table 9.2) to compare the system performance for investigation of the coordination mechanisms effectiveness. The algorithms are referred to as: baseline, selfish, centralized, and distributed.

In this work, it is assumed that batteries cannot be charged using grid power in order to reduce the trade from the grid, or from neighbors' surplus generation (this possibility is left for future research). They can only be charged by the power produced by self-owned PV systems. Moreover, we assumed that home agents also have the ability to perfectly forecast their controllable and uncontrollable appliance usage as well as the electricity generation from PV. The impact of forecasting errors is also left for future work. Another vital point of this study is to reduce electricity bills of the smart homes without changing the total electric energy consumption amount of the smart homes. Therefore, electricity appliance scheduling time is limited to the end of the scheduling horizon. Lastly, a 1-minute-based simulation time step is used to model the consumption, generation, and storage profile characteristics of smart homes in all control algorithms. However, different time step intervals are used for the communication process of the home agents, which will be explained for each control algorithm.

9.3.1 Baseline algorithm

This first control algorithm serves as a reference, to show the effectiveness of the developed control algorithms. No optimization and communication ability is defined, therefore, price data are not used for appliance scheduling or battery management

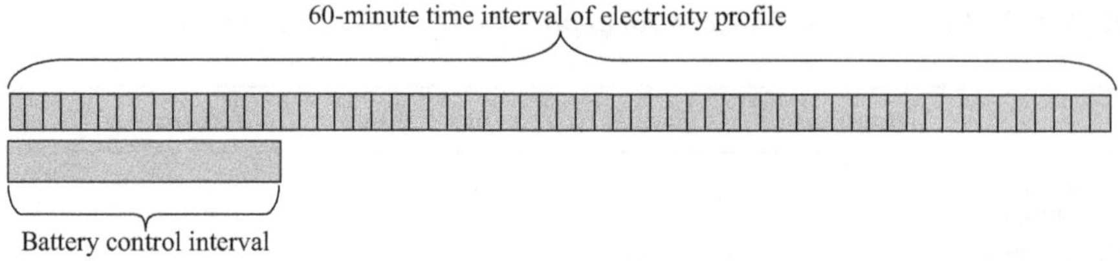

Figure 9.5 15-minute battery control interval in 1 hour

in this algorithm. With baseline, all appliances are modeled as non-deferrable loads. Home agents use extra generated electricity for self-consumption if there is battery storage in the smart home. The battery is only allowed to discharge when there is electricity consumption in the smart home, and without exceeding maximum battery discharge power. Homes without battery have no ability to control PV generation, thus they only use generated power at generation hours and feed the surplus generation back to the main grid. In this control algorithm, battery charging and discharging are decided with a 1-minute time step. Smart homes can sell surplus generation to the grid with λ_{FIT}. However, there is no automatic appliance management. The total cost of the smart homes is calculated using (9.13).

9.3.2 Selfish energy management

In selfish energy management, each home agent uses $\lambda_c(h)$, λ_{FIT}, and λ_I to optimally schedule its own electricity appliances and for battery management (charge, discharge, and idle), while taking into account user preferences. Home agents do not have any information related to their neighbors' activities; hence, there is no electric energy transfer among smart homes. In this study, a 15-minute time interval is used to control battery operations (see Figure 9.5 for 1 hour). Using a shorter time step would significantly increase the size of the optimization problem and reduce the performance of the algorithm.

During the battery control interval, the home agent decides about discharging or staying in idle mode for optimum battery management. The home agent cannot choose to charge because the battery is only able to charge the self-generated power whenever generation is higher than consumption in the smart home. Two indexes are used to control the battery operation, by comparing the generation and the consumption rates:

$$P_b(n,h) = \left\{ \begin{array}{cc} P_{pv}(n,h) - P_c(n,h), & \omega(n,k) = 0\&P_{pv}(n,h) > P_c(n,h) \\ 0, & \omega(n,k) = 0\&P_{pv}(n,h) \leq P_c(n,h) \\ P_{pv}(n,h) - P_c(n,h) & \omega(n,k) = 1 \end{array} \right\}$$

$$h = [1,2,3,\ldots,15] + (k-1)\cdot 15, \qquad k = \{1,\ldots,K\} \tag{9.14}$$

where $\omega(n,k)$ is the decision index for battery management, k is the battery control index, and $K = H/15$ is the total number of control intervals in the scheduling horizon.

Using (9.9) to (9.12), the objective function for the selfish energy management of each smart home is given by

$$\min C(n,d) = \sum_{h=1}^{H} (P_{net}(n,h) \cdot \lambda_c(h) - P_s(n,h) \cdot \lambda_{FIT} - P_{sc}(n,h) \cdot (\lambda_{FIT} + \lambda_I - \lambda_c(h)) \cdot \Delta h \tag{9.15}$$

9.3.3 Coordinated Management: distributed

In this section, the proposed coordination mechanism is presented to increase the usage of residential renewable energy inside the neighborhood area by establishing electricity trading ability among home agents. As described earlier, to enable this type of electricity trading, there should be a communication link between home agents and the aggregator agent. The case where no aggregator agent exists is not considered in this work. The coordination mechanism is divided into two categories according to decision-making process in the neighborhood: home agents and the aggregator. This section presents the distributed coordination of neighborhood area.

The distributed coordination of the smart homes requires frequent message exchange between home agents and the aggregator agent, as shown in Figure 9.6. Messages are used to inform home agents about changes in the neighborhood electricity profile with aggregate data. Home agents optimize their electricity profiles at the same time rather than using turn-based scheduling, hence, they are only informed of previous aggregate data for the neighborhood electricity profile.

The distributed optimization process is separated into several steps, as shown in Figure 9.6.

Step 1:
In the first step, the aggregator agent sends $\lambda_c(h)$, λ_{FIT}, and λ_I to the home agents. When a home agent receives these price data, it runs the optimization using (9.15) to obtain the appliance operation times and battery controls, as for selfish control. In this algorithm, however, the home agent sends two types of data: the net consumption with (9.10), and surplus generation with (9.11), to inform the aggregator agent about its electricity profile. To reduce communication requirements, rather than using the actual 1-minute information, the 30-minute average is used

$$P^m_{net}(n,l) = \sum_{l=1}^{H/L} \sum_{h=1}^{L} \frac{P_{net}(n, h + (l-1) \cdot L)}{L} \tag{9.16}$$

$$P^m_s(n,l) = \sum_{l=1}^{H/L} \sum_{h=1}^{L} \frac{P_s(n, h + (l-1) \cdot L)}{L} \tag{9.17}$$

where $l = \{1, \ldots, H/L\}$ is the index of communication data length, L is the interval value of mean data calculation, $P^m_{net}(n,l)$ is the communication data profile of $P_{net}(n,h)$, and $P^m_s(n,l)$ is the communication data profile of $P_s(n,h)$. The actual electricity profile and calculated communication data for electricity consumption are given in Figure 9.7. Home agents send $P^m_{net}(n,l)$ and $P^m_s(n,l)$ to the aggregator agent to

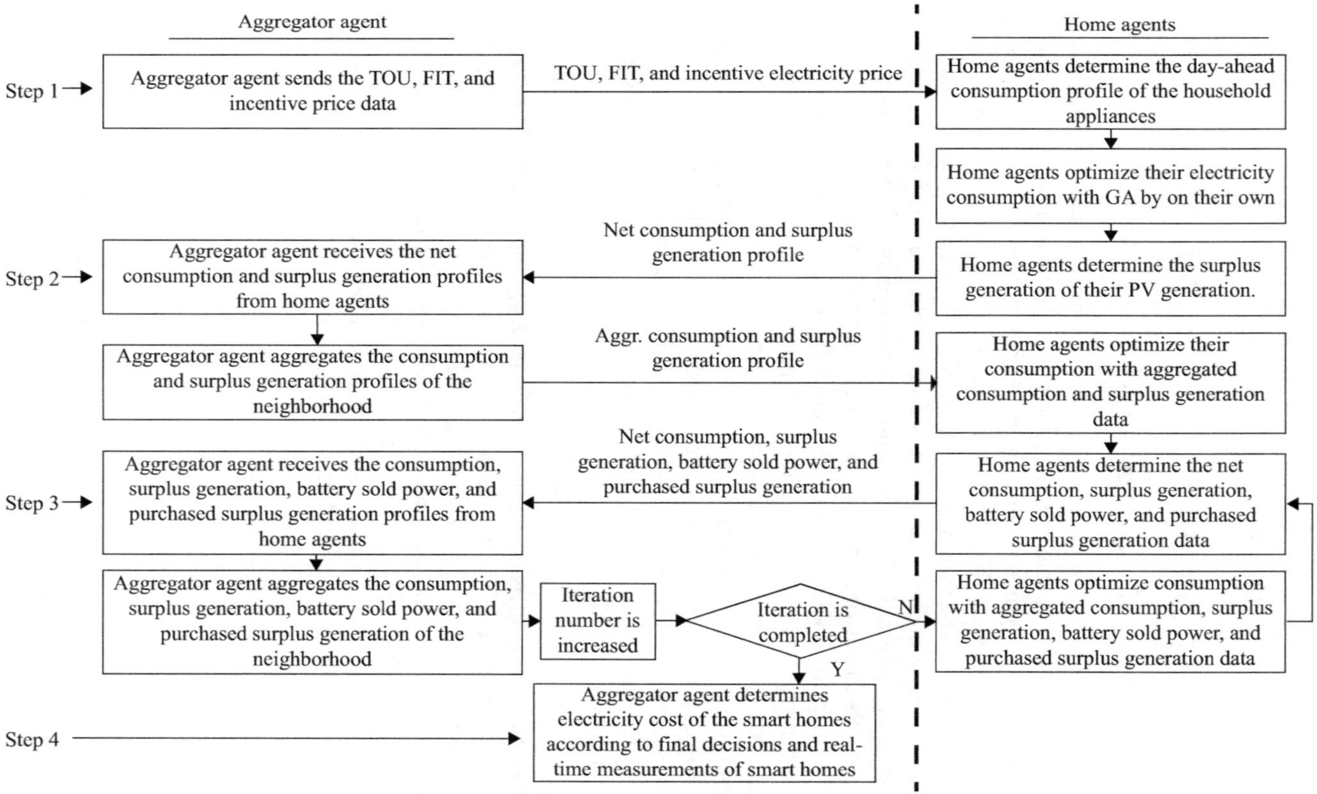

Figure 9.6 Communication flowchart of the coordinated distributed management algorithm

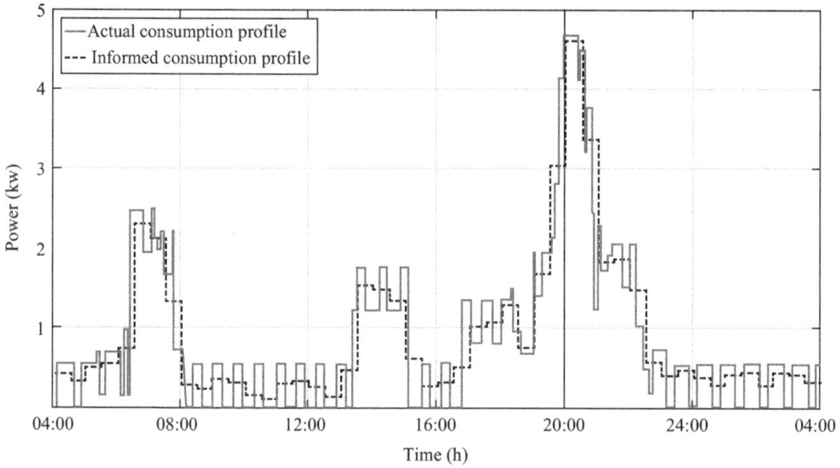

Figure 9.7 Actual (1-minute resolution, physical simulation data) and transmitted (30-minute resolution, communication data) consumption profile

inform it about their electricity profiles, and the aggregator agent, in turn, aggregates the received profiles to create a neighborhood electricity profile.

Step 2:
After the aggregator agent has created the neighborhood profile, it sends the communication data (i.e., the informed consumption profile shown in Figure 9.7) to the home agents, with $P^m_{ac}(l)$ and $P^m_{as}(l)$

$$P^m_{ac}(l) = \sum_{n=1}^{N} P^m_{net}(n, l) \tag{9.18}$$

$$P^m_{as}(l) = \sum_{n=1}^{N} P^m_{s}(n, l) \tag{9.19}$$

The home agent is informed about the neighborhood status and can decide about selling/buying electric energy to/from its neighbors. Then, to trade electricity, the home agent calculates the neighborhood electricity profile from home perspective:

$$P^m_{pac}(n, l) = P^m_{ac}(l) - P^m_{net}(n, l) \tag{9.20}$$

$$P^m_{pas}(n, l) = P^m_{as}(l) - P^m_{s}(n, l) \tag{9.21}$$

and changes the received data time index (from l to h, i.e., from a 30-minute resolution to a 1-minute resolution):

$$P_{pac}(n, h) = P^m_{pac}(n, l), \quad h = [1, \ldots, L] + (l - 1) \cdot L \tag{9.22}$$

$$P_{pas}(n, h) = P^m_{pas}(n, l), \quad h = [1, \ldots, L] + (l - 1) \cdot L \tag{9.23}$$

where $P_{pac}(n, h)$ and $P_{pas}(n, h)$ are the neighborhood consumption and generation profiles from the perspective of the home agents with a 1-minute time resolution.

Using the received price data and $P_{pac}(n,h)$ and $P_{pas}(n,h)$, home agents then optimize their electricity profile. Battery-equipped home agents also decide about the best battery operation according to conditional changes in $P_c(n,h)$, $P_{pv}(n,h)$, and $P_{ac}(n,h)$. First, home agents calculate the net electricity consumption:

$$P_{net-pac}(n,h) = P_{pac}(n,h) - P_{pas}(n,h), \tag{9.24}$$

then calculate the battery power:

$$P_b(n,t) = \begin{Bmatrix} P_{pv}(n,h) - P_c(n,h), & \omega(n,k)=0, P_{pv}(n,h)>0, P_{pv}(n,h)>P_c(n,h), \\ & P_{net-pac}(n,h)=0 \\ 0, & \omega(n,k)=0, P_{pv}(n,h)>0, P_{pv}(n,h)\le P_c(n,h), \\ & P_{net-pac}(n,h)=0 \\ P_{pv}(n,h)-P_c(n,h)-P_{ac}(n,h), & \omega(n,k)=0, P_{pv}(n,h)>0, P_{net-pac}(n,h)>0 \\ P_c(n,h)\cdot(-1), & \omega(n,k)=0, P_{pv}(n,h)=0, P_{net-pac}(n,h)=0 \\ (P_c(n,h)+P_{ac}(n,h))\cdot(-1), & \omega(n,k)=0, P_{pv}(n,h)=0, P_{net-pac}(n,h)>0 \\ P_{pv}(n,h)-P_c(n,h), & \omega(n,k)=1, P_{pv}(n,h)>0, P_{pv}(n,h)>P_c(n,h) \\ 0, & \omega(n,k)=1, P_{pv}(n,h)>0, P_{pv}(n,h)\le P_c(n,h) \\ 0, & \omega(n,k)=1, P_{pv}(n,h)=0, P_{net-pac}(n,h)=0 \\ P_c(n,h), & \omega(n,k)=1, P_{pv}(n,h)=0, P_{net-pac}(n,h)>0 \end{Bmatrix} \tag{9.25}$$

Compared to selfish battery operation, home agents control battery modes (discharge or idle) and charge whenever generation is higher than consumption. They also decide about the power sold to the neighbors $P_{b-s}(n,h)$ by using $P_{ac}(n,h)$ in (9.25), by calculating the battery discharge power. The power sold with battery discharge is then given by

$$P_{b-s}(n,h) = \begin{Bmatrix} 0, & \omega(n,k)=0, (P_c(n,h)\ge P_b(n,h)\cdot(-1)) \\ (P_b(n,h)+P_c(n,h))\cdot(-1), & \omega(n,k)=0, (P_c(n,h)<P_b(n,h)\cdot(-1)) \\ 0, & \omega(n,h)=1 \end{Bmatrix} \tag{9.26}$$

After that, home agents calculate their net consumption:

$$P_{net}(n,h) = P_c(n,h) - P_{pv}(n,h) + P_b(n,h) + P_{b-s}(n,h) \tag{9.27}$$

After using (9.10) and (9.11), home agents decide about purchasing electricity from their neighbors:

$$P_{s-p}(n,h) = \begin{Bmatrix} P_{net}(n,h), & P_{pas}(n,h) > P_{pac}(n,h)+P_{net}(n,h) \\ P_{pas}(n,h)\cdot\left(\dfrac{P_{net}(n,h)}{P_{pac}(n,h)+P_{net}(n,h)}\right), & P_{pas}(n,h)\le P_{pac}(n,h)+P_{net}(n,h) \end{Bmatrix} \tag{9.28}$$

where $P_{s-p}(n,h)$ is the purchased power from surplus generation of the neighbors. After that, home agents determine again their net consumption by adding $P_{s-p}(n,h)$:

$$P_{net}(n,h) = P_c(n,h) - P_{pv}(n,h) + P_b(n,h) + P_{b-s}(n,h) - P_{s-p}(n,h) \quad (9.29)$$

Again, (9.10) and (9.11) are used and the self-consumed power is calculated:

$$P_{sc}(n,h) = P_c(n,h) - P_{net}(n,h) - P_{s-p}(n,h) \quad (9.30)$$

At the end of the second step, home agents optimize their operation with the following objective function:

$$\min C(n,d) = \sum_{h=1}^{H} \begin{array}{l}(P_{net}(n,h) \cdot \lambda_c(h) + P_{s-p}(n,h) \cdot (\lambda_c(h) - \lambda_I/2) - P_s(n,h) \cdot \lambda_{FIT} \\ -P_{b-s}(n,h) \cdot (\lambda_{FIT} + \lambda_I/2) - P_{self}(n,h) \cdot (\lambda_{FIT} + \lambda_I - \lambda_c(h))) \cdot \Delta h\end{array} \quad (9.31)$$

After solving this optimization problem, home agents inform the aggregator agent by determining the communication data (i.e., the data that is transmitted by home agents to the aggregator agent): $P^m_{net}(n,l)$ and $P^m_s(n,l)$ with:

$$P^m_{b-s}(n,l) = \sum_{l=1}^{H/L} \sum_{h=1}^{L} \frac{P_{b-s}(n, h + (l-1) \cdot L)}{L} \quad (9.32)$$

$$P^m_{s-p}(n,l) = \sum_{l=1}^{H/L} \sum_{h=1}^{L} \frac{P_{s-p}(n, h + (l-1) \cdot L)}{L} \quad (9.33)$$

where $P^m_{b-s}(n,l)$ and $P^m_{s-p}(n,l)$ are the communication data of the sold power with battery discharge and the surplus purchased power of the smart home, respectively. In this algorithm, according to (9.31), home agents are not informed about who are buyer and seller during the decision-making for trading. This is an important part of the proposed algorithm. For instance, if a home agent buys surplus generation power, it buys directly from the aggregate neighborhood surplus profile, not individually from a seller. The same holds for a battery power seller. If a home agent discharges its battery to provide electricity for neighborhood consumption, it discharges power directly for selling to the aggregate neighborhood net consumption profile, not individually to any consumer. A home agent assumes that there is consumption/generation power at that time that can be used, but these can be changed by other home agents decision-making. Therefore, due to decision-making of all home agents at the same time, source matching is run at the end of the algorithm by the aggregator agent, using a proportional source-matching method that is not included in the objective function. Second, the incentive revenue is assumed to be shared between the buyer and seller in (9.31). Especially, in battery-equipped smart homes, the excess electricity is first used for self-consumption to earn the entire incentive (λ_I) with FIT. After that, if there is electricity that can be sold to neighbors, then home agents focus on sharing the incentive equally (i.e., $\lambda_I/2$) with each other: the seller smart home earns (FIT + $\lambda_I/2$) and the buyer smart homes earn ($\lambda_I/2$), which may be viewed as a discount on bought electricity. In smart homes with a battery, the possibility of receiving the full incentive (λ_I) is higher compared to smart homes with only PV.

Step 3:
At the end of the second step, home agents inform the aggregator agent by sending it $P^m_{net}(n,l)$, $P^m_s(n,l)$, $P^m_{b-s}(n,l)$, and $P^m_{s-p}(n,l)$, then the aggregator agent creates the neighborhood electricity profile using (9.18) and (9.19). The aggregated sold power with battery discharge and purchased power from surplus generation are given by

$$P^m_{ab-s}(l) = \sum_{n=1}^{N} P^m_{b-s}(n,l) \tag{9.34}$$

$$P^m_{as-p}(l) = \sum_{n=1}^{N} P^m_{s-p}(n,l) \tag{9.35}$$

After that, the aggregator agent sends aggregated electricity data to the home agents. The proposed algorithm iterates for a limited number of times that is arbitrarily decided. The loop is necessary to inform the home agents about the latest changes in the neighborhood profile. When home agents receive the aggregated electricity data for the neighborhood, they create their own perspective data at h domain using (9.20) with (9.22), (9.21) with (9.23). $P_{pab-s}(n,h)$ and $P_{pas-p}(n,h)$ are the neighborhood battery sold and purchased surplus generation profiles, respectively, from perspective of home agent in 1-minute time resolution:

$$P^m_{pab-s}(n,l) = P^m_{ab-s}(l) - P^m_{b-s}(n,l) \tag{9.36}$$

$$P^m_{pas-p}(n,l) = P^m_{as-p}(l) - P^m_{s-p}(n,l) \tag{9.37}$$

$$P_{pab-s}(n,h) = P^m_{pab-s}(n,l), \quad h = [1,\ldots,L] + (l-1)\cdot L \tag{9.38}$$

$$P_{pas-p}(n,h) = P^m_{pas-p}(n,l), \quad h = [1,\ldots,L] + (l-1)\cdot L \tag{9.39}$$

Using price data and the received electricity data, home agents optimize their electricity profile again. Agents use (9.24) and (9.25) for battery discharge and then decide about the available energy stored in the battery to sell for neighborhood consumption using (9.26). After that, home agents take into account other home agents decision using $P_{pab-s}(n,h)$ for selling power with battery discharge:

$$P_{b-s}(n,h) = \left\{\begin{array}{ll} P_{b-s}(n,h), & P_{net-ac}(n,h) \geq P_{pab-s}(n,h) + P_{b-s}(n,h) \\ P_{net-ac}(n,h) - P_{pab-s}(n,h), & P_{net-ac}(n,h) < P_{pab-s}(n,h) + P_{b-s}(n,h) \end{array}\right\} \tag{9.40}$$

After that, home agents use (9.27), (9.10), and (9.11) and calculate the surplus purchased power, considering $P_{pas-p}(n,h)$:

$$P_{s-p}(n,h) = \left\{\begin{array}{ll} P_{net}(n,h), & P_{pas}(n,h) > P_{pac}(n,h) + P_{net}(n,h) \\ P_{pas}(n,h)\cdot\left(\dfrac{P_{net}(n,h)}{P_{pas-p}(n,h) + P_{net}(n,h)}\right), & P_{pas}(n,h) \leq P_{pac}(n,h) + P_{net}(n,h) \end{array}\right\} \tag{9.41}$$

Home agents decide about purchasing and selling electricity, while taking into account neighboring home agents decision, and optimize their electricity cost using (9.31). At the end of the optimization, home agents determine the communication data $P^m_{net}(n,l)$, $P^m_s(n,l)$, $P^m_{b-s}(n,l)$, $P^m_{s-p}(n,l)$ and send them to the aggregator agent.

This process continues until the iteration number exceeds 3, for this work. When it does, the aggregator agent considers the latest decision of home agents as their final decision and calculates the electricity cost of smart homes. The proportional source-matching method is used to decide about the amount to trade (sold and bought) for each smart home individually. The method distributes the amount of sold and bought power according to the ratio of the smart home electricity profile with respect to the aggregated power profile. Therefore, the aggregator agent, after using (9.29), (9.10), and (9.11) in that order, calculates the amount of bought and sold energy in the neighborhood. In the formulations for the calculation of the sold and bought power, "*d*" indicates the latest decision of a home agent.

Step 4:

$$P_{s-p}(n,h) = \begin{cases} P_{net}(n,h), & P_{as}(h) \geq P^d_{as-p}(h), P^d_{s-p}(n,h) > P_{net}(n,h) \\ P^d_{s-p}(n,h), & P_{as}(h) \geq P^d_{as-p}(h), P^d_{s-p}(n,h) \leq P_{net}(n,h) \\ P_{net}(n,h), & P_{as}(h) < P^d_{as-p}(h), P_{as}(h) \cdot \left(\dfrac{P^d_{s-p}(n,h)}{P^d_{as-p}(h)}\right) > P_{net}(n,h) \\ P_{as}(h) \cdot \left(\dfrac{P^d_{s-p}(n,h)}{P^d_{as-p}(h)}\right), & P_{as}(h) < P^d_{as-p}(h), P_{as}(h) \cdot \left(\dfrac{P^d_{s-p}(n,h)}{P^d_{as-p}(n,h)}\right) \leq P_{net}(n,h) \end{cases} \tag{9.42}$$

$$P_{s-s}(n,h) = \begin{cases} P_s(n,h), & P_{as}(h) > P_s(n,h), P_{as-p}(h) = P_{as}(h) \\ P_{as-p}(h) \cdot \left(\dfrac{P_s(n,h)}{P_{as}(h)}\right), & P_{as}(h) > P_s(n,h), P_{as-p}(h) < P_{as}(h) \\ P_s(n,h), & P_{as}(h) = P_s(n,h), P_{as-p}(h) = P_s(n,h) \\ P_{as-p}(h), & P_{as}(h) = P_s(n,h), P_{as-p}(h) < P_s(n,h) \end{cases} \tag{9.43}$$

where $P_{s-s}(n,h)$ is sold power from surplus generation. According to purchased surplus generation, new values for $P^{(2)}_{net}(n,h)$ and $P^{(2)}_{net-ac}(h)$ are determined for power sold from battery discharge:

$$P^{(2)}_{net}(n,h) = P^{(1)}_{net}(n,h) - P_{s-p}(n,h) \tag{9.44}$$

$$P^{(2)}_{net-ac}(h) = P^{(1)}_{net-ac}(h) - \sum_{n=1}^{N} P_{s-p}(n,h) \tag{9.45}$$

$$P_{b-p}(n,h)=\begin{cases} P^{(2)}_{net}(n,h), & P^{(2)}_{net-ac}(h)>P^{(2)}_{net}(n,h), P^{d}_{ab-s}(h)\geq P^{(2)}_{net-ac}(h) \\ P^{d}_{ab-s}(n,h)\cdot\left(\dfrac{P^{(2)}_{net}(n,h)}{P^{(2)}_{net-ac}(h)}\right), & P^{(2)}_{net-ac}(h)>P^{(2)}_{net}(n,h), P^{d}_{ab-s}(h)<P^{(2)}_{net-ac}(h) \\ P^{(2)}_{net}(n,h), & P^{(2)}_{net-ac}(h)=P^{(2)}_{net}(n,h), P^{d}_{ab-s}(h)\geq P^{(2)}_{net-ac}(h) \\ P^{d}_{ab-s}(n,h), & P^{(2)}_{net-ac}(h)=P^{(2)}_{net}(n,h), P^{d}_{ab-s}(h<P^{(2)}_{net-ac}(h) \end{cases} \tag{9.46}$$

$$P_{b-s}(n,h)=\begin{cases} P_{ab-p}(h), & P^{d}_{ab-s}(h)=P^{d}_{b-s}(n,h), P^{d}_{b-s}(n,h)>P_{ab-p}(h) \\ P^{d}_{b-s}(n,h), & P^{d}_{ab-s}(h)=P^{d}_{b-s}(n,h), P^{d}_{b-s}(n,h)=P_{ab-p}(h) \\ P_{ab-p}(h)\cdot\left(\dfrac{P^{d}_{b-s}(n,h)}{P^{d}_{ab-s}(h)}\right), & P^{d}_{ab-s}(h)>P^{d}_{b-s}(n,h), P^{d}_{ab-s}(h)>P_{ab-p}(h) \\ P^{d}_{b-s}(n,h), & P^{d}_{ab-s}(h)>P^{d}_{b-s}(n,h), P^{d}_{ab-s}(h)=P_{ab-p}(h) \end{cases} \tag{9.47}$$

where $P_{b-p}(n,h)$ is the power purchased from battery discharge. In the equations, a detailed comparison of the electricity profile is used to decide the amount of energy that can be sold and purchased by the smart homes. Proportional source matching is used when the electricity sources are insufficient (e.g., when the generated real-time surplus generation is not enough to provide the decided purchased surplus power by the neighbor home agents). After that, the smart home cost of electricity consumption $C(n,d)$ is calculated using:

$$P_{sold}(n,h)=P_{s-s}(n,h)+P_{b-s}(n,h) \tag{9.48}$$

$$P_{purchased}(n,h)=P_{s-p}(n,h)+P_{b-p}(n,h) \tag{9.49}$$

$$P_{net}(n,h)=P_{c}(n,h)-P_{pv}(n,h)+P_{b}(n,h)+P_{sold}(n,h)-P_{purchased}(n,h) \tag{9.50}$$

Equations (9.10) and (9.11) are used for $P_{net}(n,h)$ and $P_{s}(n,h)$, then the self-consumed power is calculated:

$$P_{sc}(n,h)=P_{c}(n,h)-P_{net}(n,h)-P_{purchased}(n,h) \tag{9.51}$$

$$C(n,d)=\left(\sum_{h=1}^{H} \begin{matrix} P_{net}(n,h)\cdot\lambda_c(h)+P_{purchased}(n,h)\cdot(\lambda_c(h)-\lambda_I/2) \\ -P_{sc}(n,h)\cdot(\lambda_{FIT}+\lambda_I-\lambda_c(h))-P_{sold}(n,h)\cdot(\lambda_{FIT}+\lambda_I/2)\end{matrix}\right)\cdot\Delta h \tag{9.52}$$

With proportional source matching, every smart home has the opportunity to earn a profit, even if there is the total tradable electricity amount in the neighborhood is limited.

Another important point is that communication data are determined by taking the 30-minute average of the electricity profile. During this interval, there can be ups and downs in the electricity profile; hence, there can be mismatches between real-time data and the average. Such mismatches are handled by the proportional source-matching method. The performance of the presented algorithm can be increased by using more accurate data for communication through statistical analysis, rather than using a simple average.

It should be noted that with this algorithm, to trade electricity among smart homes, there should be excess generation in smart homes with PV due to the constant defined incentive price for all day. The algorithm can be updated to use a time-varying incentive, as for TOU, thus the efficiency of the renewable energy usage could be slightly increased in the neighborhood area by allocating battery discharge specifically to on-peak hours. The pseudo-code for the algorithm is given below.

Pseudo-code for distributed coordination, as seen from a smart home – *HDS*: home data set, *CDS*: communication data set, *ADS*: aggregated data set, APS_l: aggregated perspective data (l domain), APS_h: aggregated perspective data (h domain)

Get λ_c, λ_{FIT}, and λ_I
Optimize $Cost \rightarrow \min C(n, d)$
for $h = 1$ to H **do**
    Determine $HDS = \{P_c, P_{pv}, P_b, P_{net}, P_s, P_{sc}\}$
end for
for $l = 1$ to L **do**
Determine $CDS = \{P^m_{net}, P^m_s\}$
end for
Send *CDS* to aggregator
—
Get $ADS = \{P^m_{ac}, P^m_{as}\}$ from aggregator and transform to $APS_l = \{P^m_{pac}, P^m_{pas}\}$
Transform data from l to h domain, $APS_h = \{P_{pac}, P_{pas}\}$
Optimize $Cost \rightarrow \min C(n, d)$
for $h = 1$ to H **do**
    Determine new $HDS = \{P_c, P_{pv}, P_b, P_{net}, P_s, P_{sc}, P_{b-s}, P_{s-p}\}$
end for
for $l = 1$ to L **do**
    Determine new $CDS = \{P^m_{net}, P^m_s, P^m_{b-s}, P^m_{s-p}\}$
end for
Send *CDS* to aggregator
—

---

for *iteration* $= 1$ to 3 **do**
    Get $ADS = \{P^{m}_{ac}, P^{m}_{as}, P^{m}_{ab-s}, P^{m}_{as-p}\}$ from aggregator and transform to $APS_l = \{P^{m}_{pac}, P^{m}_{pas}, P^{m}_{pab-s}, P^{m}_{pas-p}\}$
    Transform data from l to h domain, $APS_h = \{P_{pac}, P_{pas}, P_{pab-s}, P_{pas-p}\}$
    Optimize $Cost \rightarrow \min C(n,d)$
    for $h = 1$ to H **do**
        Determine new $HDS = \{P_c, P_{pv}, P_b, P_{net}, P_s, P_{sc}, P_{b-s}, P_{s-p}\}$
    end for
    for $l = 1$ to L **do**
        Determine new $CDS = \{P^{m}_{net}, P^{m}_{s}, P^{m}_{b-s}, P^{m}_{s-p}\}$
    end for
    Send *CDS* to aggregator
end for
—

9.3.4 Coordinated Management: centralized

The second coordination mechanism is an optimization run by a central entity, here the aggregator. While home agents are frequently exchanging messages with the aggregator agent for distributed coordination, home agents send all the information about the household equipment (appliances, PV, and battery) to the aggregator agent one time as shown in Figure 9.8. The aggregator agent applies the optimization algorithm for smart homes and optimizes the neighborhood electricity profile to achieve the best overall cost reduction. After that, the aggregator agent sends the results for appliance and battery control decisions to the home agents, which, in turn, apply the control decisions.

In centralized coordination, optimization is run once by the aggregator agent, and not by the home agents themselves. However, the optimization algorithm works similar to the distributed coordination. The only difference is that the central entity focuses on the cost reduction of the neighborhood rather than of individual homes:

$$\min C(d) = \left\{ \sum_{n=1}^{N} \Big(\sum_{h=1}^{H} \begin{matrix} P_{net}(n,h)\cdot\lambda_c(h) + P_{purchased}(n,h)\cdot(\lambda_c(h) - \lambda_I/2) \\ -P_{sc}(n,h)\cdot(\lambda_{FIT} + \lambda_I - \lambda_c(h)) - P_{sold}(n,h)\cdot(\lambda_{FIT} + \lambda_I/2) \end{matrix} \Big) \cdot \Delta h \right\} \tag{9.53}$$

This centralized optimization has the advantage of being able to obtain (near) optimal solutions, as all information is available. The disadvantage of this method is its

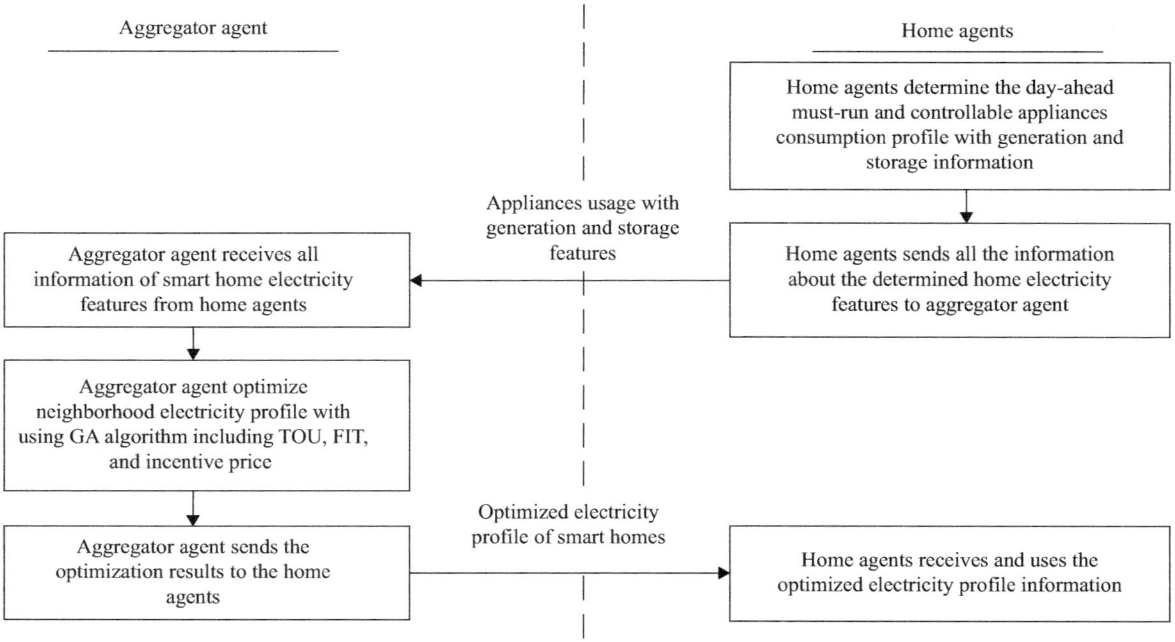

Figure 9.8 Communication flowchart of the coordinated centralized management algorithm

high computation time, due to a large number of decision variables. The pseudo-code this strategy is given below.

Pseudo-code for centralized coordination, as seen from the aggregator

---

1. **for** $n = 1$ to N **do**

2. Get all appliance, PV and battery information from the smart homes

Optimize $Cost \rightarrow \min C(d)$

end for

3. Send appliance operation times and battery control decisions to smart homes

---

9.4 Simulation Results

The four algorithms were implemented and compared in simulation, using the test system with five smart homes (two smart homes with PV and battery, one smart home with just PV, and two smart homes without PV and battery, as described in Section 9.2.3). The designed day-ahead management algorithms are run over a two-day horizon, using the rolling horizon technique shown in Figure 9.9. Home agents can improve the energy management by saving some energy in the batteries. On the first day, it is assumed that all controllable and must-run appliances are operating and in the second day it is assumed that except two controllable appliance (the washing machine and the clothes dryer, in order to reflect realistic usage patterns), others (must-run appliances and dish washer) are used. The algorithms try to reduce the electricity cost without decreasing the total electricity consumption of the smart homes during the planned scheduling horizon. The scheduling period for one day is assumed to start at 4 a.m. and end 24 hours later.

Overall, the simulation time is 2,880 minutes (48 hours), and the algorithm is run every 1,440 minutes (24 hours).

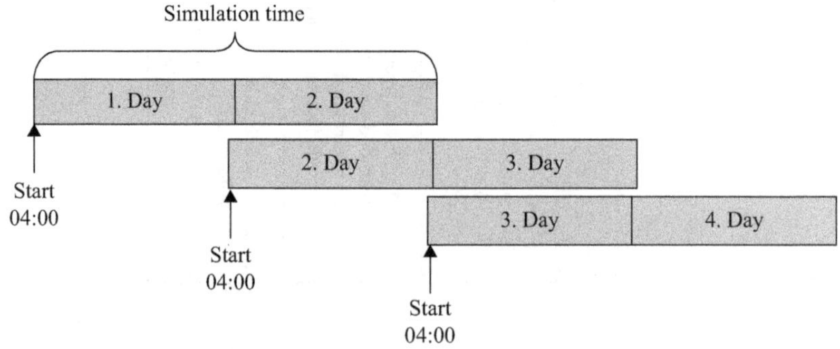

Figure 9.9 Rolling horizon optimization technique working diagram

9.4.1 Simulation setup

The simulation is performed on a dedicated desktop computer with an Intel Core i7 3.40 GHz processor and 8 GB RAM with a 64-bit Windows 7 Pro. A co-simulation setup, shown in Figure 9.10 and adapted from [25], was designed with Matlab and JADE, a MAS platform. This setup is used to model agent communication (JADE) and run the optimization algorithm (Matlab) in a simulation environment. The aggregator and home agents are modeled in JADE. The home agents are equipped with a genetic algorithm (Matlab built-in *ga* function) to run the scheduling algorithm. To enable the co-simulation, an interface was designed to transfer data between JADE and MATLAB using Transmission Control Protocol/Internet Protocol. For each agent, a different port is defined so Matlab and the agent can communicate with each other [26].

9.4.2 Overall cost comparison

The electricity cost results of the smart homes and the total neighborhood costs for the one-day length simulation are given in Table 9.3. The presented algorithms are referred to as follows: M-I: baseline, M-II: selfish, M-III: distributed, M-IV: centralized. For simplicity, the costs are given for one day. The results of M-I are taken as base case and the results of the other control algorithms are compared to this algorithm.

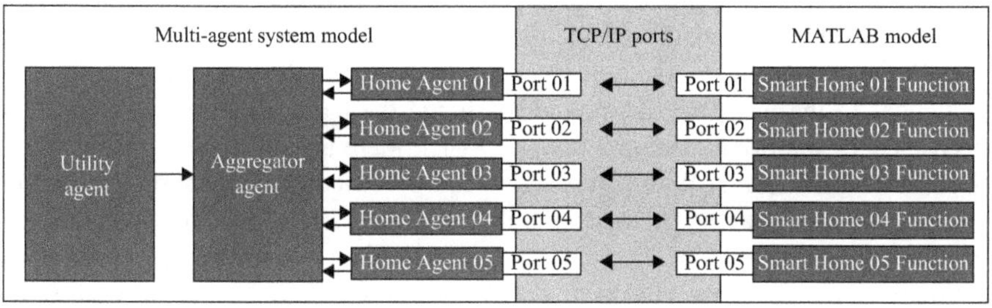

Figure 9.10 Co-simulation setup with JADE and MATLAB

Table 9.3 Smart homes and total neighborhood electricity cost

Smart homes	M-I (€)	M-II (€)	M-III (€)	M-IV (€)
**Home 01	−0.61	−0.89	−1.56	−2.36
**Home 02	−0.53	−0.90	−1.44	−2.30
*Home 03	0.71	0.43	0.40	0.35
Home 04	3.06	2.63	2.60	2.54
Home 05	3.15	2.88	2.84	2.79
Total	5.78 (ref.)	4.14 (−28.36%)	2.85 (−50.68%)	1.03 (−82.06%)

** Smart home with PV and battery.
* Smart home with just PV.

According to these results, the distributed control method shows a better performance, with a 50.68% system cost reduction compared to the base case, and an increase of performance by 21.32% compared to the selfish control method. Moreover, each home agent is able to achieve a cost reduction, which means that each smart home benefits from trading in the neighborhood area.

As expected, compared to distributed control, the centralized control method yields the best results, without violating the individual benefits. However, to solve the optimization problem, the centralized control algorithm requires approximately 30 minutes to run, while the distributed control algorithm needs less than 1 minute.

9.4.3 Detailed analysis

In this section, simulation results are analyzed by investigating power and energy values of the smart homes' profiles. In Figure 9.11, the electricity power consumption profiles of smart home 01, which has PV and battery storage systems, are shown.

At the beginning of the simulation, it is assumed that the battery storage of the smart home is fully discharged. Until the sun rises, the smart home consumes electricity by purchasing from the grid. When sufficient energy is generated by the PV system, the battery starts charging. The smart home is independent from the grid, and consumes self-generated electricity. However, with the baseline algorithm, when consumption is higher than the maximum battery discharge power (4 kW), as around 20:00, although there is enough energy in the battery for the self-consumption, the battery is unable to provide electricity, thus the home has to purchase electricity from the grid. With M-II, the home agent aims to schedule controllable electricity appliances to the low-price hours and discharge the battery adequately to provide electricity for its own consumption. With M-III, there are no changes on the home electricity consumption. With M-IV, significant electricity consumption is observed compared to other cases. The reason is that M-IV does not focus on the individual benefits of the smart home, as it aims to minimize the total electricity consumption of the neighborhood area. However, it achieves more cost reduction for each smart home than other control methods, as shown in Table 9.3.

In Figure 9.12, the electricity power consumption profiles of smart home 03 (PV only, no battery) are shown.

Due to the absence of battery storage, with M-I, the generated surplus energy is directly fed back to the grid; hence, there is no possibility to use surplus generation later in time for self-consumption. With distributed and centralized control, the purchased electricity is sold by the battery-equipped smart homes during low-price hours. So, even if a smart home has no battery storage, it can receive $\lambda_I/2$ from the discharging energy of battery-equipped smart homes, as mentioned in Section 9.3.3. As smart homes share the incentive equally, regardless of ownership, the algorithm may be considered as unfair. The compensation of investment costs (e.g., for buying the battery) is left as future work.

In Figure 9.13, the electricity power consumption profiles of the smart home 04 (no PV or battery) are shown.

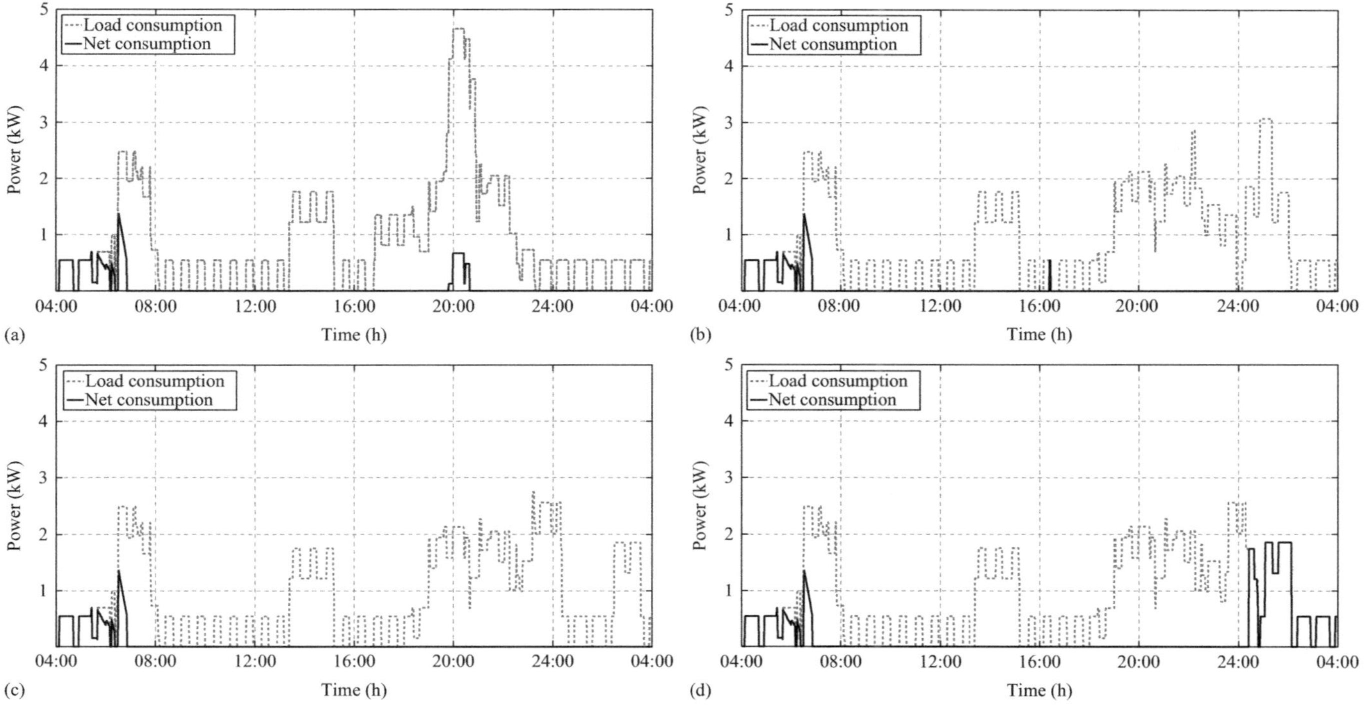

Figure 9.11 Power consumption profiles of smart home 01 (PV+battery): (a) M-I, (b) M-II, (c) M-III, and (d) M-IV

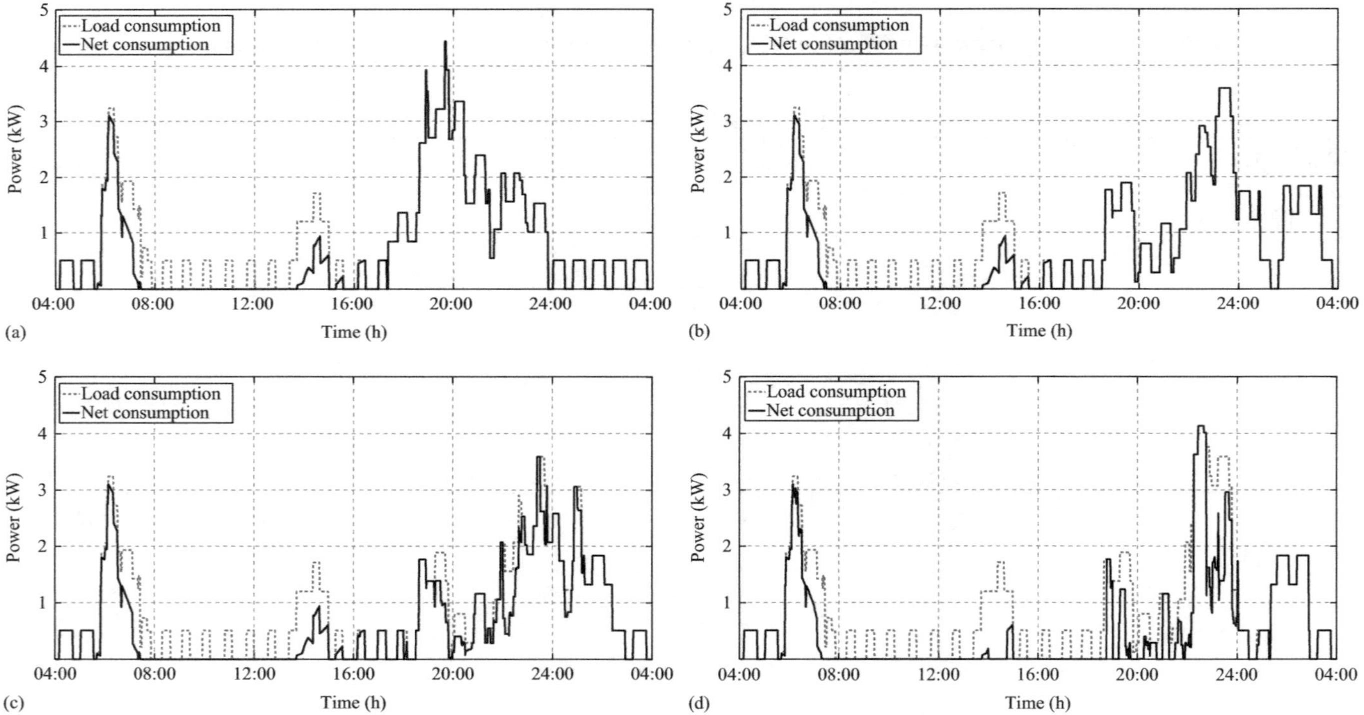

Figure 9.12 Power consumption profiles of smart home 03 (PV): (a) M-I, (b) M-II, (c) M-III, and (d) M-IV

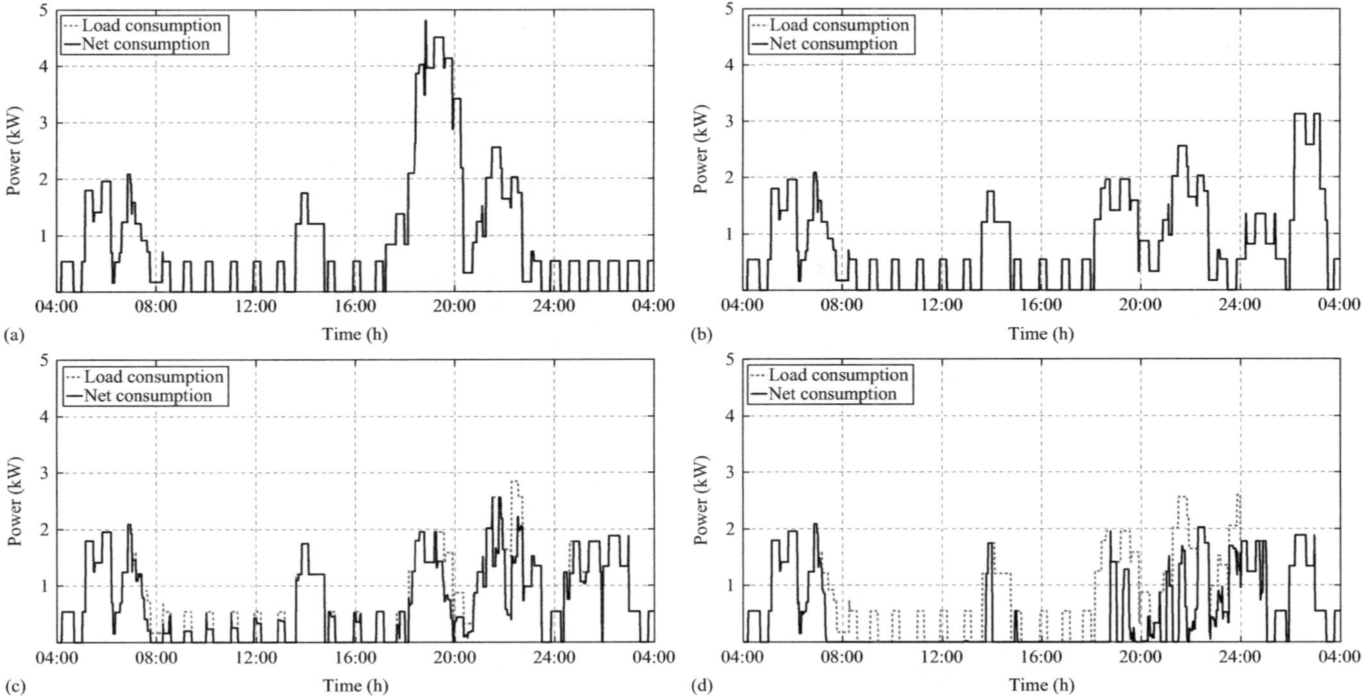

Figure 9.13 Power consumption profiles of smart home 04 (PV): (a) M-I, (b) M-II, (c) M-III, and (d) M-IV

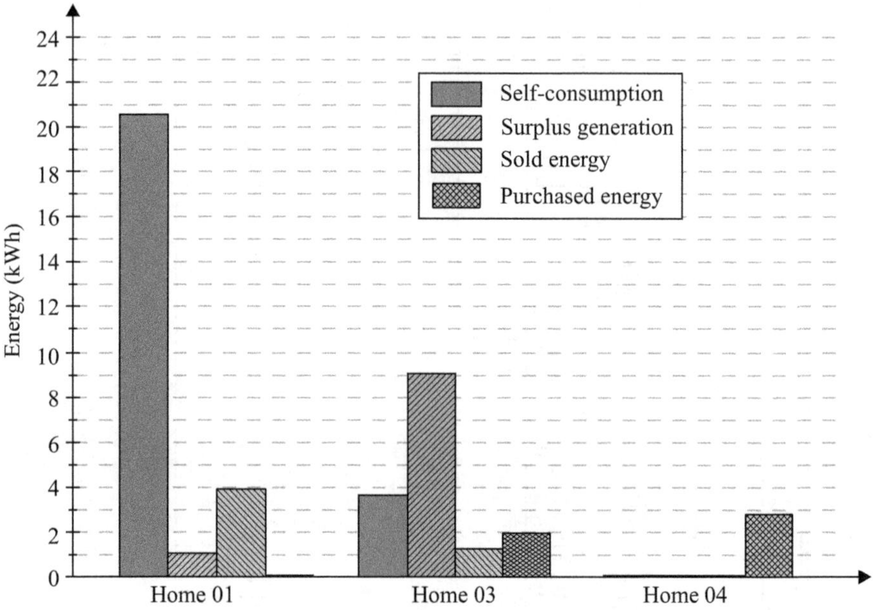

Figure 9.14 Energy analyses of smart homes 01, 03, and 04 with method III

Smart home 04 has no PV and battery installation, thus it consumes electricity directly from the grid in the base and selfish cases. With distributed and centralized control, home agent 04 has the ability to purchase electricity from neighbors with just PV and with PV and battery-equipped smart homes. During sunny hours, the smart home purchases surplus generation from other smart homes, and it also purchases electricity from the battery-equipped smart homes during night hours.

For a more detailed analysis, the electricity energy profiles of the smart homes for one day are given in Figures 9.14 and 9.15.

From this figure, it can be observed that there is a significant difference between distributed and centralized coordination with the proposed algorithm. First, the self-consumption amount of the smart home decreases in home 01, and second, the sold energy increases with the centralized control method, as there is no difference between the effects of self-consumption and energy trading on the total neighborhood electricity cost reduction. Lastly, home 04 purchased more electricity compared to home 01 and home 03 with both control methods due to the absence of self-consumption ability.

Overall, simulation results show that the proposed coordination algorithms achieve more cost reduction compared to the base cases and selfish control, and provide benefits for all smart homes no matter what type of equipment they have.

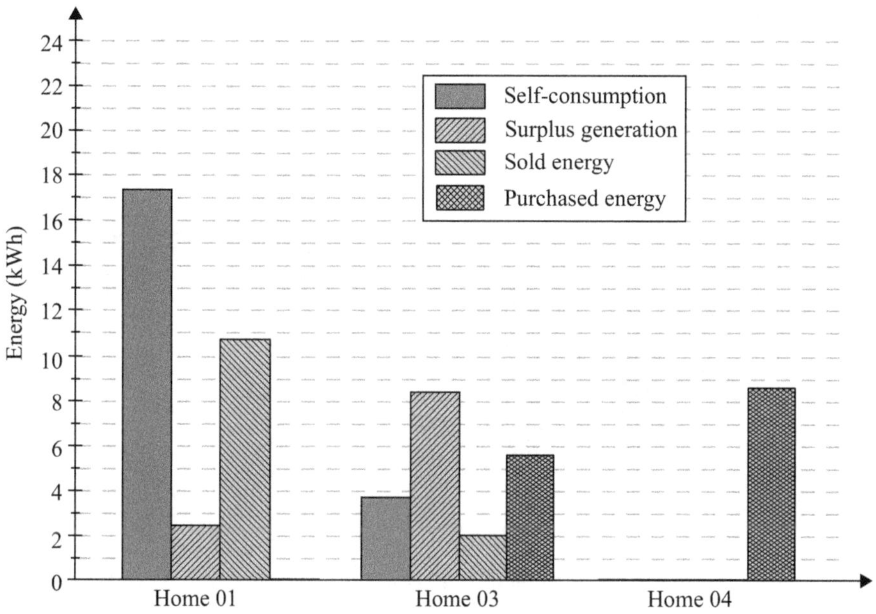

Figure 9.15 Energy analyses of smart homes 01, 03, and 04 with method IV

9.5 Discussion

9.5.1 Overview of Results

In this section, the performance of the presented algorithms is compared with two days length simulation. An overview of the results of the simulation results is given in Figure 9.16, where the total electricity cost of the smart homes is given for each smart home, and in Figure 9.17, where the aggregated neighborhood energy consumption and generation are shown.

In Figure 9.16, in the selfish control method, home agents use the optimization algorithm considering only the price data, thus based on the placement of the household loads over the time horizon, significant cost reduction can be achieved. The surplus generated energy in the smart homes is stored in the batteries, or fed back to the grid, thus the stored energy is organized to be used first during high price periods. In coordinated optimization algorithms, due to the possibility of trading energy between home agents, there is a significant cost reduction, especially for battery-equipped smart homes. It is seen that smart homes 01 and 02 switch from consumers to producers and even earn some money from this coordination. Also, smart home 03 sells its surplus PV generation. However, due to a lack of flexibility, it can sell energy only if other home agents decide to schedule their appliances to the time of surplus generation hours.

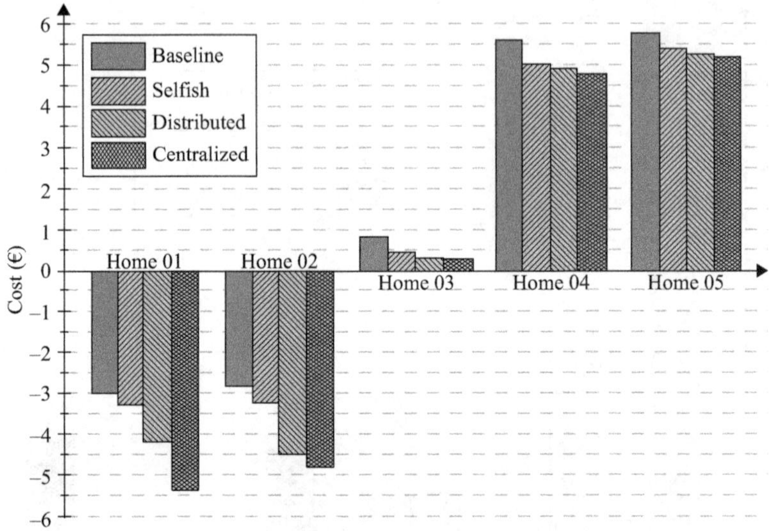

Figure 9.16 Comparison of methods with two days electricity costs of smart homes

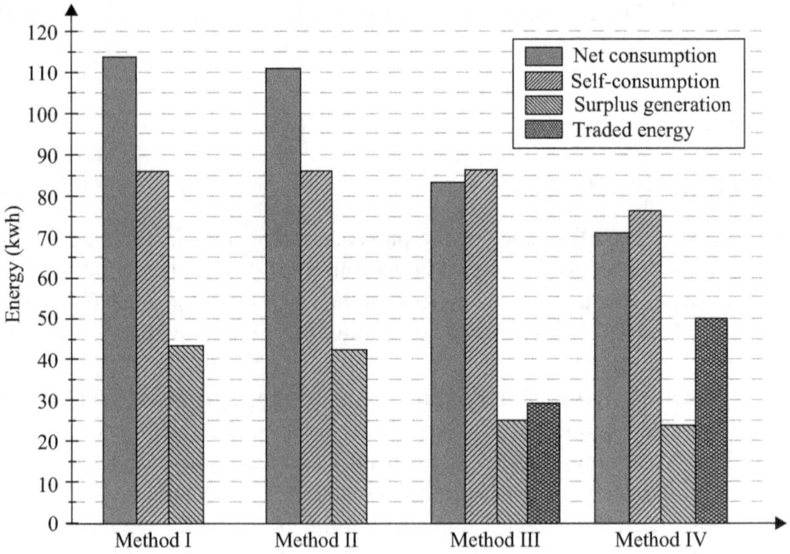

Figure 9.17 Smart homes electricity profiles comparison

In Figure 9.17, we observe a significant reduction in the net consumption of the neighborhood from the main grid. Compared to M-I, distributed control achieves a 26.63% reduction, while centralized control provides a 37.52% electricity transfer reduction from the grid. Simulation results show that by establishing a trading mechanism among smart homes, the flexibility of the neighborhood can be improved by increasing the usage of self-generated renewable energy. It should be noted that, although consumption from the grid is decreased, the total load consumption of the smart homes is unchanged. Controllable appliances operation is not shifted to the next day as mentioned before.

Generally, when both coordination approaches (distributed and centralized) are compared, it is observed that the centralized control method returns the better result. However, when the simulation times are compared, one iteration of the smart home agent in distributed coordination method takes less than 15 seconds and for 3 iterations, the total simulation time takes less than 1 minute. But with centralized coordination, the total simulation time is approximately 30 minutes. For such a small neighborhood, 30 minutes may be acceptable for day-ahead optimization, but when a neighborhood of more than 100 smart homes is considered, the simulation time could increase to unacceptable levels.

Regarding time steps, even though the communication data consider mean values for 30-minute intervals of the actual profile, the presented distributed control method showed better performance with short optimization time steps, than selfish control in 1-minute-based simulations. This means that this low time resolution in communication data does not significantly impact results. Moreover, as it is not possible to know the exact power profile because of forecasting errors, the proposed approach is able to account for a certain level of uncertainty, which originate either from forecasting errors or from taking the average of the signal.

Lastly, it should be mentioned that the results are dependent on the selected pricing scheme, whether it uses TOU, a FIT, and an incentive or not. A higher incentive would increase trading; however, this would also increase the cost burden for the collectivity. High FITs are also not in favor of such trading, as it is more interesting to sell back to the grid, rather than trade and maximize local PV generation.

Overall, the proposed algorithms have shown that coordination at the neighborhood level is able to provide benefits to each participant. Several approaches may, however, be followed, from selfish optimization to centralized optimization at the neighborhood level. A form of distributed optimization seems appropriate and promising, as it can increase the overall benefit when consumers are able to buy and sell each other energy, e.g., through an incentive price. Proper coordination strategies are thus required, with appropriate constraints on privacy, comfort, and ease of use.

9.5.2 Next steps

This study shows that there is an interesting potential for coordination mechanism in neighborhoods. However, there are a number of challenges, which are listed below, that need to be considered before such mechanisms can be further deployed.

- *Uncertainty:* the uncertainty in generation and demand forecasts was not considered in this work, for the sake of simplicity.
- *Constraints of the distribution system:* feeder and transformer capacity should be taken into account, to avoid overloading and reliability issues.
- *Rebound peaks:* as the algorithms base their decisions on prices, one may expect that large portions of the load may be shifted during the same periods of time, and could create what is known as rebound peaks [23]. Although this issue is not

significant in the proposed simulation, it needs to be accounted for in large-scale applications, as it may cause stability problems in the distribution grid.

- *Two-way battery charging:* it was assumed that batteries were only allowed to charge from self-generated electricity, due to the focus on renewable energy usage in this work. However, batteries could also be allowed to charge for neighbors surplus generation. Moreover, to further reduce electricity bills, the ability of grid charging should also be investigated.
- *Incentive, pricing scheme, and fairness:* the current scheme relies on a specific pricing scheme, with an externally funded incentive. However, the initial cost of buying some equipment such as PV panels and batteries is not considered. As a consequence, equally sharing the incentive with other homes that did not do such investments may be considered as unfair. Further investigation on incentive sharing is thus required to improve the fairness of the approach, when considering all types of costs.
- *EVs:* a high penetration of the EVs can significantly change the load profile in a neighborhood. Coordinating EV charge is another research avenue for the future. Additionally, EVs can also be used as storage devices when they are parked, using the vehicle-to-grid paradigm.
- *Annual analysis:* this chapter has presented results for one-day and two-day simulations. However, depending on time-dependent variables such as temperature, irradiance, and consumer habits, electricity load profile vary throughout the year. An annual analysis should thus be performed to determine the overall efficiency of the proposed program, and also to calculate the payback time of investments, e.g., for batteries and PV.

9.6 Conclusion

Smart grid technology enables new local energy management strategies, e.g., for neighborhoods, that can benefit both electricity providers (utilities) and consumers. While utilities are able to reduce the peak-to-average ratio of their load profile and hence reduce generation costs, end-users can also potentially reduce their electricity bills. The integration of DG and storage, turning consumers into prosumers [27], is, however, making the problem more complex. This chapter has presented an agent-based coordination mechanism for electric energy management in small-scale neighborhood areas employing TOU, FIT, and incentive pricing. Of the four presented algorithms, distributed optimization seems to be a good compromise among performance, computation needs, and privacy. Although its performance is expectably lower than for centralized optimization, it is able to leverage the possibility for home agents to buy and sell power to each other. This approach thus seems to provide a promising solution for local resource management, and may be extended to account for larger areas and more heterogeneous configurations.

References

[1] S. Nolan and M. O'Malley, "Challenges and barriers to demand response deployment and evaluation," *Applied Energy*, 2015, vol. 152, pp. 1–10.

[2] Z. Zhu, S. Lambotharan, W. H. Chin, and Z. Fan, "A game theoretic optimization framework for home demand management incorporating local energy sources," *IEEE Transactions on Industrial Informatics*, 2015, vol. 11, no. 2, pp. 353–362.

[3] M. Muratori, B.A. Schuelke-Leech, and G. Rizzoni, "Role of residential demand response in modern electricity markets," *Renewable and Sustainable Energy Reviews*, 2014, vol. 33, pp. 546–553.

[4] E. Vrettos, F. Oldewurtel, M. Vasirani, and G. Andersson, "Centralized and decentralized balance group optimization in electricity markets with demand response," in *2013 IEEE Grenoble PowerTech (POWERTECH)*, Grenoble, France, 2013, pp. 1–6.

[5] M. Juelsgaard, P. Andersen, and R. Wisniewski, "Distributed loss reduction by household consumption coordination in smart grids," *IEEE Transactions on Smart Grid*, 2014, vol. 5, no. 4, pp. 2133–2144.

[6] C. Wang and M. De Groot, "Demand response in the smart grid: a distributed computing perspective," in A. Y. Zomaya and Y. C. Lee (eds.), *Energy-efficient Distributed Computing Systems*. John Wiley & Sons, Hoboken, NJ, 2012, pp. 593–613.

[7] F. C. Schweppe, M. C. Caramanis, R. D. Tabors, and R. E. Bohn, *Spot Pricing of Electricity*. Springer, 1988, http://link.springer.com/book/10.1007%2F978-1-4613-1683-1.

[8] F. S. Bush, "Demand-response and the Advanced Metering Infrastructure," *Smart Grid: Communication-Enabled Intelligence for the Electric Power Grid.* John Wiley & Sons, Chichester, UK, 2013, pp. 235–258. [Online] http://onlinelibrary.wiley.com/book/10.1002/9781118820216

[9] M. H. Albadi and E. F. El-Saadany, "A summary of demand response in electricity markets," *Electric Power Systems Research*, 2008, vol. 78, pp. 1989–1996.

[10] M. Pipattanasomporn, M. Kuzlu, and S. Rahman, "An algorithm for intelligent home energy management and demand response analysis," *IEEE Transactions on Smart Grid*, 2012, vol. 3, no. 4, pp. 2166–2173.

[11] P. Khajavi, H. Abniki, and A. B. Arani, "The role of incentive based demand response programs in smart grid," in *10th IEEE International Conference on Environmental and Electrical Engineering (EEEIC*), Rome, Italy, 2011, pp. 11–18.

[12] S. Althaher, P. Mancarella, and J. Mutale, "Automated demand response from home energy management system under dynamic pricing and power and comfort constraints," *IEEE Transactions on Smart Grid*, 2015, vol. 6, no. 4, pp. 1874–1883.

[13] P. Yi, X. Dong, A. Iwayemi, C. Zhou, and S. Li, "Real-time opportunistic scheduling for demand response," *IEEE Transactions on Smart Grid*, 2013, vol. 4, no. 1, pp. 227–234.

[14] N. Yaagoubi and H. T. Mouftah, "A comfort based game theoretic approach for load management in the smart grid," in *IEEE Green Technologies Conference,* Denver, CO, 2013, pp. 35–41.

[15] M. Alizadeh, A. Scaglione, A. Applebaum, and K. Levitt, "Reduced-order load models for large populations of flexible appliances," *IEEE Transactions on Power Systems*, 2015, vol. 30, no. 4, pp. 1758–1774.

[16] C. Wang, Y. Zhou, J. Wu, J. Wang, Y. Zhang, and D. Wang, "Robust-index method for household load scheduling considering uncertainties of customer behavior," *IEEE Transactions on Smart Grid*, 2015, vol. 6, no. 4, pp. 1806–1818.

[17] Y. Guo, M. Pan, Y. Fang, and P. P. Khargonekar, "Coordinated energy scheduling for residential households in the smart grid," in *Third IEEE International Conference on Smart Grid Communications (SmartGridComm),* Tainan, China, 2012, pp. 121–126.

[18] S. Kahrobaee, R. A. Rajabzadeh, L. K. Soh, and S. Asgarpoor, "Multiagent study of smart grid customers with neighborhood electricity trading," *Electric Power Systems Research*, 2014, vol. 111, pp. 123–132.

[19] K. Tanaka, A. Yoza, K. Ogimi, A. Yona, T. Senjyu, T. Funabashi, and C. H. Kim, "Optimal operation of DC smart house system by controllable loads based on smart grid topology," *Renewable Energy*, 2012, vol. 39, pp. 132–139.

[20] S. D. J. McArthur, E. M. Davidson, V. M. Catterson, A. L. Dimeas, N. D. Hatziargyriou, F. Ponci, and T. Funabashi, "Multi-agent systems for power engineering applications-part I: concepts, approaches, and technical challenges," *IEEE Transactions on Power Systems*, 2007, vol. 22, no. 4, pp. 1743–1752.

[21] R. Roche, B. Blunier, A. Miraoui, V. Hilaire, and A. Koukam, "Multi-agent systems for grid energy management: a short review," in *36th Annual Conference on IEEE Industrial Electronics Society (IECON)*, Glendale, CA, 2010, pp. 3341–3346.

[22] T. M. Hansen, R. Roche, S. Suryanarayanan, A. A. Maciejewski, and H. J. Siegel, "Heuristic optimization for an aggregator-based resource allocation in the smart grid," *IEEE Transactions on Smart Grid*, 2015, vol. 6, no. 4, pp. 1785–1794.

[23] S. Shao, M. Pipattanasomporn, and S. Rahman, "Development of physical-based demand response-enabled residential load models," *IEEE Transaction on Power Systems*, 2013, vol. 28, no. 2, pp. 607–614.

[24] R. Roche, S. Suryanarayanan, T. M. Hansen, and S. Kiliccote, "A multi-agent model and strategy for residential demand response coordination," in *2015 IEEE Eindhoven PowerTech*, Eindhoven, The Netherlands, 2015, pp. 1–6.

[25] R. Roche, S. Natarajan, A. Bhattacharyya, and S. Suryanarayanan, "A framework for co-simulation of AI tools with power system analysis software," in *2012 23rd International Workshop on Database and Expert Systems Applications (DEXA)*, Vienna, Austria, 2012, pp. 350–354.

[26] F. Bellifemine, G. Caire, and D. Greenwood, *Developing Multi-Agent Systems with JADE*. John Wiley & Sons, Chichester, UK, 2007.

[27] N. Zhang, Y. Yan, and W. Su, "A game-theoretic economic operation of residential distribution system with high participation of distributed electricity prosumers," *Applied Energy*, 2015, vol. 154, pp. 471–479.

Chapter 10

Compensation of droop control in DC microgrid with multiple distributed generators

Nanfang Yang[1], *Damien Paire*[2], *Fei Gao*[2], *and Abdellatif Miraoui*[2]

Abstract

DC microgrid is a feasible and effective solution to integrate renewable energy resources, as well as to supply reliable electricity. The control objective of DC microgrids is to maintain the system's stable operation, low-voltage regulation, and proportional load sharing among the multiple distributed generators. Compared to the high-bandwidth communication-dependent master–slave control, droop control is an effective method to implement the control of DC microgrids without the requirement of communication. Droop control is an output impedance programming method, in which the output current decreases linearly with the decrease of output voltage. The load sharing is automatically achieved. However, in the real applications of low-voltage DC microgrids, the nominal reference offsets and unequal cable resistances require trade-offs to be made between voltage regulation and load sharing. Thus some compensations need to be performed so as to solve this problem.

This chapter discusses the methods to compensate the voltage error introduced by droop control as well as the unequal load sharing due to the transmission lines and the nominal voltage reference offsets. The compensation methods in the literature using low-bandwidth communication are reviewed and a unified compensation framework is proposed using the common current. In this scheme, the voltage deviation and the unequal load sharing are compensated separately. The common current is generated in each local controller by using the local module currents shared in a dedicated low-bandwidth communication line.

The contents of this chapter are organized as follows: Section 10.1 overviews and compares the active current sharing control and the droop control in DC microgrids; Section 10.2 analyses the limits of the basic droop control under the condition of nominal voltage offsets and unequal connecting cable impedances; Section 10.3

[1]GREEN Laboratory (EA 4366), Université de Lorraine, Vandoeuvre-lès-Nancy, France

[2]IRTES Laboratory (EA 7274) and FCLAB Research Federation (FR CNRS 3539), UTBM, Université Bourgogne Franche-Comté, Belfort, France

reviews and classifies the different compensation methods from the literature; Section 10.4 analyses voltage and load sharing performance of the proposed method, and investigates the boundaries of the compensation parameters to maintain system stability. In Section 10.5, some simulations are conducted in the MATLAB/Simulink environment. In Section 10.6, experimental tests are performed on a laboratory-scale test bench to verify the previous proposed theoretical analysis.

10.1 Introduction

Nowadays, microgrids are gaining more and more interest, due to their ability to integrate local distributed energy sources, energy storages, and controllable loads, as well as to improve electricity supply stability, efficiency and reduce the environmental impact. Power electronics converters are usually adopted as interfaces to connect distributed generators (DGs) with the common DC bus. The microgrid can be viewed as the system of multiple power converter interfaced DGs, and the operation of microgrids depends on the local and coordination controls of DGs.

Although most existing microgrids are in AC form, DC microgrids have been investigated in several applications, e.g., data centers [1,2], marine power systems [3–5], and residential applications [6]. The efficiency analysis shows that the loss of DC microgrids is around 15% lower than that of AC microgrids during 1 year due to the elimination of energy conversions from DC to AC and from AC to DC [7]. Besides, the adoption of DC power can also avoid the considerations of reactive power, synchronization, and the three-phase unbalance [8]. DC bus voltage is usually recommended to be 380 V DC in the USA for the connection with the 120 Vrms AC utility grid directly through a front-end rectifier [9]. The studied DC microgrids are also called as low-voltage DC microgrids to distinguish from the high-voltage DC grids.

Similar to the hierarchical structure used in the traditional power system, the hierarchical control layers can be defined for DC microgrids [10], as shown in Figure 10.1. From bottom to top, the three control levels are defined as primary control, secondary control, and tertiary control. The secondary and tertiary control located in the central controller focus on the voltage restoration and the economic and environmental optimizations of the system, whereas the primary control located in the local controller is responsible for the voltage control and load sharing [11].

The major control objective of low-voltage DC microgrids is to obtain low-voltage regulation, as well as proportional load sharing among DGs [12,13]. The control methods in the literature mainly fall into two major categories: active current sharing scheme [14] and droop-based control [10,15].

10.1.1 Active current sharing schemes

The principle of active current sharing is to generate a voltage compensation signal based on the error of individual module's instantaneous current (local current) and

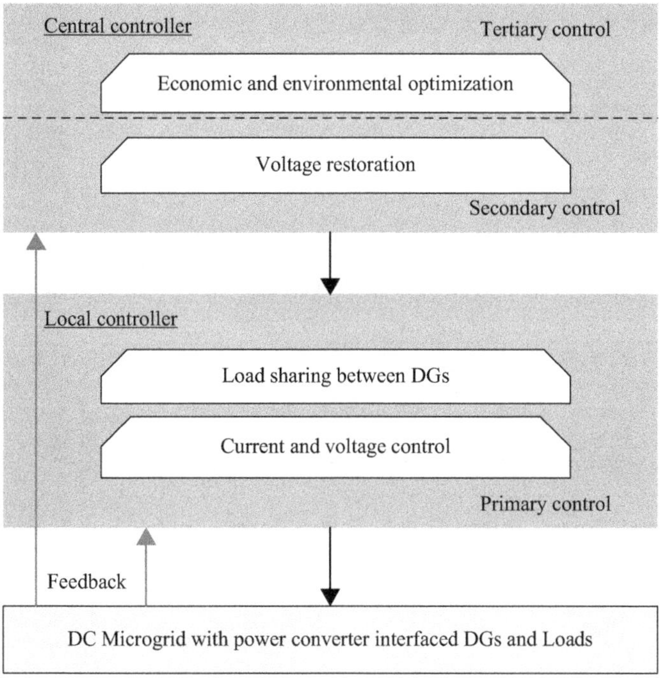

Figure 10.1 Hierarchical control structure for DC microgrids

the overall common current reference (average or maximum current through the system), and then this compensation is added to the voltage reference so as to realize proportional load sharing [16]. The mostly referred active current sharing methods are average current sharing and master–slave control. In the average current sharing, an analogue sharing bus is utilized to interconnect all the paralleling modules current signals with resistors. The average current signal from the sharing bus is fed back and is compared to the local current to generate a voltage error, and the error is used to adjust the reference voltage or current until equal load sharing is achieved.

The popular master–slave control uses one module to operate as the master, which is responsible for voltage control, while other slaves trace the output current of the master or the references given by the master, as shown in Figure 10.2. Some variations, e.g., dedicated master, rotating master, and automatic master, can reduce the dependency on a specified master module [16].

Although the master–slave control can achieve good voltage regulation and load sharing, the main drawback is that the entire system operation is highly dependent on the master unit and high-bandwidth communication [17]. The failure of the master unit or communication may result in the outage of the entire system.

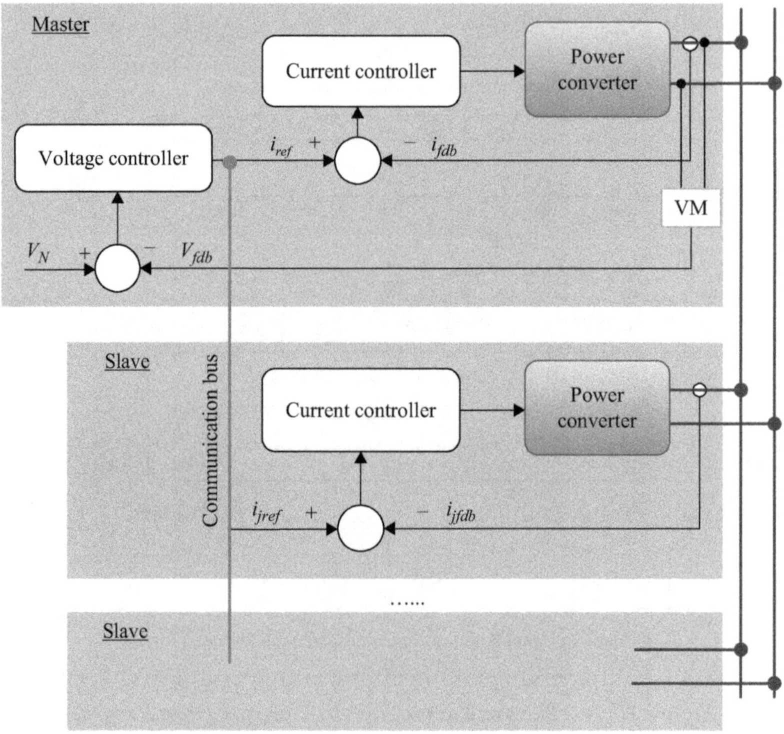

Figure 10.2 Master–slave control structure

10.1.2 Droop control technique

Droop control is an output impedance programming method, and inherently does not require communication. The basic principle is that the output voltage linearly decreases with the output current/power. This makes the connected modules to share the load automatically. In this control, all the modules participate into the voltage regulation, as shown in Figure 10.3. Therefore, disconnection or reconnection of one module has limit impact to the system operation, and the system reliability is improved compared to the active current sharing methods. Because the droop control is an output impedance programming method, i.e., open control structure, the load sharing performance is ensured by identical output impedance of the modules, and it is sensitive to the impedance of the connecting cables and the hardware of the power converters. For the droop controlled DG, usually the virtual droop resistance R_j, as shown in Figure 10.3, is relatively large compared to others, and the output impedance is dominated by the virtual droop resistance.

In AC microgrids, the real power–frequency (P–f) droop and reactive power–voltage (Q–V) droop laws are deduced under the condition that the transmission lines are mainly inductive [18]. The frequency is a global variable through the entire system, but not the voltage magnitude. The voltage drops in transmission lines will

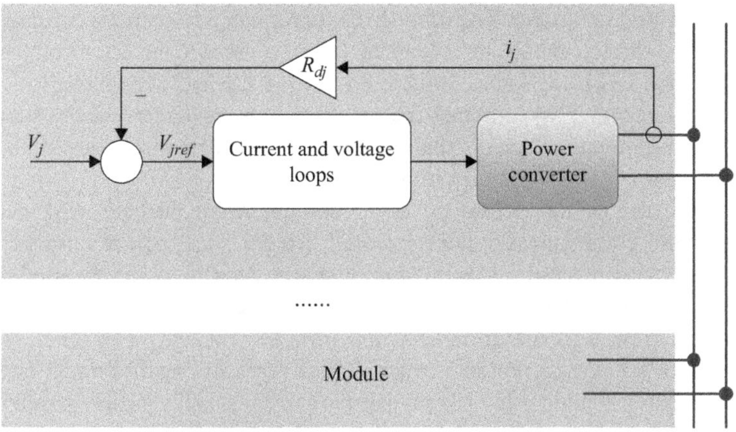

Figure 10.3 Droop control structure

lead to unequal sharing of reactive power. A similar problem is faced for the load sharing control in DC Microgrids. The power–voltage (*P–V*) droop control [19] can be implemented using DC bus signaling [20,21], in which the line resistances and voltage sensing errors are not considered, and the DC voltage is controlled in a relatively large range. However, in the low-voltage DC applications, the line resistances cannot be omitted. The connecting line parameter uncertainties will cause the unequal load sharing or circulating current among the modules like the reactive power sharing problem in low-voltage AC grids [22].

10.1.3 Compensations of droop control

The control performance of DC microgrids under droop control relies on the selection of the virtual droop resistance. The choice of virtual droop resistance needs to consider the trade-off of voltage accuracy and load sharing performance [23]. When higher droop constants are chosen, better load sharing can be obtained, but larger drop in DC bus voltage occurs.

The impact of the line voltage drops on load sharing depends on the topology of the DC microgrid, the location of the load changes, the transmission of line parameters, and voltage droop constants [12,17,24]. This makes compensation methods using full knowledge of the system not suitable.

Nominal voltage reference offsets also cause unequal load sharing. The voltage control loop of the power converter depends on the feedback sensing signal of the output voltage, and unavoidable errors in the sensed voltage signals, e.g., the quantization error due to AD conversion, may result in load sharing deterioration [22].

10.1.3.1 DC bus voltage restoration

To reduce the voltage deviations, a hierarchical structure with low-bandwidth communication is proposed in [10,25] by adjusting the nominal voltage references in a secondary central controller. The second controller can eliminate the voltage deviation, but load sharing cannot be enhanced, because the voltage error compensation feeding to each DG is the same. Another one with a supervisory control is proposed by Dragicevic *et al.* [26] to adjust the virtual droop resistances in the supervised controller. Common drawback is that the enhanced performance relies on the upper layer controller and the communication.

To avoid the central controller, some compensation methods with dedicated low-bandwidth communication are proposed. Module common peak current [27], common average current [13], and common average voltage [28] are used as common current or common voltage references to generate compensations in local control. They can be viewed as the distributed form of the secondary controller, with the common peak/average current or average voltage representing the voltage condition.

A distributed method using adaptive droop constants in the local controller is proposed in [29], which adjusts the droop constant by a fuzzy logic controller according to the output power to obtain a better voltage regulation.

10.1.3.2 Proportional load sharing

To improve the load sharing performance, the detailed unequal line parameters can be utilized to calculate the exact voltage reference for each module [30]. Although no additional communication is needed, the main drawback is the requirement of full knowledge of line parameters throughout the grid as well as the loss of flexibility. A line resistances estimation method is introduced by Khorsandi and HosseinMokhtari [31] to replace the pre-calculated values. It requires that the grid-connected mode is operated first to calculate the line resistances, and then they can be used in the islanding mode. The nominal voltage references mismatch can also be represented as circulation currents under the no-load condition. An iterative method is adopted to adjust the nominal voltage references during the no-load condition, so as to obtain zero circulation currents [22]. To adopt this method, all the DGs must start from the no-load condition at the same time. Another similar process is also proposed to adjust the droop constants during the initial process without load [32].

These lead to the loss of plug-and-play capability and also the effect of transmission line on load sharing is not considered. In [28], a distributed compensation structure with dedicated low-bandwidth communication is proposed. The voltage and current information are shared with each module; the load sharing error is compensated by the error of local current and the common current. The error goes through a PI controller or sliding-mode-control [33,34] to generate load sharing compensation. A similar structure is adopted to generate voltage compensation using the error between average voltage and module instantaneous voltage. Although this complicated structure can resolve the impact of transmission line, the determination of the two PI compensation controllers is really a great challenge.

In summary, most of the literatures investigate either the voltage regulation or load sharing in low-voltage DC microgrids. Although the method proposed in [28] considers both the voltage regulation and the load sharing, the multiple PI controller parameters make it not easy to be implemented, and also the tuning of the controllers is not investigated.

Table 10.1 Performance comparison of different droop-based control methods

	Voltage regulation	Load sharing	Communication dependency
Master–slave control	Good	Good	High
Basic droop control	Normal	Unsatisfactory	No
Hierarchical structure	Good	Unsatisfactory	Low

In this chapter, a mixed compensation structure in the local controller using common current through the system is proposed to enhance both voltage regulation and load sharing performance. The common current is designed as the average local current in per unit through the whole system, which is generated in the local controller, with the aid of a dedicated low-bandwidth communication. The required dedicated low-bandwidth communication can be implemented with CAN or power line communication [35,36] to avoid an additional line setup. The voltage regulation, load sharing performance, and communication dependency of some major control methods are listed and compared in Table 10.1.

10.2 Analysis of the basic droop control

According to the *Thévenin theorem*, the power converter interfaced DGs can be modeled as an imperfect voltage source (an ideal voltage source with inner resistors in series), when droop control is applied [37]. Then, the DC microgrids can be represented as the paralleling of multiple imperfect voltage sources. The relationship between output current and voltage reference can be expressed by

$$V_j - V_{MG} = i_j \left(R_{oj} + R_{cj}\right) \tag{10.1}$$

where V_j denotes the equivalent nominal voltage reference of the jth module, V_{MG} denotes the microgrid voltage, as well as the voltage at load point, i_j denotes the injected current to DC bus by the module, R_{oj} denotes the equivalent output resistance, and R_{cj} denotes the connecting line resistance. The subscript $j = 1, 2, \ldots$ indicates the jth module connected to the DC microgrid. It can be seen from the equation that the current/power injected to the DC bus depends on the voltage deviation and output resistance R_j. The output current/power can be adjusted by the nominal voltage references or output resistances.

The equivalent circuit of a two-module DC microgrid in shown in Figure 10.4 the equivalent output resistor R_j is composed of the built-in resistor r_j and virtual droop resistor R_{dj}. It can be expressed by

$$R_j = r_j + R_{dj} \tag{10.2}$$

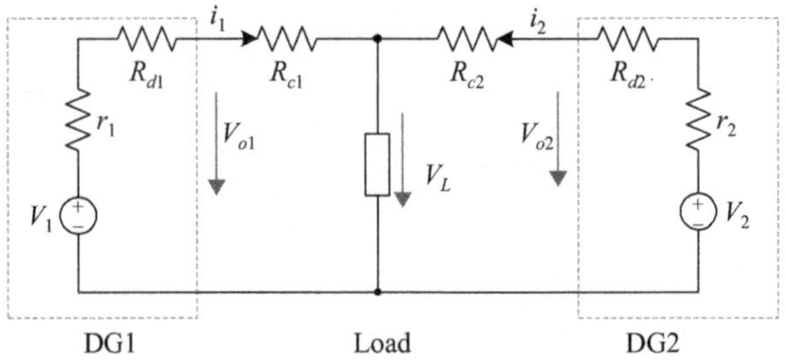

Figure 10.4 Equivalent circuit of a DC microgrid with two distributed generation units. © 2015 Elsevier. Reprinted with the permission from Reference 38

Then, the jth module output current can be deduced from the equivalent circuit, as follows:

$$i_j = \frac{V_j - V_{MG}}{R_j + R_{cj}} = \frac{V_j - V_{MG}}{r_j + R_{dj} + R_{cj}} \tag{10.3}$$

The built-in resistances are determined by the structure and physical parameters of the power converters. They are not identical in different modules. This is one source of the unequal sharing, but usually the built-in resistances are relatively small compared to the virtual droop resistances. When the built-in resistances are omitted, the steady-state load sharing error between the two modules is given by (10.4)

$$\Delta i_{12} = i_1 - i_2 = \frac{(R_{d2} + R_{c2})(V_1 - V_{MG}) - (R_{d1} + R_{c1})(V_2 - V_{MG})}{(R_{d1} + R_{c1})(R_{d2} + R_{c2})} \tag{10.4}$$

The load sharing error relies on the output resistances and nominal voltage references. Equal load sharing $i_{err} = 0$ can be achieved only under the condition of identical output resistances and accurate voltage nominal references. Small nominal voltage offsets introduced by voltage feedback sensing signal, and connecting cable resistances may result in significant performance deterioration, and these effects are analysed in the following subsections.

10.2.1 Nominal voltage reference offset

The nominal voltage references offset are usually introduced by the physical implementation of voltage measurement [13]. A small sensed voltage error may lead to a significant load sharing error, especially when the virtual droop resistances are relatively small. The influence can be demonstrated by Figure 10.5a. When two modules are connected in parallel to supply power for a constant current load, then the load sharing error of the two modules is given by

$$\Delta i_{12} = \frac{(R_{d2} + R_{c2})(V_n + \delta V_1 - V_{MG}) - (R_{d1} + R_{c1})(V_n + \delta V_2 - V_{MG})}{(R_{d1} + R_{c1})(R_{d2} + R_{c2})} \tag{10.5}$$

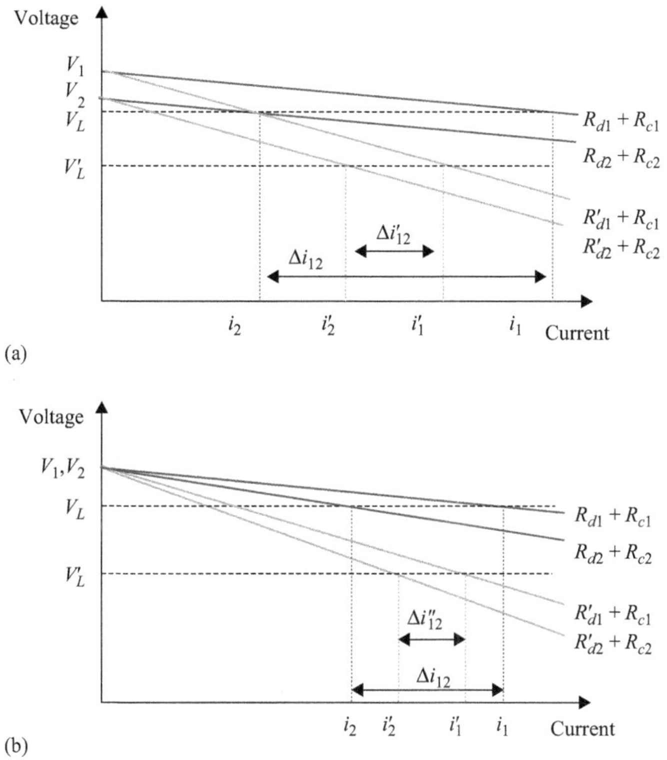

Figure 10.5 The influences of unequal nominal voltage references and cable resistances on the load sharing and DC bus voltage regulation. The droop resistances ($R'_{d1} = R'_{d2}$) > ($R_{d1} = R_{d2}$). (a) The nominal voltage references are different, the cable resistances are the same $R_{c1} = R_{c2}$ and (b) The nominal voltage references are assumed to be equal and the cable resistances are different $R_{c1} < R_{c2}$. © 2015 Elsevier. Reprinted with the permission from Reference 38

where V_n is the nominal microgrid voltage, δV_j is the jth module's nominal voltage offset, and the module's nominal reference voltage $V_j = V_n + \delta V_j$.

When per unit system is adopted, the normalized droop resistances are selected to be the same, i.e., $R_{d1} = R_{d2} = R_d$. If the line resistances are equal $R_{c1} = R_{c2} = R_c$, the load sharing error and DC bus voltage drop can be expressed as

$$\Delta i_{12} = \frac{\delta V_1 - \delta V_2}{R_d + R_c} \tag{10.6}$$

$$\Delta V_{MG} = \frac{1}{2}[\delta V_1 + \delta V_2 - (R_d + R_c)i_L] \tag{10.7}$$

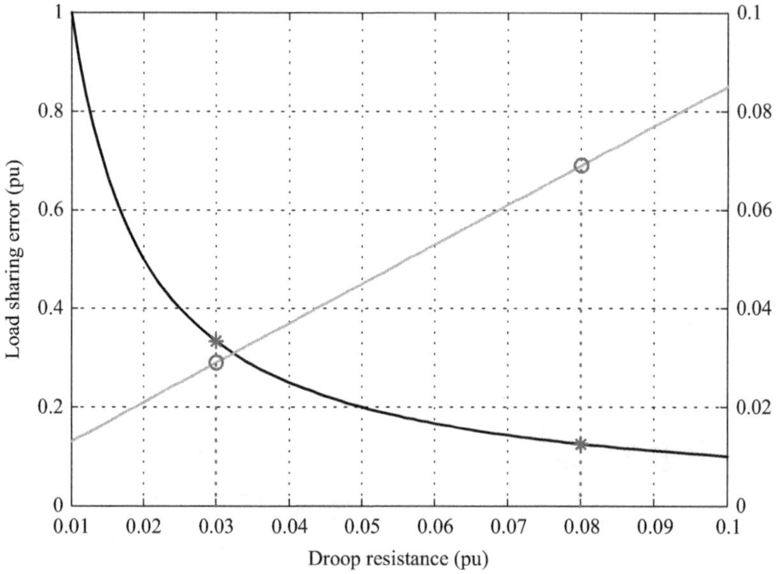

Figure 10.6 The load sharing error and DC bus voltage drop in relationship to virtual droop resistance and the voltage drop to droop resistance, when 1% nominal voltage reference offset between two modules is considered

where i_L is the load current. It can be seen from Figure 10.5b that higher droop resistance $R'_d > R_d$ leads to better load sharing but worse DC bus voltage performance. The load sharing error and voltage drop are shown in (10.8) and (10.9), respectively.

$$\Delta i'_{12} = \frac{\delta V_1 - \delta V_2}{R'_d + R_c} < \Delta i_{12} \tag{10.8}$$

$$\Delta V'_{MG} = \frac{1}{2}[\delta V_1 + \delta V_2 - (R'_d + R_c)i_L] > \Delta V_{MG} \tag{10.9}$$

According to (10.8), when the nominal voltages offset between the two modules is 1% ($\delta V_1 = 1\%, \delta V_2 = 0$), droop resistance is chosen as $R_d = 0.03$pu and the line resistances $R_c = 0.01$pu. The resulted load sharing error will be as large as 0.25pu, when a 0.8pu load is connected. If the droop resistances increase to $R'_d = 0.08$pu, then the current error deceases to 11.1% of the rated current. But the DC bus voltage drop increases from less than 2.7% to about 6.7%, this may not be acceptable. The load sharing error and DC bus voltage drop in relationship to droop resistance are shown in Figure 10.6. It indicates that the influence of unequal nominal voltage can be reduced with higher droop gain, but the performance of voltage regulation becomes worse.

10.2.2 Unequal cable resistances

Due to geographic locations of DGs, the line resistances may be comparable to droop resistances; this will cause significant load sharing errors. It can be demonstrated in Figure 10.5b. Similarly, the effect of unequal line resistances can be reduced by relative higher droop resistances, but the voltage regulation performance may be violated. If the nominal voltage reference error is not considered ($V_1 = V_2 = V_n$), the load sharing error deduced from (10.4) is

$$\Delta i_{12} = \frac{(R_{d2} + R_{c2})(V_n - V_{MG}) - (R_{d1} + R_{c1})(V_n - V_{MG})}{(R_{d1} + R_{c1})(R_{d2} + R_{c2})} \tag{10.10}$$

When the droop resistances are selected to be same, $R_{d1} = R_{d2} = R_d$ in per unit system, the load sharing error and DC bus voltage drop can be expressed as follows:

$$\Delta i_{12} = \frac{(R_{c2} - R_{c1})(V_n - V_{MG})}{(R_d + R_{c1})(R_d + R_{c2})} \tag{10.11}$$

$$\Delta V_{MG} = V_n - \frac{R_d + R_{c1}}{2R_d + R_{c1} + R_{c2}} i_L \tag{10.12}$$

When higher droop resistors $R'_d > R_d$ are adopted, the load sharing error and voltage drop become:

$$\Delta i''_{12} = \frac{(R_{c2} - R_{c1})(V_n - V_{MG})}{(R'_d + R_{c1})(R'_d + R_{c2})} < \Delta i_{12} \tag{10.13}$$

$$\Delta V''_{MG} = V_n - \frac{R'_d + R_{c1}}{2R'_d + R_{c1} + R_{c2}} i_L > \Delta V_{MG} \tag{10.14}$$

Then, a better load sharing is achieved, but worse voltage performance accompanies.

In brief, basic droop control benefits high reliability and easy implementation because it does not require any communication. On the other hand, this scheme is an open loop technique to individually program the output impedance of each module. Thus, a trade-off must be made between the load sharing and output voltage regulation. Some compensations with the aid of low-bandwidth communication can be applied to adjust droop resistance or nominal voltage reference, so as to achieve good voltage regulation and load sharing simultaneously.

10.3 Compensation of the basic droop control

The problem of the basic droop control is the conflict between voltage regulation and load sharing. The inherent limits make that it cannot realize lower voltage regulation and equal load sharing simultaneously. The voltage across the DC microgrid is not constant, thus cannot be used as the global variable like frequency in AC microgrid. Thus, low-bandwidth communication is usually introduced to share the voltage or current information, so as to generate the common reference. The low-bandwidth communication is used as an auxiliary to the basic droop control, such that the system performance can be enhanced with communication and the system still can work without communication or when communication fails.

Some researches are conducted to improve the performance of basic droop control technique with low-bandwidth communication, e.g., using a hierarchical structure to restore the voltage deviation in a centralized secondary controller [10], generating voltage deviation compensations in the local controller by average current [13], or the error between common voltage reference and the local output voltage [39]. According to their objectives, these compensation strategies can be classified into three categories: voltage restoration methods, load sharing compensation methods, and mixed methods.

10.3.1 Voltage restoration methods

The hierarchical control structure is proposed by Guerrero *et al.* [10] to restore the voltage deviation (referred as Type AI), as shown in Figure 10.7a. The low-bandwidth communication line between the secondary controller and local controllers is used to transmit the voltage compensation from the secondary control to local control. The secondary controller compares the voltage reference with measured voltage at the point of common coupling (PCC) or point of load, and the voltage error goes through a voltage regulator (e.g., PI) to generate the global voltage compensation. The voltage compensation is transmitted to all local controllers through a low-bandwidth communication.

The voltage deviation of droop control can also be compensated locally by voltage error [39] with the aid of a dedicated communication line (referred as Type AII), as shown in Figure 10.7b. The output voltage information of all connected modules is shared with a dedicated low-bandwidth communication line. Each module receives the voltage information to generate common average voltage (common voltage). The error between the nominal voltage reference and the average output voltage passes through a voltage compensation controller (i.e., PI regulator) to generate the global voltage compensation. This compensation method can be viewed as the distributed form of Type AI, in which the average output voltage is used to represent the load voltage and the voltage restoration control is dedicated into each module.

Not only the average output voltage but also the common average output current (common current) can be used to design the voltage drop compensation in local control structure [13], which is shown in Figure 10.7c (referred as Type AIII). In this scheme, the current information of all connected modules is shared through the dedicated low-bandwidth communication bus. The common current is generated in each module in the local controller, and then the voltage compensation is generated by the product of common current and a compensation coefficient. Besides, the maximum module current could be an alternative to be used to generate voltage compensation.

The aforementioned Types AI and AII use the nominal voltage as reference to regulate the load voltage or common voltage, thus the load voltage or common voltage can achieve the nominal voltage reference without static error. However, the design of the voltage control, e.g., the tuning of PI controller is not an easy task; the dynamic of the compensation should be much slower so as to avoid any coupling with the voltage control loop. Although Type AIII cannot totally eliminate the voltage derivation, it

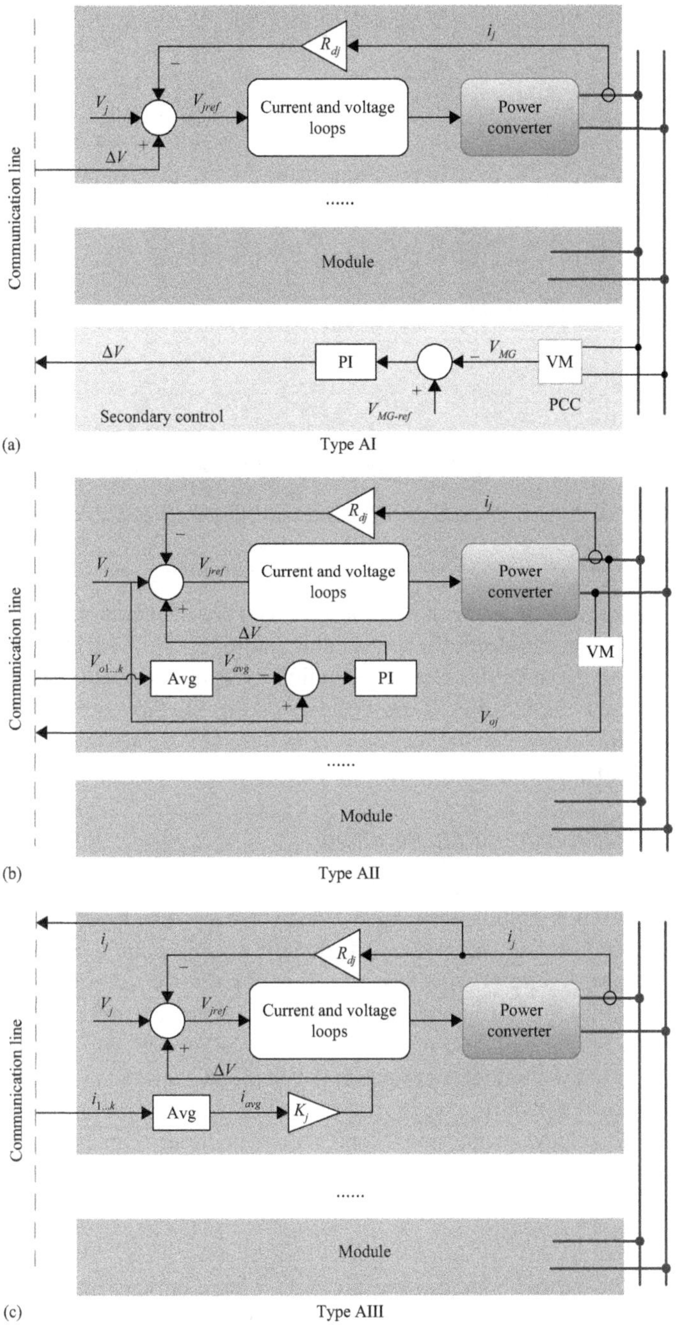

Figure 10.7 Voltage compensation methods of the basic droop control. (a) Type AI, (b) Type AII, and (c) Type AIII. © 2015 Elsevier. Adapted with the permission from Reference 38

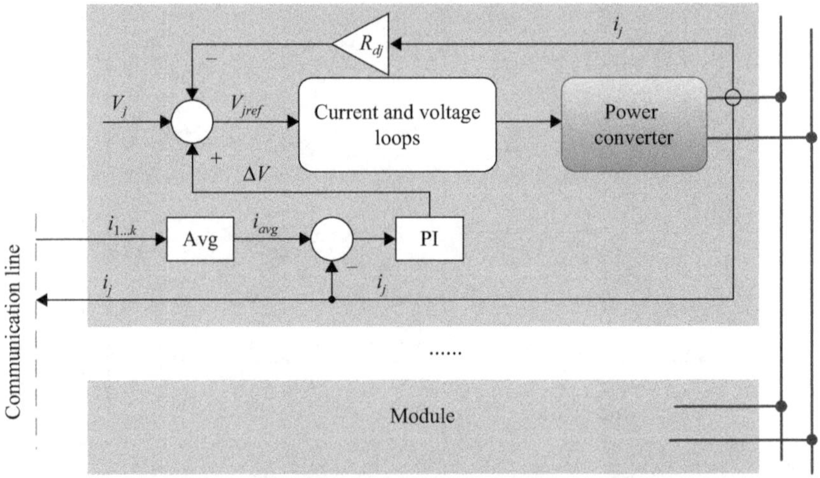

Figure 10.8 Load sharing compensation method (Type B) of the basic droop control

can be significantly reduced with a properly selected compensation coefficient. The design of the compensation coefficient is rather intuitive. Compared to Type AI, Types AII and AIII can enhance the system reliability by using dedicated communication instead of the central controller. It must be noticed that the three compensation methods generate global voltage compensation, which is same for each module, thus only the voltage performance is improved and not the load sharing.

10.3.2 Load sharing compensation

The load sharing compensation cannot be realized by global compensation; individual compensation in each module is required. A dedicated low-bandwidth communication line is introduced among the connected modules to share the current information [39]. As shown in Figure 10.8 (referred as Type B), the shared current information is used to generate the normalized average common current reference, and a load sharing compensation regulator is used to control the module current to follow this reference. Thus, all the modules can achieve equal load sharing. An alternative using the maximum common current reference can achieve similar performance.

Type B can achieve proportional load sharing without error when PI control is adopted, but the voltage performance keeps the same as the basic droop control.

10.3.3 Mixed methods

The above-discussed methods can only realize voltage compensation or load sharing compensation; thus, mixed methods should be introduced to achieve these both compensations. The bus voltage can be restored by the global voltage compensation generated by voltage or current information, while the load sharing compensation requires current information. Therefore, two mixed methods can be deduced from the compensation analysis: Type CI, as shown in Figure 10.9a, uses the shared voltage

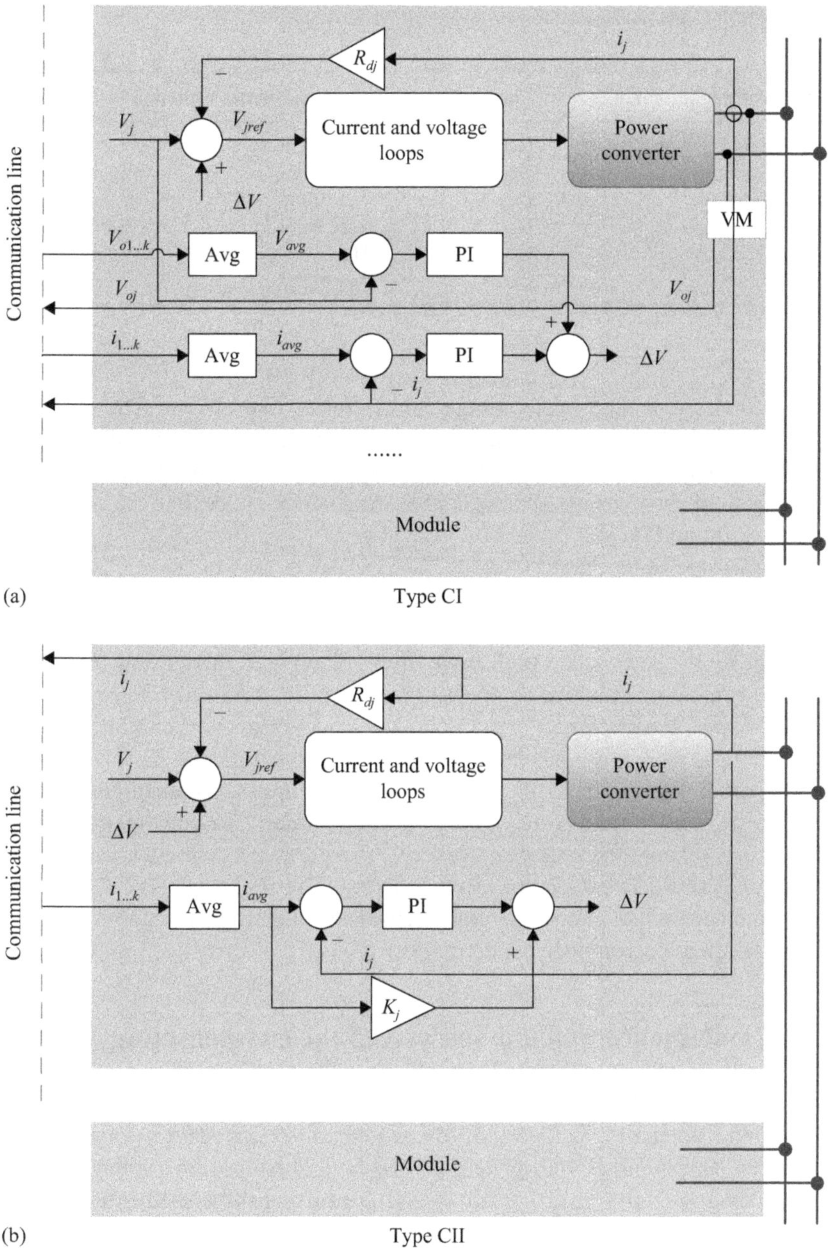

Figure 10.9 Mixed compensation methods of the basic droop control. (a) Type CI and (b) Type CII. © 2015 Elsevier. Adapted with the permission from Reference 38

Table 10.2 Comparisons of the compensation methods

Type	Voltage performance	Load sharing	Communication burden	Complexity
AI	Good	Depend	Light	Medium
AII	Medium	Depend	Medium	Medium
AIII	Medium	Depend	Medium	Medium
B	Depend	Good	Medium	Medium
CI	Good	Good	Heavy	Complicated
CII	Good	Good	Medium	Medium

information to generate global voltage compensation like in Type AII, while using the shared current information to generate load sharing compensation like in Type B [39]; Type CII, as shown in Figure 10.9b, uses the shared current information to compensate both the bus voltage and the load sharing performance, like the combination of Types AIII and B [38].

Type CI inheriting the performance of Type AII is able to achieve zero error in bus voltage control, while Type CII can only improve the control of DC bus voltage with the adjustment of compensation coefficient. However, Type CI requires the communication line to share both the output voltage and the current information, and the parameters of the two compensation controllers are needed to be tuned. The proposed Type CII can reduce the communication burden by using only shared current information, as well as the number of compensation controllers.

The compensation methods are listed in Table 10.2 and compared by voltage performance, equal load sharing, communication burden, and complexity. Although Type CI can achieve the best performance, it requires the highest communication burden, and has the most complicated structure. The proposed Type CII can be a better candidate achieving comparable performance using medium communication burden and medium complicated structure.

10.4 Implementation and analysis of the compensation

The DC bus voltage drop is derived from the droop control in the local controller, while the unequal load sharing is the result of unequal output impedance, connecting line resistances, and nominal voltage reference offset. Therefore, the voltage and load sharing compensation can be separately designed using the aforementioned mixed compensation method Type CII.

As shown in Figure 10.9, a dedicated low-bandwidth communication is introduced to share the local current through the whole system. Then, the local controller can generate normalized average current reference (normalized common current) by

$$i_{avg} = \frac{\sum i_j I_{nj}}{\sum I_{nj}} \tag{10.15}$$

where i_{avg} is the normalized average current reference, which represents the load condition of the whole system, I_{nj} is the rated current of the jth module, and i_j is the instantaneous module current.

10.4.1 Voltage deviation compensation

In basic droop control, the output voltage drops linearly with the output current/power, and this leads to the voltage deviation. The most intuitive method is to compensate it by the load current or common average current as described in Type AIII, such that the voltage deviation can be reduced. The voltage compensation is given by

$$\Delta V_j' = K_j i_{avg} \tag{10.16}$$

where K_j is the voltage compensation coefficient to restore bus voltage, and it should be selected with the same normalized value for each module. Its value should be smaller than the droop resistance to retain the droop control function [13] and also selected properly to maintain the system stability (detailed discussion in Section 10.4.3).

When the DC microgrid is simplified to be a source-load system, the voltage error V_{drop} reduces with the increase of the compensation coefficient as shown in (10.17), if the nominal voltage reference offset is not considered

$$V_{drop} = R_j i_j - K_j i_{avg} \tag{10.17}$$

The normalized common current i_{avg} equals to the normalized local current i_j, and the bus voltage performance can be improved. For example, when the compensation coefficient is half of the virtual droop resistance, the voltage error would decrease to half.

10.4.2 Load sharing compensation

Load sharing errors come from the differences of the modules, thus requiring compensation individually in each module rather than globally. Similar to the voltage compensation, the normalized common current reference is compared with the local current, and then the error feeds into a PI to generate the load sharing compensation. The local current in each module is controlled to follow the common current reference such that equal load sharing is achieved. The load sharing compensation is given by

$$\Delta V_j'' = \frac{K_{Pj}s + K_{Ij}}{s}\left(i_{avg} - i_j\right) \tag{10.18}$$

where s is the integrator, K_{Pj}, K_{Ij} are the parameters of load sharing compensator in the jth module. In steady state, the normalized local current can be controlled well by following the common reference, and ideally with no load sharing error. The dynamic of the compensation loop needs to be designed much slower than the voltage control loop to avoid any interactions between them. For example, if the voltage loop bandwidth is 100 rad/s, then the bandwidth of the compensation loop usually is selected no higher than 20 rad/s.

10.4.3 Stability analysis

The adoption of the voltage compensation coefficient in Type AIII or Type CII will reduce the effect of droop control, and may affect the stability of the system, especially when a constant power load (CPL) is connected [40]. Therefore, a model of the DC microgrid is required to conduct the stability tests, such that the boundaries of the compensation coefficient can be determined.

The droop-controlled DG can be represented by an imperfect voltage source using the *Thévenin theorem*. If the cable resistance and inductance can be combined with the equivalent circuit of the DG, then it becomes a perfect voltage source with an equivalent source resistor and an equivalent inductor in series. Then, the DC microgrid can be represented by multiple DG equivalent circuits in parallel as shown in Figure 10.10. V_j is the nominal voltage reference of the jth DG, R_j is the combination of droop resistance and the line resistance, L_j is the connecting line inductance, and C_{dc} denotes the combination of source output capacitance and load input capacitance.

Consider the microgrid supply power to a CPL. The relationship of the load current and the load voltage in ideal CPL is written by

$$I_L = \frac{P_{CPL}}{V_L} \tag{10.19}$$

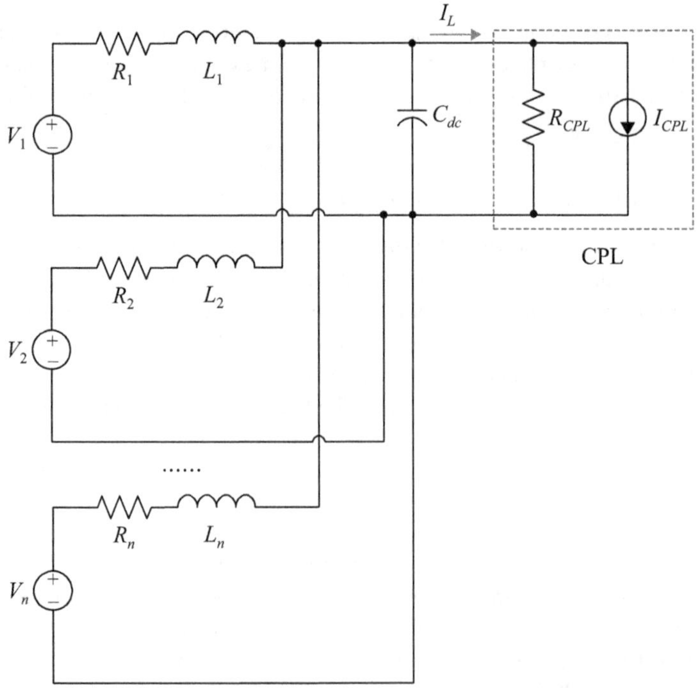

Figure 10.10 The equivalent circuit of a DC microgrid with n DGs

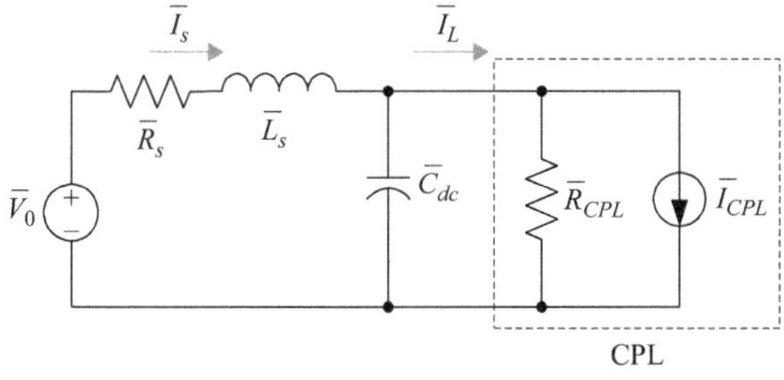

Figure 10.11 Reduced-order model of the DC microgrid

where I_L is the current absorbed by the CPL and V_L is the load voltage. Linearizing the CPL model using the Taylor expansion at the operation load voltage V_e approximates the load current as

$$I_L \approx I_{CPL} + \frac{V_L}{R_{CPL}} = 2\frac{P_{CPL}}{V_e} - \frac{P_{CPL}}{V_e^2}V_L \tag{10.20}$$

The resulting equivalent circuit of CPL is composed of an equivalent negative resistor $R_{CPL} = -V_e^2/P_{CPL}$ and an equivalent current sink $I_{CPL} = 2P_{CPL}/V_e$ in parallel.

Using the arithmetic mean value to the represent the distributed parameters of individual DGs, the equivalent circuit of the microgrid can be reduced to an equivalent DG supplying power to the CPL [41]. The equivalent circuit is shown in Figure 10.11, and the linearized state-space model is given by

$$\frac{d}{dt}\begin{bmatrix} \overline{I}_s \\ V_L \end{bmatrix} = \begin{bmatrix} -\frac{\overline{R}_s}{\overline{L}_d} & -\frac{1}{\overline{L}_d} \\ \frac{1}{\overline{C}_{dc}} & -\frac{1}{\overline{C}_{dc}\overline{R}_{CPL}} \end{bmatrix}\begin{bmatrix} \overline{I}_s \\ V_L \end{bmatrix} + \begin{bmatrix} \frac{1}{\overline{L}_d} & 0 \\ 0 & -\frac{1}{\overline{C}_{dc}} \end{bmatrix}\begin{bmatrix} V_N \\ \overline{I}_{CPL} \end{bmatrix} \tag{10.21}$$

where $\overline{I}_s$ is the average current supply by the equivalent DG, $\overline{R}_s$ denotes the equivalent combined average virtual droop resistance and average cable resistance, and $\overline{L}_s$ denotes the equivalent average cable inductance. If we consider the microgrid has n DGs, the load parameters also need to be converted into equivalent average parameters as follows:

$$\begin{cases} \overline{R}_{CPL} = n \times R_{CPL} \\ \overline{I}_{CCL} = \dfrac{I_{CCL}}{n} \\ \overline{I}_{CPL} = \dfrac{I_{CPL}}{n} \\ \overline{C}_{dc} = \dfrac{C_{dc}}{n} \end{cases} \tag{10.22}$$

The introduction of voltage compensation changes the value of the equivalent source resistance, thus a small-signal stability test is conducted to search the margin of the equivalent source resistance.

Table 10.3 The circuit parameters to conduct the stability analysis

	Equiv. resistance	Equiv. inductance	Equiv. capacitance	Equiv. load
Value	0.5 Ω	200 μH	0.2 mF	1000 W

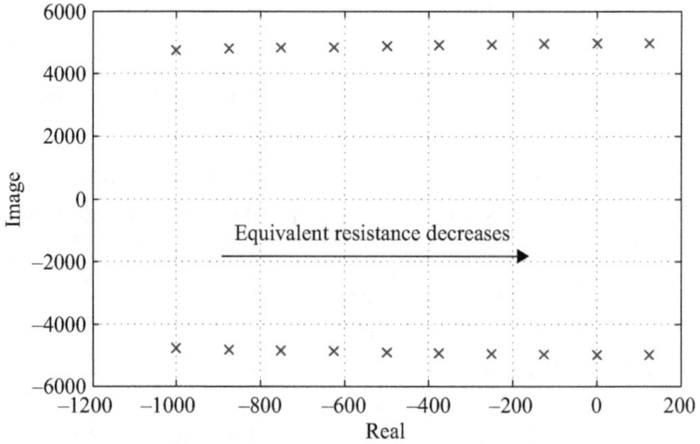

Figure 10.12 The eigenvalue traces with a decrease of equivalent source resistance

Using the linearized model of CPL, the DC microgrid can be viewed as a linear time-invariant system. The global asymptotic stability can be analyzed by the locations of the state matrix eigenvalues. When all the eigenvalues have negative real parts, the system is stable in the sense of Lyapunov. If any of the eigenvalues has positive real part, the system is unstable. The equivalent source resistance decreases from 0.1 Ω to 0.02 Ω (20% of the original value). The other parameters used for the analysis are listed in Table 10.3.

The traces of the eigenvalues are shown in Figure 10.12 with variable equivalent source resistance. The eigenvalues gradually approach the right-hand plane, with the decrease of the equivalent resistance, and the real part of the eigenvalues becomes positive when the equivalent resistance reaches 0.03 Ω. Therefore, the minimum value that can be applied is around 30% of the original value. The voltage compensation coefficient needs to be limited no larger than 70% of the droop resistance, so as to maintain the system stability.

10.5 Simulation

10.5.1 Simulation setup

The DC microgrid with three modules is modeled and simulated in MATLAB/ Simulink, as shown in Figure 10.13. The rated parameters of the three modules

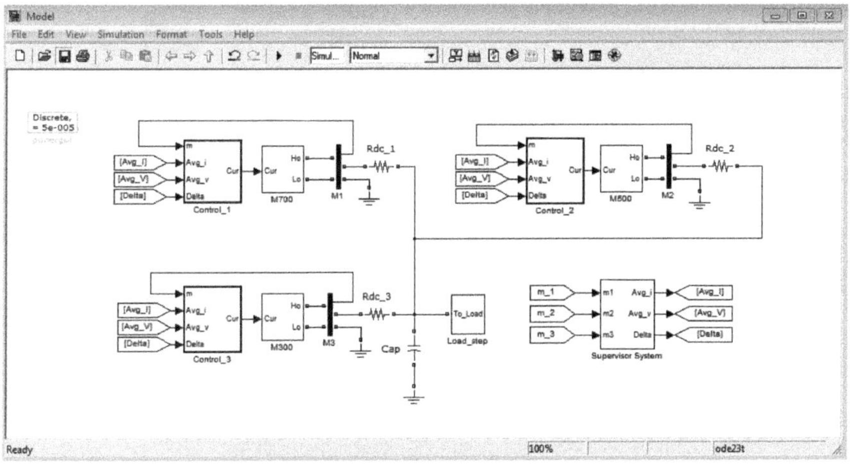

Figure 10.13 Simulation model in MATLAB/Simulink

Table 10.4 Parameters of the DC microgrid

Module	Rated voltage (V)	Rated current (A)	Line resistance (mΩ)	Voltage offset (%)
M700	70	10	10	0.5
M500	50	10	30	0
M300	60	5	0*	0.25

*The value is smaller than 1 mΩ during experiments.

are 700 W/70 V, 500 W/50 V, and 300 W/60 V. The three modules are connected to the common DC bus via boost-type DC–DC converters. The local control is usually realized by cascaded control with inner current control and outer voltage control. The dynamics of the current control is much higher than outer loop, thus the current controlled DG is modeled as controllable current source in the simulation model. The DC bus voltage adopted is 100 V to keep the same configuration as the experimental test, in which the bus voltage is limited by the test bench. The other parameters of the three modules used for simulation are listed in Table 10.4.

It is assumed that the three distributed generation modules are geographically distributed, and the module M300 is located near the load, others are connected to the load with cables about 1 m and 3 m for modules M700 and M500, respectively. Single conductor cable (5.5 mm^2) is used as the connecting cable, and the resistors are designed to be 0.01 Ω, 0.03 Ω, to consider the influences of transmission lines [7]. To simulate the influence of unequal nominal references, the nominal reference offsets for these three modules are manually set to 0.5%, 0%, and 0.25%.

The inner current loop can be implemented by peak current modulation or average current control using PI control or sliding mode control [42] to obtain a first-order

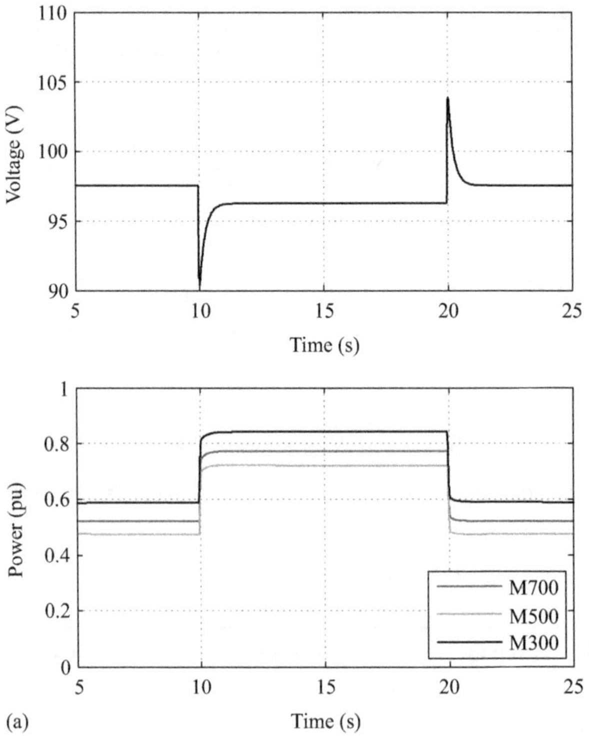

Figure 10.14 Simulation results of different compensation methods. The droop resistance is 0.05pu, while the voltage compensation coefficient adopted in Types AIII and CII is 0.02pu. (a) Basic droop control.

current response. The time constant of current loop is usually as small as one or several control periods. In the simulation, the inner current loop is simplified to be a current source, and then the equivalent models of the DGs become controllable current sources paralleled with an output capacitor. A PI controller is used for voltage control loop, and the droop resistors are set to 0.05pu, to achieve 5% voltage regulation theoretically. Per unit system is adopted, as it is convenient to compare the performance of the DGs with different rated powers and also it makes tuning of control parameters easier. The base value of voltage is selected as 100 V for all the DGs, while the base current is chosen as the rated output current of the individual DG and the base resistance is the ratio of base voltage over base current.

10.5.2 Simulation results

The aforementioned basic droop control, hierarchical structure with the secondary central controller (Type AI), dedicated voltage compensation using the common voltage (Type AII), dedicated voltage compensation with the common current (Type AIII),

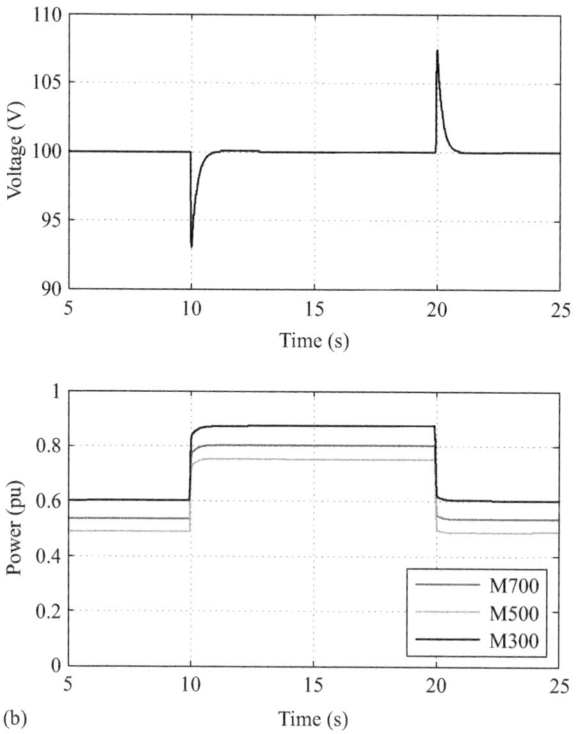

Figure 10.14 (Continued)
(b) Type AI – Hierarchical structure with secondary control.

load sharing compensation with the common current (Type B), voltage compensated using the common voltage and load sharing compensated by common current (Type CI), and the proposed compensation method (Type CII) using the common current are simulated to compare their steady-state performances. A step load is applied to test the performance of different compensation methods. The load steps up from 800 to 1200 W at $t = 10$s and then steps back at $t = 20$s, the results are shown in Figure 10.14.

Without communication, the voltage drop is determined by the droop resistance, thus heavy load results to larger voltage drop in basic droop control. As shown in Figure 10.14a, the load cannot be shared proportionally due to the unequal line impedances as well as the nominal voltage reference offsets. With the introduction of secondary voltage control at PCC, Type AI can obtain precise bus voltage control without error at the reference point no matter heavy or light load is applied, as seen from Figure 10.14b. However, the load sharing error among the three modules cannot be compensated. When the voltage compensation is dedicated at each local controller, this is Type AII, as shown in Figure 10.14c, the bus voltage is controlled without error similar as Type AI and no improvement in the load sharing performance. With the help

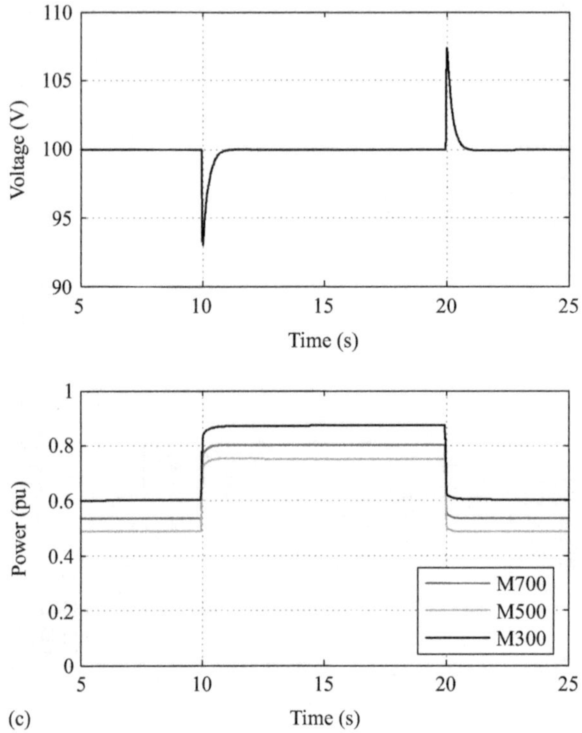

Figure 10.14 (Continued)
(c) Type AII – Dedicated voltage compensation using common voltage.

of dedicated communication line, the common current (normalized global average current) is generated in each module. In Type AIII, the droop effect can be partially compensated by using the average current, thus the voltage drop is reduced while the load sharing performance is same as that of the basic droop control, as shown in Figure 10.14d.

On the contrary, Type B can only obtain equal load sharing not voltage performance improvement. Similar to the Type AIII, the common current is generated in each module in Type B. But this common current is used as a reference, and local currents are controlled to follow this reference. Thus proportional load sharing is achieved as shown in Figure 10.14e. Both the common current and voltage references are generated with the aid of dedicated communication line in Type CI. It is like the combination of Type AII and Type B, thus it can achieve good performances in both voltage and load sharing, as shown in Figure 10.14f. In the proposed Type CII, the generated common current is used to compensate the bus voltage as Type AIII and the load sharing error as Type B. It can enhance both voltage and load sharing performances, as shown in Figure 10.14g, during different load conditions, but with a simple compensation structure and lighter communication burden.

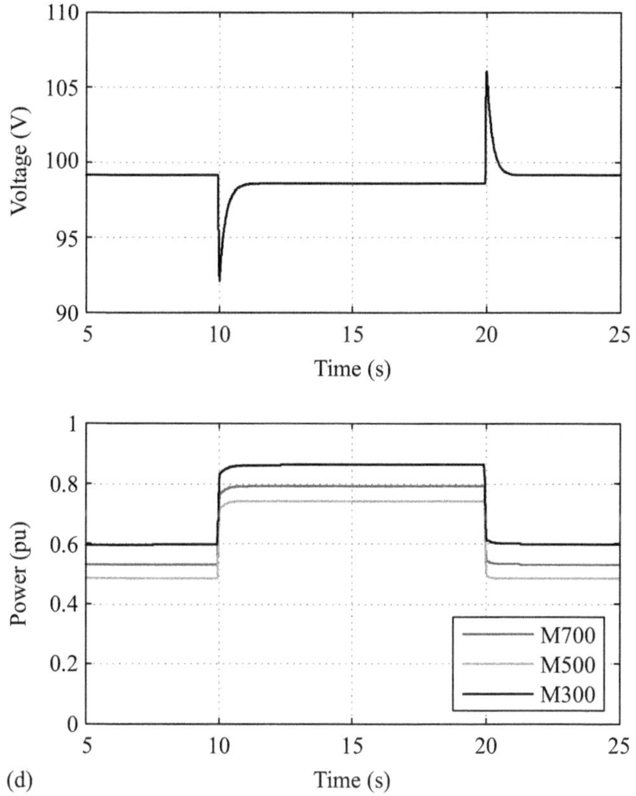

Figure 10.14 (*Continued*)
(d) Type AIII – Dedicated voltage compensation using common current.

10.5.3 Evaluation of the compensation methods

To evaluate the performances of the compensation methods, some metrics are introduced to achieve quantitative comparison, i.e., the voltage regulation index (VI) and the load sharing index (LSI). VI is utilized to reflect the overall DC-bus voltage deviation from the nominal voltage, which is defined by

$$\mathrm{VI} = \sqrt{\frac{\sum_{t=1}^{T}(V_t - V_N)^2}{T}} \tag{10.23}$$

where T is the number of the sampling points to be considered and V_N is the DC-bus nominal voltage. A zero VI value indicates a perfect voltage regulation without error, and higher VI means larger DC bus voltage deviation.

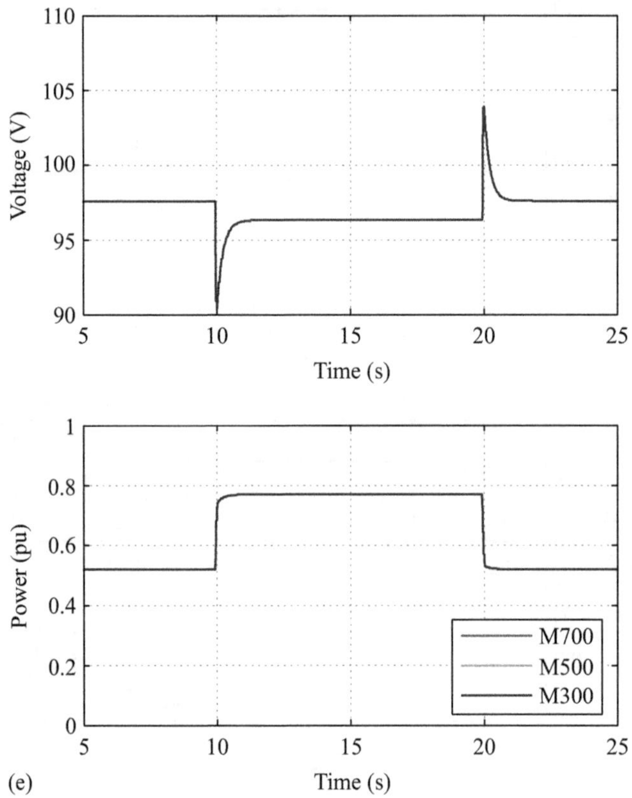

Figure 10.14 (*Continued*)
(e) Type B – Load sharing compensation using common current.

Similar to the VI, the LSI is built to measure the error between the local output current and the common current, which is defined by

$$\mathrm{LSI} = \sqrt{\frac{\sum_{j=1}^{n}\left(\frac{\sum_{t=1}^{T}\left(I_{j,t}-I_{N,t}\right)^2}{T}\right)}{n}} \tag{10.24}$$

where n is the number of DGs, $I_{j,t}$ is the sampling current of the jth DG, and $I_{N,t}$ is the common average current of the n DGs through the microgrid.

The performances of different compensation methods are evaluated by VI and LSI, the results are given in Figure 10.15. The blue bars show the VI, whereas the dark red bars give the LSI. Types AI, AII, and CI can achieve best voltage performance, i.e., lowest VI. Types B, CI, and CII achieve best load sharing performance, i.e., lowest LSI around 0.0001. Overall best performance is achieved by Type CI, and followed by CII, which has simpler structure and flexibility to adjust the voltage regulation.

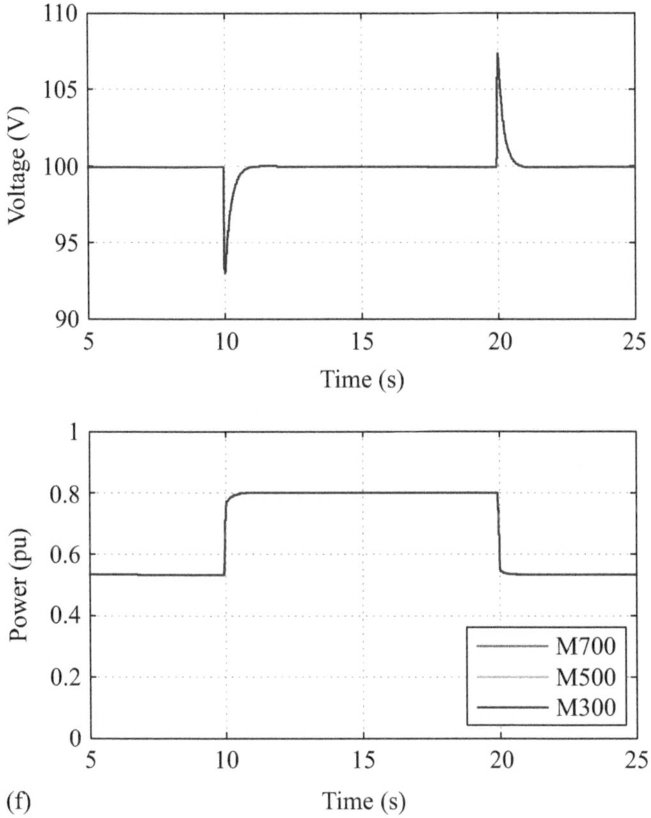

Figure 10.14 *(Continued)*
(f) Type CI – Mixed compensation using common current and voltage.

10.6 Experimental verification

10.6.1 Experiment setup

A laboratory-scale low-voltage DC microgrid, with three DGs, as shown in Figure 10.16, is constructed to test the compensation algorithms. The three DGs are connected in parallel to the common DC bus via boost-type DC–DC converters. A similar control structure using droop control with compensation is adopted for each DG. The parameters of the three DGs and the connecting cable parameters are the same as that adopted in simulation, listed in Table 10.4. M300 is directly connected to the load with a cable of which the resistance is smaller than 1 mΩ, while the others are connected in distance.

The schematic diagram of the experimental platform is shown in Figure 10.17, in which the output voltages and input currents are sensed using voltage and current

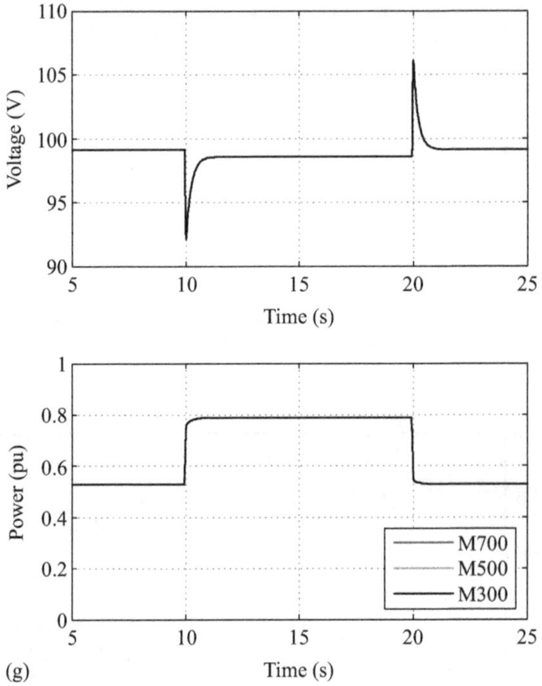

Figure 10.14 (Continued)
(g) Type CII – Mixed compensation using common current.

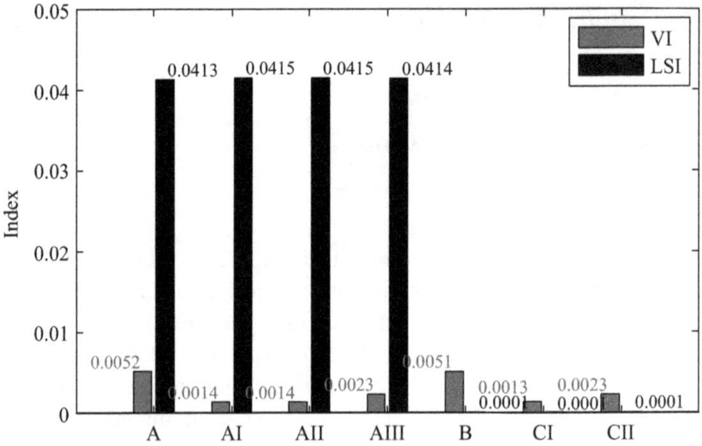

Figure 10.15 Quantitative comparison of the compensation methods

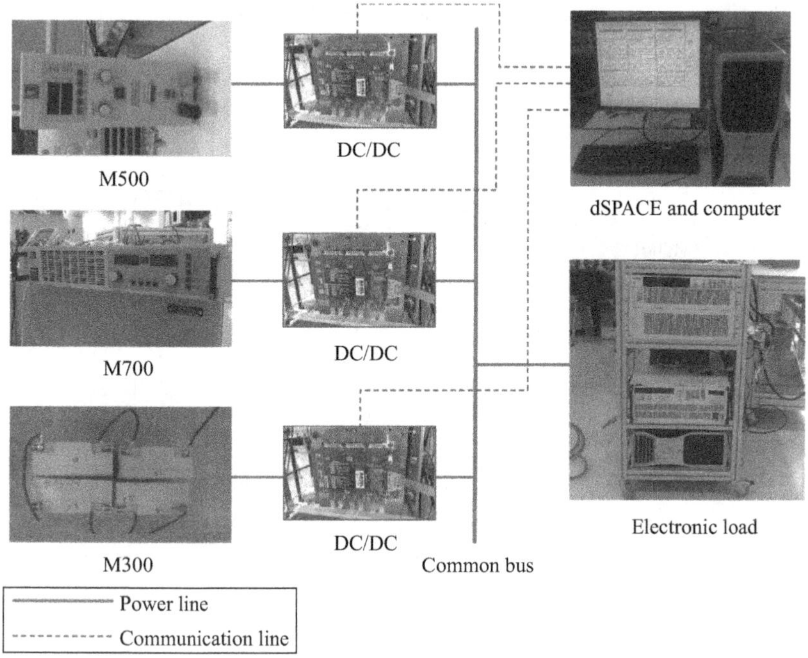

Figure 10.16 Laboratory setup of the DC microgrid. ©2015 Elsevier. Reprinted with the permission from Reference 38

sensors, and then current and voltage analogue signals go through low-pass filters (cut-off frequency 5 kHz). The filtered signals are then fed into the dSPACE DS1104 board to be converted to digital signals. It should be noticed that the adoption of per unit system allows the use of input current. This reduces the requirement of output current sensors because the input current sensor is essential for the inner current control loop.

The control algorithms of the three modules are implemented in a real-time platform DS1104. The control implementation includes four sub-units: the inner current loop with the time constant of 0.2 ms, voltage control loop operates every 1.0 ms, the droop technique unit, and the compensation unit each with a time constant of 10 ms. Both the current control and the voltage control are implemented by PI controllers. A five-point moving average filter is applied to the output current signals, before they are used in droop control sub-units and as well as for sharing with the low-bandwidth communication. The technique requires the common current reference for every module to generate voltage and load sharing compensations. In this test bench, the digital low-bandwidth communication is not implemented but simulated with a time delay of 10 ms when receiving the current/power information. The PWM signals are generated by TMS320F240 chip in DS1104 to control the DC–DC converters.

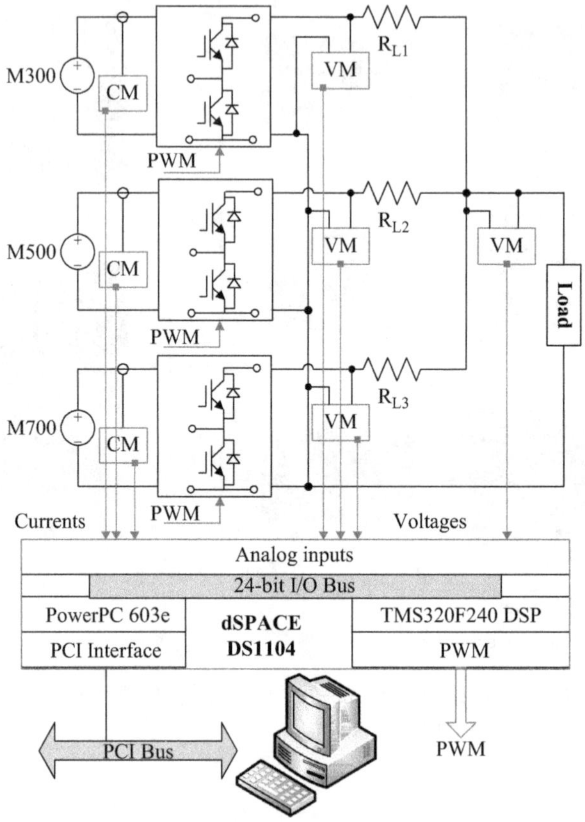

Figure 10.17 The schematic diagram of the laboratory-scale DC microgrid. ©2015 Elsevier. Reprinted with the permission from Reference 38

10.6.2 Load steps analysis

To verify the performance of the proposed method and compare with other methods, resistive load steps are utilized to conduct the experimental tests. The resistive load steps up from 400 to 600 W at around $t = 13$s and steps back at around $t = 50$s. Because the load steps are manually implemented, the step timings for different tests are not exactly the same.

The microgrid voltage performance and load sharing using the previously discussed compensation methods are shown in Figure 10.18. The same conclusion can be drawn by comparing these results. Moreover, it should be noticed that heavier load may cause some modules to go into current limit control mode without load sharing compensation. Under such circumstance, the secondary controller will become ineffective. The proposed method can overcome this and operates well during a wider range of load changes.

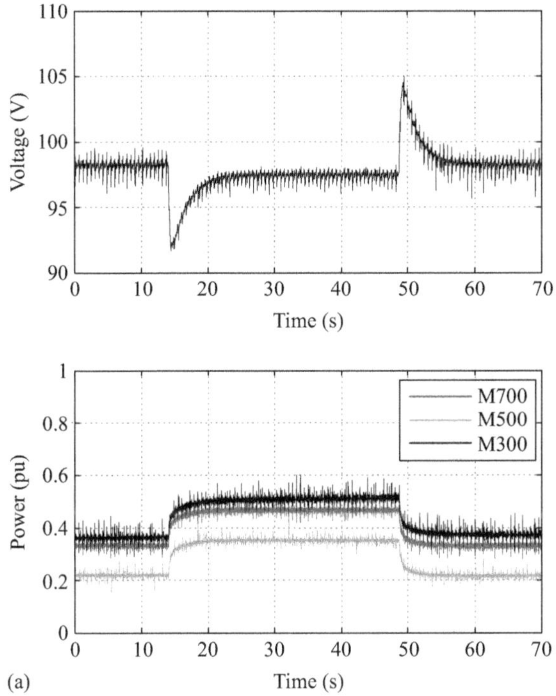

Figure 10.18 Experimental results of the voltage and load sharing performances under step load. The adopted droop resistance is 0.05pu, and the voltage compensation coefficient is 0.02pu in Types AIII and CII. (a) Basic droop control.

10.7 Conclusion

The basic droop control technique requires trade-offs between voltage regulation and load sharing, especially when the nominal voltage reference errors and unequal cable resistances are considered in low-voltage DC microgrids. Although the hierarchical structure with secondary voltage control can be used to restore the voltage drops, it has no improvement in load sharing performances. In this chapter, different compensation methods with low-bandwidth communication are classified and compared. A simple mixed compensation method using the common current is proposed to improve both voltage and load sharing performances. Small-signal stability tests are conducted using the reduced-order model to determine the boundaries of the compensation coefficients. Simulation and experimental results have verified the proposed analysis. However, it should be noticed that these compensation methods just deal with the steady-state performance and not the dynamic performance, which still needs further investigation.

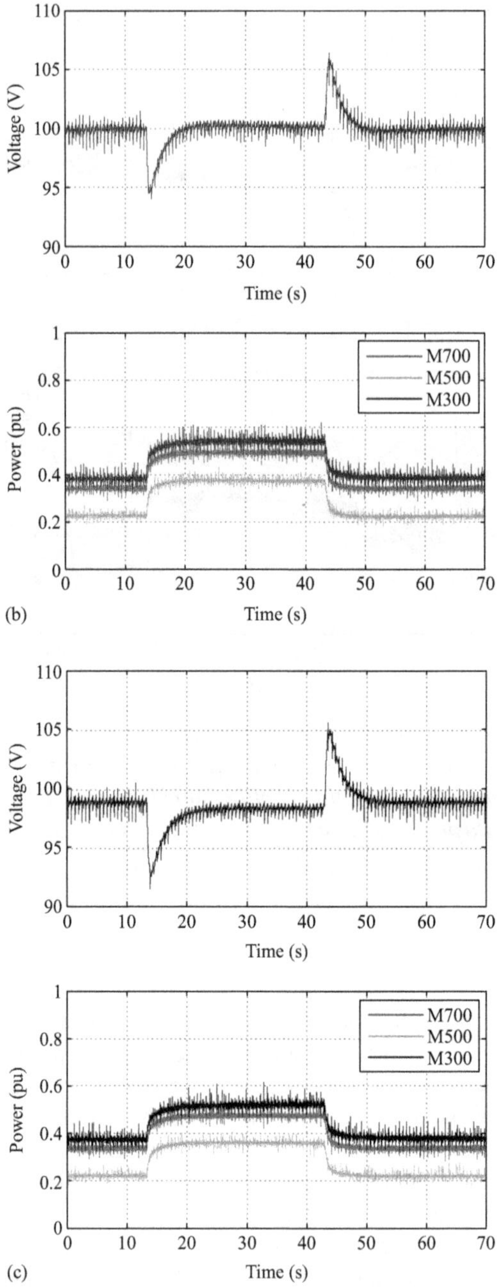

Figure 10.18 (*Continued*)
(b) Type AI – Hierarchical structure with secondary control. (c) Type AIII – Dedicate compensation using common current.

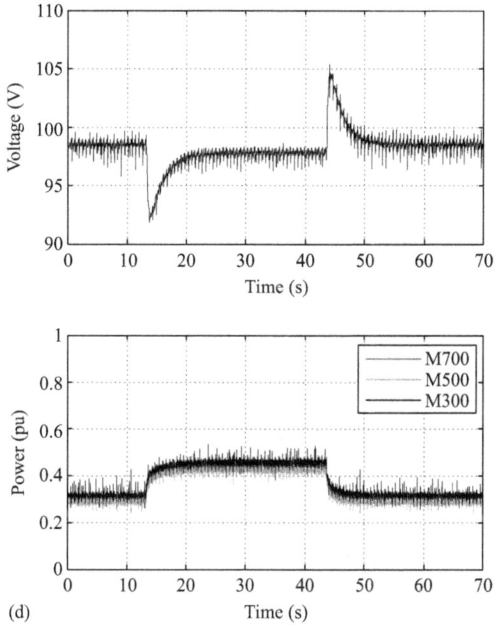

(d)

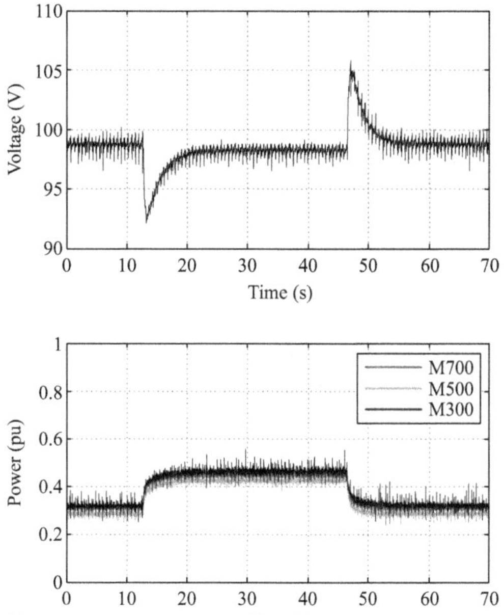

(e)

Figure 10.18 (*Continued*)
(d) Type B – Load sharing compensation using common current.
(e) Type CII – Mixed compensation using common current.

References

[1] D. Salomonsson and A. Sannino, "Comparative Design and Analysis of DC-Link-Voltage Controllers for Grid-Connected Voltage-Source Converter," in *Proceedings of IEEE Industry Applications Society Annual Meeting*, 2007, no. 2, pp. 1593–1600.

[2] D. Salomonsson, L. Söder, and A. Sannino, "An adaptive control system for a DC microgrid for data centers," *IEEE Transactions on Industrial Applications*, 2008, vol. 44, no. 6, pp. 1910–1917.

[3] M. Bash, R. Chan, and J. Crider, "A medium voltage DC testbed for ship power system research," in *Proceedings of IEEE Electric Ship Technologies Symposium*, 2009, pp. 560–567.

[4] K. Jia, M. Sumner, E. Christopher, T. Bi, and D. Thomas, "Advanced DC zonal marine power system protection," *IET Generation, Transmission & Distribution*, February 2014, vol. 8, no. 2, pp. 301–309.

[5] B. Zahedi and L. Norum, "Modeling and simulation of all-electric ships with low voltage DC hybrid power systems," *IEEE Transactions on Power Electronics*, 2013, vol. 28, no. 10, pp. 4525–4537.

[6] H. Kakigano, Y. Miura, and T. Ise, "Low-voltage bipolar-type DC microgrid for super high quality distribution," *IEEE Transactions on Power Electronics*, December 2010, vol. 25, no. 12, pp. 3066–3075.

[7] H. Kakigano, M. Nomura, and T. Ise, "Loss evaluation of DC distribution for residential houses compared with AC system," in *Proceedings of International Power Electronics Conference*, 2010, pp. 480–486.

[8] T. Ise, "Advantages and circuit configuration of a DC microgrids," slides presented in Symposium on Microgrids, Montreal, 2006.

[9] I. Cvetkovic, D. Boroyevich, and F. Lee, "Future home DC-based renewable energy nanoGrid system," in *Proceedings of Conference on Grand Challenges in Modeling & Simulation*, 2010, pp. 1–7.

[10] J. M. Guerrero, J. C. Vasquez, J. Matas, L. G. De Vicuña, and M. Castilla, "Hierarchical control of droop-controlled AC and DC microgrids – a general approach toward standardization," *IEEE Transactions on Industrial. Electronics*, 2011, vol. 58, no. 1, pp. 158–172.

[11] N. Yang, D. Paire, F. Gao, and A. Miraoui, "Power management strategies for microgrid-A short review," in *Proceedings of IEEE Industry Applications Society Annual Meeting*, 2013, pp. 1–9.

[12] S. Anand and B. Fernandes, "Steady state performance analysis for load sharing in DC distributed generation system," in *Proceedings of International Conference on Environment and Electrical Engineering*, 2011, no. 3, pp. 1–4.

[13] S. Anand, B. G. Fernandes, and J. Guerrero, "Distributed control to ensure proportional load sharing and improve voltage regulation in low-voltage DC microgrids," *IEEE Transactions on Power Electronics*, April 2013, vol. 28, no. 4, pp. 1900–1913.

[14] S. K. Mazumder, M. Tahir, and K. Acharya, "Master-Slave Current-Sharing Control of a Parallel DC-DC Converter System Over an RF Communication Interface," *IEEE Transactions on Industrial Electronics*, January 2008, vol. 55, no. 1, pp. 59–66.
[15] B. Irving and M. Jovanovic, "Analysis, design, and performance evaluation of droop current-sharing method," in *Proceedings of IEEE Applied Power Electronics Conference and Exposition*, 2000, pp. 235–241.
[16] S. Luo, Z. Ye, R.-L. Lin, and F. C. Lee, "A classification and evaluation of paralleling methods for power supply modules," in *Proceedings of IEEE Power Electronics Specialists Conference*, 1999, vol. 2, pp. 901–908.
[17] T. M. Haileselassie and K. Uhlen, "Impact of DC line voltage drops on power flow of mtdc using droop control," *IEEE Transactions on Power Systems*, August 2012, vol. 27, no. 3, pp. 1441–1449.
[18] J. Guerrero and J. Matas, "Wireless-control strategy for parallel operation of distributed-generation inverters," *IEEE Transactions on Industrial Electronics*, 2006, vol. 53, no. 5, pp. 1461–1470.
[19] B. K. Johnson, R. H. Lasseter, F. L. Alvarado, and R. Adapa, "Expandable multiterminal DC systems based on voltage droop," *IEEE Transactions on Power Delivery*, 1993, vol. 8, no. 4, pp. 1926–1932.
[20] J. Schonberger, R. Duke, and S. D. Roud, "DC-bus signaling: a distributed control strategy for a hybrid renewable nanogrid," *IEEE Transactions on Industrial Electronics*, 2006, vol. 53, no. 5, pp. 1453–1460.
[21] K. Sun, L. Zhang, and J. M. Guerrero, "A distributed control strategy based on DC bus signaling for modular photovoltaic generation systems with battery energy storage," *IEEE Transactions on Power Electronics*, 2011, vol. 26, no. 10, pp. 3032–3045.
[22] S. Anand and B. G. Fernandes, "Modified droop controller for paralleling of dc–dc converters in standalone dc system," *IET Power Electronics*, July 2012, vol. 5, no. 6, pp. 782–789.
[23] J. B. Wang, "Study of cable resistance and remote-sensing scheme in parallel DC/DC converter system via primary droop current-sharing control," *IET Power Electronics*, July 2012, vol. 5, no. 6, pp. 885–898.
[24] P. Karlsson and J. Svensson, "DC bus voltage control for a distributed power system," *IEEE Transactions on Power Electronics*, November 2003, vol. 18, no. 6, pp. 1405–1412.
[25] M. Hua and H. Hu, "Multilayer control for inverters in parallel operation without intercommunications," *IEEE Transactions on Power Electronics*, 2012, vol. 27, no. 8, pp. 3651–3663.
[26] T. Dragicevic, J. Guerrero, J. Vasquez, and D. Skrlec, "Supervisory control of an adaptive-droop regulated DC microgrid with battery management capability," *IEEE Transactions on Power Electronics*, 2014, vol. 29, no. 2, pp. 695–706.
[27] J. Kim, H. Choi, and B. Cho, "A novel droop method for converter parallel operation," *IEEE Transactions on Power Electronics*, 2002, vol. 17, no. 1, pp. 25–32.

[28] X. Lu, J. Guerrero, K. Sun, and J. Vasquez, "An improved control method for DC microgrids based on low bandwidth communication with DC bus voltage restoration and enhanced current sharing accuracy," *IEEE Transactions on Power Electronics*, 2014, vol. 29, no. 4, pp. 1800–1812.
[29] H. Kakigano, Y. Miura, and T. Ise, "Distribution voltage control for DC microgrids using fuzzy control and gain-scheduling technique," *IEEE Transactions on Power Electronics*, May 2013, vol. 28, no. 5, pp. 2246–2258.
[30] J. B. Wang, "Parallel DC/DC converters system with a novel primary droop current sharing control," *IET Power Electronics*, 2012, vol. 5, no. 8, pp. 1569–1580.
[31] A. Khorsandi and M. A. HosseinMokhtari, "A decentralized control method for a low-voltage DC microgrid," *IEEE Transactions of Energy Conversion*, 2014, vol. 29, no. 4, pp. 793–801.
[32] A. Erdogan and M. Aydemir, "Application of adaptive droop method to boost converters operating at the output of fuel cells," in *Proceedings of International Conference on Electrical and Electronics Engineering*, 2009, no. 1, pp. I-321–I-325.
[33] S. K. Mazumder, A. H. Nayfeh, and A. Borojevic, "Robust control of parallel DC-DC buck converters by combining integral-variable-structure and multiple-sliding-surface control schemes," *IEEE Transactions on Power Electronics*, May 2002, vol. 17, no. 3, pp. 428–437.
[34] Y. Chen and J. Liang, "Paralleling magamp-postregulator modules with sliding-mode-control method," *IEEE Transactions on Industrial Electronics*, June 2006, vol. 53, no. 3, pp. 974–983.
[35] Z. Fu and W. Qu, "Implementation of a current share method for paralleling of DC/DC converter without control interconnection," in *Proceedings of International Conference on Electrical Machines and Systems*, 2003, pp. 394–397.
[36] T. Dragicevic, J. M. Guerrero, and J. C. Vasquez, "A distributed control strategy for coordination of an autonomous LVDC microgrid based on power-line signaling," *IEEE Transactions on Industrial Electronics*, 2014, vol. 61, no. 7, pp. 3313–3326.
[37] Y. Huang and C. Tse, "Circuit theoretic classification of parallel connected DC-DC converters," *IEEE Transactions on Circuits Systems I Regular Paper*, 2007, vol. 54, no. 5, pp. 1099–1108.
[38] N. Yang, D. Paire, F. Gao, A. Miraoui, and W. Liu, "Compensation of droop control using common load condition in DC microgrids to improve voltage regulation and load sharing," *International Journal of Electrical Power Energy Systems*, 2015, vol. 64, pp. 752–760.
[39] X. Lu, J. Guerrero, K. Sun, and J. Vasquez, "Hierarchical control of parallel AC–DC converter interfaces for hybrid microgrids," *IEEE Transactions on Smart Grid*, 2013, vol. 5, no. 2, pp. 683–692.
[40] A. Khaligh and A. Emadi, "Mixed DCM/CCM pulse adjustment with constant power loads," *IEEE Transactions on Aerospace Electronics Systems*, April 2008, vol. 44, no. 2, pp. 766–782.

[41] Andre Pires Nobrega Tahim, Daniel J. Pagano, E. Lenz, and V. Stramosk, "Modeling and stability analysis of islanded DC microgrids under droop control," *IEEE Transactions on Power Electronics*, 2015, vol. 30, no. 8, pp. 4597–4607.
[42] Y. Gurkaynak and A. Khaligh, "Control and power management of a grid connected residential photovoltaic system with plug-in hybrid electric vehicle (PHEV) Load," in *Proceedings of IEEE Applied Power Electronics Conference and Exposition*, 2009, no. 1, pp. 2086–2091.

Chapter 11

Co-operative responsive electric vehicles for social-economic dispatch

Mehdi Rahmani-andebili[1] *and Ganesh Kumar Venayagamoorthy*[1,2,3]

Abstract

Nowadays, plug-in electric vehicles (PEVs) are widely promoted to replace conventional internal combustion-based vehicles to reduce dependency on fossil fuels and decrease greenhouse gases emissions. This is likely to create a huge stress to the power system and create new problems in its operation due to uncontrolled power and energy transactions by PEVs. Nevertheless, the stress can be avoided by proper coordination and cooperation of responsive PEVs (REVs) with utilities and system operators. This chapter presents a study on the integration of power system generation assets, REVs, and behavior of REV owners as a cyber-physical-social power system (CPSPS). Sustainability criteria are applied in a CPSPS framework to formulate a social-economic dispatch with REVs (SED-REVs) problem. This involves the participation of REVs through modeling of the behavior of REV owners and the load demands of a given power system. Two types of modeling are carried out, namely, first, modeling the participation percentage of REVs as their responsiveness level with respect to incentive schemes in a SED-REVs problem and second, stochastic modeling of the presence of REVs in SmartParks (parking lots for PEVs with capability to carry out energy transactions). The variability and stochastic nature of the REVs behavior and load demand is addressed by applying a heuristic optimization methodology to solve the SED-REVs problem.

[1]The Holcombe Department of Electrical and Computer Engineering, Clemson University, Clemson, SC, USA
[2]Real-Time Power and Intelligent Systems Laboratory, Clemson University, Clemson, SC, USA
[3]Eskom Centre of Excellence in HVDC Engineering, University of KwaZulu-Natal, Durban, South Africa

Nomenclature

Indices and sets

$g \in S^G$	generation unit
$Model \in S^{Model}$	responsiveness model of the REVs with respect to the introduced incentive scheme
$t \in S^T$	time in hour related to operation period
$t \in S^\tau$	time in hour related to optimization period in the forward-looking objective function
$x_f(\cdot) \in S^f(\cdot)$	forecasted value for the uncertain state
$x_f^{Narr}(\cdot)$	forecasted value for a number of arriving PEVs to SmartParks
$x_f^{Ndep}(\cdot)$	forecasted value for a number of departing PEVs from SmartParks
$x_f^{SOC}(\cdot)$	forecasted value for state-of-charge of the REVs
$x_f^D(\cdot)$	forecasted value for demand level

System parameters and variables

$A^B(\cdot)$	effective ampere-hour throughput of REV battery
A_{Tot}^B	total cumulative ampere-hour throughput of REV battery in its life cycle
C^{REV}	capacity of REV battery
$Cost_{Daily}(\cdot)$	daily operation cost of the system as the objective function of the problem
$Cost_{FL}(\cdot)$	forward-looking objective function of the problem
$Cost_{Hourly}(\cdot)$	hourly operation cost of the system
$Cost^{Em}(\cdot)$	CO_2 emission cost of the thermal power plants
$Cost^{IBM}(\cdot)$	cost of incentive-based management of the REVs fleet
$Cost^{Risk}(\cdot)$	risk cost of load not supplied of the system
$Cost^{Fuel}(\cdot)$	fuel cost of the thermal power plants
$Cost^{LL}(\cdot)$	total cost of the REVs batteries life loss
$Cost_{REV}^{LL}(\cdot)$	cost of a REV battery life loss
$D(\cdot)$	system load demand
$Em(\cdot)$	CO_2 emission level produced by the thermal power plants
$ELNS(\cdot)$	total expected load not supplied of the system
$LNS(\cdot)$	load not supplied of the system due to outage of a generation unit
$N^{REV}(\cdot)$	number of the REVs parked at SmartParks
$N^{PEV}(\cdot)$	number of the PEVs parked at SmartParks
$N^{arr}(\cdot)$	number of arriving PEVs to SmartParks

$N^{dep}(\cdot)$	number of departing PEVs from SmartParks
$Prob^{Outage}$	outage probability of a generation unit
$\underline{P^G}, \overline{P^G}$	minimum and maximum power limits of the thermal power plant
$P^G(\cdot)$	power of a thermal power plant
$P^{REV}(\cdot)$	power of the REV battery
$\lvert P^{REV}(\cdot)\rvert$	magnitude of power of the REV battery
$\underline{P^{REV}}, \overline{P^{REV}}$	minimum and maximum power limits of the REV battery
$P_{Tot}^{REV}(\cdot)$	total power of the REV batteries
$Price^B$	price of the REV battery
RUR^G, RDR^G	ramp up rate and ramp-down rate of the thermal power plant
$\underline{SOC^{REV}}$	minimum allowable state-of-charge of the REV battery
$\overline{SOC^{REV}}$	maximum allowable state-of-charge of the REV battery
$SOC^{PEV}(\cdot)$	state-of-charge of the REV battery
$UORR^G$	generation unit outage replacement rate
$VOLL(\cdot)$	value of lost load of customers
$\alpha_1^F, \alpha_2^F, \alpha_3^F$	fuel cost coefficients of the thermal power plant
$\alpha_1^E, \alpha_2^E, \alpha_3^E$	CO_2 emission coefficients of the thermal power plant
β^E	CO_2 emission cost factor
λ^B	effective weighting factor concerned with life loss cost of the REV battery
ξ_{Model}	participation percentage of the REVs in the SED-REVs problem considering the model
ψ	value of charging cost discount
$\pi(\cdot)$	electricity price
μ_{er}	mean of prediction errors related to the uncertain state
σ_{er}	standard deviation of prediction errors related to the uncertain state

SA algorithm parameters and variables

N^{SA}	number of generations of new solution at every stage
r_k^{SA}	random number in the range of $[0,1)$ at stage k
y_k^{SA}	binary variable as indicator of acceptance of new solution at stage k
ρ_k^{SA}	adaptive probability for acceptance of new solution at stage k
η^{SA}	coefficient for gradually decreasing temperature of molten metal
ε_k^{SA}	internal energy of molten metal at stage k
$\theta_{Initial}^{SA}$	initial temperature of molten metal
θ_k^{SA}	temperature of molten metal at stage k

11.1 Introduction

A recent study demonstrates that almost 27% of the total energy consumption and 33% of greenhouse gases emissions in the world are related to the transportation sector [1]. The carbon dioxide (CO_2) emissions from the transportation sector were reported to be 24% in 2005, the second highest from the energy industry [2]. In order to address the environmental concerns resulting from conventional vehicles, the usage of plug-in electric vehicles (PEVs) is promoted. Studies show that advancements in battery technology have accelerated PEVs utilization in some developed countries in the last decade [3,4]. However, replacing conventional vehicles on the road with PEVs is likely to create a stress to the power system and create new problems such as system overload and oscillations, and cause spikes in energy market prices due to uncontrolled charging/discharging by the PEVs [5,6]. Nevertheless, the above mentioned problems can be avoided by proper coordination and cooperation of responsive PEVs (REVs) with utilities and system operators. REVs are PEVs willing to follow schedules for their energy transactions as per requests from electric power system operators. The coordination and cooperation will primarily address social-economic aspects. Creating a sustainable and reliable energy and transportation system is a societal responsibility, and therefore influences and is influenced by our social behavior. In other words, coordination and cooperation of the REVs aiming at a social-economic dispatch (SED-REVs) can minimize or eliminate the undesirable consequences of PEVs, and bring about increased social welfare through economic, environmental, and reliability benefits.

It has been reported that private vehicles are at parking lots in idle state for more than 90% of daytime [7]. This reality indicates that even partial replacement of the conventional vehicles with the PEVs can bring about a huge potential for participation of the REVs in the SED-REVs problem. Several studies have investigated the presence of PEVs in various activities. In [8], the feasibility of utilizing Ontario's grid for charging PEVs has been analyzed applying a zonal model of Ontario's transmission network and base-load generation capacities for the period of 2009–2025. Reference [9] proposes a model for evaluation of the performance of vehicle-to-grid (V2G) for microgrid energy management. In [10], PEVs have been utilized to support smart grids by offering frequency regulation as the ancillary service.

In [11], emissions of thermal plants have been converted into an emission cost function and integrated into the cost function for the economic dispatch problem in order to address the environmental concern caused by operation of fossil fuel-based thermal plants. In [12–15], in addition to emissions, ramp-rate limits of thermal power plants have been taken into consideration in the economic dispatch problem. In [16], PEVs and renewable energy sources have been considered for creating a sustainable integrated electricity and transportation infrastructure to minimize cost and emissions from energy use.

However, in the above-mentioned studies, a detailed sustainability study related to the cooperation and behavior of REVs in the economic dispatch problem has not been considered. In this study, based on the sustainability criteria discussed in [17], the participation of REVs is modeled to solve the SED-REVs problem in a cyber-physical-social infrastructure. Herein, the stochastic presence of the REVs in the SmartParks (parking lots for PEVs with capability to carry out energy transactions) including their arrival time, departure time, and state-of-charge (SOC) of the REVs batteries is modeled. Also, the responsiveness level of the REVs in respect to the introduced incentive scheme as their participation percentage in the SED-REVs problem is modeled. Herein, the SED-REVs problem as a nonlinear, non-convex, and stochastic optimization problem is solved applying a heuristic technique.

The rest of the chapter is outlined as follows. Section 11.2 briefly describes the social-economic sustainability indicators or measures for the SED-REVs problem. In Section 11.3, the responsiveness model of the REVs to the incentive scheme is presented. In Section 11.4, the problem of SED-RESs is defined and formulated. The optimization methodology applied to solve the problem of SED-RESs is described in Section 11.5. Typical results and discussions are presented in Section 11.6. Finally, the conclusion is given in Section 11.7.

11.2 Sustainability indictors

11.2.1 Social sustainability indicator

As mentioned above, the growing environmental problem caused by emissions from use of energy can be minimized by responsive societal behavior. Depleting fossil fuels and emitting hazardous pollutant into the atmosphere have many negative impacts on the planet. These undesirable effects include climate change and environmental pollution [17,18]. There is a negative impact on the health of people from the combustion of fossil fuels in thermal power plants that mainly puts out CO_2 [17,19,20]. Herein, minimizing CO_2 emissions released by thermal power plants as a result of the cooperation of REVs in the SED-REVs problem is taken into account by considering the value of the penalty for CO_2 emission.

PEVs depend heavily on their batteries to operate and it is known that the disposal of batteries is not environmentally friendly. Therefore, it is important to integrate the life of the battery in the SED-REVs problem. Furthermore, increasing the reliability of the delivered electricity to the customers, due to co-operation of the REVs in the SED-REVs problem, can be considered as the important aspect of the social sustainability indicator. Herein, the risk of load not supplied to the customers is taken into account and integrated in the objective function of the SED-REVs problem through the value of lost load (VOLL) of the customers. The VOLL is the value of the customers' damage due to disruption in their electricity service.

11.2.2 Economic sustainability indicator

The co-operation of REVs in the SED-REVs problem must have an overall economic benefit. In other words, minimizing the value of the total operation cost of the system that includes the fuel cost of the thermal power plants, the cost of CO_2 emission produced by the thermal power plants, the cost of risk of load not supplied of the system, the cost of incentive-based management of the REVs fleet, and the REVs

Table 11.1 Participation models of the REVs in the SED-REVs problem as the function of discount on charging cost [21]

	Participation percentage of the REVs in the SED-REVs problem
Linear model	$\xi_{Lin} = \psi$
Exponential model	$\xi_{Exp} = 100 \times e^{M \times (\psi/100-1)},\ M \gg 1$
Power model	$\xi_{Pot} = 100 \times \left(\frac{\psi}{100}\right)^n$
Logarithmic model	$\xi_{Log} = 100 \times ln\left(\frac{\psi}{100} \times (exp(1) - 1) + 1\right)$

battery life loss cost due to V2G and grid-to-vehicle (G2V) is considered as economic sustainability indicator.

11.3 Responsiveness model of the REVs with respect to incentive scheme

Herein, the responsiveness level of the REVs with respect to the value of discount on charging cost (ψ) is defined as their participation percentage (ξ) in the SED-REVs program. The value of discount is introduced by the independent system operator (ISO) to the REVs. The value of discount on charging cost is determined using the models presented in [21]. In the given models that include linear, logarithmic, exponential, and power mathematical functions, the value of ξ is calculated based on the value of ψ as the introduced scheme of incentive-based management of the REVs fleet. As can be seen in Table 11.1, these models have been designed to yield 100% for participation percentage of the REVs in the SED-REVs program under free charging and none under zero discount on the charging cost. The graphic illustration of the presented models can be seen in Figure 11.1.

11.4 SED-REVs problem

Figure 11.2 illustrates the typical cyber-physical-social system (CPSS) framework considered in this study that comprises a monitoring and decision-making system located in a utility control center or at an ISO site. Figure 11.2 also shows the physical assets including the power generation system with some thermal power plants, the communication network, and the installed SmartParks with PEVs parked. The communication network is used to connect the utility control center or ISO to the thermal power plants and online REVs. In other words, the ISO announces the incentive schemes to the REVs parked in the SmartParks and estimates their responsiveness level. After that, the ISO/utility control center computes the generation level of every

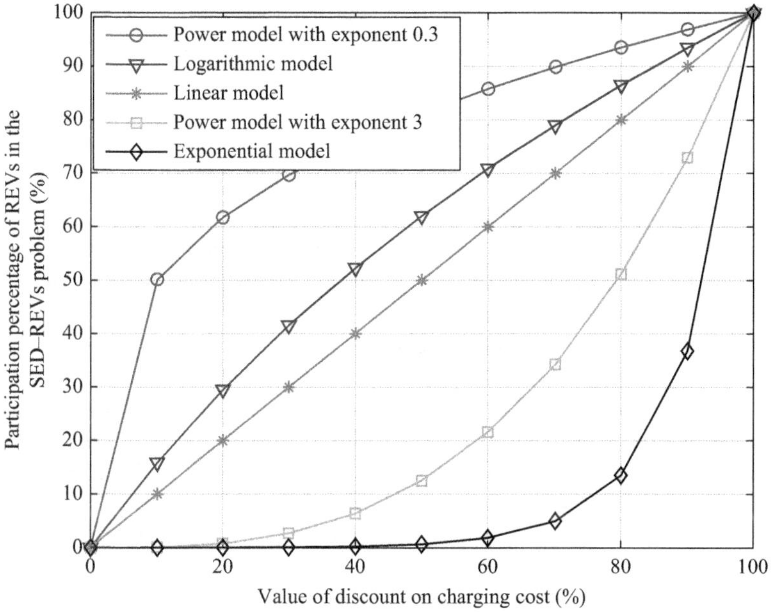

Figure 11.1 Curves related to participation percentage of the REVs in the SED-REVs problem with respect to value of discount on charging cost [21]

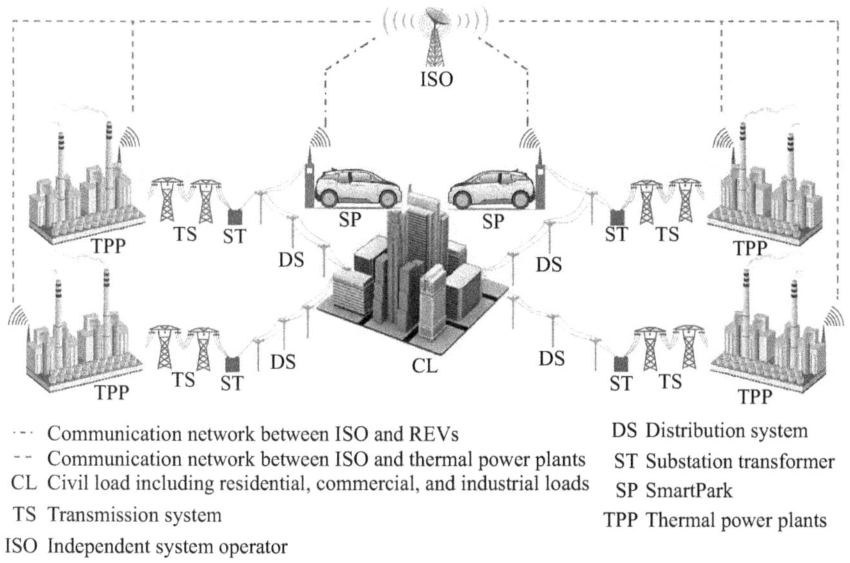

Figure 11.2 The cyber-physical-social system (CPSS) under study that comprises the communication network, the physical assets, and a computing and decision-making center located in a utility control center or ISO

thermal power plant while considering the participation percentage of the REVs in the SED-REVs problem, and then informs the results to the thermal power plants. In the following, the objective function of the SED-REVs problem with different cost terms, and also the constraints of the problem are presented and described.

11.4.1 Objective function of the SED-REVs problem

The daily operation cost of the system that must be minimized is defined as the objective function of the SED-REVs problem considering social-economic sustainability dimensions given in (11.1–11.3). As can be seen in (11.2), the daily operation cost of the system is the sum of the hourly operation cost of the system over the operation period. The hourly operation cost of the system presented in (11.3) includes five cost terms that are related to the fuel cost of the thermal power plants, the CO_2 emission cost of the thermal power plants, the life loss cost of the REVs batteries, the cost of incentive-based management of REVs fleet, and the risk cost of load not supplied to the system. Herein, the decision variables of the SED-RESs problem include the power of every thermal power plant, the power of every REV, and the value of discount on charging cost

$$min\left\{Cost_{Daily}\right\} \tag{11.1}$$

$$Cost_{Daily} = \sum_{t\in S^T} Cost_{Hourly}(t) \quad S^T = \{1,\ldots,Nt\} \tag{11.2}$$

$$\begin{aligned} Cost_{Hourly}(t) = {} & Cost^{Fuel}(t) + Cost^{Em}(t) + Cost^{LL}(t) + Cost^{IBM}(t) \\ & + Cost^{Risk}(t), \quad \forall t \in S^T \end{aligned} \tag{11.3}$$

11.4.1.1 Fuel cost of the thermal power plants

The fuel cost of every thermal power plant is considered as a quadratic polynomial. The total fuel cost function of the thermal power plants of the system is presented in the following equation:

$$\begin{aligned} Cost^{Fuel}(t) = \sum_{g\in S^G} \left(\alpha^F_{1,g} \times \left(P^G_g(t)\right)^2 + \alpha^F_{2,g} \times \left(P^G_g(t)\right) + \alpha^F_{3,g}\right), \quad \forall t \in S^T \\ S^G = \{1,\ldots,Ng\} \end{aligned} \tag{11.4}$$

11.4.1.2 CO_2 emissions cost of the thermal power plants

The CO_2 emission level of every thermal power plant is considered a quadratic polynomial. CO_2 is the primary greenhouse gas pollutant, accounting for nearly three-quarters of global greenhouse gas emissions and 84% of US greenhouse gas emissions [22]. The total CO_2 emissions produced by the thermal power plants in the system are given in (11.5). Also, the CO_2 emission cost function of the system is presented in (11.6).

$$Em(t) = \sum_{g \in S^G} \left(\alpha^E_{1,g} \times \left(P^G_g(t)\right)^2 + \alpha^E_{2,g} \times \left(P^G_g(t)\right) + \alpha^E_{3,g} \right), \quad \forall t \in S^T \quad (11.5)$$

$$Cost^{Em}(t) = \beta^E \times Em(t), \quad \forall t \in S^T \quad (11.6)$$

11.4.1.3 Life loss cost of the REVs batteries

The value of life loss cost of a REV's battery is determined based on the effective ampere-hour throughput of the battery (A^B) due to the V2G action, as can be seen in (11.7) [23]. Herein, A^B_{Tot} is the total cumulative ampere-hour throughput of the battery in its life cycle, $Price^B$ is the price of the battery installed in the REV, and λ^B is the effective weighting factor used in the model introduced in [23]. In the presented model, the value of the effective weighting factor has a nonlinear relationship with the SOC of the vehicle's battery. The value of life loss cost of REVs batteries is presented in (11.8). The total number of the REVs parked at SmartParks is determined using (11.9)

$$Cost^{LL}_{REV}(t) = \frac{\lambda^B \times A^B(t)}{A^B_{Tot}} \times Price^B, \quad \forall t \in S^T \quad (11.7)$$

$$Cost^{LL}(t) = Cost^{LL}_{REV}(t) \times N^{REV}(t), \quad \forall t \in S^T \quad (11.8)$$

$$\begin{aligned} N^{REV}(t) &= \xi_{Model} \times N^{PEV}(t), \forall t \in S^T, \quad \forall Model \in S^{Model} \\ S^{Model} &= \{Lin, Exp, Pot, Log\} \end{aligned} \quad (11.9)$$

11.4.1.4 Cost of incentive-based management of the REVs fleet

In this study, managing the REVs fleet is done by motivating them to participate in the SED-REVs problem by introducing different schemes of charging cost discount. The value of hourly cost of incentive-based management of the REVs fleet due to considering the discount on the charging cost is presented in (11.10). Herein, the battery of every REV that participates in the SED-REVs problem is fully charged considering the value of discount on the charging cost.

$$\begin{aligned} Cost^{IBM}(t) = &\left(\frac{100 - SOC^{REV}(t)}{100} \right) \times C^{REV} \times \frac{\psi}{100} \\ &\times \pi(t) \times N^{REV}(t), \quad \forall t \in S^T \end{aligned} \quad (11.10)$$

11.4.1.5 Risk cost of load not supplied of the system

Herein, the value of the expected load not supplied of the system is defined as the risk level of the system presented in (11.11). The outage probability of every thermal power plant is determined using (11.12), considering the related unit outage replacement rate. The value of the hourly cost related to the hourly risk of load not supplied of the system is given in (11.13). As can be seen in (11.13), the value of hourly risk cost of the system depends on the hourly value of risk level and VOLL of the customers

$$ELNS(t) = \sum_{g \in S^G} Prob_g^{Outage} \times LNS_g(t), \quad \forall t \in S^T \tag{11.11}$$

$$Prob_g^{Outage} = \frac{UORR_g^G}{1 - UORR_g^G} \prod_{g \in S^G} \left(1 - UORR_g^G\right), \quad \forall g \in S^G \tag{11.12}$$

$$Cost^{Risk}(t) = ELNS(t) \times VOLL(t), \quad \forall t \in S^T \tag{11.13}$$

11.4.2 Constraints of the SED-REVs problem

11.4.2.1 System power balance limit

This constraint that must be held at every time step of the operation period is presented in (11.14). Also, the total power generated or consumed by the REVs fleet is given in (11.15). Herein, the power of every REV is considered positive if the REV is doing V2G and this power is supposed to be negative if the REV is doing G2V.

$$\sum_{g \in S^G} P_g^G(t) + P_{Tot}^{REV}(t) = D(t), \quad \forall t \in S^T \tag{11.14}$$

$$P_{Tot}^{REV}(t) = P^{REV}(t) \times N^{REV}(t), \quad \forall t \in S^T \tag{11.15}$$

11.4.2.2 Generation units' power limits

The maximum power and minimum power constraints of every thermal power plant at every time step of the operation period are presented in the following equation:

$$\underline{P_g^G} \le P_g^G(t) \le \overline{P_g^G}, \quad \forall t \in S^T, \forall g \in S^G \tag{11.16}$$

11.4.2.3 Generation units' ramp-up rate and ramp-down rate limits

The ramp-up rate and ramp-down rate constraints of every thermal power plant at every time step of the operation period are presented in (11.17) and (11.18), respectively,

$$\left(P_g^G(t+1) - P_g^G(t)\right) \le RUR_g^G, \quad \forall t \in S^T, \forall g \in S^G \tag{11.17}$$

$$\left(P_g^G(t) - P_g^G(t+1)\right) \le RDR_g^G, \quad \forall t \in S^T, \forall g \in S^G \tag{11.18}$$

11.4.2.4 REVs batteries' power limits

The output power limit and input power limit of every REV's battery in the V2G and G2V actions are presented in (11.19) that must be held at every time step of the operation period. The magnitude of power of every REV's battery must be within the minimum and maximum allowable limits

$$\underline{P^{REV}} \le \left|P^{REV}(t)\right| \le \overline{P^{REV}}, \quad \forall t \in S^T \tag{11.19}$$

11.4.2.5 REVs batteries' SOC limits

In order to prolong the life time of the REVs batteries, the battery of every REV must not be discharged more than the allowable depth of discharge. Moreover, the battery of every REV may have an upper limit for its SOC. Thus, the SOC of the battery of every REV at every time step of the operation period must be in the defined and allowable minimum and maximum SOC limits presented in

$$\underline{SOC^{REV}} \leq SOC^{REV}(t) \leq \overline{SOC^{REV}}, \quad \forall t \in S^T \tag{11.20}$$

11.5 Proposed optimization technique

In this section, a stochastic optimization algorithm for solving the constrained optimization problem formulation described in Section 11.4 for SED-REVSs is described.

11.5.1 Stochastic optimization

Based on the proposed formulation, at every time step of the operation period (t), the SED-REVs problem is solved for the updated optimization time horizon $(t+1, \ldots, t+N\tau)$; however, only the dispatch signals of the target time $(t+1)$ are accepted as the decision signals, as can be seen in Figure 11.3. Therefore, the objective function of the problem presented in (11.1) is modified and changed into the forward-looking objective function presented in (11.21). As can be seen in (11.22), the forward-looking objective function is the sum of the hourly operation cost of the system over the updated optimization time horizon

$$min\,\{Cost_{FL}(t)\} \tag{11.21}$$

$$Cost_{FL}(t) = \sum_{t \in S^\tau} Cost_{Hourly}(t) \quad S^\tau = \{t+1, \ldots, t+N\tau\} \tag{11.22}$$

The flowchart of the proposed methodology that uses simulated annealing (SA) algorithm as its optimization tool is illustrated in Figure 11.4. As can be seen, at first, the system parameters along with the forecasted system states are received. Then, uncertainties of the predicted data are addressed using the scenario-based approach. After that, the problem is optimized stochastically by the SA algorithm. Next, values

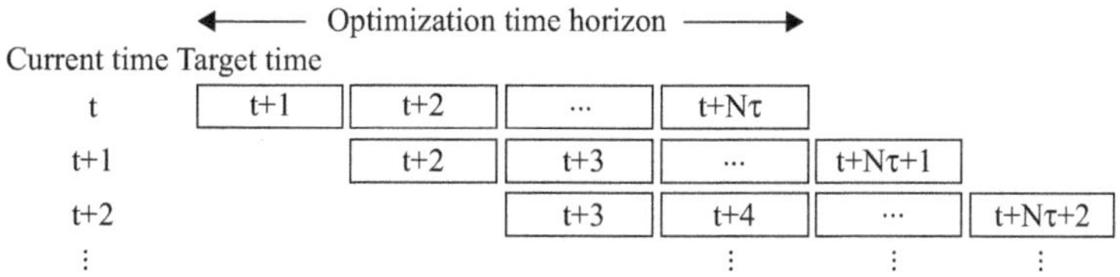

Figure 11.3 Concept of adaptability and dynamism in the proposed optimization technique

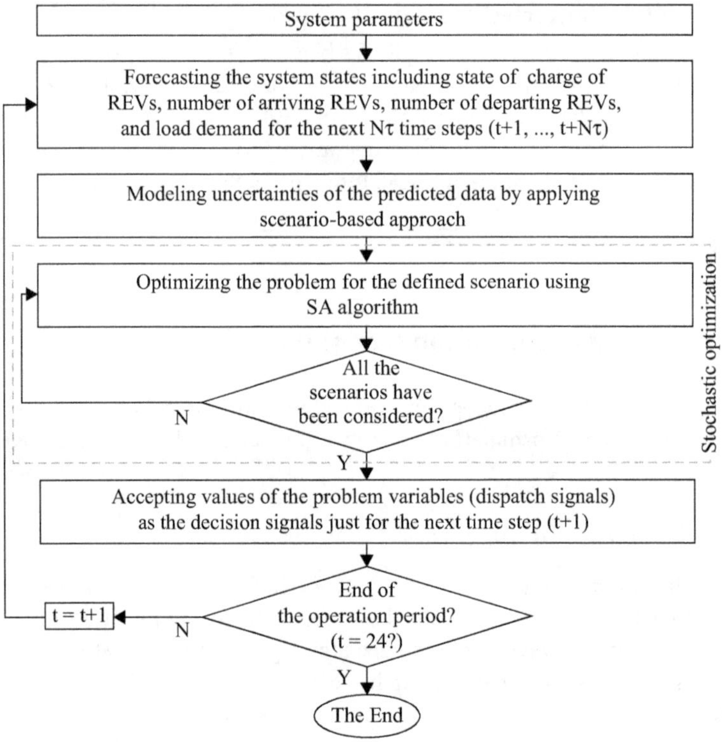

Figure 11.4 The flowchart that illustrates procedure of the proposed optimization technique

of the problem variables for the target time (next time step) are accepted as the decision signals. This process is repeated for every time step (hour) of the operation period (day).

11.5.2 Forecasting uncertain states of the problem

Herein, the number of arriving PEVs to the SmartParks, the number of departing PEVs from the SmartParks, the value of SOC of the REVs, and the value of load demand as the uncertain states of the system for the next $N\tau$ time steps are predicted using a feedforward neural network trained with the Levenberg-Marquardt back-propagation algorithm available in MATLAB. The set of forecasted values of the uncertain states over the optimization time horizon is presented in

$$S^f(t) = \left\{x_f^{Narr}(t), x_f^{Ndep}(t), x_f^{SOC}(t), x_f^{D}(t)\right\}, \quad \forall t \in S^\tau \tag{11.23}$$

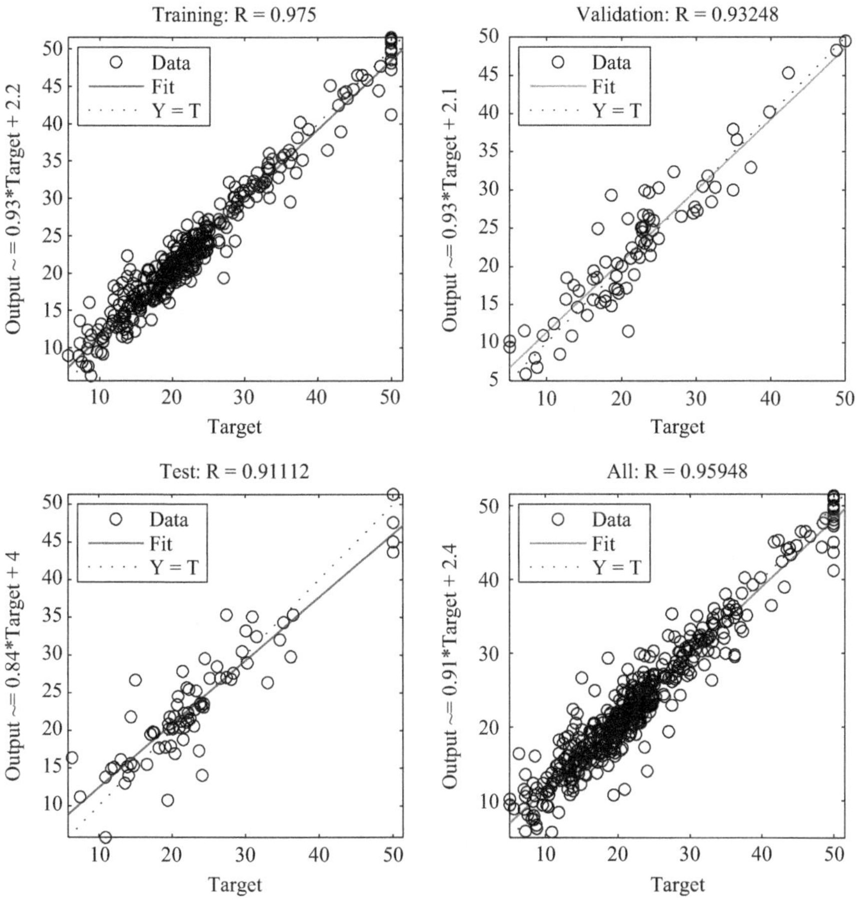

Figure 11.5 Prediction errors measured by comparing value of difference between target and output data

11.5.3 Modeling uncertainties of the forecasted data

In this study, in order to address the uncertainties of the predicted data, the scenario-based approach is applied [24,25]. Herein, in the validation and testing processes of the data forecasting, the output data are compared with the target data and values of errors of the predictions are measured (Figure 11.5), and then they are fitted on an appropriate probability density function curve (Figure 11.6). In this problem, it is observed that the predication errors can be approximated by a Gaussian probability density function with an appropriate mean (μ_{er}) and standard deviation (σ_{er}) [25]. After that, the curve is divided into three areas to define three distinct values for the prediction inaccuracy with the probabilities about 0.1587 p.u., 0.6826 p.u., and 0.1587 p.u. related to $\mu_{er} - 2\sigma_{er}$, μ_{er}, and $\mu_{er} + 2\sigma_{er}$, respectively. These values are

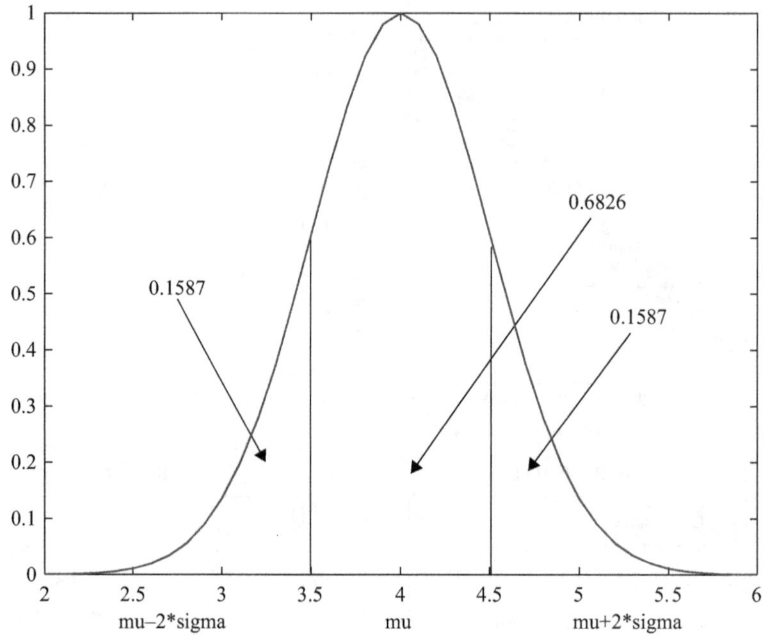

Figure 11.6 Gaussian probability density function related to the fitted prediction error of an uncertain state of the system

the most suitable amounts for the prediction inaccuracies if it is intended to divide the Gaussian probability density function into three sections. Therefore, at every time step, every uncertain state of the system (x) has three different values with different probabilities that are presented in (11.24–11.27). Based on this concept, several effective scenarios are defined for the values of every uncertain state over the optimization time horizon, which are graphically illustrated in Figure 11.7. Herein, every scenario presents the considered values for the uncertain state of the problem at every time step of the optimization time horizon.

$$\begin{cases} N^{arr}(t) = x_f^{Narr}(t) + \mu_{er}^{Narr} - 2\sigma_{er}^{Narr} \\ N^{arr}(t) = x_f^{Narr}(t) + \mu_{er}^{Narr} \\ N^{arr}(t) = x_f^{Narr}(t) + \mu_{er}^{Narr} + 2\sigma_{er}^{Narr} \end{cases}, \quad \forall t \in S^{\tau} \tag{11.24}$$

$$\begin{cases} N^{dep}(t) = x_f^{Ndep}(t) + \mu_{er}^{Ndep} - 2\sigma_{er}^{Ndep} \\ N^{dep}(t) = x_f^{Ndep}(t) + \mu_{er}^{Ndep} \\ N^{dep}(t) = x_f^{Ndep}(t) + \mu_{er}^{Ndep} + 2\sigma_{er}^{Ndep} \end{cases}, \quad \forall t \in S^{\tau} \tag{11.25}$$

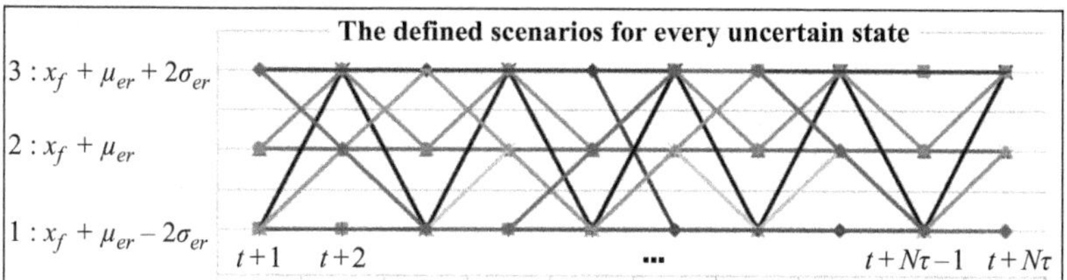

	1	2	3	4	5	6	7	8	9	10
1	3	3	3	3	3	1	1	1	1	1
2	1	1	1	1	1	3	3	3	3	3
3	2	2	2	2	2	2	2	2	2	2
4	1	3	1	3	1	3	1	3	1	3
5	1	2	1	2	1	2	1	2	1	2
6	2	3	2	3	2	3	2	3	2	3
7	1	1	1	1	1	1	1	1	1	1
8	3	3	3	3	3	3	3	3	3	3
9	1	2	3	2	1	2	3	2	1	2
10	3	2	1	1	2	3	3	2	1	1

Figure 11.7 The defined scenarios for the probabilistic values of every uncertain state of the system over the optimization time horizon

$$\begin{cases} SOC^{REV}(t) = x_f^{SOC}(t) + \mu_{er}^{SOC} - 2\sigma_{er}^{SOC} \\ SOC^{REV}(t) = x_f^{SOC}(t) + \mu_{er}^{SOC} \\ SOC^{REV}(t) = x_f^{SOC}(t) + \mu_{er}^{SOC} + 2\sigma_{er}^{SOC} \end{cases}, \quad \forall t \in S^{\tau} \tag{11.26}$$

$$\begin{cases} D(t) = x_f^{D}(t) + \mu_{er}^{D} - 2\sigma_{er}^{D} \\ D(t) = x_f^{D}(t) + \mu_{er}^{D} \\ D(t) = x_f^{D}(t) + \mu_{er}^{D} + 2\sigma_{er}^{D} \end{cases}, \quad \forall t \in S^{\tau} \tag{11.27}$$

11.5.4 SA algorithm as the optimization tool

In this study, the SA algorithm as the optimization tool is applied to solve the *SED-REVs* problem. Other optimization algorithms such as particle swarm optimization could be applied to solve this problem [26]. Herein, the value of the forward-looking objective function is defined as the value of internal energy of molten metal (ε^{SA}) and then it is tried to minimize the value of this energy. In the following, different steps for applying the SA algorithm are presented and described.

Step 1: *Primary data*
Setting controlling parameters of the SA algorithm: These parameters include the initial temperature of the molten metal ($\theta_{Initial}^{SA}$), the number of generating new random solutions at every stage (N^{SA}), and the value of coefficient for gradually decreasing the temperature of the molten metal (η^{SA}).

Parameters of the system: Values of all the system parameters and the predicted states of the system considering their uncertainties included in the defined scenarios are obtained. Moreover, the responsiveness model of the REVs and value of charging cost discount are determined.

Initial solution: A random solution for the problem variables is generated as an initial solution.

Step 2: *Generating an acceptable solution*

Generating new solution: A random solution for the problem variables is generated in the vicinity of the old one.

Checking problem constraints: All the problem constraints are checked for the optimization time horizon and if they are correct, the value of the internal energy of the molten metal is measured and the algorithm goes on, otherwise, the process is continued from Step 2.

Checking SA acceptance criterion: The SA acceptance criterion is presented in (11.28). Based on this principle, the problem solution resulted in decreased internal energy of the molten metal is always accepted; however, the solution with increased value of the internal energy is accepted just by an adaptive probability presented in (11.29). The value of this adaptive probability is reduced as the molten metal is cooled.

$$y_k^{SA} = \begin{cases} 1, & \varepsilon_{k+1}^{SA} < \varepsilon_k^{SA} \text{ or } \varepsilon_{k+1}^{SA} \geqslant \varepsilon_k^{SA}, r_k^{SA} > \rho_k^{SA} \\ 0, & \varepsilon_{k+1}^{SA} \geqslant \varepsilon_k^{SA}, r_k^{SA} \leq \rho_k^{SA} \end{cases} \tag{11.28}$$

$$\rho_k^{SA} = e^{-\left(\varepsilon_{k+1}^{SA} - \varepsilon_k^{SA}\right)/\theta_k^{SA}} \tag{11.29}$$

Step 3: *Checking number of the iteration for the current temperature*

If the number of iterations in the current temperature is not equal to the predefined value (N^{SA}), the process is continued form Step 2; otherwise, the temperature of the molten metal is decreased based on the following equation:

$$\theta_{k+1}^{SA} = \eta^{SA} \times \theta_k^{SA} \tag{11.30}$$

Step 4: *Concluding*

Checking temperature of the molten metal: The temperature of the molten metal is measured and if the molten metal is frozen, the optimization process is terminated; otherwise, the process is continued from Step 2.

Introducing outcomes: The consequences include optimal values for the problem variables.

Herein, the presented procedure is repeated for every responsiveness model of the REVs and for every value of charging cost discount. The value of the discount is modified with a 10% step. Then, the optimal value of the charging cost discount for every participation model of the REVs in the SED-REVs is explored based on the minimum daily operation cost of the system.

Table 11.2 The technical data of the thermal power plants [27]

Generator	1	2	3	4	5	6	7
α_1^F ($/MW²h)	0.69	0.68	0.68	0.89	0.91	0.99	0.98
α_2^F ($/MWh)	16.50	19.70	22.26	27.74	25.92	27.27	27.79
α_3^F ($)	680	450	370	480	660	665	670
α_1^E (ton/MW²h)	0.01	0.1	0.02	0.03	0.04	0.05	0.06
α_2^E (ton/MWh)	0.16	0.19	0.22	0.27	0.25	0.27	0.28
α_3^E (ton)	1.13	1.32	1.43	1.51	1.54	1.85	1.91
$\underline{P^G}$ (MW)	20	15	10	20	10	10	10
$\overline{P^G}$ (MW)	300	200	180	140	120	80	80
RUR^G,RDR^G	70	70	60	50	40	40	40

11.6 Numerical study and results analysis

11.6.1 Initial data

Technical data of the generation units including their fuel cost coefficients, CO_2 emission coefficients, power limits, and ramp-up rate and ramp-down rate are presented in Table 11.2 [27]. Furthermore, the value of penalty for CO_2 emission is assumed about $10 per ton based on the California Air Resources Board auction of greenhouse gases emissions [28].

The total number of PEVs is considered in this study 20,000. The rated power and capacity of battery of every REV are assumed to be about 50 kW and 50 kWh, respectively. Moreover, the minimum and maximum allowable limits for SOC of the battery of every REV are considered to be 30% and 100%, respectively. The price of a battery installed in a REV and the value of the total cumulative ampere-hour throughput of a battery during its life cycle is considered to be $10,000 and 1,000,000 ampere-hour, respectively. Moreover, the charging and discharging voltages of the REVs batteries are assumed to be about 400 V.

Table 11.3 presents forecasted values, and also the mean and standard deviation of the prediction errors related to every uncertain state of the REVs behavior for every hour of the operation period. Also, the values of the forecasted demand level and the corresponding mean and standard deviation of the prediction errors for every hour of the operation period are given in Table 11.4. Additionally, the values of the controlling parameters of the SA algorithm including $\theta_{Initial}^{SA}$, N^{SA}, and η^{SA} are considered to be about 900 centigrade, 90 times, and 0.9, respectively.

11.6.2 Studying the system with a predefined scheme of incentive

Herein, 60% discount on the charging cost, as the default scheme of incentive, is introduced to the REVs and the problem is studied. In other words, the charging cost

Table 11.3 The forecasted values, mean, and standard deviation of the prediction errors related to every uncertain state

Hour	Connected PEVs (%)			Connection duration (hour)			State-of-charge (%)		
	x_f^{Narr}	μ_{er}^{Narr}	σ_{er}^{Narr}	x_f^{Ndep}	μ_{er}^{Ndep}	σ_{er}^{Ndep}	x_f^{SOC}	μ_{er}^{SOC}	σ_{er}^{SOC}
1	0.00	0.00	0.00	7	0	0	20	10	8
2	0.00	0.00	0.00	6	0	0	20	10	8
3	0.00	0.00	0.00	5	0	0	20	10	8
4	0.00	0.00	0.00	4	0	0	20	10	8
5	0.00	0.00	0.00	3	0	0	20	10	8
6	0.00	0.00	0.00	2	0	0	20	10	8
7	0.00	0.00	0.00	1	0	0	20	10	8
8	0.01	0.00	0.00	22	1	1	72	10	8
9	0.01	0.00	0.00	22	1	1	72	10	8
10	0.02	0.00	0.00	21	1	1	72	10	8
11	0.02	0.00	0.00	20	1	1	72	10	8
12	0.03	0.00	0.00	19	1	1	72	10	8
13	0.02	0.00	0.00	18	1	1	55	10	8
14	0.01	0.00	0.00	17	1	1	55	10	8
15	0.01	0.00	0.00	16	1	1	55	10	8
16	0.02	0.00	0.00	15	1	1	55	10	8
17	0.04	0.00	0.00	14	1	1	37	10	8
18	0.03	0.00	0.00	13	1	1	37	10	8
19	0.09	0.01	0.01	12	1	1	37	10	8
20	0.13	0.01	0.01	11	1	1	37	10	8
21	0.15	0.01	0.01	10	1	1	37	10	8
22	0.15	0.01	0.01	9	1	1	37	10	8
23	0.14	0.01	0.01	8	1	1	20	10	8
24	0.07	0.01	0.01	7	1	1	20	10	8

Table 11.4 Values of the forecasted demand level and the corresponding mean and standard deviation of the prediction errors

Hour	x_f^D (MW)	μ_{er}^D (MW)	σ_{er}^D (MW)	Hour	x_f^D (MW)	μ_{er}^D (MW)	σ_{er}^D (MW)
1	454.9	20	10	13	672.0	20	10
2	460.3	20	10	14	699.1	20	10
3	449.5	20	10	15	739.8	20	10
4	449.5	20	10	16	743.2	20	10
5	454.9	20	10	17	712.7	20	10
6	438.6	20	10	18	830.1	20	10
7	454.9	20	10	19	937.3	20	10
8	468.5	20	10	20	973.2	20	10
9	597.4	20	10	21	980.0	20	10
10	644.9	20	10	22	961.7	20	10
11	672.0	20	10	23	855.2	20	10
12	665.2	20	10	24	791.4	20	10

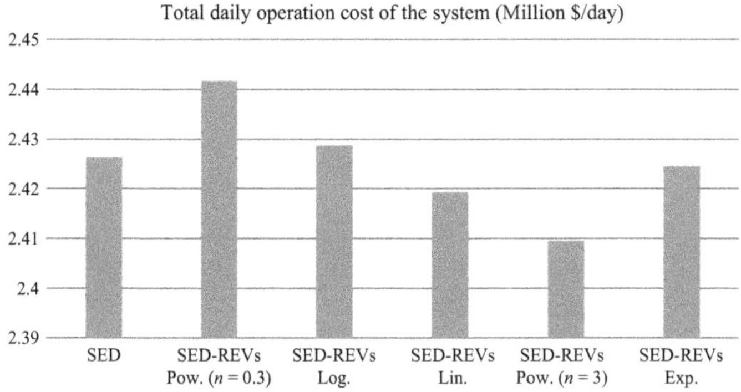

Figure 11.8 Value of total daily operation cost of the system for the SED and SED-REVs problems considering different participation models and the predetermined incentive scheme (Million $/day)

Table 11.5 Simulation results of the SED and SED-REVs problems considering different participation models and the predetermined incentive scheme

		SED-REVs problem				
	SED problem	**Power model with exponent 0.3**	**Logarithmic model**	**Linear model**	**Power model with exponent 3**	**Exponential model**
Value of discount (%)	0	60	60	60	60	60
Participation level (%)	0	85.79	70.85	60	21.60	1.83
Risk level (MW/day)	28.59	4.57	5.24	6.30	16.19	27.29
CO_2 emission level (ton/day)	45,025	43,899	43,972	44,041	44,494	44,981
Incentive-based management cost ($/day)	0	91,596	75,645	64,059	23,061	1955
Life loss cost of the REVs' batteries ($/day)	0	37,362	33,429	28,856	11,631	969
Risk cost ($/day)	47,537	5,182	7,360	9,378	27,081	45,577
CO_2 emission cost ($/day)	450,251	438,995	439,722	440,413	444,942	449,816
Fuel cost ($/day)	1,928,472	1,868,523	1,872,583	1,876,585	1,902,771	1,926,184
Total operation cost ($/day)	2,426,261	2,441,660	2,428,741	2,419,293	2,409,488	2,424,503

of the REVs is 40% of the real charging cost. Figure 11.8 illustrates the total daily operation cost of the system for the SED problem, and also for the SED-RESs problem considering different responsiveness models including power model with exponent 0.3, logarithmic model, linear model, power model with exponent 3, and exponential model. As can be seen, the participation of RESs in the SED-REVs in some cases has a negative effect, since the total daily operation cost of the system has been increased. Detailed results of the problem simulation are presented in Table 11.5. As can be seen,

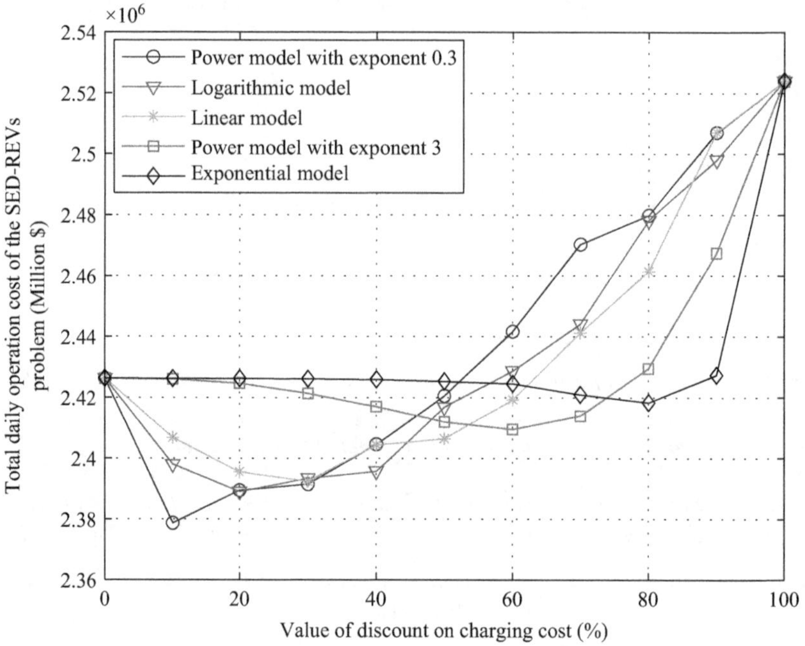

Figure 11.9 The curves concerned with total daily operation cost of the SED-REVs problem with respect to the value of discount on charging cost considering different participation models

due to the difference in the responsiveness level of the REVs with different models, participation percentages of the REVs in the SED-REVs problem are not equal. In other words, the exponential model and power model with exponent 0.3 has the lowest and highest participation percentage in the SED-REVs problem.

11.6.3 Investigating optimal scheme of incentive

Herein, the SED-REVs problem is simulated for every participation model and for all possible values of the charging cost discount with a 10% step. The total daily operation cost curves with respect to charging cost discount are illustrated in Figure 11.9. As can be seen, increasing the value of charging cost discount for raising motivation of the REVs is not always beneficial since the total daily operation cost curves do not have a purely descending trend. In other words considering a higher discount on the charging cost brings about extra cost, since the total daily operation cost of the SED-REVs problem is more than the total daily operation cost of the SED problem without participation of the REVs. Thus introducing a high value of incentive to the REVs does not benefit but also has a detrimental impact. Furthermore, there is just one optimal

Table 11.6 Simulation results of the SED and SED-REVs problems considering different participation models and the related optimal incentive scheme

		SED-REVs problem				
	SED problem	**Power model with exponent 0.3**	**Logarithmic model**	**Linear model**	**Power model with exponent 3**	**Exponential model**
Optimal discount (%)	0	10	20	30	60	80
Optimal participation level (%)	0	50.11	29.53	30	21.60	13.53
Risk level (MW/day)	28.59	9.16	13.24	12.86	16.19	20.29
CO_2 emission level (ton/day)	45,025	44,176	44,389	44,350	44,494	44,650
Incentive-based management cost ($/day)	0	8,918	10,512	16,014	23,061	19,265
Life loss cost of the REVs' batteries ($/day)	0	28,255	16,132	16,654	11,631	7,373
Risk cost ($/day)	47,537	14,229	22,133	21,496	27,081	33,900
CO_2 emission cost ($/day)	450,251	441,768	443,898	443,509	444,942	446,506
Fuel cost ($/day)	1,928,472	1,885,547	1,896,403	1,894,729	1,902,771	1,911,247
Total operation cost ($/day)	2,426,261	2,378,719	2,389,080	2,392,405	2,409,488	2,418,293

value on the charging cost discount for every responsiveness model. Moreover, the optimal value of the incentive for every participation model is different. Therefore, considering an incidental scheme of incentive and introducing it to the REVs would not be advantageous.

Detailed results of the problem simulation are presented in Table 11.6. As can be seen, although considering the optimal scheme of incentive imposes extra expenses due to REVs batteries life loss cost and incentive-based management cost, the total daily operation cost of the SED-REVs problem for every model is decreased compared with the result of the SED problem.

As can be seen in Table 11.6 and Figure 11.10, the power model with exponent 0.3 is the most appropriate model from economic point of view, since the related total daily operation cost of the system is minimum. In addition, this model is considered as the desirable model from social and environmental viewpoints, as the minimum risk and CO_2 emission levels are obtained with this model.

Figure 11.11 illustrates the curve related to optimal participation of the REVs in the SED-REVs problem with respect to value of optimal charging cost discount. This curve is the outcome of regression of the discrete data regarding the optimal participation percentage and the optimal charging cost discount for different models

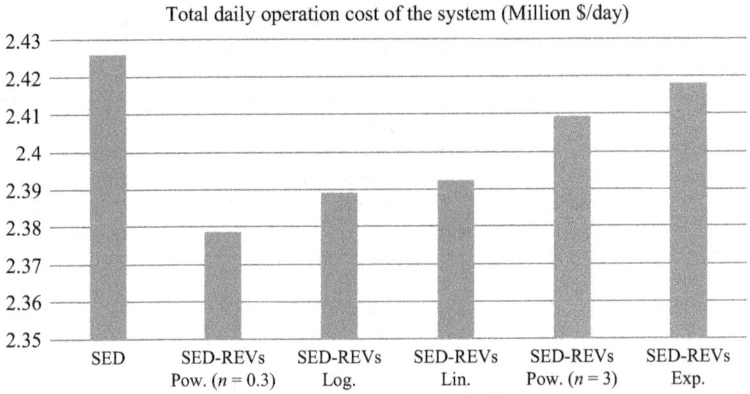

Figure 11.10 Value of total daily operation cost of the system for the SED and SED-REVs problems considering different participation models and the related optimal incentive scheme (Million $/day)

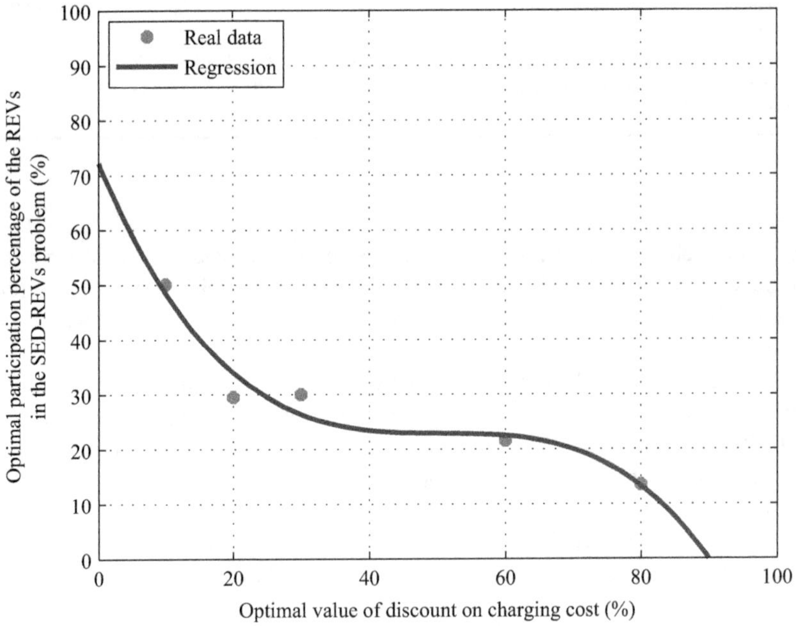

Figure 11.11 The curve related to the optimal participation of the REVs in the SED-REVs problem with respect to the optimal value of charging cost discount

presented in Table 11.6. Based on this, the crossing point of every participation model with this curve gives the optimal value of charging cost discount and optimal participation percentage of the REVs in a SED-REVs problem.

11.7 Conclusion

This chapter has presented a study on the integration of power system generation assets, REVs, and behavior of REV owners as a cyber-physical-social power system (CPSPS). A SED-REVs problem was formulated to enhance sustainability through a CPSPS framework. This involved the participation of REVs through modeling of the behavior of REV owners and the load demands of a given power system. The variability and stochastic nature of the REVs behavior and load demand was addressed by applying a heuristic optimization methodology to arrive at optimal dispatch for the SED-REVs problem.

It was observed that considering a default incentive scheme and introducing it to the REVs may not be beneficial for every participation model of the REVs in the SED-REVs problem. Thus, the incentive scheme should be designed optimally. Moreover, it was demonstrated that the optimal value of incentive for every participation model from economic viewpoint is not identical. Therefore, the behavior of the REVs fleet regarding their participation in the SED-REVs problem with respect to the introduced incentive scheme should be studied and modeled precisely, as this issue can significantly affect the result of the problem. After investigating the optimal scheme of incentive and introducing it to the REVs, it was observed that optimal cooperation of the REVs in the SED-REVs problem with any responsiveness model is advantageous from an economic, environmental, and reliability point of views, and consequently from a social welfare viewpoint.

Acknowledgment

This work was supported in part by the US National Science Foundation (NSF) under grants 1312260 and 1232070 awarded to Dr Venayagamoorthy. Any opinions, findings, and conclusions or recommendations expressed in this material are those of the authors and do not necessarily reflect the views of National Science Foundation.

References

[1] S. F. Tie, and C.W. Tan, "A review of energy sources and energy management system in electric vehicles," *Renewable and Sustainable Energy Reviews*, April 2013, vol. 20, pp. 82–102.

[2] G. K. Venayagamoorthy and G. Braband, "Carbon reduction potential with intelligent control of power systems," *Proceedings of the* 2008 *IFAC World Congress*, Seoul, South Korea, July 6–11, 2008.

[3] Canadian Automobile Association, "Electric vehicles: what you need to know." [Online]. Available: http://electricvehicles.caa.ca/, accessed March 2016.

[4] BC Hydro, "Integrated Resource Plan-Appendix 2A-2012 Electric Load Forecast." [Online]. Available: https://www.bchydro.com/content/dam/BCHydro/customer-portal/documents/corporate/regulatory-planning-documents/integratedresource-plans/current-plan/0200a-nov-2013-irp-appx-2a.pdf, accessed March 2016.

[5] L. P. Fernandez, T. G. San Roman, R. Cossent, C. M. Domingo, and P. Frias, "Assessment of the impact of plug-in electric vehicles on distribution networks," *IEEE Transactions on Power Systems*, February 2011, vol. 26, no. 1, pp. 206–213.

[6] P. Mitra and G. K. Venayagamoorthy, "Wide area control for improving stability of a power system with plug-in electric vehicles," *IET Generation, Transmission & Distribution*, October 2010, vol. 4, no. 10, pp. 1151–1163.

[7] W. Kempton, "Vehicle to grid power," FERC, 2007. [Online]. Available: https://www.ferc.gov/media/news-releases/2007/2007-3/10-22-07-v2g.pdf, accessed March 2016.

[8] A. Hajimiragha, C.A. Caizares, M. W. Fowler, and A. Elkamel, "Optimal transition to plug-in hybrid electric vehicles in Ontario, Canada, considering the electricity-grid limitations," *IEEE Transactions on Industrial Electronics*, February 2010, vol. 57, no. 2, pp. 690–701.

[9] C. Battistelli, L. Baringo, and A. J. Conejo, "Optimal energy management of small electric energy systems including V2G facilities and renewable energy sources," *Electric Power Systems Research*, November 2012, vol. 92, pp. 50–59.

[10] J. R. Pillai, and B. Bak-Jensen, "Integration of vehicle-to-grid in the Western Danish power system," *IEEE Transactions on Sustainable Energy*, January 2011, vol. 2, no. 1, pp. 12–19.

[11] P. Venkatesh, R. Gnanadass, and N. P. Padhy, "Comparison and application of evolutionary programming techniques to combined economic emission dispatch with line flow constraints," *IEEE Transactions on Power Systems*, May 2003, vol. 18, no. 2, pp. 688–697.

[12] C. M. Huang and Y. C. Huang, "A novel approach to real-time economic emission power dispatch," *IEEE Transactions on Power Systems*, February 2003, vol. 18, no. 1, pp. 288–294.

[13] P. C. Chen and C. M. Huang, "Biobjective power dispatch using goal-attainment method and adaptive polynomial networks," *IEEE Transactions on Energy Conversion*, December 2004, vol. 19, no. 4, pp. 741–747.

[14] T. Niknam, F. Golestaneh, and M. S. Sadeghi, "θ-Multiobjective teaching-learning-based optimization for dynamic economic emission dispatch," *IEEE Systems Journal*, June 2012, vol. 6, no. 2, pp. 341–352.

[15] Y. Xia, S. G. Ghiocel, D. Dotta, D. Shawhan, A. Kindle, and J. H. Chow, "A simultaneous perturbation approach for solving economic dispatch problems with emission, storage, and network constraints," *IEEE Transactions on Smart Grid*, December 2013, vol. 4, no. 4, pp. 2356–2363.

[16] A. Y. Saber and G. K. Venayagamoorthy, "Plug-in vehicles and renewable energy sources for cost and emission reductions," *IEEE Transactions on Industrial Electronics*, April 2011, vol. 58, no. 4, pp. 1229–1238.

[17] E. Santoyo-Castelazo and A. Azapagic, "Sustainability assessment of energy systems: integrating environmental, economic and social aspects," *Journal of Cleaner Production*, October 2014, vol. 80, pp. 119–138.

[18] J. B. Guinee, M. Gorree, R. Heijungs, G. Huppes, R. Kleijn, and A. de Koning, "Life cycle assessment: an operational guide to the ISO standards – Part 2a." Ministry of Housing, *Spatial planning and environment (VROM) and centre of environmental science (CML)*, Netherlands, 2001.

[19] G. Dorini, Z. Kapelan, and A. Azapagic, "Managing uncertainty in multiple-criteria decision making related to sustainability assessment," *Clean Technologies and Environmental Policy*, February 2011, vol. 13 no. 1, pp. 133–139.

[20] L. Stamford and A. Azapagic, "Sustainability indicators for the assessment of nuclear power," *Energy*, October 2011, vol. 36, no. 10, pp. 6037–6057.

[21] M. Rahmani-andebili and G. K. Venayagamoorthy, "SmartPark placement and operation for improving system reliability and market participation," *Electric Power Systems Research*, June 2015, vol. 123, no. 6, pp. 21–30.

[22] U.S. Environmental Protection Agency (EPA), "Learn about carbon pollution from power plants," [Online]. Available: http://www.epa.gov/cleanpowerplan/learn-about-carbon-pollution-power-plants, accessed February 2016.

[23] D. P. Jenkins J. Fletcher, and D. Kane, "Lifetime prediction and sizing of lead-acid batteries for micro generation storage applications," *IET Renewable Power Generation*, September 2008, vol. 2, no. 3, pp. 191–200.

[24] V. S. Pappala, I. Erlich, K. Rohrig, and J. Dobschinski, "A stochastic model for the optimal operation of a wind-thermal power system," *IEEE Transactions on Power Systems*, May 2009, vol. 24, no. 2, pp. 940–950.

[25] M. Rahmani-andebili, "Spinning reserve supply with presence of plug-in electric vehicles aggregator considering compromise between cost and reliability," *IET Generation, Transmission & Distribution*, December 2013, vol. 7, no. 12, pp. 1442–1452.

[26] Y. del Valle, G. K. Venayagamoorthy, S. Mohagheghi, J. C. Hernandez, and R. G. Harley, "Particle swarm optimization: Basic concepts, variants and applications in power systems," *IEEE Transactions on Evolutionary Computation*, April 2008, vol. 12, no. 2, pp. 171–195.

[27] M. Rahmani-andebili and G. K. Venayagamoorthy, "Stochastic and adaptive optimization for combined economic and emission dispatch with renewables," *Proceedings of the IEEE Computational Intelligence Application in Smart Grid (CIASG) Symposium*, Cape Town, South Africa, December 8–10, 2015.

[28] U.S. Energy Information Administration (EIA), "California's first greenhouse gas emissions auction sells near minimum price." [Online]. Available: http://www.eia.gov/todayinenergy/detail.cfm?id=9310, accessed January 2015.

Part 4

Social aspects and implementations

Chapter 12

If you build it, will they come? Getting consumers on board with the future of the smart grid

Patricia A. Aloise-Young[1], *Jennifer E. Cross*[2], *and Perla K. Sandoval*[1]

Abstract

Inside a smart home, smart appliances operate automatically. Outside, a smart meter receives and transmits information to the utility. In an instant, the smart grid optimizes the use of distributed resources to ensure that an uninterrupted supply of power is delivered to the home. This is the vision that energy technology experts have held of the smart grid, but there is one key piece of this picture missing – *people*. However, the goals of the smart grid can only be achieved with the cooperation of the consumer. Moreover, one of the defining characteristics of the smart grid is that it enables informed participation by customers. Thus, the consumer is integral to the success of the future of the smart grid. This chapter provides a roadmap for interdisciplinary collaboration between technology experts and social scientists, and provides concrete suggestions for getting consumers on board with the future of the smart grid.

12.1 Introduction

Inside a smart home, smart appliances decide when and how to operate. Outside, a smart meter communicates with the utility. Meanwhile, the smart grid optimizes the use of distributed resources to deliver power to the home. This is the vision that energy technology experts have held of the smart grid, but there is one key piece of this picture missing – *people.*

The smart grid is a multi-dimensional cyber-physical-social system and in this chapter we will discuss the importance of considering people's behaviors, cognitions, and preferences in the design and implementation of smart homes, smart meters, and distributed generation. Moreover, we will highlight the contribution that social science can make to that process by reviewing case studies of previous smart grid failures and suggest how social science could have made these efforts more successful. One

[1]Dept. of Psychology, Colorado State University, CO, USA
[2]Dept. of Sociology, Colorado State University, CO, USA

of the themes that will recur throughout this discussion is that designing and implementing systems that support the vision of the smart grid require attention to the barriers and opportunities at several levels: individual, organizational, and community. Social scientists theorize about each of these levels, their unique opportunities, and constraints on individual and group behaviors, as well as their distinct levers for fostering change. Across numerous fields (e.g., anthropology, economics, organizational studies, psychology, and sociology), social scientists have been investigating and testing programs designed to foster energy conservation and energy efficiency. For example, while smart meters are installed on individual homes and must overcome barriers for individual homeowners, adoption and use involves regulatory components (community barriers) and layoffs of meter readers (organizational barriers) as well.

The goals of the smart grid can only be achieved with the cooperation of the consumer. Moreover, one of the defining characteristics of the smart grid is that it enables informed participation by customers [1]. Thus, the consumer is integral to the success of the future of the smart grid. This chapter provides a roadmap for interdisciplinary collaboration between technology experts and social scientists, and provides concrete suggestions for getting consumers on board with the future of the smart grid.

12.2 Smart homes

On April 23, 1990, *Design News* [2] proclaimed "Smart Houses are on the way." Twenty-five years later, smart homes are only just beginning to see mainstream acceptance. Harris Poll conducted an online survey in 2015 with 4,043 adults [3] which revealed that 28% had at least one smart feature (defined as "products or tools that aid in controlling a home's functions such as lighting, temperature, security, safety, and entertainment, either remotely by a phone, tablet, computer or with a separate automatic system within the home itself") in their homes. But even now, smart homes' entry into the mainstream is in doubt. Why has it already taken so long and why is the smart home still not a sure bet in 2015?

In one of IEEE's earliest papers on the topic of smart homes, Valtchev and Frankov [4] defined a smart home as "a house or living environment that contains the technology to allow devices and systems to be controlled automatically" (p. 126). If we contrast that definition with the one used in the Harris Poll, we can see that what has changed is the addition of the ability of home functions to be controlled, *by the user*, via phone, tablet, or computer. It is this added functionality which is finally bringing the smart home into the mainstream. What the market shows is that users want more control over their environment, not less. They want to be able to remotely control the functions of their home. As a result, smart phones have emerged as an important enabling technology for smart homes and it was not until 2013 [5] that the majority of American adults owned a smartphone.

Although price, and a variety of technological elements such as networking capabilities have been obstacles to smart home uptake [6], as early as the mid-1990s some researchers [7 and Meyer and Schulze (1996) as cited in 8] had identified another

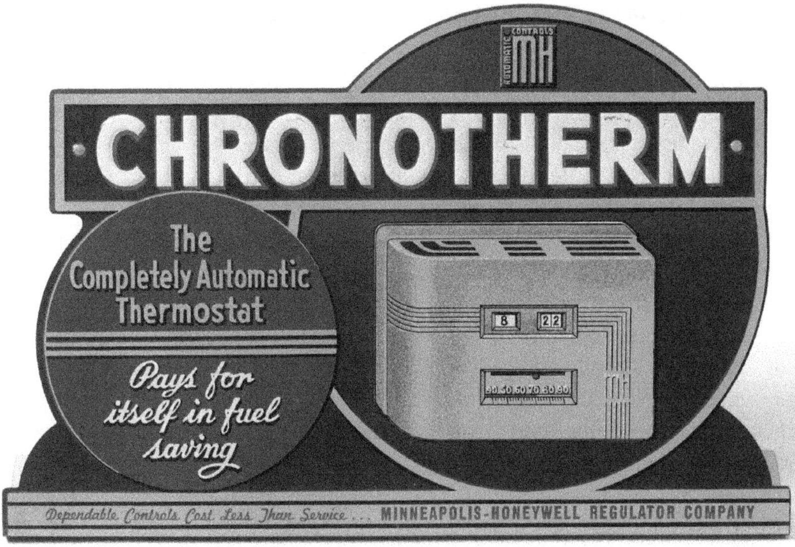

Figure 12.1 Metal stand-up counter sign advertising the Minneapolis-Honeywell Regulator Company's Chronotherm Automatic Thermostat. Created by Donaldson, Art Sign between 1930 and 1945. © Minnesota Historical Society [13], reprinted, with permission

obstacle to the progress of the smart home: designers were paying relatively little attention to the needs of consumers. Social scientists have explored the meaning of *home*, and have documented its key characteristics: a sense of possession and control, an expectation of privacy, personal freedom, and a sense of security [9]. Thus, any technology or activity proposed for residential spaces must support personal freedom, privacy, control, and safety. There were several features that users wanted from smart homes, including remote control, ease of operation, and help with managing everyday tasks, that the smart home was not delivering. Not only were smart home features being designed without asking users what they wanted, but very little usability research was being conducted to determine whether consumers could utilize the technology that was being developed [7].

12.2.1 Smart thermostats and smart appliances

Thermostats provide an excellent example. Arguably, the first smart home feature was the programmable thermostat (see Figure 12.1). The theory was automation: set it and forget it; the reality is that the vast majority of users override the programmed settings. A variety of sources, including the US Department of Energy (DOE), advertise to consumers that programmable thermostats can result in noticeable reductions in their energy bills [10], which is true. However, field research shows while programmable thermostats can reduce energy used for heating by as much as 65%, in practice 75% of households fail to achieve it [11]. Consumers are purchasing these thermostats expecting savings and instead, usability studies [12] show that users find the thermostats difficult to operate and incompatible with their needs. As a result, they fail to achieve energy savings. This caused DOE to remove the ENERGY STAR label from programmable thermostats in 2009.

Even in an industry usually driven by consumer research, household appliances, smart appliances have failed to deliver the features that consumers want, landing smart appliances on popular technology and social media site Mashable's 2011 list of 14 tech flops. For instance, in a 1998 article in the IEEE Spectrum [14], two members of the GE R&D team discussed the design of smart appliances. They said,

> Along with these advances, however, has arisen concerns over the user-interface. Most appliances present the user with a range of dials or buttons on which to enter set points for various process parameters; in a washer, these would be water level, load size and fabric type. In fact, most users rarely ever change the settings. (p. 36, emphasis added)

GE had given users a set of features that they either did not want or were intimidated by. Similarly, in 2013 an LG executive bragged in a press release that with their smart machines you could "start a load of laundry while driving home from work" [15] which is not viewed as a worthwhile feature by consumers [16].

In a white paper about the future of the smart grid [17], the Association of Home Appliance Manufacturers focused heavily on the cost of energy, concluding that current residential tariffs and rate structures are not sufficient to incentivize consumer participation in Smart Grid technologies. However, we would argue that rather than relying on rate structures such as time of use, or rebates to support interest in smart homes, the industry should be focused on providing consumers with the functionality they really want. User-centered design [18] and task-centered design [19] both suggest that in order to design a usable system the users' characteristics need to be taken into account during the design process. We extend that view further to argue that design teams need to be interdisciplinary because the people who are best qualified for some steps in the process are not technology experts, but rather social scientists.

12.2.1.1 User-centered design

One of the fundamentals of creating a design with good usability is understanding the behavior of different types of users. For instance, Greenberg [20] suggests that designers need to conduct observations of and/or interviews with end-users before designing prototypes (Figure 12.2). Undergraduate degree programs in Psychology and Sociology train students to be able to create and administer interviews and surveys with sound psychometric qualities. Similarly, these students have been taught how to code observational data so that the results will be valid and reliable. Moreover, these students often have additional tools at their disposal such as focus groups, participatory ethnographies, and diary studies. In contrast, a review of Human–Computer Interaction degree requirements reveals that it is relatively rare for students to receive any training in these research skills. Since the final design will only be as good as

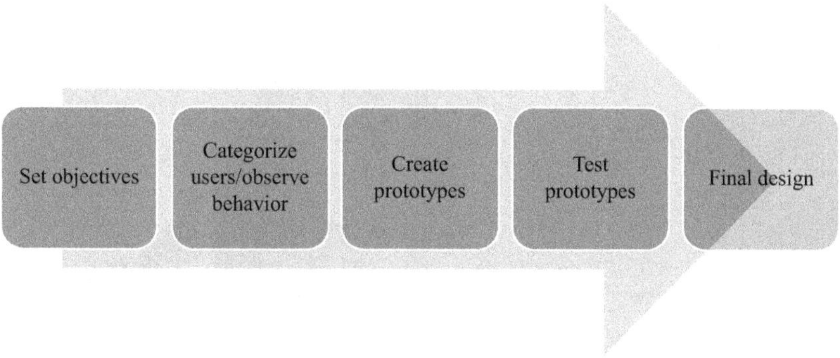

Figure 12.2 Design process incorporating user behavior

the quality of the data that are collected in this step, it is critical that the person doing the observations/interviews be trained in social science methodology. Given that designers have little background in social science it is rare to see designs such as this thermostat [21] based on user behavior and social science theory.

In addition to informing the design, the step of testing prototypes will also require observing behavior. In essence, the entire process itself is a psychological experiment. That is, representative users must be recruited, groups assigned to different conditions, extraneous variables accounted for, and behavior observed to draw valid conclusions. One of the key components to this process, and the marketing process that will occur after the product is designed, is understanding how to segment the population. This is important for both the design process and advertising the final product, and is another area where social scientists can lend their expertise.

12.2.1.2 Diffusion of innovations

As high-tech products are introduced into the marketplace, there is a well-documented process, called diffusion of innovations [22] that occurs. According to the diffusion of innovations model, there are five market segments: innovators, early adopters, early majority, late majority, and laggards. There is a qualitative difference between the innovators/early adopters and the other segments (indicated in Figure 12.3 as the chasm). Members of the early and late majority are more risk averse than are innovators and early adopters. Innovators and early adopters are more tech-savvy. Their purchase decisions are driven by a desire to be on the cutting edge and they are willing to spend more for products. Members of the early and late majority are more reliant on social proof for their purchase decisions, so that good word of mouth from the innovators and early adopters is key for a product to jump the chasm [23].

The expansion and contraction in the smart security camera market is an example of the Diffusion of Innovations curve and the different market segments that exist within the overall population. In June 2015, sales associates of the American real estate franchise Coldwell Banker were asked how interested their buyers were in a variety of smart home features including: appliances, entertainment, lighting,

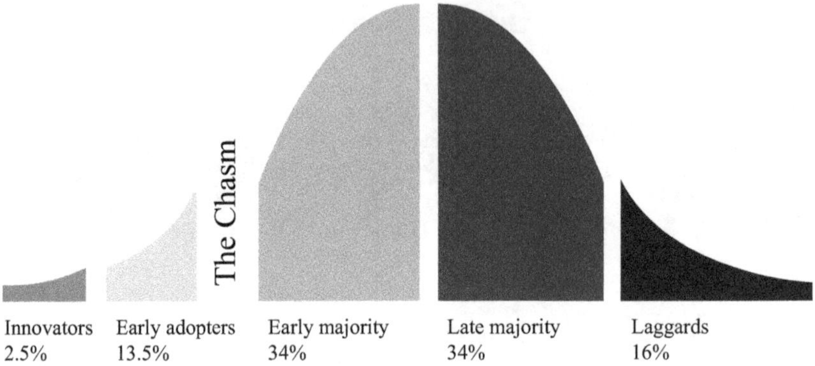

Figure 12.3 Diffusion of innovations curve [24]

security, and temperature control [25]. Security was reported to be the feature that most buyers expressed interest in. Argus Insights reports that the market for these products showed robust growth in 2014 but that this has slowed dramatically in the first quarter of 2015 [26]. The current market contraction is associated with consumers expressing frustration that security products are complicated and difficult to set up and use. This is because the market has moved beyond the innovators and early adopters and is trying to now capture the early majority. However, unlike tech-savvy early adopters who can overcome barriers encountered during setup and operation, mainstream consumers simply return the items. This is another mistake that destines some products to failure. Businesses assume that the diffusion of innovations model applies only to marketing. However, it also has to be taken into consideration in the design process. The features that are important to early adopters are not the same as the features that sell products to the majority. Moreover, when the majority is not considered in the design process it can have cascading effects: flaw in the design process → bad customer experiences → negative reviews on online sites → declining sales (because the majority requires social proof). In the case of smart security cameras, reviews indicating that these items are difficult to set up, incompatible with home networks, etc. this could seriously stall the growth of the industry.

12.2.2 Summary

In summary, the driving force behind the design of smart homes and other green technologies should not be, What is technologically possible? (and then, once we have designed it – How do we sell it?) but rather, What added functionality do users find valuable? (and then, How can we make it a reality?). This is analogous to the systems engineering approach, where the starting point for the design process is needs analysis [27]. Recently, Honeywell's Lyric thermostat was designed in collaboration with Frog design using the principles of user-centered design [16]. It is too early to tell whether it will deliver better results for the customer. The NEST thermostat was designed with usability in mind, and in a side-by-side test it did deliver better energy savings than a traditional programmable thermostat [28], despite the fact that the NEST was not originally designed for energy savings. Customer satisfaction with both products is high, and Honeywell and NEST are generating healthy profits as the smart thermostat market shows significant growth in North America and Europe. This provides optimism for the future of smart homes.

12.3 Smart meters

Smart meters are a key piece in the smart grid puzzle. The smart grid is only possible with two-way communication between the elements of the grid. Moreover, in laying out its goals for the smart grid, the US government has identified "empowerment of consumers with enhanced information to save energy, ensure privacy, and shrink bills" [29] as one of its four core goals, and smart meters have been touted as a promising mechanism for achieving that goal. However, despite the importance of smart meters and their potential benefit to consumers, initial attempts to install them did not go smoothly. We discuss what went wrong with the initial wave of smart meter deployments and whether smart meters are delivering the lasting energy savings that had been anticipated.

In 2013, the US Energy Administration reported that nearly 52 million smart meters had been installed and activated in the United States [30]. There was significant growth in smart meter deployments associated with the 2009 American Recovery and Reinvestment Act (ARRA). This stimulus package designated a large portion of funding to the modernization of the US energy grid. However, one of the earliest and most influential smart meter deployments in the United States began in 2007 in California. Pacific Gas and Electric (PG&E) is one of the largest electricity providers in the United States with over 5 million customers, but PG&E quickly became known for smart meters.

12.3.1 Smart meter backlash

Unexpectedly, the deployment experienced a wide array of negative consequences such as threats of class action lawsuits, customer complaints of inaccurate bills and sudden price increases, allegations that the radio frequency transmissions emitted by smart meters would lead to cancer and other negative health effects, and widescale protests that delayed meter installations [31,32]. Such was the influx of complaints that the California Public Utility Commission conducted an independent investigation to investigate concerns of smart meter accuracy and health dangers [31]. Anti-smart meter websites were created encouraging customers to write to policy-makers with a series of concerns including privacy and security issues that resonated with many law-makers. Protests spread to other parts of the country where smart meter deployments had already been approved and some deployments (e.g., Baltimore Gas & Electric) were put on hold while the rest of the country watched to see what would happen in California. There was even a 2013 anti-smart meter documentary produced entitled *Take back your power* [33].

12.3.2 Shortcomings of the PG&E deployment

Governments and utilities around the world were investing heavily in smart meters to provide both the utilities and their customers with "advanced approaches to energy production, transmission, and consumption" [34, p. 790]. For utility companies, smart meters gave them the value proposition of managing more efficiently energy use during peak loads through electrical load micromanagement [35]. Customers would also be able to take greater control of their energy bill using the detailed information that smart meters make possible [36]. Moreover, according to the California PUC, "[smart meters] provide customers with greater control over their electricity use when coupled with time-based rates" [37]. In theory, this sounds consistent with the smart grid's goal of empowering users and the users should be excited about it, so what happened?

By themselves, smart meters provide more detailed information to the utility, but they do not, on their own, provide more detailed information to the user. Smart meters need other supporting technologies in order to be of value to the consumer [34]. Many utilities have chosen to make the more fine grained data from smart meters available to consumers via a web portal such as this one shown in Figure 12.4 from Fort Collins Utilities [38] which provides the user with 15-minute interval data when it is posted to the user's billing account (once per day).

Similarly, PG&E has a webportal that enables households to see their historical energy use data in various formats such as graphs that compare monthly energy use as well as energy use compared to other households [39]. The idea behind webportals such as these is to give consumers the information that they need to change their behavior and save money/conserve energy. However, there are two types of feedback that utility companies can offer via smart meters, indirect and direct feedback [40]. Indirect feedback is information about energy use provided through billing and account services that details energy use information in a static format and) that is temporally separated from the behavior (i.e., delayed). PG&E's user account website would be an example of indirect feedback. The key difference between indirect and direct feedback is that direct feedback allows the user to interact with it. Real-time feedback, such as electricity usage rendered on an in-home-display (IHD), is an example of direct feedback. Users can plug and unplug appliances and see the impact on their usage – they can learn from doing [40]. Thus, the informational value of indirect feedback, the type presented to PG&E customers, is much smaller than from direct feedback. Most of the current webportals do not package information in ways that are actionable for customers, nor based on the best social science on how to provide feedback that will impact behavior. For example, even the 15-minute interval data presented in Figure 12.4 only shows consumers their past use. It does not help them understand why their use is high nor how it could be lowered. In this figure, the utility customer can see her daily use, but not what causes it to be high or low. For example, the customer would have to record, "I ran my dishwasher at 6:30 a.m." and then check her dashboard the next day to determine how much electricity her dishwasher consumed. In addition, the site tells the user that her use is "down 19%" and gives her an overlay of weather data, but does not provide her with an option for

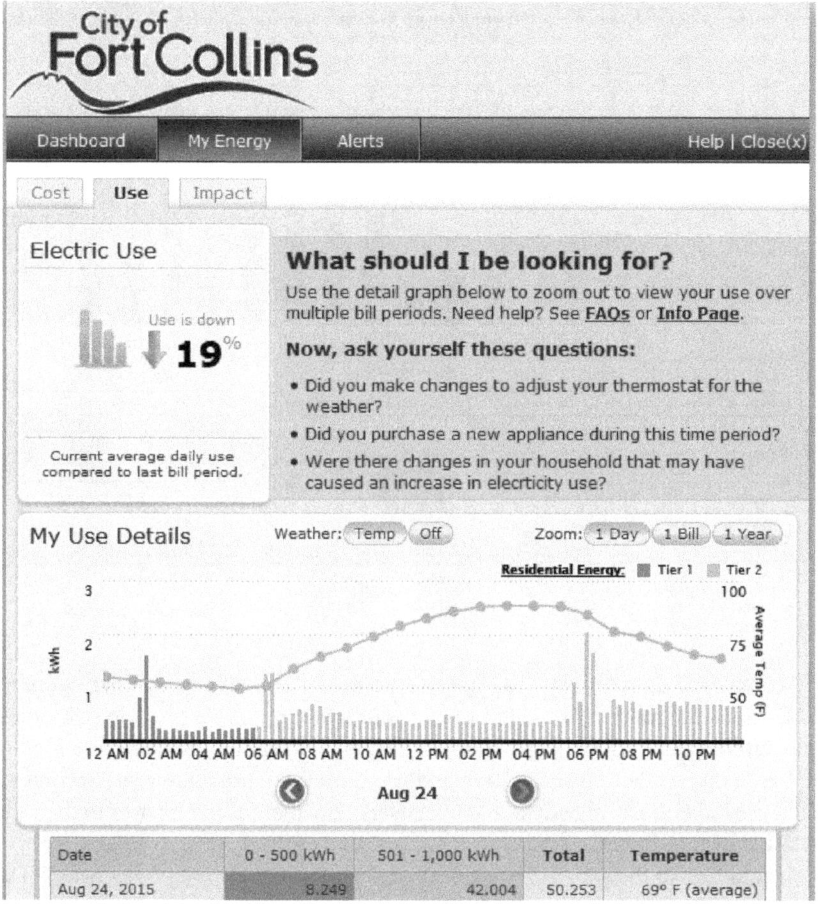

Figure 12.4 Example web portal interface with 15-minute interval data

weather normalizing her usage. Her inability to separate the large seasonal variations on her usage from the relatively smaller effects of her behavior will severely limit her understanding of what actions to take. However, one positive quality of this this site is that it provides customers with an option to receive alerts when their usage exceeds a user-set level. Thus, they can predict if they will hit the highest price tier later in the month and adjust their usage accordingly. The portal itself also tells users their expected bill for the month, given their current average daily use.

Smart meters have been identified as crucial to changing behavior because feedback is key to the success of any behavioral intervention. The closer the feedback occurs to the behavior, the more impact it will have – in the case of energy conservation we can see a linear trend between how quickly people get feedback about their usage and how much energy they save [40]. However, historical comparisons

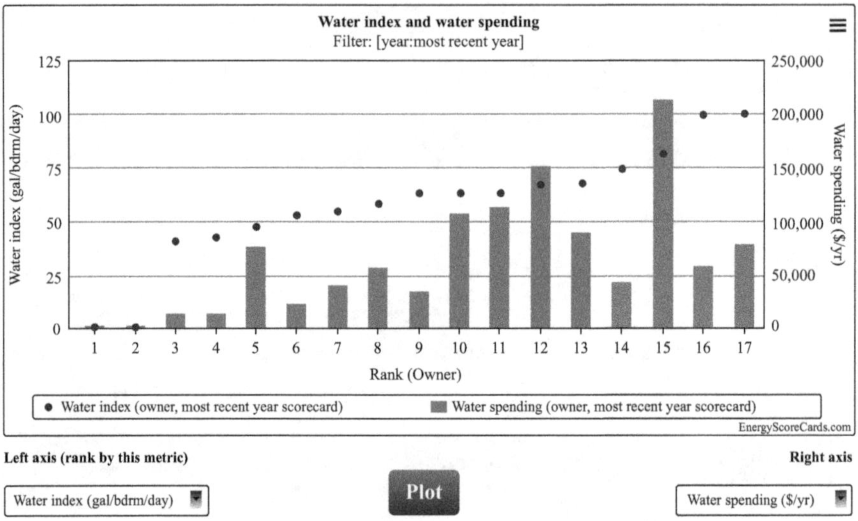

Figure 12.5 Web portal showing social comparison information

(i.e., comparing an individual to himself) are not as effective as comparing an individual to his neighbors. Thus, the most useful energy portals provide two types of comparative data. First they provide an analysis of actual use compared to predicted use, based on weather normalization and use during the previous year. Second, they provide social comparative data that tells users how their use compares to their peers such as shown in Figure 12.5.

In addition to the shortcomings of its webportal, there were issues with PG&E's rate structure that also challenged its smart meter deployment. After installing smart meters, PG&E (with the approval of the California PUC), adopted a rate structure that incorporates seasonal, tiered, TOU rates. For example, this means that a household would be charged a different amount for running their dishwasher depending on the month (seasonal), day of the week and time of day (TOU), and whether the household had consumed 100 or 1500 kWh in that billing cycle (tiered). This is how PG&E explains it to the user (see also Figure 12.6):

> With Time of Use, you still need to conserve energy throughout the month to avoid and react to higher priced tiers, but the price you pay also varies throughout the time of day and season. This works great if you can use air conditioning and major appliances in the morning or evening during the summer. Rates are lowest in the morning, late evening and weekends from May through October, and at all times outside 5-8PM on weekdays from November through April. The graphic below shows Time of Use pricing (Tier 2) over one year. See the next section to understand Time of Use and Tiers.

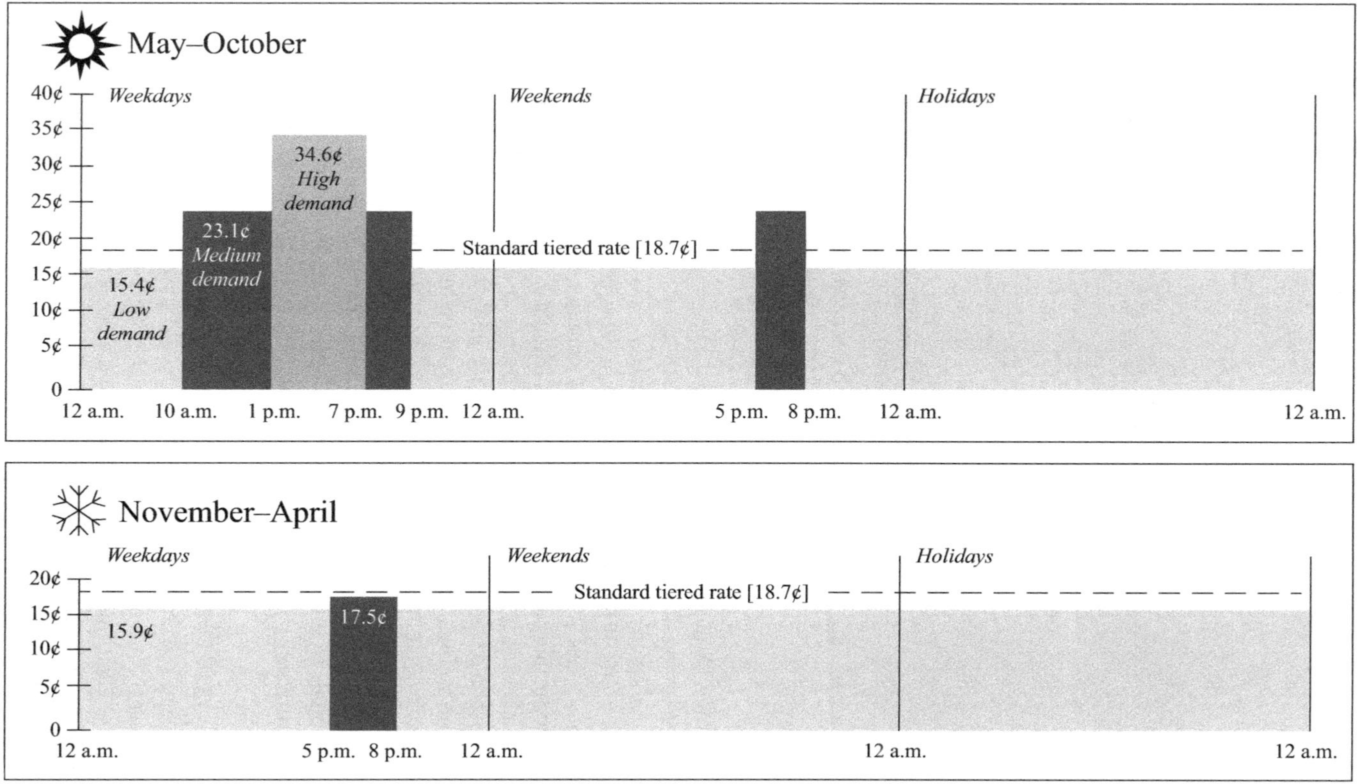

Figure 12.6 Tier 2 residential TOU electricity pricing from www.pge.com

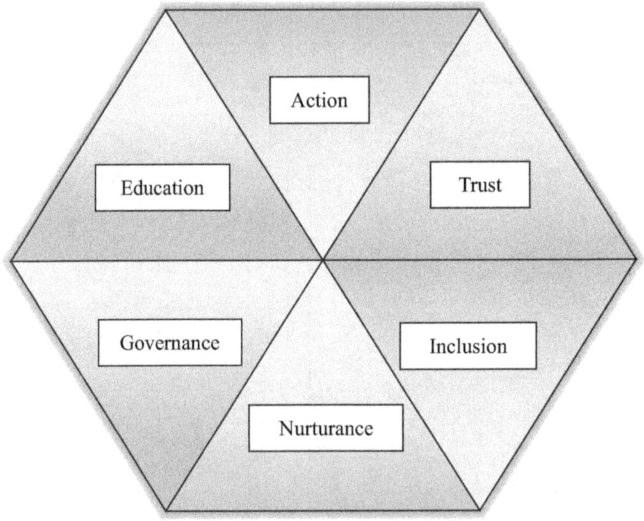

Figure 12.7 EATING approach to community engagement

A rate structure with this level of complexity will leave the average user bewildered and rather than giving the user more control, will make the user feel helpless.

Once a smart meter is installed, it immediately provides valuable information to the utility, but by itself the smart meter does not provide more detailed information to the user, nor does it provide users with an interface to increase their control. Most utility companies deployed smart meters without all the supporting technology that would enable the best feedback to the customer and the capacity for increased control of household energy use in response to available feedback. This is an excellent example of how a single technology has barriers at multiple levels. At the individual level, residents want information and control. Whilst smart meters have this capacity, they were not deployed at the large scale with these features fully functional. This is primarily due to barriers at the organizational level, that is, utilities did not have the resources (capital, IT infrastructure, and personnel) to deploy smart meters with the supporting hardware (IHDs) or interactive apps desired by customers.

12.3.2.1 Keys to successful community engagement

What does social science suggest that PG&E could have done differently to make the smart meter rollout more successful? First, they failed to engage their customers before going to their homes and installing the meters. This resulted in a grassroots effort spontaneously forming to fight PG&E. Sarkissian and Hofer [41] have laid out the components of a successful community engagement strategy called the EATING approach (Figure 12.7), and as we will discuss, the PG&E smart meter rollout lacked several of these key components.

In their book *Kitchen table sustainability*, Sarkassian and Hofer describe their experiences with numerous successful and unsuccessful community-wide initiatives [41]. They identify six participatory planning strategies that are critical for successful community engagement: education, action, trust, inclusion, nourishment, and governance. In *education*, we ask: What information and knowledge do we need to communicate to members of the community? It is important that the relevant information be communicated in a way that is accessible to the audience (without using jargon) and that the definition of what is important is based on the values and concerns of the community. The component of *action* addresses the frustration that community members feel when they are given a forum for input but perceive that little comes of it. In a successful community engagement strategy, community members are made aware of how their input will be used, and that community input has the potential to affect decision-making. The third component, *trust*, may be the most important because trust between the groups feeds into other elements of the engagement process. For example, the education component can be affected by the perceived trustworthiness of information. Conflicting claims may be made by different groups and there may be no way to discriminate between them except on the basis of trust. Thus, successful engagement processes require the development of trust between diverse groups, sometimes with competing interests. Trust can be developed, but it must be consciously nurtured through interaction. *Inclusion* is the next component of a successful community engagement strategy. It can be tempting to only communicate with people who share our views or who are culturally similar to us because this is easy. Inclusion is a justice issue (all those who are affected by the issue should have a voice) and an empowerment issue (by giving people a voice, you give them a sense of agency). These are both particularly salient in the area of the smart grid where underserved groups (e.g., the poor, rural) are disproportionately affected by energy policy. The component of *nourishment* addresses the issue of burnout and suggests that to sustain engagement over the long term that people need to engage with nature and their community. The final principle is *governance* and is composed of four main principles: (1) accountability, (2) transparency, (3) strategic adaptability, and (4) participation. The processes by which a powerful entity (whether it is a government or corporation) works should be transparent and the entity should be accountable to the members of the community for not only the final decision, but the process itself. Strategic adaptability suggests that governance begin with a shared vision of the community, whereas participation refers to inclusivity in decision-making. Thus, this component ties into several of the other components.

12.3.2.2 Lessons learned from PG&E

The EATING model provides a theoretical framework for understanding suggestions [42] for how the PG&E deployment could have been more successful. Beginning with the component of education, it is well recognized that PG&E did a poor job of educating its customers about the meters. Although providing the right information is always important, smart meters were a new technology and were unknown to consumers. In 2009, the Boston Consulting Group surveyed 1,253 residents of Texas and

California in zip codes where smart meters had been deployed and found that more than 50% of respondents had never heard of smart meters [43]. Not only did this undermine PG&E's engagement efforts by failing to capture consumers who would have been positively disposed toward the meters, if they had been adequately educated, but it also created the opportunity for less reputable information sources to fill the void.

Lack of trust was extremely problematic for PG&E. They helped to create mistrust by changing rates concurrently with the meter installation, and by introducing TOU rates that were so complex as to be impossible to understand by the ordinary user. As a result, when consumers' bills went up, they blamed the change they understood, the new meter. This fed into some consumers' conspiracy theories that the meters were part of a profit motive by the utility to raise revenue. Moreover, many consumers who called to complain about their bills reported being told that the sudden increase was due to an unusually hot summer. This relates to the EATING component of action in which community members become frustrated when they are given an opportunity to provide input (in this case customer complaints) but perceive that little comes of it. It should be noted that the PG&E website now guarantees that they will check the meter and communicate results to a customer within seven days of a customer's request.

In addition to the misstep of failing to educate their consumers regarding what smart meters are and why they should care about them, PG&E missed the opportunity to have their consumers really benefit from the meters by not giving them access to real-time data. However, in this respect, PG&E is not alone. To date, utilities have been unwilling to invest in giving their customers IHDs. Since it is an expensive proposition, we examine whether consumers are likely to stay engaged with IHDs over the long haul and whether IHDs can help utilities meet other goals, such as reducing peak loads, which might convince them to invest in the technology.

Long-term studies of IHDs are lacking. However, in the research conducted to date it does appear that interest in IHDs decreases over time. For example, Houde *et al.* found that households receiving real-time feedback showed initial reductions in energy usage, but returned to baseline after 5 weeks [44]. A year-long study in the UK provides a somewhat more optimistic picture [45]. Interviews with 11 homeowners one year after they had received IHDs revealed that the monitors had had a profound effect on levels of awareness about energy consumption (e.g., "We have changed and I think we've changed forever, because we don't think about it these days. It's part of life now to make sure that you're not using too much power, and we're all aware of it." p. 129). Of the 11 residents, 8 reported that they were still using the IHD after one year. However, they all reported that their use of the device was considerably lower than it had been at the beginning of the year, e.g.,

> As it's more of a novelty you definitely check it more, but I think it's definitely part of the background now. It's become part of the fabric really… and also it doesn't really change that much anymore, so I don't feel I have a need to keep monitoring it. I sort of feel I have a reasonable view of what it's going to be anyway. (p. 128–9)

The interviews also revealed how the IHDs were used to save energy. Specifically, there were three main behavioral results: "Using it hot" – the IHDs alerted homeowners to unusual levels of usage which prompted them to switch off unused appliances; "making considered decisions" – IHDs helped homeowners identify energy hogs (i.e., particularly wasteful appliances or behaviors), and then to use them less, or replace them; and "spillover effects" – discussions about energy consumption and conservation strategies with family and friends. Considered decisions were especially likely to occur only when homeowners first got the IHD, whereas the other two behaviors were more lasting.

In another study of the effects of IHDs, Jessoe and Rapson found that real-time energy usage data can dramatically increase the price elasticity of demand. Specifically, households that received real-time pricing and usage information on an IHD demonstrated a $3\times$ conservation effect compared with households that received advance warning (via email, text, or phone, determined by user preference) of price changes alone [46]. Although this study lasted for only one month, this suggests that IHDs could be used to increase the impact of demand response programs or to help consumers navigate complex rate structures like the one used by PG&E.

12.3.3 Summary

Smart meters got off to a rough start with the PG&E deployment, but this crucial smart grid technology has weathered the storm. Many utilities benefitted from watching the struggles of the early adopters and have done a better job of educating their customers about the importance of smart meters. Smart meter technology will continue to raise questions about control, privacy, and safety. Smart meters will be the most successful when residents feel that the meters contribute to their sense of home by providing control and their concerns for safety (health), security, and privacy have been acknowledged and addressed. The initial controversies around smart meters are understandable when viewed through the lens of a social scientist: smart meters created a new threat to privacy and personal freedom without providing the benefits of real-time feedback or increased control. Moreover, because the utilities primarily invested in the part of the equation that benefited them, their promises of energy savings have been largely unrealized [39].

12.4 Distributed generation

Recently, NREL [47] published a report showing that it was technologically possible to support 80% renewables on the grid. However, getting to 80% renewables is not only a major technological challenge, it will also require convincing policy-makers, financiers, utility executives, and in some cases taxpayers that investing in the infrastructure to support this change to the grid is warranted and will pay off (financially or otherwise) in the end. Thus, as energy technology experts consider the problem of how to optimize distributed generation sources to provide reliable power to households, it is equally important that social scientists work to increase the distributed

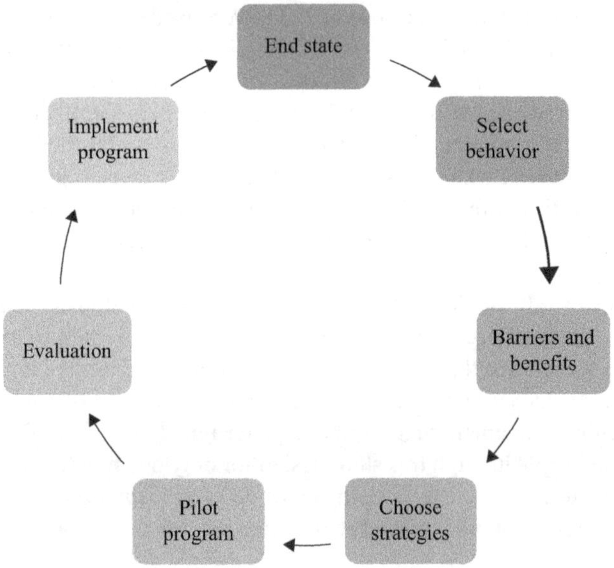

Figure 12.8 Community-based social marketing

generation assets available to the grid. As we discuss in the following section, there are barriers at the community, organizational, and individual levels that need to be understood and overcome. Moreover, there are motivations, or levers for change, at each of these levels that can be utilized to help move us toward our goal.

12.4.1 Community-based social marketing

Community-based social marketing (CBSM) [48] is a data-driven process (see Figure 12.8) for promoting behavior change that can be very useful in this effort. In CBSM, we begin with the end state (what we are trying to accomplish) which in this example is increasing distributed generation on the grid. We then examine the behaviors that can get us to the end state. There are usually a number of behaviors related to the end state for us to choose from. For example, the behaviors that can increase distributed generation include: (a) increasing the number of homeowners with rooftop solar (PV), (b) increasing (utility and homeowner) participation in community solar gardens, (c) increasing the number of solar collectors, (d) convincing electric vehicle (EV) owners (private or fleet) to discharge vehicle batteries to the grid during peak hours, and recharge them overnight, (e) increasing the number of offshore wind turbines, and (f) increasing the number of onshore wind farms.

The next step in the CBSM process is to select which of these behaviors we are going to target with our behavior change efforts. This is based on three criteria: (a) probable impact, that is, the relationship between the individual behavior and the end state (e.g., targeting PV is more attractive in areas that receive more sunshine), (b) malleability of the behavior (expected probability of success for a behavior change program), and (c) current penetration of the behavior. That is, we look for a high impact behavior that relatively few people are currently engaging in, but that we have reason to believe they would be willing to engage in. Ideally, this is based on previous research using a similar behavior change program. If such evidence does not exist, then the potential audience would be surveyed or interviewed to assess their willingness to undertake the behavior. Once we have selected the behavior or behaviors (depending on the size of the program) that show the most promise, we identify the target audience(s) for that behavior, and we begin to conduct focus groups and/or individual interviews with stakeholders to identify barriers and benefits for the target behaviors: What are the obstacles that stand in the way of engaging in the behavior? What motivates individuals to undertake the behavior of interest? Each of the potential behaviors related to the end state will have a unique set of barriers and benefits associated with it, as is demonstrated in the following example.

In 2009, Kowal [49] interviewed 19 residents of Fort Collins, Colorado: 13 who had purchased PV (6 with no rebate and 7 with a rebate) and 6 respondents who were interested in purchasing PV but had not done so. The sample was typical of early adopters: 16 respondents held a Master's degree or higher, and of the 14 who were willing to provide their household income, responses ranged between $50,000 (the approximate US median household income) and $200,000, with an average of $95,571. The majority was male and worked or was otherwise involved in an industry related in some way to PV including electricians, engineering, architecture, and sciences related to renewable energy. Not surprisingly, the participants reported being very knowledgeable about solar power.

Four categories of motivations for purchasing PV emerged from participants' interviews: environmental, financial, leadership, and energy independence. Examples of statements that participants made that fell into each category are given in Table 12.1. In addition, six categories of barriers for purchasing PV emerged in the interviews

Table 12.1 Motivations for PV adoption

Motivations	Quotes
Environmental	"It was the environmentally appropriate thing to do" "It reduces coal mining and global warming" "We are quite aware of the environmental damage of gas and pollution"
Financial	"There is a reduction in the utility bills" "I won't ever have to worry about an electric bill"
Leadership	"I want to set an example for energy efficiency and renewable energy" "...role model to the rest of the community" "...role model professionally and personally"
Energy independence	"...to start making my own energy as opposed to buying it" "Now I believe in global warming, we need the alternatives, so it's energy independent, it's global warming and I'm a technology geek so it's interesting to see this stuff develop" "...to cut ties from the utility company"

Table 12.2 Barriers to PV adoption

Barriers	Quotes
Upfront costs	"First cost is high, there's some federal money, but cost is high…" "Cost was a factor" "The cost was the biggest obstacle in my decision making"
Long payback	"There's no payback" "Payback for me is like thirty years" "I don't expect that this will pay back"
Billing	"Net metering wasn't established at the time" "I did not get more panels because [net metering] wasn't guaranteed" "Sometimes get billed for electricity and don't get credit"
Siting issues	"The house has a weird roof orientation. I'm not a perfect candidate and had to work with the existing infrastructure" "It is a challenge in old town. There are mature trees, there is a large tree on our neighbour's lot, it shades all our solar on winter mornings" "There's a tree right there, those two trees need to go and this one (points to neighbour's property)" "There IS a chimney and tree in the way"
Aesthetics	"[Aesthetics] were a consideration" "…picked a lot facing north so you couldn't see them" "I'm glad they're not sails, I'm glad they're flush, I had just gotten the house re-roofed. It's not as attractive, but oh well" "Some of the houses we saw had PV panels that were very ugly, and we were concerned" "We might not have done it if the only option was to have some ugly structure supporting it..I think aesthetics is an important aspect…"
Maintenance	"Cleaning snow off" "Had to clean snow off, they get a little dirty" "There might be some maintenance issues down the road" "It's another system you have to worry about"

including three financial barriers, upfront costs, long payback, and billing, and three non-financial barriers, siting issues, aesthetics, and maintenance (see Table 12.2).

Using these data as a starting point, we can see how the barriers and benefits for each behavior related to different forms of generation will share commonalities, but be unique. For instance, there are no upfront costs, siting issues, or aesthetic concerns associated with asking EV owners to discharge their batteries to the grid during peak hours. However, other motivations/barriers that were mentioned in regard to rooftop PV will likely overlap such as billing and maintenance issues. EV owners might be motivated by the compensation they receive for discharging their batteries (e.g., feed in tariffs) and be concerned about how repeatedly discharging their batteries would affect the performance of and the potential need to replace their EV battery (i.e., barrier of maintenance). Individual homeowners who are considering investing in community solar gardens (PV panels hosted by a community partner with grid-connected subscribers) do not have to consider siting issues, aesthetics, or maintenance, whereas a landowner considering a wind farm would.

Research shows that although many of the barriers and benefits obtained with a particular sample can be generalized, some are contextually specific with regard to both time and place. For example, although the motivations for purchasing rooftop PV by homeowners in the UK [50] and Wisconsin [51] were very similar to those reported above, UK early adopters appear to be less concerned about aesthetics. Moreover, Schelly [51] directly compared Colorado and Wisconsin homeowners' motivations relative to the respective states' electricity policies and found that Wisconsin's statewide feed-in tariff impacted both uptake of PV and subsequent conservation behaviors. Consequently, in the CBSM model it is important to assess local barriers and benefits, even if a literature exists from which to extrapolate.

Once the barriers and benefits have been established, the appropriate behavior change strategies can be selected. An important consideration is whether the barriers/benefits occur at the individual level, the organizational level or the community level. For example, if we consider the aesthetics of PV, a homeowner might only wish to install solar panels only if they can be installed on the rear of the home, so as not to interfere with the "curb appeal" of the home (individual barrier). Alternatively, an HOA or Homeowners' Association (organizational barrier) might have an approval process for rooftop PV with specific appearance requirements such as that solar panels match the color of the roof, and that the color of the trim be black, brown, gray or non-reflective metallic (e.g., Oak Bluffs Estates, Texas). Finally, state codes (community barrier) may restrict how far above the roofline the solar panel may extend. The CBSM framework is based in social psychology and more often focuses on individual barriers and benefits than policy change. In many cases, including but not limited to PV adoption, even when the end-user or a homeowner is the target, CBSM must be integrated with change strategies at other levels (organizational, community). Without buy-in from key organizational players, like HOAs or local utilities, even well-developed CBSM strategies will be less than successful. The design and deployment of the new technologies that will enable the smart grid will require not only integrated design efforts, but also change management strategies alongside CBSM strategies and community engagement efforts [52–54].

At the individual level, some barriers and benefits may be financial, but it is a mistake to assume that individuals make decisions like businesses do – on a rational bottom line. Each behavior is likely to have a mix of motivations, both economic and non-economic, just as the PV study exemplified. McKenzie-Mohr [55] outlines a number of strategies to address these motivations such as goal-setting, commitment, competitions, prompts, and convenience. Persuasive messages are often developed as part of CBSM programs. The placement of those messages, in time and space, is important. The audience needs to receive the message at the decision point, the place and time when they are deciding whether or not to engage in the behavior. In order to deliver an effective message at the decision point, we need to have a thorough understanding of the behavior and decision-making process underlying it. Social diffusion (e.g., social media) is one mechanism for spreading the message, and as we have already discussed, the early and late majority are particularly affected by social proof. Similarly, social norms, the knowledge of what most people do, and what most people approve of, is a powerful persuasive force that is often used in CBSM. These tools are combined with the knowledge of the target audience's barriers and benefits to design a specific behavior change intervention.

In the next phase of the project, the intervention is piloted on a small sample. Ideally, program participants exposed to the new program are compared with controls (i.e., by conducting a randomized experimental test). Too often, inexperienced interventionists will skip this step and move straight to a full-scale intervention only to discover after the fact that what they designed was ineffective. Pilot testing allows us not only to determine that our program works, but why. If certain elements are superfluous and can be eliminated, then the program can move forward with full implementation more cost-effectively.

12.4.2 Summary

CBSM is a promising tool as we pursue our goal of increased renewable penetration on the grid. However, behavior change initiatives could be instrumental in other aspects of the equation as well. For example, the NREL models demonstrating that it was possible to get to 80% renewables by 2050 were predicated on zero growth in demand. Clearly, both routine and one-time behaviors (i.e., buying habits) have a profound impact on electricity demand. Finally, the optimization of resources is far less challenging (and expensive) if customers can be persuaded to cooperate with demand response programs. Thus, there are wide-ranging opportunities for social science to assist in meeting the goals of the smart grid.

12.5 Conclusions

In the *2010 Smart Grid System Report* [1], DOE described a characteristic of the smart grid:

> Consumers become an integral part of the electric power system. They help balance supply and demand and support reliability by modifying the way they use and purchase electricity. These modifications come as a result of consumers having choices that motivate different purchasing patterns and behavior. These choices involve new technologies, new information about consumers' electricity use, and new forms of electricity pricing and incentives. (p. 16)

In order to develop technology and consumer products that empower residents to substantially reduce their electricity consumption in the home, to produce energy from renewable sources, and reduce our impact on the planet, we must do more to integrate the insights and tools from social science at every stage from design, to usability studies, to adoption and marketing. As we described above, a sense of home is founded on people feeling that they have personal freedom, privacy, safety, security, and control over their home. New technologies will not succeed unless they do two things: (1) contribute to an individual's sense of home and (2) provide empowering and enabling technology that addresses customer needs.

Across the different levels of the smart grid, we have described five common mistakes in the development and deployment of new technology and products:

1. Ignoring consumers' need for control.
2. Failing to observe human behavior and understand customers' needs.
3. Delivering information in a form that it inaccessible or unappealing.
4. Promising without delivering.
5. Failing to educate and engage customers.

Correcting these mistakes requires incorporating social science methodology into each step of the technology development process, ideally with an interdisciplinary team. Consideration, and subsequent development of strategies to overcome barriers at the individual, organizational, and community levels will be required. As we described, any new product or technology may need to overcome barriers at more than one level in order to be successful.

We outlined a number of approaches for avoiding these common mistakes in future smart grid endeavors including:

1. User-centered design: a systematic process for product development that integrates customer feedback and knowledge from various disciplines in the process of design [18].
2. Community-based social marketing: a research-based strategy for selecting behaviors and developing programs that empowers individuals to adopt new behaviors [55].
3. Models for organizational change: change management is an established best practice for organizations undergoing a substantial shift in their business practices [52–54].
4. Participatory planning strategies: a detailed process for engaging a community of people in sustainability initiatives (the EATING approach) [41].

It is important to recognize that, just as in other fields, these approaches need to be implemented systematically and rigorously in order to be effective. Unfortunately, it is all too common for programs to run where a few of the steps of CBSM or participatory planning have been cherry picked. Then, when the results fall short of expectations, people dismiss these approaches. The challenge is not that we do not have *best practice* strategies and processes, but rather that organizations and businesses either do not know about them, or fail to commit the resources to the utilization of best practices.

Because it can be difficult for those outside the field to discriminate between a properly designed/executed program and one that superficially sounds like it has all the right buzz words, it is important to focus on the qualifications of the person or organization administering the program. Specifically, do they have training and expertise in social science methodology? Meeting aggressive goals for greenhouse gas reductions is going to require that technology experts and social scientists work together in the same way that we hope the user will work with the grid.

References

[1] U.S. Department of Energy, *2010 Smart Grid System Report*. Washington, DC, February 2012. [Online]. Available: http://energy.gov/sites/prod/files/2010%20Smart%20Grid%20System%20Report.pdf [Last accessed on: August 31, 2015].

[2] "Smart Houses On The Way." *Design News*, April 23, 1990. [Online], vol. 46, no. 8, p. 15. Available from: Academic Search Premier, Ipswich, MA. [Last accessed on: August 31, 2015].

[3] Coldwell Banker Real Estate LLC, *Coldwell Banker Real Estate Survey Finds the Technology of Tomorrow Helps Many Smart Homes Sell Faster Today*, Maddison, NJ, March 2015. [Online]. Available: https://www.coldwellbanker.com/press-release/smarthomes [Last accessed on: August 31, 2015].

[4] D. Valtchev and I. Frankov, "Service gateway architecture for a smart home," *IEEE Communications Magazine*, 2002, vol. 40, no. 4, pp. 126–132.

[5] A. Smith, (*Smartphone Ownership 2013*, Pew Research Center, June, 2013. [Online]. Available from: http://www.pewinternet.org/2013/06/05/smartphone-ownership-2013/ [Last accessed on: August 31, 2015].

[6] A. Zipperer, P. A. Aloise-Young, S. Suryanarayanan, *et al.*, "Electric energy management in the smart home: perspectives on enabling technologies and consumer behavior," *Proceedings of the IEEE*, November 2013, vol. 101, no. 11, pp. 2397–2408.

[7] D. Gann, J. Barlow, and T. Venables, *Digital Futures: Making Homes Smarter*, Chartered Institute of Housing, Coventry, 1999.

[8] Frances K. Aldrich, "Smart homes: past, present and future," in *Inside the Smart Home*, R. Harper, Ed. Springer-Verlag London, 2003, pp. 17–39.

[9] D. Seamon, "Phenomenology, geography and geographical education," *Journal of Geography in Higher Education*, 1979, vol. 3, no. 2, pp. 40–50.

[10] U.S. Department of Energy, "Tips: Programmable Thermostats," 2015. [Online]. Available from: http://www.energy.gov/energysaver/tips-programmable-thermostats [Last accessed on: August 31, 2015].

[11] B. Urban and C. Gomez, "A case for thermostat user models," in *Proceedings of the 13th Conference of International Building Performance Simulation Association*, 2013.

[12] A. Meir, C. Aragon, B. Hurwitz, *et al.*, "How people actually use thermostats," in *Proceedings of* 2010 *ACEEE Summer Study on Energy Efficiency in Buildings, Chapter 2*, pp. 193–206. [Online]. Available from: http://escholarship.org/uc/item/3vd5q0cp [Last accessed on: August 31, 2015].

[13] Honeywell Regulator Company, "Metal Stand-up Counter Sign Advertising the Minneapolis Honeywell Regulator Company's Chronotherm Automatic Thermostat," Created by Donaldson Art Sign between 1930 and 1945, 2015. [Online]. Available from: http://www.mnopedia.org/multimedia/honeywell-chronotherm-counter-sign [Last accessed on: August 31, 2015].

[14] V. Badami and N. Chbat, "Home appliances get smart," *IEEE Spectrum*, 1998, vol. 35, no. 8, pp. 36–43.

[15] F. Manjoo, "A fridge with a touch screen! A washer that connects to the internet! Why are smart appliances so stupid?" January 11, 2013. [Online]. Available from: http://www.slate.com/articles/technology/technology/2013/01/smart_appliances_a_washer_that_connects_to_the_internet_and_other_amazingly.html [Last accessed on: August 31, 2015].

[16] L. Jackson, "How visual design influences our behaviour and changes our world," Presented at *Behavior, Energy and Climate Change Conference*, Washington, DC, December 9, 2014.

[17] Association of Home Appliance Manufacturers, "Smart Grid White Paper: The home appliance industry's principles & requirements for achieving a widely accepted smart grid," Washington, DC, December 2009. [Online]. Available from: https://www.aham.org/index.php?ht=a/GetDocumentAction/i/44191 [Last accessed on: August 31, 2015].

[18] R. Hartson and P. Pyla, *The UX Book: Process and Guidelines for Ensuring a Quality User Experience.* Morgan Kaufmann, Burlington, MA, 2012.

[19] C. Lewis and J. Riemann, "Task centered user interface design: a practical introduction," 1st ed. Boulder: University of Colorado, 1993. [Online]. Available from: tp.cs.colorado.edu/pub/cs/distribs/clewis/HCIDesign-Book/ [Last accessed on: August 31, 2015].

[20] S. Greenberg, "Working through task-centered system design," in *The Handbook of Task Analysis for Human–computer Interaction*, vol. 2, D. Diaper and N. Stanton, Eds. Lawrence Erlbaum Associates, Mahwah, NJ, 2004, pp. 49–66.

[21] R. Vastamäki, I. Sinkkonen, and C. Leinonen, "A behavioural model of temperature controller usage and energy saving," *Personal & Ubiquitous Computing*, 2005, vol. 9, no. 4, pp. 250–259. doi:10.1007/s00779-004-0326-3

[22] E. M. Rogers, *Diffusion of Innovations*. The Free Press New York, 1962.

[23] D. E. Rosen, J. E. Schroeder, and E. F. Purinton, "Marketing high tech products: lessons in customer focus from the marketplace," *Academy of Marketing Science Review*, 1998, vol. 6, pp. 1–17.

[24] "Academia Britanica Cuscatieca." [Online]. Available: https://ibbandm.wikispaces.com/4.3+Product, 2016. [Last accessed on: January 31, 2016].

[25] Coldwell Banker Real Estate LLC, "Americans ready for the smart home: results of the Coldwell Banker and CNET Smart Home Survey. Coldwell Banker Real Estate LLC," August 2015. [Online]. Available: https://www.coldwellbanker.com/content/pdfs/survey_final.pdf [Last accessed on: August 31, 2015].

[26] Argus Insight. "New report by Argus Insights reveals that of all Home Automation devices, security cameras are most talked about and despite investments in Nest and Dropcam by Google, and SmartThings by Samsung, consumer demand still lags," Los Gatos, CA, June 2015. [Online]. Available: http://www.argusinsights.com/connected-home-release/ [Last accessed on: August 31, 2015].

[27] B. S. Blanchard and W. J. Fabrycky, *Systems Engineering and Analysis*, 5th ed. Prentice Hall Upper Saddle River, NJ, 2011.

[28] Cadmus Group. "Indiana: Smart thermostat pilot studies." [Online]. Available: http://www.cadmusgroup.com/case-studies/indiana-smart-thermostat-pilot-studies/ [Last accessed on: August 31, 2015].

[29] Department of Energy, "Smart Grid," 2015 [Online]. Available: http://energy.gov/oe/services/technology-development/smart-grid [Last accessed on: August 31, 2015].

[30] US Energy Information Administration, "How many smart meters are installed in the United States, and who has them? FAQ," 2015 [Online]. Available: http://www.eia.gov/tools/faqs/faq.cfm?id=108&t=3 [Last accessed on: August 31, 2015].

[31] W. Koch, "Smart home meters draw consumer complaints," *USA Today*, 2015. Available: http://content.usatoday.com/communities/greenhouse/post/2010/07/smart-meters-draw-complaints/1#.Vdx6u8mZ484 [Last accessed on: August 31, 2015].

[32] S. Michels, How smart are smart meters? *PBS Newshour*, July 20, 2012. [Online]. Available: http://www.pbs.org/newshour/rundown/how-smart-are-smart-meters/ [Last accessed on: August 31, 2015].

[33] *Take back your power*. Australia, Canada, USA, Josh del Sol UK, 2014.

[34] T. Krishnamurti, D. Schwartz, A. Davis, *et al.*, "Preparing for smart grid technologies: a behavioral decision research approach to understanding consumer expectations about smart meters," *Energy Policy*, February 2012, vol. 41, pp. 790–797.

[35] S. S. S. R. Depuru, L. Wang, and V. Devabhaktuni, "Smart meters for power grid: Challenges, issues, advantages and status," *Renewable and Sustainable Energy Reviews*, 2011, vol. 15, no. 6, pp. 2736–2742. doi: 10.1016/j.rser.2011.02.039

[36] S. Darby, "Smart metering: what potential for householder engagement?" *Building Research & Information*, 2010, vol. 38, no. 5, pp. 442–457.

[37] California Public Utilities Commission, "The benefits of smart meters," 2015 [Online]. Available: http://www.cpuc.ca.gov/PUC/energy/Demand+Response/benefits.htm [Last accessed on: August 31, 2015].

[38] Fort Collins Utilities, "Monitor my use." [Online]. Available: http://www.fcgov.com/utilities/sustainability-leadership/advanced-meter-fort-collins/monitor-my-use, 2015. [Last accessed on: August 31, 2015].

[39] C. Mooney, "Why 50 million smart meters still haven't fixed America's energy habits." *Washington Post*, January 29, 2015. [Online]. Available: http://www.washingtonpost.com/news/energy-environment/wp/2015/01/29/americans-are-this-close-to-finally-understanding-their-electricity-bills/ [Last accessed on: August 31, 2015].

[40] A. K. Ehrhardt-Martinez, K. Donnelly, and J. A. Laitner, "Advanced metering initiatives and residential feedback programs: a meta-review for household electricity-saving opportunities," *Energy*, 2010, vol. 123(6), pp. 128.

[41] W. Sarkissian and N. Hofer, *Kitchen Table Sustainability: Practical Recipes for Community Engagement with Sustainability*. Earthscan London, 2009.

[42] K. Fehrenbacher, "Lesson learned from the PG&E Smart Meter Suit: it's a communication problem." *Gigaom.com*, November 19, 2009 [Online]. Available: https://gigaom.com/2009/11/19/lesson-learned-from-the-pge-smart-meter-suit-its-a-communication-problem/ [Last accessed on: August 31, 2015].

[43] Boston Consulting Group, "The smart-meter opportunity: The US consumer perspective," May 2010.

[44] S. Houde, A. Todd, A. Sudarshan, J. A. Flora, and K. C. Armel, "Real-time feedback and electricity consumption: a field experiment assessing the potential for savings and persistence," *Energy Journal*, 2013, vol. 34, no. 1, pp. 87–102.

[45] T. Hargreaves, M. Nye, and J. Burgess, "Keeping energy visible? Exploring how householders interact with feedback from smart energy monitors in the longer term," *Energy Policy*, 2013, vol. 52, pp. 126–134.

[46] K. Jessoe and D. Rapson, "Knowledge is (Less) Power: experimental evidence from residential energy use," *American Economic Review*, 2012, vol. 104, no. 4, pp. 1417–1438. doi:10.1257/aer.104.4.1417

[47] T. Mai, D. Sandor, R. Wiser, and T. Schneider, "Renewable electricity futures study. executive summary," National Renewable Energy Laboratory, Golden, CO, Tech. Rep. NREL/TP-6A20-52409, 2012. doi:10.2172/1063076

[48] D. McKenzie-Mohr, "Fostering sustainable behavior through community-based social marketing," *American Psychologist*, 2000, vol. 55, no. 5, pp. 531–537.

[49] D. Kowal, "An exploratory study of the adoption of residential solar photovoltaic systems in Fort Collins, Colorado." Unpublished Manuscript, 2014.

[50] A. Faiers and C. Neame, "Consumer attitudes towards domestic solar power systems," *Energy Policy*, September 2006, vol. 34, no. 14, pp. 1797–1806.

[51] C. Schelly, "Residential solar electricity adoption: what motivates, and what matters? A case study of early adopters," *Energy Research & Social Science*, 2014, vol. 2, pp. 183–191.

[52] B. Doppelt, *Leading Change Toward Sustainability*, Greenleaf Sheffield, UK, 2003.

[53] J. P. Kotter, "Leading change: why transformation efforts fail," *Harvard Business Review*, 1995, vol. 73, no. 2, pp.59–67.

[54] A. Hoffman and R. Henn, "Overcoming the social and psychological barriers to green building," *Organization & Environment*, 2008, vol. 21, no. 4, pp. 390–419.

[55] D. McKenzie-Mohr, *Fostering Sustainable Behavior: An Introduction to Community-based Social Marketing*. 3rd ed. New Society Publishers, Gabriola Island, British Columbia, 2011.

Chapter 13

Risks, threats and mitigation strategies for SCADA systems

Helge Janicke[1], *Allan Cook*[1], *Andrew Nicholson*[1] *and Kevin Jones*[1]

Abstract

Industrial Control Systems are at the heart of our critical infrastructures. Supervisory Control and Data Acquisition Systems (SCADA) are providing engineers with visibility and control of large-scale distributed control system infrastructures. Whilst these systems have in the past been isolated, today's technological advancements and changes in business needs mean that they are increasingly integrated with other networks that are connected to the Internet and infrastructures. This chapter provides an introduction to SCADA systems and reviews and analyses the resulting risks, threats arising from cyber attacks. The remainder of the chapter then investigates strategies to mitigate some of the risks that result from the increasing exposure of SCADA systems through prevention, attribution and incident response.

13.1 Introduction

Industrial Control Systems (ICS) are at the heart of the prosperity of industrialised nations. They operate and control our Critical National Infrastructures (CNI), such as electricity, gas or water distribution and waste treatment. Not only are our infrastructures dependent on these technologies, but also the industries that form the basis of our economy. Everything from import and export logistics relying on rail transport and harbour control systems to production and automated assembly lines ICS operating continuously with a minimum of disruption to their services are essential to our economy.

ICS come in many forms. Supervisory Control and Data Acquisition (SCADA) systems predominantly are deployed in distributed settings where controls are distributed and data is communicated centrally for decision making by supervisory

[1]Software Technology Research Laboratory, Faculty of Technology, De Montfort University, The Gateway, Leicester, UK

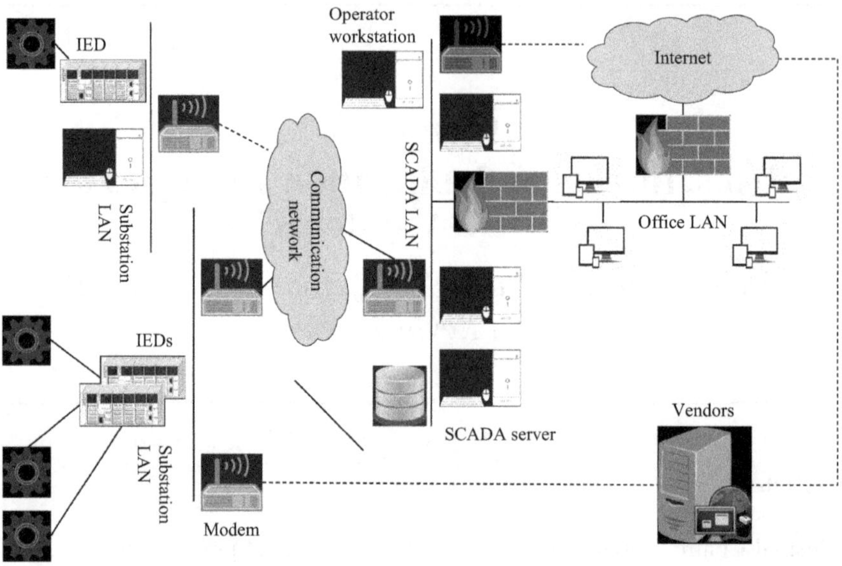

Figure 13.1 An electric power transmission SCADA network. Adapted from [1]

circuits or control engineers. These distributed systems can span very large areas, for example in electric power transmission networks where it is not feasible to implement physical controls effectively or where it is too expensive to maintain permanent staff to operate the systems. In many cases these systems are also operating in hazardous environments which make their maintenance difficult and the permanent stationing of personnel impossible, with oil rigs and deep sea explorations being examples. An example electric power transmission system is illustrated in Figure 13.1.

Whilst traditionally SCADA systems have been deployed in isolation of other network infrastructures, in recent years there has been the move to connect these systems to the wider business networks and thus increasing their exposure to cyber attacks. The increase in connectivity has been mainly driven by business needs to gather real-time information on their production and a move to increasingly provide goods and services just-in-time reducing operating costs induced by warehousing and storage. Business needs for faster information and reductions in control staff eroded the previous isolation of the business and control networks. Architectures that physically separated these networks, referred to as air-gaps, became too difficult to maintain and administrate; as a consequence air-gaps have gradually been replaced with firewalls and other less efficient means of separation. This has led to a situation in which it is common to find control system interfaces publicly exposed on the Internet and readily be accessible through web interfaces that can be discovered with ease through search engines such as Shodan [2]. Whilst in many cases this seems to be due to a lack of awareness of cyber security issues of control systems, the exposure to the Internet is only one of the threats that cyber attacks pose to our well-being.

The distributed nature of SCADA systems makes the deployment of physical controls difficult. As the functioning of the control systems relies on the interconnectivity of the various network components, gaining access to a single node of the control network may be sufficient to launch an effective cyber attack that can cascade through the entire infrastructure. In the case of electricity transmission networks this means that potentially thousands of kilometres of deployed substation equipment requires adequate physical protection. This would be prohibitively expensive and difficult to achieve even if sufficient resources were available. Experience has shown that many attacks are not perpetrated by external adversaries, but are due to insider attacks launched by disgruntled employees, rogue contractors or elaborate industrial sabotage [3]. Against attacks initiated or aided by these, normally trusted, individuals many of our technical protections are inadequate. This makes cyber defence as much of a social and people issue as one rooted in technology itself.

In this chapter we explore the threats posed by cyber attacks on ICS, covering both technological and social aspects that need to be considered in a holistic defence. To better understand the context in which security control can be deployed we will introduce the reader to the widely accepted Purdue model [4] that describes control systems in six tiers, ranging from the high-level business IT systems to the deployment of physical devices such as sensors and actuators that affect the physical realm. Using this structure we then explore the suitability of security tools normally deployed in our IT systems in the underlying control system infrastructures and highlight some of the challenges that need to be overcome, both technologically as well as from an organisational process perspective.

13.1.1 Components of industrial control systems

ICS is a general term that encompasses a family of process automation technologies, including Supervisory Control and Data Acquisition (SCADA) systems and Distributed Control Systems (DCS). These systems use Programmable Logic Controllers (PLC) or similar Remote Terminal Units (RTU) and Intelligent Electronic Devices (IED) such as numeric protection relays, to manage electromechanical equipment in either local or distributed environments. Their application covers a range of industrial sectors and critical infrastructures essential to the operations of any industrialised nation-state. The Purdue model, described in Figure 13.2, describes a reference architecture for control hierarchy that has become the standard within ICS. It describes six levels within an organisation managing an ICS as depicted in Figure 13.2.

13.1.1.1 Level 5, the enterprise network

An organisation's corporate network encompassing managed access to the Internet with centralised IT functions together with the customer-facing services. This layer is frequently remotely accessible, although in reality this feature is also found at lower levels due to the dispersion of ICS assets over wide geographic areas as operational efficiencies dictate that lower-level ICS components must communicate with this layer to make business relevant information accessible. At the enterprise level, the

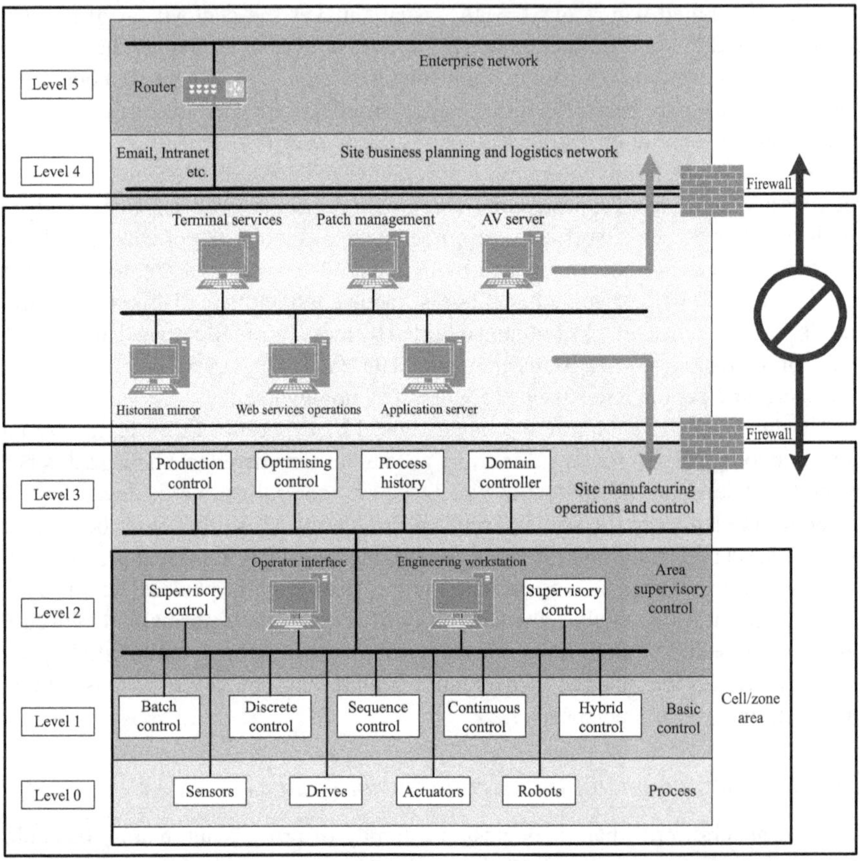

Figure 13.2 The Purdue model for control hierarchy. Adapted from [4]

differences between operations that run ICS or SCADA systems and other industry sectors are minimal. Most of the common network protocols and services are present at this level, such as mail through SMTP and IMAP/POP3 services, HTTP/HTTPS, DNS and underlying TCP, UDP and ARP protocols that run the network based on an IP network stack. The use of commodity IT at this level of the architecture provides an attack surface based on established cyber attack tools and techniques such as "spear phishing," a technique that delivers malware in an email targeted at an individual that purports to be sent for legitimate business reasons [5]. According to Wu [6], the corporate network is a well understood attack-vector from which further access to the underlying control system infrastructure can be gained. This view is widely shared and understood to be one of the major threats to ICS and SCADA environments [7], as businesses also manage the interaction with their supply chains using data provided by the real-time ICS connection managed through the lower levels [8].

13.1.1.2 Level 4, the Logistics network

Similar to the enterprise level, the logistics network also relies on standard IT services and is often seen as an extension to Level 5. The activities on the logistic network are mostly outward facing, with around 90% of traffic being routed to the Internet or to data-centres. However, much of the data processed at this level is sourced from layer 3 of the Purdue model and thus contains pertinent information about the control network itself. Whilst access to these information does not provide a direct threat to the integrity of the ICS and associated control processes, it can contain important information that is gathered as part of reconnaissance or industrial espionage. At this level Internet access and external email is available, but also data relating to the underlying control system is being processed through Manufacturing Execution Systems (MES) and Enterprise Resource Planning (ERP) tools. Historians and other logging mechanisms are storing some process critical events and data for auditing purposes, often mirrored or replicated from Level 3.

The Purdue model segregates the ICS network from the Internet-facing network between Levels 4 and 3 using a "De-Militarised Zone" (DMZ) that implements a pair of firewalls, one at the boundary of Level 3, the other at the boundary of Level 4, to inspect traffic between them and block inappropriate messages [9]. However, as tools such as Shodan [2] and Byres [10] note, this is often not implemented in the real-world deployment of ICS infrastructures. According to Byres, nearly 40% of all reported ICS incidents involve data being transmitted from the corporate network to the ICS. Whilst this is reasonably old data, evidence from Kaspersky [11] shows that in major critical infrastructures still over 80% of systems are connected, when believed to be isolated. This is aggravated by the use of protocols such as OPC which are designed to ease the integration of many proprietary DCS and PLC systems. OPC, according to Luallen [12], is the least frequently monitored protocol but a well-known pivoting point for attackers or reconnaissance and subsequent exploitation of ICS. OPC is a critical component with more than 25% of end users surveyed by Kaspersky [13] considering the loss of OPC communication to lead to a halt of production.

13.1.1.3 Level 3, the site manufacturing operations and control

This is the highest level of control equipment referred to as Operational Technology (OT) [14]. Historically it sat at the highest level of an ICS stack before connecting them to the business networks of Levels 4 and 5 became common practice. The site manufacturing operations and control layer is designed to manage end-to-end operational functions and support the production work flows. Sitting at the boundary of traditional ICS and business networks, this level was intended to provide a clear abstraction between Levels 4 and 2. However, in practice it is often used as a staging area for deploying changes into the operational environment and data being passed up to the enterprise (Levels 4 and 5) systems through the DMZ.

The applications at Level 3 are primarily based on standard computing equipment and operating systems (Unix/Linux-based or Microsoft Windows), and for this reason usually support standard Ethernet and IP networking protocols. This level typically encompasses standard IT hardware such as PCs, along with services such as DHCP,

LDAP, DNS, and file servers. As such, these elements are vulnerable to the same cyber attacks as business systems

Garitano [15] highlights that the OT at this layer has many of the ICS characteristic performance requirements found at lower levels. This reliance on real-time communication and processing of data means that any technology that introduces delay and increases latency or jitter can lead to unstable processes with serious consequences for the ICS as a whole. This has implications when considering the introduction of traditional security mechanisms such as firewalls and anti-virus solutions into these environments.

13.1.1.4 Level 2, area supervisory control

This level manages the supervisory control at site-level and is responsible for the end-to-end production. It communicates directly with controllers situated at Level 1 and shares data with the enterprise through Level 3. Commonly used DCS manage assets and performance within the site through Level 1 devices using closed-loop feedback to mediate groups of localised controllers that share the overall tasks of carrying out an entire production process. The DCS is the main interface for control engineers and provides engineering workstations to supervise and control the area or manufacturing process, as well as the ability to reconfigure the logic of the control devices at Levels 1 and 0.

13.1.1.5 Level 1, basic control, and Level 0, process

Although logically abstracted, it is often difficult to separate the functionality provided at Levels 1 and 0. Level 1 comprises controllers that direct and manipulate the manufacturing process by interacting with the Level 0 devices (e.g., I/O, sensors, and actuators).

Control devices operating at Levels 1 and 0 typically use operating systems unfamiliar to IT professionals.

The operation of these devices is normally configured through engineering workstations situated at Level 2 and typically involves changing the control logic of the device and other configuration parameters, such as poll-rates. Controllers at this level implement basic control loops fed by sensory information from real-time devices at Level 0, which result in instructions being fed to other real-time devices to actuate physical industrial machinery. At this level both hardware and software are optimised for performance and use proprietary technologies. It is not uncommon for these devices to remain in service for between 10 and 20 years. The impact on production availability and risk of changes to certified safety processes means that updates and patches to this equipment are infrequent, and as a result security vulnerabilities can persist for long periods of time.

The major threats to control systems components at this level are characterised as response injection, command injections, or denial of service attacks [16]. Given the relative lack of authentication, injection attacks are relatively easy to achieve on many control systems once access to the network has been established. The comparably low processing power and the real-time properties of many control systems make denial of service attacks or degradation of service attacks comparatively straightforward to achieve. For example Shayto *et al.* [17] demonstrated how an Internet Control Message Protocol (ICMP) ping targeting a fragile, misconfigured TCP/IP stack resulted in a loss of service on a control device.

Whilst the Purdue reference model is not used to govern ICS implementations, it reflects the general architectural principles adopted whereby the control of industrial equipment is managed in a layered hierarchy that is logically, if not physically, separated from the management of the industrial facility and its business processes. Importantly, it defines the areas of an industrial control architecture where IP-based protocols transition to legacy serial communications.

This section reviewed some of the components that make up typical ICS and explained their connectivity and function in the overall structure of the Purdue model. We also discussed some of the threats that are targeting these components and what perpetrators may achieve through compromising the various layers. The following will look at some of the currently available technologies for protecting ICS and how they correspond to the layers of the Purdue model.

13.2 Security technologies and their role in ICS protection

In this section we will look at a number of technologies that are currently available in the protection of traditional IT systems and explore the challenges that prevent their adoption in ICS, and in particular the lower layers of the Purdue model. Not all of these challenges have been overcome and significant issues remain.

13.2.1 Level 5

Given the similarity to traditional IT architectures at this level, the protection mechanisms and controls used for standard IT are adequate. However, there is evidence to suggest that whilst these technologies are widely available, their implementation has been inadequate. The US Department of Homeland Security [18] states that the majority of incidents between 2009 and 2011 could have been identified if traditional IT security mechanisms and techniques had been properly deployed. Managed network configurations, the filtering of known bad IP addresses or domains through ingress/egress filtering and authentication and access controls would have reduced the attack surface significantly. The mechanisms at the OT levels are typically less sophisticated than those deployed in the wider enterprise, and any direct integration between them allows a direct path to the generally insecure devices that manage the physical processes of the operation or facility.

Work by Stouffer *et al.* [9] suggests the use of separate technologies in the corporate and OT networks increases containment. This is particularly of interest as the number of users that require access at the enterprise level is significantly higher than that requiring access to the underlying OT network. Through the separation of security controls, phishing, spear-phishing attacks and the effect of malware infections can be contained and their impact limited.

13.2.2 Level 4

The use of IP-based protocols and standard operating systems in Level 4 means that traditional IT security mechanisms are adequate. Lebanidze [19] discusses instituting practices such as proper network segmentation and firewalling, regular security patching, up-to-date anti-virus software that runs regularly, security-aware software development and acquisition processes as appropriate. Given Level 4 sits at the boundary to the OT, best practice with respect to cyber security must be followed in order to prevent unauthorised access to the underlying control system. This includes assurances that level 4 does not provide an egress route for data from the OT to inappropriate external entities and should include traffic monitoring and other anomaly detection techniques to combat industrial espionage. On the other hand interactions between this level and the control networks is required. Galloway & Hancke [20] highlight that engineers involved in the design and maintenance of control networks must be familiar with both traditional enterprise concerns, such as network security, as well as traditional industrial concerns such as determinism and response time. Stouffer *et al.* [9] recommends that when designing a network architecture for an ICS deployment, it is necessary to separate the ICS network from the corporate network. However, they also acknowledge that a connection is required between the ICS and corporate networks and propose the deployment of an intermediate DMZ in order to marshal the communications between IT and OT.

13.2.3 Level 3

Whilst most of the Level-4 and Level-5 components are located in areas that traditional IT security mechanisms can protect, the systems situated at Level 3 are increasingly close to production floors and in many organisations the stovepipe administrative structures, where IT and OT security staff are separated, can interfere with effective security management. Level 3 is the first layer that mixes IT and OT and whilst the use of standard IP protocols is prevalent, the typical traffic that is observable differs significantly. A common problem at this level is the temptation to utilise out of the box IT security controls as they are ill suited to the industrial requirements of this level [15]. One of the issues is that legacy protocols are tunnelled through TCP/IP connections for reasons of cost and efficiency, leading to many of the well-known security vulnerabilities against TCP/IP such as ARP spoofing and cache poisoning propagating within the OT as well as in the higher layers, ultimately leading to an increase in the complexity of the security mechanisms required.

Intrusion Detection Systems, which are effective at Levels 4 and 5, are presented with challenges at this level and need to be considered with care as they can introduce latency that may impact the safety systems designed to bring the OT to a safe state when certain boundary conditions are met. This is a fundamental component of safety critical systems, to include safeguards that lead to a graceful shutdown or other actions that ensure that physical damage is limited as much as possible. The triggering of a safety system is not a normal system behaviour that can be observed and learned by traditional IDS systems that operate on profiling of normal system behaviour to detect anomalies, however it is an intrinsically benign function of the OT. This means that security mechanisms deployed in OT must be aware of the processes themselves.

13.2.4 Level 2

Hadžiosmanović *et al.* [22] propose that IDS and other traditional security measures cannot, by themselves, detect or mitigate process-related threats in this level and offers that the identifiers for threats that intend to target the process under control, rather than the control system, can only be detected by analysing the data at a higher semantic level. The scale of contemporary ICS presents significant challenges for the configuration of monitoring tools that capture process events through system logs. Logs and monitoring data, whilst useful for the analysis of failures and post-incident investigations, need to be adequately protected. Falsified monitoring data may leave control engineers ignorant of system malfunctions until it is not possible to recover the system. This can be achieved, for example, through well-known man-in-the-middle attacks that allow an adversary to intercept the data streamed from control components such as PLCs to Human Machine Interfaces (HMI). Similarly falsified data-streams from a compromised component can also mislead IDS [23]. Successful ICS protection and analysis will require a more detailed analysis due to the complexities and single purpose nature of many of these systems. This means that advanced modelling [24] and automated discovery tools [25] should be used to facilitate a rigorous security analysis that takes into account the processes, operator interactions as well as sub-system interdependencies [26] (Table 13.1).

Table 13.1 Differences between CIS and CPS intrusion detection [21]

ICS	CPS
An ICT Intrusion Detection Systems (IDS) monitor host- or network-level activity (e.g., an HTTP request or a web server)	A CPS IDS monitors the physical processes (and hence the laws of physics) that govern the behaviour of physical devices that make certain behaviours more likely to be seen than others
An IT IDS monitors user-triggered activities, leading to unacceptably high false-positive rates due to the unpredictability of user behaviours.	A CPS IDS monitors activities that are frequently automated and time driven in a closed-loop setting, thus providing some regularity and predictability for behaviour monitoring
An ICT IDS deals with mostly non-zero-day attacks rendering knowledge-based detection effective	A CPS IDS deals with zero-day or highly sophisticated attacks, rendering knowledge-based detection ineffective
An ICT IDS often does not have to deal with legacy components, making behaviour specification of the physical processes governing legacy components unnecessary	A CPS IDS often must deal with legacy technology, making behaviour-specification detection an effective technique by precisely specifying the physical processes governing the behaviour of legacy components

13.2.5 Level 1 and Level 0

Beresford [27] explains how some PLCs run x86 Linux with all processes having root privileges. If commands can be injected it is possible to launch a command shell running as root, proving complete control of the device and therefore the industrial equipment it manages. The lack of protocol encryption or authentication proved the device susceptible to spoofing, session hijacking and denials of service. The CockpitCI Consortium [28] reported that on PLCs running the VxWorks operating system they observed a diagnostic service running on a port that would execute instructions without any form of authentication. The paper also described how it was possible to use the Modbus protocol to force single and multiple coil values by manipulating the Modbus function codes in order to switch off industrial equipment, whilst simultaneously suppressing the status updates from the equipment, denying the system operator or DCS the awareness of the true state of the control system. In similar work Rrushi & Campbell [29] explained how memory addresses and shellcode, injected by Modbus traffic into PLC memory, could be made to appear as valid coil and/or register values by exploiting mappings between the addresses of data items defined by Modbus and the actual address locations in which they were stored. The traffic, although malicious, was fully conformant with the Modbus protocol specification, highlighting the need for context-aware intrusion detection within OT networks.

Whilst expanding on the threat of ICS reconnaissance, Gao & Morris [30] demonstrated that it is possible to use the device identification services of Modbus to determine the PLC run status, vendor name, product code, major and minor revisions, vendor URL, product name, model name, user application name and other device-specific information, all of which can be used to search freely available ICS vulnerability databases.

Barbosa *et al.* [31], Goldenberg & Wool [32] and Kleinmann & Wool [23] highlight how devices in Levels 1 and 0 produce periodic traffic patterns using protocols that are insecure by design and are therefore susceptible to replay attacks, hijacking, spoofing and manipulation. As the devices in question have limited processing resources and are inherently proprietary they will remain vulnerable until the vendors implement on-board security services, and even then the ICS users would still have to accept the risk of upgrading their infrastructure. Traffic inspection, transport layer security and device whitelisting all potentially impact on latency, and as many protocols in use are not based on IP, would require specific hardware and/or drivers in order to facilitate this.

Accepting the current limitations of the devices and protocols, security mechanisms in Levels 1 and 0 lend themselves to procedural security management and configuration control. Therefore, whilst not specifically designed for ICS, the concepts of Information Security Management Systems as defined in ISO 27001 offer the procedural basis for mitigations in these layers.

In this section we presented a number of the security risks and threats to the physical infrastructure that is managed by SCADA systems connected to enterprise networks. Whilst we focused on the technical aspects of securing the infrastructure at the various abstraction levels of the Purdue architecture, other risks cannot be ignored. One significant problem is that of a failure to patch infrastructures on a regular basis. Standard IT systems are regularly updated to close vulnerabilities, and the IT infrastructure at Levels 5 and 4 is designed to facilitate these updates. In the lower levels of the Purdue model the situation is very different. Whilst patches are released by vendors of the control systems, research by Byres [33] shows that they are only deployed in the system is a small minority of cases. The reasons for these are many, but one of the most important is the need to guarantee the availability of services. Patches that require a shutdown of production are a high risk to the organisation, firstly because the shutdown of large infrastructures is expensive and also carries the risk of complications when the systems are restarted. This is aggravated by uncertainties about the effect the patch has on the overall system as only few infrastructures maintain adequate testing facilities that can be used to validate the patch before deployment. Another aspect is the risk-averse culture in which these systems have been developed, deployed and maintained over several decades. In environments where changes to operational systems could have unforeseen consequences, the reluctance to introduce software or architectural changes is apparent. Unfortunately, a vulnerable system is inherently at risk, and will continue to operate until this vulnerability is exploited by an adversary, making the risk difficult to quantify. In addition responsibilities to manage the risk are often not clearly divided or even identified. Many vulnerabilities are found in individual components residing on the lower layers of the Purdue model. Vendors of these components will provide patches and test them against the known specifications of that component, however when deployed these components are integrated into larger subsystems and then deployed and operated by public or private enterprises it is often unclear how overall security or operations are affected.

13.3 Managing threats to ICS/SCADA

So far we have concentrated on the technical controls that could provide increased protection against threats to control systems. Now we shall consider how the attribution of attacks and post-incident forensics can manage the threats to an operational system. For being able to attribute attacks it is useful to consider what the motivations and capabilities of attackers are. Robinson [34] gives an overview of the current threat-landscape for ICS/SCADA systems.

13.3.1 Attribution of attacks against ICS/SCADA

Attribution of attacks, the means by which an attacker or attack source is identified, is beneficial to serve as a deterrent for future attacks, improve defensive techniques and provide a basis for interrupting attacks that are in progress [35]. Response to cyber attacks, such as prosecution, political sanctions, attack back, can be considered by decision makers when attack actors are known and attribution is accurate. The relevance of attribution has been questioned and, especially in an enterprise environment, the focus is often on resilience rather than identifying the attack actor. However, in ICS/SCADA environments, attacks are more likely to be against specific targets, and with more significant consequences, e.g., damage to machinery and the potential for harm and loss of life. Therefore in this environment attribution is far more pertinent.

Attribution of attacks is a multi-faceted and difficult problem and becomes even more challenging when considering ICS/SCADA environments which often mix both IT and OT. Techniques themselves may have limitations, requiring widespread deployment or being legally or ethically questionable. In addition, there are various ways that attack actors can perform attribution avoidance to subvert or confuse attribution techniques, potentially implying false blame against innocent parties. For example, attackers may obscure their identity through mechanisms designed to protect anonymity on the Internet [36]. Proxies, TOR (The Onion Router) and onion-routed networks are specifically designed to defeat attribution techniques such as traceback. TOR provides anonymity by establishing a connection to an encrypted network of volunteer routing devices that do not offer any reverse path reconstruction services [37]. Less sophisticated methods of anonymity include the use of SOCKS or VPN proxies. SOCKS proxies allow a HTTP/HTTPS point-to-point connection with a proxy server. Whilst it is technically possible to trace IP traffic back to a SOCKS proxy server and examine the connection logs to establish the originating addresses, often this is prevented by the location of the server itself. The operators of proxies are often located in countries that do not subscribe to transnational cybercrime initiatives and have no incentive to make this attribution data available [36]. VPN proxies allow traffic to be encrypted and tunnelled through a series of routing devices to appear at an exit node elsewhere in the world, usually through a NATed IP address. The same log inspection techniques and issues apply equally to VPNs. Cybercrime syndicates may rent out space in data centres in countries that have strict privacy laws and set up their own VPN infrastructure.

A wide range of technology-based techniques exist to facilitate attribution, primarily focusing on enterprise IT environments. Wheeler and Larsen [38] provided a baseline survey of technical attribution techniques (Figure 13.3) and a taxonomy that has been used as primary literature in a number of studies since its initial publication [35,39–41]. This section reviews the developments in three key areas of Wheeler and Larsen's research, those being traceback techniques, honeypot technology and malware analysis. We frame these three technical attribution techniques in the context of ICS/SCADA environments. We also discuss "intelligence-led attribution," which involves following an intelligence approach using non-technical techniques.

13.3.2 Traceback techniques

Traceback is a class of methods that encompasses techniques by which the traffic from a target device is recursively stepped-back through its routing path to its originating source device, as shown in Figure 13.4. The figure shows three paths that represent possible attack paths from suspected attackers. Traceback creates an attack graph showing the intermediate devices that the attack passed through.

Kuznetsov *et al.* [43] aggregated the significant research in this area into three distinct approaches. The first category included manual methods of traffic tracing, and required the routing device to support input debugging as well as constraining the

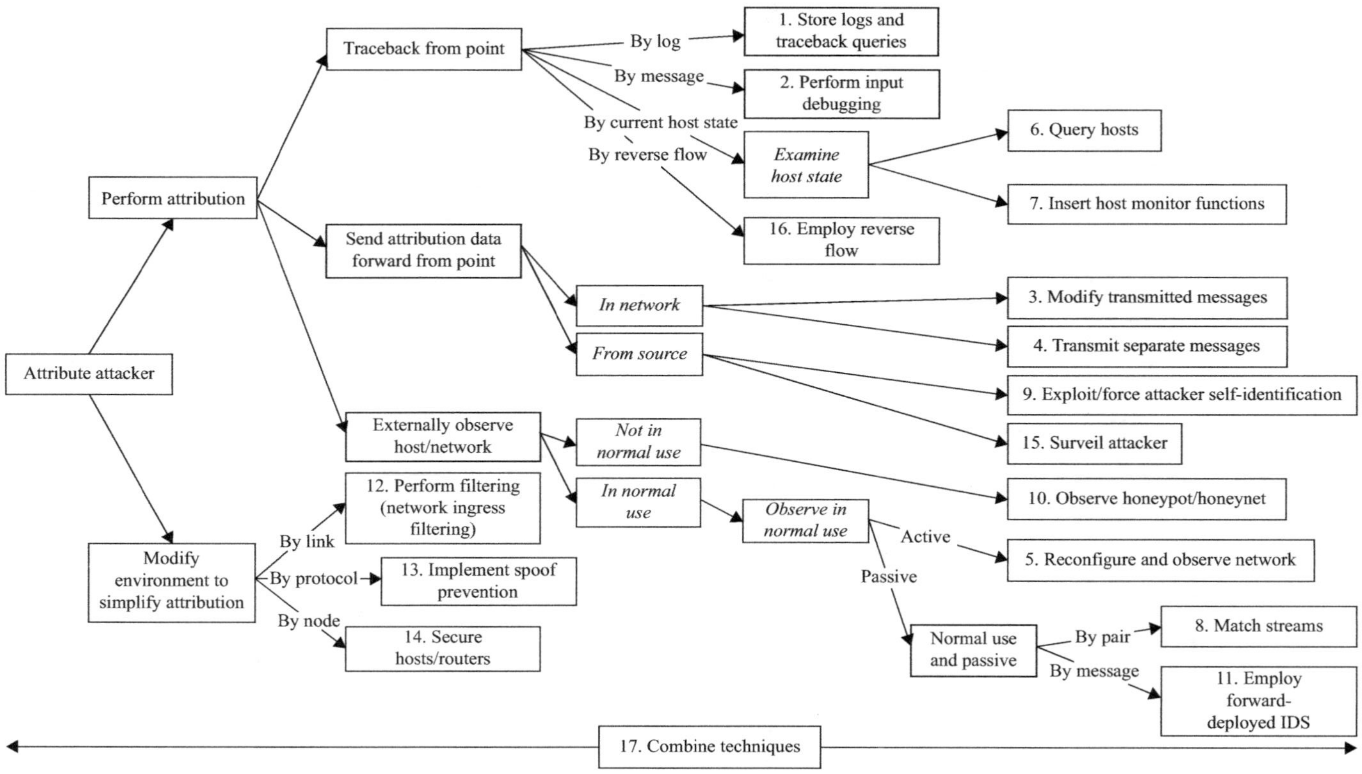

Figure 13.3 Wheeler and Larsen attribution technique taxonomy [38]

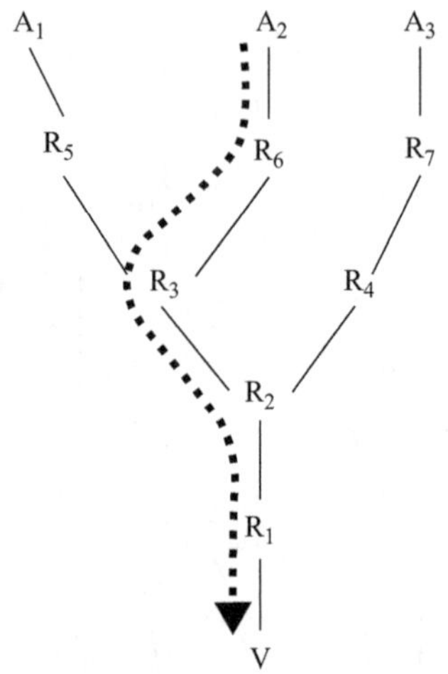

Figure 13.4 Theoretical attack graph [42]

period of analysis to the duration of the attack itself. The second category spanned logging techniques, whereby routers persist information about the traffic they have encountered. These are impractical due to storage requirements. One variant, however, the Source Path Isolation Engine (SPIE) techniques of Snoeren *et al.* [44], capable of tracing the route of a single packet through SPIE-compliant routers, could resolve storage issues by only collecting hashes of the packets. Whilst this reduces the storage overhead, Gao and Ansari [45] highlight that the computational requirements increased as a consequence. The third category included the various methods of probabilistic packet marking (PPM), and ICMP traceback (iTrace).

PPM, originally proposed by Savage *et al.* [42], and extended by Song and Perrig [46] and Belinky and Ansari [47], uses packet marking to sample a number of packets with path data, so that should a target device receive a sufficient volume of such packets it could reconstruct the entire path back to the source. Marking information is stored in unused or infrequently used packet header fields, such as the 16-bit Identification field. Savage *et al.* [42] suggest that 75 packets is sufficient when the path length is 10 and the number of attackers is small. When the number of attackers is large, this technique becomes ineffective; thousands of packets are required and convergence time increases.

Traceback methods typically require a modification to the network infrastructure over which they will operate, and it is questionable how cost-effective this would be given the scale of the modern Internet. More important, however, is that all traceback techniques, including a hybrid model proposed by Gong and Sarac [48], fail to address the nature of contemporary, multi-stage attacks described by Clark and Landau [49] whereby intermediary devices are coerced by malware to infiltrate one computer to use as a platform to attack a second etc., in an ongoing process of originator obfuscation. At best, they will only attribute the attack to a coerced device.

Traceback techniques suffer from a number of problems that mean deployment in the Internet environment is unlikely. These techniques provide direct artefacts such as the source IP address, however since IP addresses may be associated with compromised machines, this is of little use. It is only useful if the owner of the IP address endpoint is willing to fully cooperate and allow forensic investigation of their machine(s).

Traceback is also intrusive, requiring infrastructure changes for deployment and packet/router modifications, additional traffic or additional storage and processing requirements. Furthermore, the onus of who should manage these aspects is unclear. Finally, traceback techniques may introduce new attack vectors. For example, packet logging produces additional traffic and could cause a DDoS attack in itself.

Traceback as an attribution technique suffers in enterprise IT environments as it only provides a source address or attack path to a source address. This can ultimately be foiled when an attack actor uses any type of stepping stone, e.g., Tor, VPN, proxy or compromised system. There could be any number of stepping stones that add degrees of separation between the adversary and the victim. To make matters more complex, the attack path could be routed through multiple countries and legal jurisdictions. This problem is still inherently true when deployed in SCADA systems. Additionally, if a traceback goes beyond the domain control of the organisation, then there must continue to be traceback capability in other domains to attribute any further. When using the Internet as an example, every ISP would need to support traceback. Finally, traceback is a network-based attribution technique. Should malware propagate through removable media, then it would not be identified by this technique.

13.3.3 Honeypots

A honeypot is a system, or set of systems, where vulnerable software and services are hosted in order to allow attack activities to be monitored and logged. They approach the issue of attribution of attacks differently to traceback methods, by observing an attack in situ. Unlike traceback methods, honeypots have seen varied use within ICS/SCADA environments. Pothamsetty and Franz [50], created the first publicly acknowledged SCADA honeypot, to "determine the feasibility of building a software framework to simulate a variety of industrial networks and devices." They found that there was a lack of information relating to SCADA vulnerabilities and attacks. A technical deliverable was produced, a SCADA honeypot based on a low interaction honeypot, Honeyd. Honeyd simulates many network protocols such as HTTP, SMTP and FTP. Honeyd can be extended to simulate more network protocols using simple scripts. Franz and Pothamsetty created scripts to simulate the SCADA services of a Modicon Quantum device with HTTP, FTP, Telnet and Modbus services. They also created a Java applet, "StatusApplet.java," which could be accessed via a web server and simulated the status of a SCADA field device. The technical implementation of this honeynet was primitive and at a proof of concept stage. Subsequently little effort was placed on concealing the honeypot status. For example, the action event on the HTML forms reads "action='honeyd-feedback.py'", an indicator that the SCADA system is actually a honeypot.

Researchers at Digital Bond [51] expanded upon the work of Franz and Pothamsetty [50] with two VMWare virtual images. One image contained a SCADA honeypot based on Pothamsetty and Franz's work and another image contained a gateway to monitor activity, collect data and prevent outbound connections from compromised honeypots. The images also included a collection of Snort IDS preprocessors and plugins specifically for SCADA protocols. What made this work unique was that the gateway image could be placed in front of either the SCADA honeypot or a real non-production PLC. The latter configuration was important because it enabled a physical SCADA device to be used as a honeypot.

Rrushi and Campbell [29] continued the trend of using real devices when they proposed "reactor mirage theory." Their proposal used deception to detect intrusions against the nuclear power sector. Their prototype made active decisions to draw adversaries towards a honeypot which used real industrial devices. By populating the environment with deceptive systems they increased the possibility of an adversary targeting a non-production system. Similarly, by using real devices as deception systems and creating simulated activity (Modbus protocol traffic), they increased realism and decreased the possibility of an adversary discovering that they are interacting with a honeypot. Despite these benefits, the costs associated with deploying many real devices for deception purposes is high and any increase of network traffic directly or indirectly on a production SCADA network should be approached with close scrutiny.

In another academic proposal, Valli [52] described a SCADA forensics framework which combined the Snort IDS with two low interaction honeypots; Honeyd and Nepenthes. Known SCADA exploits were replayed in a controlled lab to create network IDS rules which would then influence configurations for the two honeypots. However, it is unclear if this initial proposal received further attention.

The SCADA honeypot proposals discussed so far focused on the assumption that attacks are network borne. In 2012 students at Bonn University in Germany, led by Sebastian Poeplau, created Ghost USB, a honeypot which emulates USB devices to counter the threat of malware that propagated by removable media [53]. Currently, Ghost USB supports 32-bit Windows XP/7 and is supported by the Honeynet Project. The project has obvious uses for SCADA systems; Stuxnet was presented as a use case for Ghost USB, since the malware propagated by USB devices connected to SCADA engineer workstations. This tool could be deployed on production systems with little cost and is readily available.

Wilhoit [54] attempted to address the attribution of attacks specifically against ICS by employing a set of honeypots that advertised themselves as an operational system with PLCs attached. The honeypot architecture ran the BeEF framework (Browser Exploitation Framework Project) to embed a script into web pages that was executed every time an attacker compromised the site authentication. The script determined the geographical location of the attacker, as well as capturing further information for statistical analysis. The study managed to identify the locations of the attackers, and made a high-level assessment of the perceived intent of the attacks, stating that if "an attack was targeted in nature, for instance, but did not compromise the operation of a target ICS device, the attackers' motivation could be espionage or information gathering. If an attack, however, compromised the operation of a target ICS device, depending on how badly it was affected then the motivation could be considered destructive in nature." There would be merit in considering this research in the context of the Miller and Rowe [3] taxonomy.

Most recently the Honeynet Project has announced and released Conpot, which aims to simplify the process of setting up SCADA honeypots [55]. Conpot currently supports Modbus and SNMP, however the developers intend to add support for other protocols. The tool simulates the Siemens S7-200 PLC. Conpot feeds into HPFeeds, a data sharing platform which is used by a number of the Honeynet Project honeypots.

Spitzner [56] highlighted the limitations of honeypots due to the narrow field of view available to them, and that it only allows a focus on attacks against specific targets (i.e., the honeypot). He highlighted that whilst the data capture can be very rich, it does not encompass all of the surrounding behaviour that may occur outside of the honeypot that may indicate the wider events associated with an attack.

In order to try and address this issue, Wagener *et al.* [57] adopted high-interaction, self-adapting honeypots that introduced simulated failures into the interactions to repeatedly attract attackers and lure them into revealing as much information about themselves as possible. The study notes that attackers showed a level of determination to achieve their perceived objectives and that we assume that attackers are rational and follow a specific goal during attacks, implying a focus on modus operandi.

13.3.4 Malware analysis

Malware, in its various forms, virus, worm, trojan, adware, spyware, backdoors and rootkits, are analysed to identify characteristics that can be used to attribute. Malware analysis in enterprise IT can be split into two areas: behavioural analysis and code analysis.

Behavioural analysis examines the way that malware interacts with the environment. Malware might make changes to the registry, create new processes, hide files, execute other binaries, contact command-and-control servers, cover tracks by deleting evidence of its modifications (as Stuxnet did), disable security protections, record user interaction (e.g., keylogging), harvest sensitive data, exfiltrate data, attempt to update, pivot to other systems, establish back doors and more. A controlled sandbox environment is usually created to examine this behaviour.

Code analysis is concerned with examining the code that makes up the malware. Source code for malware might be available, although it is unlikely. Otherwise, reverse engineering and debugging take place. Reverse engineering involves restoring the malware's binary machine code to human-readable assembly code, using tools such as IDA Pro and OllyDbg. The code can then be executed in a debugger to step through the instructions, inspect register contents, identify embedded strings and set breakpoints to figure out exactly what the malware does.

Practitioners used reverse engineering against the Stuxnet malware [58]. They identified clues in the code, such as binary compile times, suspicious variable names, registry keys that appear to be dates and directory names that might be biblical names. Some or all of these clues could have been false flags; data that was purposely crafted to implicate another entity as the malware authors. Symantec consulted the expertise of SCADA practitioners, such as Ralph Langner, in order to understand the effects that the Stuxnet malware had on the Siemens PLCs. This again highlights the diverse skill sets required for the SCADA environment and the necessity for security professionals to work closely with SCADA engineers.

13.3.5 An intelligence-led attribution approach

A number of non-technical investigatory techniques may offer alternative or complementary approaches to assigning attribution to a cyber attack. We label this approach "intelligence-led" techniques.

As technical attribution techniques offer limited and varying degrees of actionable data, Carr [59] proposed that the "one thing you can count on is that someone has to pay for the necessities of virtual combat. Therefore, one sound strategy in any cyber investigation is to follow the money trail created by the necessary logistics of organizing a cyber attack domain registration, hosting services, acquisition of software, bandwidth, and so on." He highlighted that although false identities are often used when registering and acquiring services, the increased use of social media and the increasing size of individual and corporate digital footprints allows for a forensic examination of online presence and identity may reveal such deceptions. Gantz *et al.* [60] estimated that approximately 45 GB of data existed for every person on the planet. They also discussed the analysis of "digital shadows," that ambient content data created by traffic cameras, use of ATMs, online transactions etc.

An analysis of alleged Chinese computer attack behaviour [61] resulting from a reported seven years of covert observation offered an insight into the scale and complexity of attacks on ICS. Targets included transportation, navigation, engineering, food and agriculture, chemicals, energy, aerospace and mining – all areas where ICS were likely to be used. Its attribution of the observed attack behaviour to China was based upon a mix of technical measures and intelligence data gathering and analysis. In particular the report focused on commonalities between attack methods, consistencies in naming conventions and comparative analysis of malware.

Both Fireeye [62] and Shivraj [63] described the consistency in attack behaviour observed from common sources. Fireeye leveraged their position as a supplier of commercial security products to gather and analyse Advanced Persistent Threat (APT) callback traffic and events in order to establish patterns of behaviour and command and control traffic. Shivraj [63] defined the stages of contemporary APT behaviour with a focus on SCADA attacks and illustrated how common malware approaches can be applied to ICS targets with limited alteration required, at least at the early stages of an attack. The combined findings of both papers could be potentially combined to provide an indication of attack attribution and a tangible assessment of where the target is in the attack cycle, and therefore what preventative measures may be appropriate as a consequence.

Langner [64], in his investigation of the Stuxnet malware, was unable to provide any substantive evidence to attribute the originator of the code, but did find significant indicators as to the evolution of the software and its intended effects. In particular, he highlighted the level of industrial process and control system knowledge required to develop the malware, and speculated as to the high level of testing that would have been required to prove the payload prior to its release. The necessity for the malware to traverse the traditional IT layers of the target environment before its compromise of the ICS to damage the physical elements of the system under control gave rise to a complex piece of software. Langner [64] believed that to develop Stuxnet required nation-state resources. Although he provided no irrefutable evidence for this, he presented a compelling argument based on the complexity of the development undertaken. Knake (2010), took a more pragmatic and empirical view when testifying to the US House of Representatives on the cyber threat, stating that at the uppermost level of threat, that of a nation-state, the issue of attribution is simplified as "there are a limited number of actors capable of carrying out such attacks." It is perhaps worth considering the requisite capability of an actor when attempting to assign attribution to a covert attack.

This concept of nation-state capability was extended by Geers *et al.* [65] in an attempt to characterise the motivations and nature of state-sponsored cyber attacks. In a discussion of cyber warfare, the paper proposed that "[a] cyber attack is best understood not as an end in itself, but as a potentially powerful means to a wide variety of political, military, and economic goals." In this context, an analysis of the intent of an attack would perhaps elucidate which nation-state(s) would benefit from the outcome of the attack, and from this we could derive motive. Whilst not an attribution method in itself, it would allow for an investigation into attribution to be focused on likely perpetrators.

In a post-Stuxnet analysis, Bencsáth *et al.* [66] undertook comparative analyses of malware in their investigation of the Duqu, Flame and Gauss executables and highlighted that Duqu shared "striking similarities with Stuxnet" and proposed that there were indications that the three malware tools were part of the same family, suggesting at least a partial common source.

Accepting that there are inherent problems with absolute attribution of cyber attacks, Kalutarage *et al.* [67] proposed a probabilistic approach based on Bayesian methods. The methodology divided the problem into two smaller domains; evidence fusion and aggregation (described as "accumulation"), and the subsequent analysis (described as the "anomaly definition"). The accumulation allowed for the incorporation and use of many Bayesian approaches and prepared the anomaly definition to allow the analysis of attacker activity patterns within a series of node profiles. The data used in the experimentation came from a series of logging techniques and appeared to be entirely IP-based. However, there appeared to be nothing in the methodology that would preclude the use of serial data or Historian records from an ICS. In the context of intelligence-led attribution analyses, there may be some valuable research to be undertaken in the field of probabilistic attribution.

13.4 Post-incident forensics in ICS/SCADA

Post-incident forensics are closely linked with attribution as they offer an additional opportunity to identify the source of an attack and potentially prevent its recurrence.

Table 13.2 Expected lifespan of data [69]

Type of data	Lifespan
Registers, peripheral memory, caches, etc.	Nanoseconds
Main memory	Ten nanoseconds
Network state	Milliseconds
Running processes	Seconds
Disk	Minutes
Floppies, backup media, etc.	Years
CD-Roms, printouts, etc.	Tens of years

It also allows the operators of an ICS or SCADA system to restore their systems to a functioning state,and identify and fix other affected systems.

There are two forensic approaches commonly used: traditional forensics and live forensics. Traditional forensics, also known as "postmortem forensics," was once the best practice and de facto approach for forensic analysis [68]. Using this method, upon finding a compromised system, or system suspected to be compromised, the system is immediately, and ungracefully, powered down to prevent further malware propagation. This does not allow the system to shut down safely and volatile data is lost. Hard disks are then placed into a write-blocker device, so that data can only be read and integrity is preserved. A bit-level forensic copy is created and then forensic analysis takes place on the copy. The benefit of this approach is that data is less likely to be modified and therefore reliable if used as evidence. However, this approach is less appealing for a number of reasons, especially as volatile data is lost, as shown in Table 13.2.

More recently best practice has shifted towards live forensics [68]. In this approach, rather denying power, the system is left in its current state. In live forensics data acquisition takes place whilst the system is operational. In traditional IT systems, tools are used to capture running processes, RAM memory, browsing history and more, in the order of volatility. Performing live forensics on an operational machine in a SCADA environment poses significant challenges; accidentally causing the machine to crash could be catastrophic. Ahmed [70] discusses this issue and suggests using fail-over systems to allow for live forensic analysis to take place.

However, in all likelihood, any post-incident investigator will be competing with recovery efforts which will almost certainly destroy evidence. There is also clearly a logistics concern when performing SCADA forensics. Field devices could be located many miles away, perhaps on different continents, or perhaps in difficult to reach places, such as on the ocean floor. Physically reaching these devices may not be possible. Similarly, some devices may not have disks or persistent data stores, or may only have serial interfaces and therefore not be compatible with contemporary forensics tools. The diversity of devices that a forensic investigator can encounter in the SCADA environment is far wider than that of the traditional IT domain. Traditional IT systems have a lifespan of 3–5 years, perhaps 10 at most, whilst 10 years for a SCADA systems would not be considered anywhere near the end of its life. However, as PLCs and other SCADA devices continue to move towards commercial-off-the-shelf (COTS) hardware and software, the forensic analysis of SCADA systems becomes standardised and potentially simpler.

Another field of forensics used in enterprise IT environments is network forensics. This approach involves two stages: collecting network messages and then analysing network messages. Existing infrastructure such as switches and routers can be configured to collect messages, or extra equipment can be deployed, e.g., a network tap device. By logging messages, analysis can take place during an incident or post-incident. Collection of data is relatively straightforward. An organisation must identify points in the network were they wish to collect network data. Mahmood *et al.* [71] describes traditional network analysis problems and network sniffer deployment in a SCADA environment. An area that requires further consideration is when traditional communication channels other than standard Ethernet are used. Industrial Ethernet, which can require specifically designed routing and switching devices, or older serial communications based on RS232 or RS485 may require specialist devices to gather traffic. Also, the use of communications bearers other than cable or fibre-optic, such as radio or GSM links, present additional traffic capture challenges.

Forensics is primarily a practitioner-led field with research taking place as and when it is required. In a recent effort to outline a research agenda for this field, SCADA forensics was identified as a predominant theme [72]. The following points were identified as near future research for forensics in SCADA systems: (i) Collection of evidence in the absence of persistent memory, (ii) Hardware-based capture devices for control systems network audit trails, (iii) Honeypots for control systems as part of the investigatory process, (iv) Radio frequency forensics and (v) Intrusion detection systems for control systems.

13.5 Conclusion

The protection of ICS and SCADA systems is complicated, with a broad range of security issues across every layer of the architecture. The discussion of these systems has taken place within the logical framework of the Purdue reference model. However, the reality of many ICS is that they do not follow such architectural demarcations, with systems and protocols crossing intangible layers at will. Connectivity to the Internet allows reconnaissance activity around an ICS as well as an opportunity for malware delivery. Spear phishing is as much of a threat to ICS as to general IT networks. Inadequate controls within a corporate network can provide an environment in which malware can propagate and extend its reach within an organisation. The range of attack vectors grows within an ICS as vendor engineers access control equipment either remotely or locally, with little control over the security of their devices or network connectivity.

The mix of IT and OT within a control system limits the extent to which traditional IT security mechanisms can adequately defend such environments. Adoption of defence-in-depth approaches that combine existing security tools with monitoring systems native to control devices offer the potential to mitigate a number of the risks, but some will inevitably remain. No single attribution method provides enough substantive evidence to identify the perpetrator of the attack on a control system sufficiently to support a prosecution, and therefore serve to further reduce risk, but it is possible that a combination of techniques offers the potential to build a probabilistic model that, combined with post-incident forensic techniques, may improve attack attribution, or at least increase the level of deterrence.

References

[1] G. Dan, H. Sandberg, M. Ekstedt, and G. Björkman, "Challenges in power system information security," *Security Privacy, IEEE*, July 2012, vol. 10, no. 4, pp. 62–70.

[2] Shodan, "Shodan search engine," [Online]. Available: http://www.shodanhq.com, Accessed March 3, 2015.

[3] B. Miller and D. Rowe, "A survey SCADA of and critical infrastructure incidents," in *Proceedings of the 1st Annual conference on Research in Information Technology*. ACM, 2012, pp. 51–56.

[4] T. J. Williams, "The purdue enterprise reference architecture," *Computers in Industry*, 1994, vol. 24, no. 2, pp. 141–158.

[5] C. Wueest, *Targeted attacks against the energy sector*, Symantec Security Response, Mountain View, CA, 2014.

[6] T. Wu, J. F. P. Disso, K. Jones, and A. Campos, "Towards a SCADA forensics architecture," in *Proceedings of the 1st International Symposium on ICS & SCADA Cyber Security Research 2013*. BCS, 2013, pp. 12–21.

[7] E. Byres. "Using ANSI/ISA-99 standards to improve control system security." The Industrial Ethernet Book. [Online]. Available: http://www.iebmedia.com/ethernet.php?id=8460&parentid=74&themeid=255&hft=68&showdetail=true&bb=1&PHPSESSID=vk9tck0p6rq69jt6d3l6co7lh7, Accessed August 4, 2016.

[8] E. Udassin, "Control system attack vectors and examples: field site and corporate network," in *Proceedings of S4 SCADA Security Conference*, 2008.

[9] K. Stouffer, J. Falco, and K. Scarfone, *Guide to industrial control systems (ICS) security*, NIST special publication, Gaithersburg, MD, pp. 800-8-82, 2011.

[10] E. Byres, D. Leversage, and N. Kube, Security incidents and trends in SCADA and process industries, *The Industrial Ethernet Book*, 2007.

[11] Kaspersky, "Industrial control systems and network protection using kaspersky endpoint security for business." [Online]. Available: http://media.kaspersky.com/pdf/Brochure_KESB_IndustrMode_Eng_WEB.pdf, 2014.

[12] M. Luallen, "Breaches on the rise in control systems: a SANS survey." [Online]. Available: http://www.sans.org/reading-room/whitepapers/analyst/breaches-rise-control-systems-survey-34665, April 2014.

[13] Kaspersky, "Cyberthreats to ICS systems – you don't have to be a target to become a victim." [Online]. Available: http://media.kaspersky.com/en/business-security/kaspersky-ics-protection-whitepaper.pdf?icid=it-IT:ent-content, 2014.
[14] H. Patzlaff, "D7. 1 preliminary report on forensic analysis for industrial systems," EU, Crisalis Project, Tech. Rep., [Online]. Available: http://www.crisalis-project.eu/sites/crisalis-project.eu/files/crisalis_deliverable-D7.1.pdf, 2011.
[15] I. Garitano, R. Uribeetxeberria, and U. Zurutuza, "A review of scada anomaly detection systems," in *Soft Computing Models in Industrial and Environmental Applications, 6th International Conference SOCO 2011*. Springer, 2011, pp. 357–366.
[16] T. Morris and K. Pavurapu, "A retrofit network transaction data logger and intrusion detection system for transmission and distribution substations," in *Power and Energy (PECon), 2010 IEEE International Conference on*. IEEE, 2010, pp. 958–963.
[17] R. Shayto, B. Porter, R. Chandia, M. Papa, and S. Shenoi, "Assessing the integrity of field devices in modbus networks," in *Critical Infrastructure Protection II*. Springer, 2008, pp. 115–128.
[18] U. D. of Homeland Security, "US-CERT incident response summary report 2009–2011." [Online]. Available: https://ics-cert.us-cert.gov/sites/default/files/documents/ICS-CERT%20Incident%20Response%20Summary%20Report%20(2009-2011)_S508C.pdf, Accessed August 4, 2016.
[19] E. Lebanidze, "Guide to developing a cyber security and risk mitigation plan." [Online]. Available: https://www.smartgrid.gov/files/CyberSecurityGuideforanElectricCooperativeV11-21.pdf, NRECA/ Cooperative Research Network, Tech. Rep., 2011.
[20] B. Galloway and G. P. Hancke, "Introduction to industrial control networks," *Communications Surveys & Tutorials, IEEE*, vol. 15, no. 2, pp. 860–880, 2013.
[21] R. Mitchell and I.-R. Chen, "A survey of intrusion detection techniques for cyber-physical systems," *ACM Computing Surveys (CSUR)*, 2014, vol. 46, no. 4, p. 55.
[22] D. Hadžiosmanović, D. Bolzoni, and P. H. Hartel, "A log mining approach for process monitoring in scada," *International Journal of Information Security*, 2012, vol. 11, no. 4, pp. 231–251.
[23] A. Kleinmann and A. Wool, "Accurate modeling of the siemens s7 SCADA protocol for intrusion detection and digital forensic," *Journal of Digital Forensics, Security and Law*, 2014, vol. 9, no. 2, pp. 37–50.
[24] M. Cheminod, L. Durante, L. Seno, and A. Valenzano, "Automated analysis of access policies in industrial plants," in *Proceedings of the 2nd International Symposium on ICS & SCADA Cyber Security Research 2014*, ser. ICS-CSR 2014. UK: BCS, 2014, pp. 43–52. [Online]. Available: http://dx.doi.org/10.14236/ewic/ics-csr2014.6, 2014.
[25] A. Wedgbury and K. Jones, "Automated asset discovery in industrial control systems – exploring the problem," in *3rd International Symposium for ICS & SCADA Cyber Security Research 2015, ICS-CSR 2015, 17–18 September*

2015, University of Applied Sciences Ingolstadt, Germany, ser. Workshops in Computing, H. Janicke and K. Jones, (Eds). BCS, [Online]. Available: http://ewic.bcs.org/content/ConWebDoc/55101, 2015.

[26] L. Lemaire, J. Lapon, B. De Decker, and V. Naessens, "A sysml extension for security analysis of industrial control systems," in *Proceedings of the 2nd International Symposium on ICS & SCADA Cyber Security Research 2014*. BCS, 2014, pp. 1–9.

[27] D. Beresford, "Exploiting siemens simatic s7 plcs," *Black Hat USA*, 2011.

[28] C. Consortium, "Cybersecurity on scada: risk prediction, analysis and reaction tools for critical infrastructures," *Project Proposal, PART B*, 2011.

[29] J. Rrushi and R. Campbell, "Detecting cyber attacks on nuclear power plants," in *Critical infrastructure protection II*. Springer, Boston, MA, 2008, pp. 41–54.

[30] W. Gao and T. H. Morris, "On cyber attacks and signature based intrusion detection for modbus based industrial control systems," *Journal of Digital Forensics, Security and Law*, 2014, vol. 9, no. 1, pp. 37–56.

[31] R. R. R. Barbosa, *Anomaly detection in SCADA systems: a network based approach*. University of Twente, Netherlands, 2014.

[32] N. Goldenberg and A. Wool, "Accurate modeling of modbus/tcp for intrusion detection in SCADA systems," *International Journal of Critical Infrastructure Protection*, 2013, vol. 6, no. 2, pp. 63–75.

[33] E. Byres and D. Pederson, "Understanding opc and how it is deployed." Digital Bond. [Online]. Available: https://ics-cert.us-cert.gov/sites/default/files/recommended_practices/OPC%20Security%20WP1.pdf, 2007.

[34] M. Robinson, "The SCADA threat landscape," in *1st International Symposium for ICS & SCADA Cyber Security Research 2013, ICS-CSR 2013, 16–17 September 2013, Leicester, UK*, ser. Workshops in Computing, H. Janicke and K. Jones, (Eds) BCS, [Online]. Available: http://ewic.bcs.org/content/ConWebDoc/51166, 2013.

[35] J. Hunker, B. Hutchinson, and J. Margulies, "Role and challenges for sufficient cyber-attack attribution," *Institute for Information Infrastructure Protection*, 2008.

[36] G. Will and P. John, *Cyber crime and espionage*. Syngress, 2011.

[37] R. Dingledine, N. Mathewson, and P. Syverson, "Tor: The second-generation onion router," in *Proceedings of the 13th Conference on USENIX Security Symposium – Volume 13*, ser. SSYM'04. Berkeley, CA: USENIX Association, pp. 21–21. [Online]. Available: http://dl.acm.org/citation.cfm?id=1251375.1251396, 2004.

[38] D. A. Wheeler and G. N. Larsen, "Techniques for cyber attack attribution," DTIC Document, Tech. Rep., 2003.

[39] M. Dacier, V.-H. Pham, and O. Thonnard, "The wombat attack attribution method: some results," in *Information systems security*. Springer, Berlin, Heidleberg, 2009, pp. 19–37.

[40] M. Bishop, C. Gates, and J. Hunker, "The sisterhood of the traveling packets," in *Proceedings of the 2009 workshop on New security paradigms workshop*. ACM, 2009, pp. 59–70.

[41] M. C. Libicki, *Cyberdeterrence and cyberwar*. Rand Corporation, Santa Monica, CA, 2009.

[42] S. Savage, D. Wetherall, A. Karlin, and T. Anderson, "Practical network support for ip traceback," *ACM SIGCOMM Computer Communication Review*, 2000, vol. 30, no. 4, pp. 295–306.

[43] V. Kuznetsov, H. Sandstrom, and A. Simkin, "An evaluation of different ip traceback approaches," in *Information and Communications Security*. Springer, Heidelberg, Germany, 2002, pp. 37–48.

[44] A. C. Snoeren, C. Partridge, L. A. Sanchez, *et al.*, "Hash-based ip traceback," *ACM SIGCOMM Computer Communication Review*, 2001, vol. 31, no. 4, pp. 3–14.

[45] Z. Gao and N. Ansari, "Tracing cyber attacks from the practical perspective," *Communications Magazine, IEEE*, 2005, vol. 43, no. 5, pp. 123–131.

[46] D. X. Song and A. Perrig, "Advanced and authenticated marking schemes for IP traceback," in *INFOCOM 2001. Twentieth Annual Joint Conference of the IEEE Computer and Communications Societies. Proceedings. IEEE*, 2001, vol. 2, IEEE, pp. 878–886.

[47] A. Belenky and N. Ansari, "Ip traceback with deterministic packet marking," *IEEE communications letters*, 2003, vol. 7, no. 4, pp. 162–164.

[48] C. Gong and K. Sarac, "A more practical approach for single-packet ip traceback using packet logging and marking," *Parallel and Distributed Systems, IEEE Transactions on*, October 2008, vol. 19, no. 10, pp. 1310–1324.

[49] D. D. Clark and S. Landau, "The problem isn't attribution: it's multi-stage attacks," in *Proceedings of the Re-architecting the Internet Workshop*. ACM, 2010, p. 11.

[50] V. Pothamsetty and M. Franz, "Scada honeynet project: building honeypots for industrial networks," Critical Infrastructure Assurance Group (CIAG), Cisco Systems Inc. 2008. [Online]. Available: http://scadahoneynet.souceforge.net/, Accessed August 4, 2016.

[51] PROFIBUS/PROFINET. Digital Bond. [Online]. Available: http://www.digitalbond.com/scadapedia/protocols/profibusprofinet/, 2014.

[52] C. Valli, "SCADA forensics with snort IDS," in *Proceedings of WORLDCOMP2009, Security and Management 2009*. CSREA Press, Las Vegas, NV, 2009, pp. 618–621.

[53] S. Poeplau and J. Gassen, "A honeypot for arbitrary malware on USB storage devices," in *Risk and Security of Internet and Systems (CRiSIS), 2012 7th International Conference on*. IEEE, 2012, pp. 1–8.

[54] K. Wilhoit, "ICS, SCADA, and non-traditional incident response," Trend Micro, 2013.

[55] T. H. Project, "Conpot." [Online]. Available: https://www.honeynet.org/node/1047, November 2013.

[56] L. Spitzner, *Honeypots: tracking hackers*. Addison-Wesley Longman Publishing Co., Inc., Boston, MA, 2002.

[57] G. Wagener, A. Dulaunoy, T. Engel and R. State, "Self adaptive high interaction honeypots driven by game theory," in *Stabilization, safety, and security of distributed systems*. Springer, 2009, pp. 741–755.

[58] N. Falliere, L. O. Murchu, and E. Chien, "W32. stuxnet dossier," *White paper, Symantec Corp., Security Response*, 2011, vol. 5, pp. 13–22.

[59] N. B. Carr, "Development of a tailored methodology and forensic toolkit for industrial control systems incident response," PhD dissertation, Monterey, California: Naval Postgraduate School, 2014.

[60] J. F. Gantz and C. Chute, "The diverse and exploding digital universe: an updated forecast of worldwide information growth through 2011," in *IDC analyze the future*. IDC, 2008. [Online]. Available: http://www.ifap.ru/library/book268.pdf, Accessed August 4, 2016.

[61] M. I. Center, "Apt1: Exposing one of china's cyber espionage units," *Mandian.com*, 2013.

[62] FireEye, "The advanced cyber attack landscape." [Online]. Available: https://www2.fireeye.com/WEB2013ATLReport.html, 2013.

[63] A. Shivraj, "Cyber threat evolution with a focus on SCADA attacks." [Online]. Available: http://digitalstrategies.tuck.dartmouth.edu/digital/assets/images/Project2011_AnantShivray.pdf, May 2011.

[64] R. Langner, "To kill a centrifuge," *Langner Group*, 2013. [Online]. Available: http://www.langner.com/en/wp-content/uploads/2013/11/To-kill-a-centrifuge.pdf, Accessed August 4, 2016.

[65] K. Geers, D. Kindlund, N. Moran, and R. Rachwald, "World war c: understanding nation-state motives behind today's advanced cyber attacks," Technical report, FireEye, Tech. Rep., 2014.

[66] B. Bencsáth, G. Pék, L. Buttyán, and M. Felegyhazi, "The cousins of stuxnet: Duqu, flame, and gauss," *Future Internet*, 2012, vol. 4, no. 4, pp. 971–1003.

[67] H. K. Kalutarage, S. Shaikh, Q. Zhou, A. E. James, "Sensing for suspicion at scale: A Bayesian approach for cyber conflict attribution and reasoning," in *Cyber conflict (CYCON), 2012 4th international conference on*. IEEE, 2012, pp. 1–19.

[68] F. Adelstein, "Live forensics: Diagnosing your system without killing it first," *Communications of the ACM*, February 2006, vol. 49, no. 2, pp. 63–66. [Online]. Available: http://doi.acm.org/10.1145/1113034.1113070

[69] D. Farmer and W. Venema, *Forensic discovery*. Addison Wesley, Upper Saddle River, NJ, 2004.

[70] I. Ahmed, S. Obermeier, M. Naedele, and G. G. Richard III, "Scada systems: challenges for forensic investigators," *Computer*, 2012, vol. 45, no. 12, pp. 44–51.

[71] A. N. Mahmood, C. Leckie, J. Hu, Z. Tari, and M. Atiquzzaman, "Network traffic analysis and SCADA security," in *Handbook of Information and Communication Security*. Springer, Heidelberg, Germany, 2010, pp. 383–405.

[72] K. Nance, B. Hay, and M. Bishop, "Digital forensics: defining a research agenda," in *System Sciences, 2009. HICSS'09. 42nd Hawaii International Conference on*. IEEE, 2009, pp. 1–6.

Chapter 14

Occupant-engaged fast demand response for commercial buildings

Zhen Song[1], *Xianjun Sam Zheng*[1] *and Sanjeev Srivastava*[1]

Abstract

Demand response (DR) refers to the active participation by retail customers in electricity markets, seeing and responding to prices as the prices change over time [1]. Occupants are excluded from traditional commercial building DR control loops. However, our experimental results suggest that motivated occupants can achieve significant load shaving with proper information. The analysis indicates more load shaving potentials with financial incentives.

During the DR events, utility companies send DR requests to commercial building facility managers (FMs), who often send emails to occupants with generic instructions. Unless the buildings are equipped with advanced hardware, office building FMs have limited means to reduce significant peak load in the DR periods. We present a software-based, occupant-engaged fast DR system for commercial office buildings. Our collaborative DR (cDR) module is built upon our collaborative energy management and control (cEMC) platform. Through a web portal, occupants can submit a preferred temperature range and schedule.

In this chapter, we present a novel occupant-engaged collaborative DR system from the perspective of cyber-physical-social systems (CPSSes). Traditional building automation systems (BASes) are designed based on thermal and mechanical requirements, without considering the impact of occupants' psychological motivations, such as peer pressure, social recognition, and gaming experiences. The cDR system is designed with cybernetics and social factor in mind. From the cybernetics perspective, we introduced a *semantic building data model* to enable micro-zoning scenarios. From the social behavior perspective, (1) we developed a *zonal virtual energy meter* algorithm to split the whole building energy consumption into individual occupants and (2) provide an embedded social network, in order to encourage an energy competition game. The effectiveness of the energy game is validated by real building experiments. In addition, we developed a game-theoretical *optimal incentive design*

[1]Siemens Corporation, Corporate Technology, 755 College Road East, Princeton, USA

(OID) algorithm to allocate real financial incentives to individual occupants, and conducted a simulation-based study.

During field experiments on a mid-size office building at Pittsburgh, in 2014 summer, we observed up to 1.6% load reduction for the traditional email-based DR method, and 11–15% load reduction using our cDR system without compromising on comfort. Up to 56.7% load was shed, with an acceptable loss of comfort.

14.1 Introduction

14.1.1 Introduction to demand response

Demand response (DR) refers to the active participation by retail customers in electricity markets, seeing and responding to prices as the prices change over time [1]. DR is one approach for demand side management (DSM). In the legacy power grid system, electricity demand and supply capacity are often unbalanced. On the demand side, the energy users were not motivated to improve the efficiency and reduce emissions of the whole grid. On the supply side, utility companies and independent system operators (IOSs) were not capable to communicate and influence end-users [2]. They are under contractual obligation to provide electricity to the end-users under any condition. Much electricity generation capacities are built for short periods of peak demands.

The social and economic benefits of DR are significant. DR can improve the efficiency of the whole power grid, avoid building peaking power plants, reduce consumer energy bills, reduce utility marginal costs, and achieve significant environmental and economic benefits. We hereby illustrate the potential benefits with a brief overview of today's electricity market in the USA, which will motivate our problem descriptions in the following chapters.

From the economic and environment perspectives, the ideal demand curve shall be perfectly flat, such that the generation capability is used 100% at any time. Therefore, the utility companies can generate "just enough" electricity at constant rate, and all the electricity is consumed by the demand with minimal loss. In practice, short period peak demand is often much higher than average load. A significant benefit of DR is to avoid building more power plants to meet the peak demand that run just *several hours per year* [1]. To meet the peak load, utilities may either run expensive peaking power plants, which are featured with short start up time (measured in minutes) in exchange for high energy cost. The coal or nuclear power plants suffer from slow startup time (hours to half a day), and must be started long before the expected peak load. Therefore, their low energy costs are largely sacrificed. If 3% of the USA peak demands can be shaved, i.e., shifted to off peak hours, 20 GW can be reduced [1,3], which is equivalent to 40 median size coal power plants totaling about 500 MW in capacity. According to the DOE, up to 19% of peak load, or 199 GW, of the USA may be shaved by 2019, if the maximum potential of DR is fully implemented [3]. As shown in Figure 14.1, the peak load of New York is often about 70% higher than the average load [5].

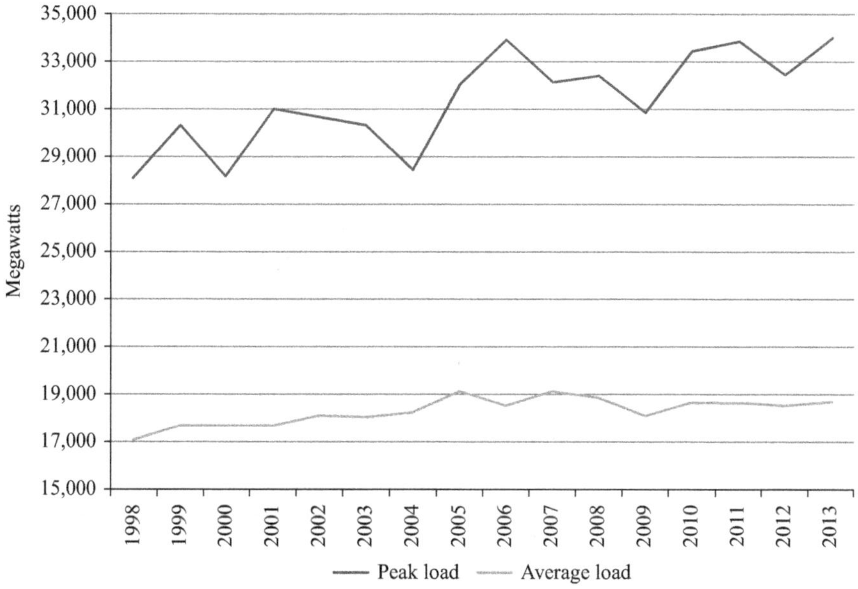

Figure 14.1 New York State peak vs. averaged load [2]

In the USA, and many other developed countries, the importance of DR will increase in the foreseeable future. More renewable energy sources will be available to reduce the average load. However, peak load may not decrease, as shown in the California load prediction curve in Figure 14.2. There are different DR contracts, such as incentive-based, tariff-based programs [6,7], etc. The detailed pricing model depends on utilities or aggregators, and can be rather complicated [6,7]. In terms of time scale, DR can be day-head or in short period [1], as shown in Figure 14.3. The short period DR must be supported by an automated system. This chapter focuses on this scenario.

14.1.2 Commercial buildings as cyber-physical-social systems

14.1.2.1 Human-in-the-loop architecture

Cyber-Physical-Systems (CPS) integrate the dynamics of the physical processes with those of software and communication, providing abstractions, modeling, design, and analysis techniques for the integrated whole [8]. For some systems, human social behaviors are intertwined with the cyber and physical world. These are Cyber-Physical-Social Systems (CPSSes). In this chapter, we consider commercial buildings as CPSSes. In the past decades, researchers have explored novel design paradigms from different perspectives, such as microprocessor design [9] or computer networking. A typical middle-sized commercial building, such as an office, may host 100–500 occupants, with 2–5 roof top units (RTUs), a boiler, and a chiller to serve about 15,000 sqft (1,393.5 m^2) of operation area [10]. Large buildings, such as a hotel, may be of 40,000 sqft (3,716.1 m^2) area, serve thousands of occupants, with

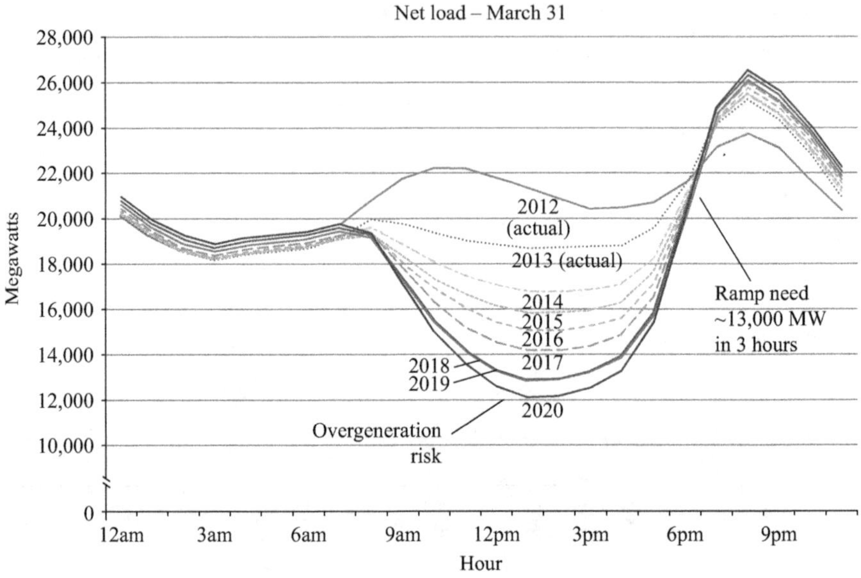

Figure 14.2 Net load curve of California [11]

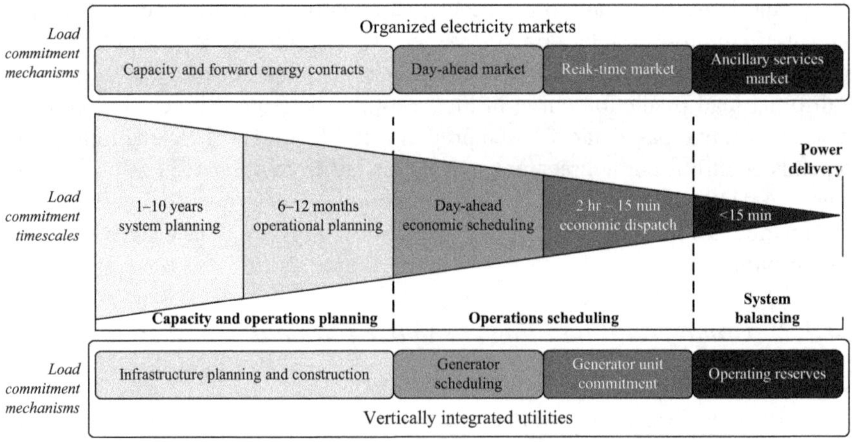

Figure 14.3 DR time scale [3]

more complicated Heating, Ventilation, Air Conditioning (HVAC) and electric systems. Modern buildings are controlled by Direct Digital Control (DDC) systems, i.e., networked computers. The occupants' behavior has major impacts on the building operation and energy usage.

From the perspective of CPSSes, commercial buildings are extended beyond mechanical systems. Figure 14.4 illustrates the proposed system architecture, where we consider occupants as actuators and sensors. The Human–Machine Interface

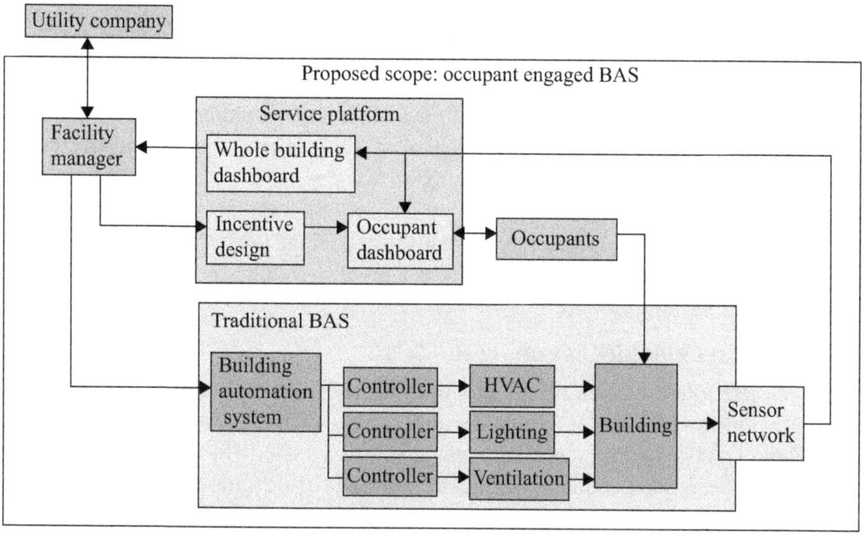

Figure 14.4 BAS from the CPSS perspective

(HMI), i.e., the occupant dashboard, is the major control signal for occupants, with a zonal energy dashboard as a complementary feedback mechanism. The occupant dashboard aggregates and enables visualizing information from two sources: (1) Data collected by sensors, either from the existing sensors in the HVAC system or by retrofitting additional sensors. With enough sensors, it is possible to split the whole building energy consumptions to individual zones; (2) Social activities from their peers or FMs. Peer pressure can either motivate occupants to engage in DR activities or, if improperly designed, discourage the occupants. Human psychology is *subjective*, *uncertain*, and *time varying*. We argue that novel algorithms to connect the subjective psychological factors and objective physics laws are of vital importance to ensure occupants' tendency to adopt environmental friendly resolutions.

14.1.2.2 Cyber, physical and social factors

In the following paragraphs, we describe the architecture from several different perspectives. From the cybernetic perspective, buildings are power grids that are instrumented with networked smart devices, such as BASes, RoofTop Units (RTUs), Air-Handling Units (AHUs), Terminal Equipment Controllers (TECs) for individual Variable Air Volume (VAV) boxes, boilers, chillers, and meters. The control and sensing logics are spread across the equipment and operate on different *modes*. The term mode in the HVAC industry is corresponding to the *state* of a Finite State Machine (FSM) in the computer science domain. For example, each RTU has heating, cooling, and swing season modes. The VAV boxes have day and night modes, in addition to heating and cooling modes.

From the physics perspective, buildings are thermo-dynamic systems as well as electric systems and are governed by physics laws. For energy-efficiency purposes, the operations must be optimized with the physics constraints in mind. For example, the air-conditioning system cannot support significant temperature differences across adjacent thermal zones.

The social factor is ignored by traditional BAS design methodologies, but it is the emphasis of this chapter. The paramount goal of a building is to service its occupants, and a large portion of the energy is utilized for the purpose of thermal comfort. However, traditional methodologies focus on *single* person psychology model, and social behaviors are largely ignored. With more occupants connected to the Internet or intranet, we argue that the social behavior is a pivoting point toward the new building domain CPSS paradigm.

14.1.2.3 Psychological comfort models

As a concrete example, we review the definition of thermal comfort in the age of Internet. Traditionally, HVAC engineers quantify the thermal comfort using Predicted Mean Vote (PMV) or Predicted Percentage of Dissatisfied (PPD). Based on experiments with groups of Danish occupants, Fanger [12] developed the PMV thermal comfort model:

$$\begin{aligned} PMV = {} & (0.303e^{-0.036M} + 0.028)((M - W) - 3.96e^{-8} f_{cl}((t_{cl} + 273)^4 \\ & - (t_r + 273)^4 - f_{cl} h_c (t_{cl} - t_a) - 3.05(5.73 - 0.007(M - W) - p_a) \\ & - 0.0014M(34 - t_a)) \end{aligned}$$

And

$$f_{cl} = \begin{cases} 1 + 0.2 I_{cl}, & I_{cl} < 0.5clo \\ 1.05 + 0.1 I_{cl}, & I_{cl} \geq 0.5clo \end{cases}$$

$$R_{cl} = 0.155 I_{cl}$$

$$h_c = 12.1 V^{0.5}$$

where

- f_{cl} : clothing factor.
- h_c : convective heat transfer coefficient.
- I_{cl} : clothing insulation [clo].[1]
- M : metabolic rate [W/m^2] 115 for all scenarios.
- p_a : vapor pressure of air [kPa].
- R_{cl} : clothing thermal insulation.
- t_a : air temperature [°C].
- t_{cl} : surface temperature of clothing [°C].
- t_r : mean radiant temperature [°C].
- V : air velocity [m/s].
- W: external work (assumed = 0).

PPD is a function of PMV, and it is defined as

$$PPD = 100 - 95e^{(-(0.3353PMV^4 + 0.2179PMV^2))}$$

[1]The clothing insulation is measured in clo, where 1 clo = 0.155 K $\cdot$ m^2 $\cdot$ W^{-1}.

The PMV and PPV capture the scientific relationship from many physical parameters to human psychological comfort level, measured by percentage. They, however, can be too complicated for most engineering practices. Based on PMV, the American Society of Heating Refrigerating and Air-conditioning Engineers (ASHRAE) developed a 7-level thermal sensation scale system and captured the summer and winter comfort zones on a psychometric chart [12]. The ASHRAE method is more practical and widely adopted in engineering practices. In theory, engineers can quantify the expected percentage of satisfied occupants, given a particular indoor temperature and humidity configurations. The PMV is a static thermal comfort model that ignored many potential impact factors, such as clothing, radiation, outdoor temperature, etc. In reality, occupants of a particular building often show diverse thermal preferences as compared to those subjects in the PMV experiments. More dynamic thermal comfort models have been discussed by other researchers [13], but these approaches are all based on predefined survey methods.

PMV was developed in the 1960s, long before the Internet and many Information Technology (IT) systems. Although PMV still provides solid scientific guidance, there are more practical engineering methods for comfort control. The PMV model has more scientific implications than engineering impacts. For example, it is *impractical* for an FM to track the clothing thermal insulation of hundreds of occupants on daily bases. Today's standard practice is to simply define a narrow band of temperature ranges, such as 70–72°F in summers.[2] Today, most occupants in commercial buildings have access to the Internet or the intranet. In this work, we ask occupants to type in their preferences directly via web interfaces. Working independently, several research groups proposed and experimented occupant-engaged thermal control around the same time [14–16], but with a *different design philosophy* in mind. These researchers introduced different temperature set point *voting* mechanisms based on *unregulated voting*, and observed about 10% energy savings in specific experiments [14,15].

14.1.2.4 Motivation of the proposed methodology

We present a method based on software *arbitration*, where many temperature, ventilation set points are computed based on optimization processes. Although social study may indicate energy-saving effects due to room (zonal) temperature set point adjustments, we did not adopt this approach, for the following reasons:

(1) We intend to *achieve significant energy savings without comfort loss*: Our method does not encourage occupant to sacrifice comfort by adjusting the temperature set points. Our method achieves energy savings using occupants' schedule. Although it is scientifically feasible for occupants to be comfortable even with comfort loss due to peer pressure, we do not consider the objective comfort enhancement. In this chapter, we consider thermal comfort as an *objective* metric, which can be measured by sensors.

(2) We utilized VAV box *micro-zoning* for energy savings and load shaving: In fact, we allow larger temperature bands in order to improve comfort, potentially with

[2] As we will explain later in this chapter, due to the static nature of the BAS control logics, even the limited temperature set points are not always achievable.

more energy consumptions. However, the industry's heuristic is that comfort is affordable, i.e., 1°F adjustment corresponds to 1% energy consumption change,[3] assuming reheat coil in the VAV boxes are not triggered due to the temperature adjustments.

(3) Our approach features several layers to achieve *energy saving* and *comfort enhancements* simultaneously: We introduce *occupant calendar* for energy savings, and utilized *softThermostat* for comfort enhancement [17,18]. The VAV box and RTU settings are adjusted on real-time, based on temperature and ventilation arbitrations.

(4) We establish social energy games based on our *energy split* algorithm: With built-in social network features, we encourage occupants to compare their normalized zonal energy savings and load shavings with their peers. The energy split algorithm gives occupants feedbacks on the impacts of their behavior, thereby encouraging energy awareness. Previous studies on energy games [19,20] are not based on energy split algorithms.

14.1.3 Occupant-engaged DR

In today's commercial buildings, DR events are often sent to Facility Managers (FMs), or the BAS managed by the FM, from the utility. An FM is responsible for *accepting* or *rejecting* the DR request, based on the estimated load shaving capability. In today's standard practices, if the DR request is accepted, the FM usually broadcasts the message via emails. Occupants cannot reply to offer feedback, since one FM is often responsible for serving hundreds of occupants, sometimes thousands. Additionally, it is impractical for the FM to *track* occupants' requirements and *validate* their performances. For example, the VAV box settings shall be adjusted, in theory, every time when an occupant leave or enter a thermal zone. In a mid-size commercial building, there might be hundreds of thermal zones and several hundreds of occupants. One FM cannot track every occupant manually. Existing techniques, such as Direct Load Control (DLC) or RTU set points adjustment, allow FMs to shave peak loads. However, without sufficient information, it is difficult for FMs to decide how and when to shave the loads.

The architecture of the traditional scenario is shown in Figure 14.5a. In this scenario, occupants have little control over the comfort settings and there are limited means (phone calls and emails) for them to express their experience and comfort needs to FMs. Meanwhile, existing BASes provide no feedback to the occupant about his energy use behavior and how he or she should behave to support FMs to achieve DR targets.

The one-way, top-down process of building control could lead to two opposite consequences in DR events. Under one circumstance, FMs who implement aggressive energy policies often achieve energy-saving goals with the sacrifice of occupants' comfort, even disrupting occupants from their current work. Under most of other circumstances, FM usually have to relax the energy policies to avoid occupant complains, and often miss the opportunity to achieve the DR target from utility, therefore

[3]This is an estimation method used by mechanical engineers based on their experience.

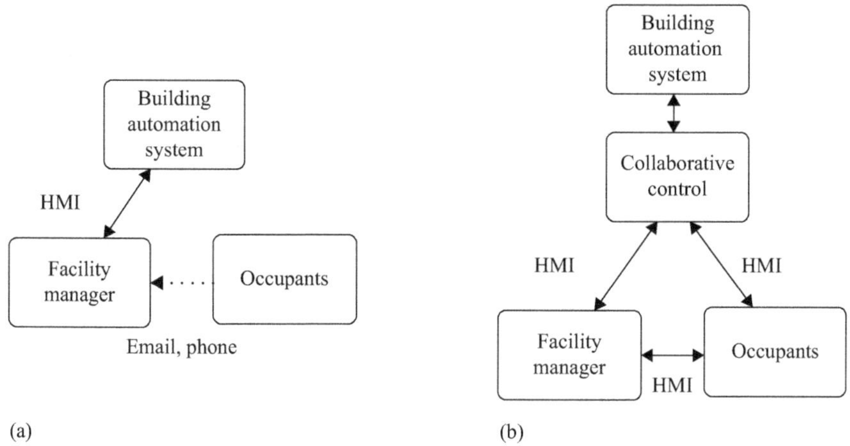

Figure 14.5 Traditional BAS (left) vs. cDR architectures (right)

lose incentives. In brief, accurate and fast DR is difficult without proper feedback information from the occupants.

Figure 14.5b shows the communication flow of our proposed collaborative Energy Management Control (cEMC) software, which has a component to enable a collaborative Demand Response (cDR) scenario. Occupants are engaged in the whole process of the DR event, in order to achieve more load shaving without comfort loss. Each occupant has access to a web portal to directly communicate with the BAS. In case of conflictive requirements, the software automatically solves the conflicts with its software arbitrator. There are three types of conflicts: (1) Conflicts among different occupants, i.e., *inconsistent requirements*, (2) Occupants' requirements may be in conflict with the HVAC system, or other mechanical and electronic systems, i.e., *unfeasible requirements*; (3) Occupants' preferences may be in conflict with the FM energy policy, i.e., *energy-inefficient requirements*. The outputs of the cEMC systems are a number of set points *without conflicts* with the BAS.

In order to actively engage occupants, the occupant web portal was designed to provide most *relevant information* and an *intuitive* HMI to reduce their learning and using efforts. In traditional DR, occupants are either informed about the *total building energy consumption* and are not provided with any energy context information at all. In the cDR scenario, we split the total building energy consumption into individual zone, with a new algorithm. This novel *zonal energy dashboard* allows each occupant to relate his/her behavior with the energy impacts. More importantly, the combination of our *occupant web portal* and our *zonal energy dashboard* algorithm enables *energy competition games*. Technically, FM can split the DR incentive [19,20], such as rebate or discount, to motivate occupants to participate in the DR events.

The system is designed to enable three methods for collaboration at different levels:

(1) *Software arbitration*: Occupants can configure the web portal according to their needs, without considering other constraints. The software arbitrator component automatically resolves inconsistent requirements, unfeasible requirements, and

energy-inefficient requirements. For example, if the temperature set points among different zones are higher than what the HVAC system can support, the *temperature arbitrator* will adjust the set points to ensure that they are feasible, consistent, and comply with FM's regulations.

(2) *Social networking and energy games*: If the conflict cannot be resolved by the software arbitrator, occupants can discuss within the embedded social network. For example, if the occupants in a thermal zone have different temperature preferences, the arbitrated set point is the average of the set points. In the occupant web portal, one can read the energy consumption of their individual zone, and the ranking of their zone.

(3) *FM arbitration*: If the occupants cannot reach an agreement with the aforementioned methods, they shall raise the issue to the FM. This manual arbitration is the last option. In our field experiments, it never happened.

14.2 Collaborative, occupant-engaged fast demand responses

14.2.1 The software architecture of cEMC

FMs receive notification of fast DR requests 15 minutes to several hours ahead of those events. As a comparison, "day ahead DR" notifications are issued by 5 p.m. of the day ahead of the event [10]. Automatic control is an enabling technology for fast DR scenarios. It is impossible for an FM to manually investigate hundreds of occupants' shedable, non-urgent, electricity demands in a couple of minutes. If an FM accepts the fast DR request yet failed to achieve the load sheding target, then financial penality may be applied. In this chapter, we present the cDR system to assist the FM. We must highlight that *energy efficiency* (EE) is the foundation for DR technologies. Our cDR system is built upon our cEMC platform. In terms of software implementation, cDR is one of the components of cEMC. Featured with Service-Oriented Architecture, the components in cEMC are loosely coupled and communicate via web services.

The cEMC software is a vendor-independent software platform which extends the functionality of an existing BAS by means of active occupant engagement in energy management and comfort control. Figure 14.6 shows the system architecture, where communication between cEMC and the BAS is provided by an industry standard open communication protocol, BACnet. At the center of the cEMC system is a device called the "cEMC controller," which sits on the network to interconnect building occupants, the FMs, and the BAS. The cEMC controller applies techniques for both occupancy-based energy management and micro-zoning building environmental control to maximize or optimize building EE and to manage occupant-engaged DR events. Field engineers type in building-specific information, i.e., the topological relationships between floors and the zones, etc. in the building configuration database. The configuration data establish a metamodel for *hardware abstraction*. The application logics can be wrapped as *building-independent apps*, such as temperature arbitration app or vetilation arbitration app. Each occupant has a *personal agent* to track his/her personal preferences. Each thermal zone has a *zone agent* to calculate zone set points, based on the outputs from the personal agents. The collaborative

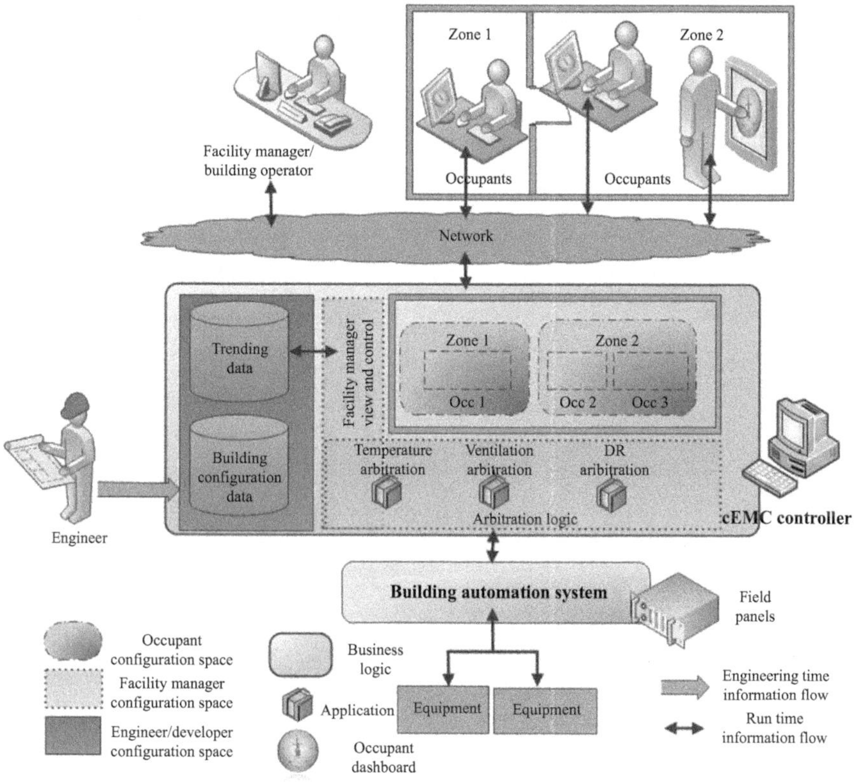

Figure 14.6 The system diagram for cEMC

HMIs provide an holistic view to present the energy impacts of their behaviors and decisions. For example, instead of showing the whole building energy consumption, the occupant HMI visualizes the energy consumption of the individual zone. In addition, the occupants can directly submit their personal comfort preferences and office schedule, and receive action items for DR events. Simultaneously, the FM can manage and respond to such occupant requests, configure energy policy as rules, and issue DR commands from the FM Dashboard.

14.2.2 Design for engaging human–machine interface: Occupants' Dashboard and FM's HMI

User experiences are of vital importance to win occupants' cooperation. We designed and developed the occupant HMI for building collaborative control that allows interactive communications and collaborations between occupants and FMs. Through this HMI, FMs can refine energy policies in ways that optimize energy saving and improve occupants' comfort. The resulting Occupant Dashboard includes a "softThermostat" and a DR portal. The softThermostat allows occupants to set up their comfort preference as a range, input their out-of-office schedules, command their desired

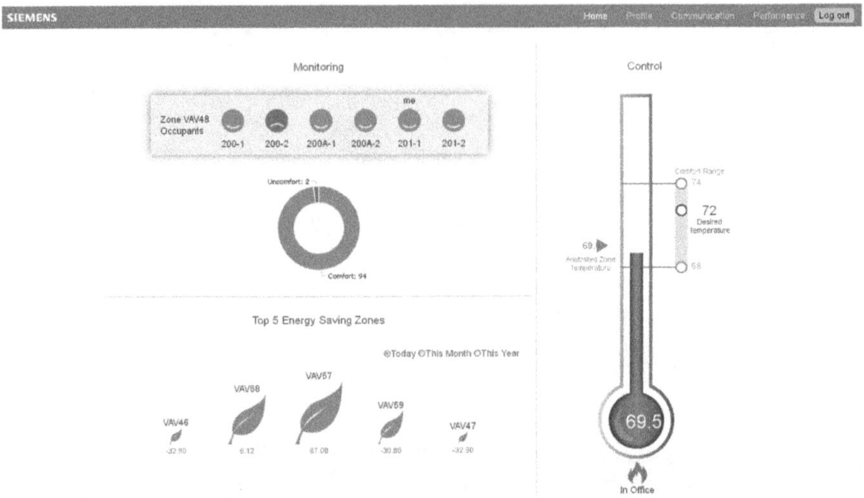

Figure 14.7 The softThermostat home view

zone temperature settings, and monitor their environmental conditions and individual energy performances. The DR portal displays DR information from FMs including detailed DR instructions customized for the different groups of occupants. The FM Dashboard supports FMs by allowing them to monitor building energy performance, occupant comfort, and zone-level energy use, and to initiate DR events.

The homepage of the Occupant's Dashboard shown in Figure 14.7 includes a comfort/energy monitoring panel (on the left side) and a control panel with a display that looks like a mercury-in-glass thermometer (on the right side). Previous research [21] has shown that intuitive design and visualization of real-time energy consumption patterns can trigger more ecologically responsible behavior from building occupants. An occupant can set his/her preferred temperature range (set from 68°F to 74°F in this figure) and desired temperature (set at 72°F in this figure) using the "Control" panel, which also shows sensor measured current zone temperature in the thermometer bulb (reading 69.5°F in the figure). This design style is a novel balance between skeuomorphism and flat design. The key idea of skeuomorphism is to create HMI mimicking daily objects, such as the traditional mercury thermometer, therefore reducing the learning effort for users. Although digital objects are not physical objects, they display information as if they were familiar devices. In our case, a typical user naturally reads the number "67.5" at the bottom of the thermometer as room temperature, even though there is no descriptive text, such as "°F" or "room temperature."

The design challenge in our case has been to properly communicate novel concepts that are not seen in standard room thermostats. The concept of temperature and ventilation arbitration is new to occupants. Traditional thermostats are shared among occupants and do not have calendars. In order to communicate with occupants, we

Figure 14.8 Occupant zone view

bind these new objects with familiar experiences. The inputs and outputs are split to different sides of the thermometer. All the inputs are on the right hand side with circle handles as visual indicators to invite occupants to draw them, and they can be dragged. On the left hand side, the only output is indicated with a triangle. When users change their input on the HMI, a short animation is rendered and the triangle moves to the proper location, i.e., the zone temperature set point *after arbitration*. This indicates that a cause–result relationship exists between the user input and the system output. In practice, we observe that users quickly distinguish the inputs and outputs, and they often asked how the system calculated the arbitrated temperature. Without either a manual or instructions, most users can identify the "arbitrated temperature" as the key feature of this thermostat and that its value is calculated by some underlying algorithm. These are the important concepts we want to communicate with the users via our intuitive design.

Figure 14.9 is the calendar view of each occupant, where the repetitive working schedule and out-of-office calendar are shown side by side. Occupants need to input their routine schedules information when they first sign onto the system. After this initial set up, they can make quick changes or override the calendar easily with only a few mouse clicks. Occupant calendars are used by the cDR controller to relax the zone temperature settings and reduce air flow when the occupants are out of their normal work space. In order to improve energy-efficient behaviors, we also introduce a personal energy performance view, shown in Figure 14.10, where the occupants are informed of the energy consequences of their behaviors and can take appropriate actions to eliminate energy waste.

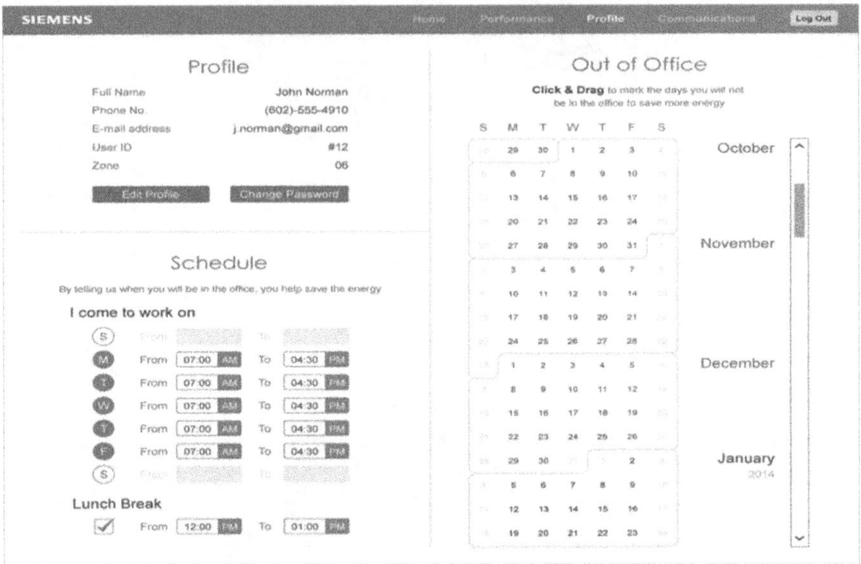

Figure 14.9 Occupant calendar view

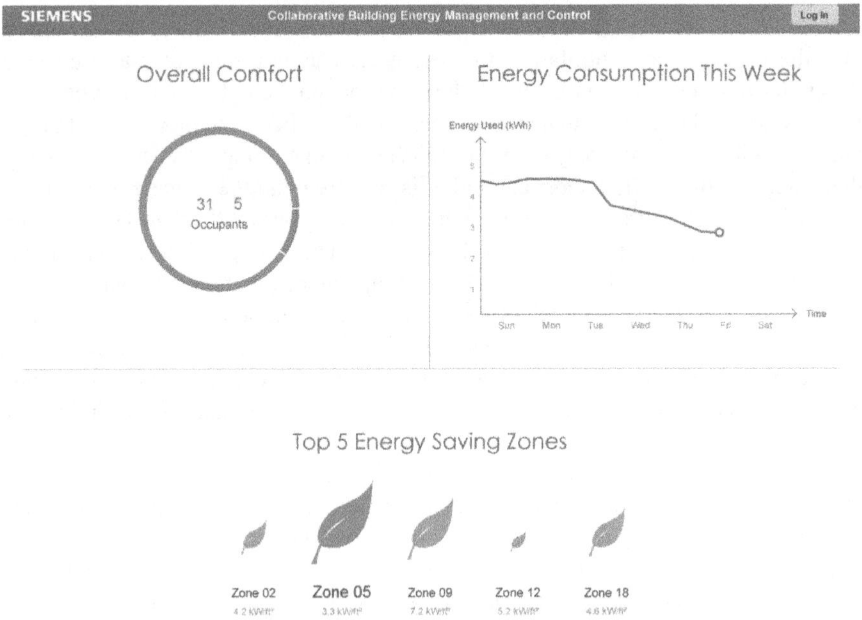

Figure 14.10 Occupant overall comfort and energy performance view

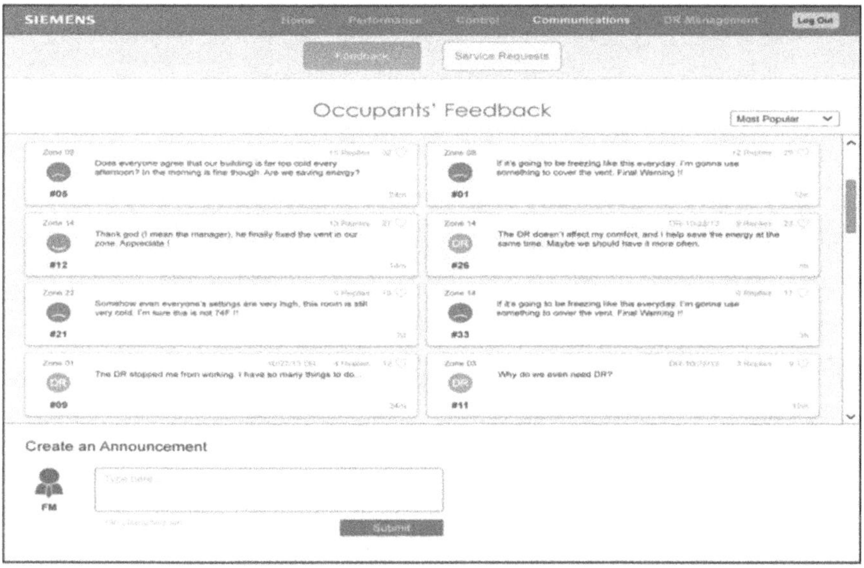

Figure 14.11 Embedded social network portal

Based on occupants' inputs of preference and schedule, both the Occupant's and the FM's HMIs can also visualize the energy-efficiency information as calculated by the cEMC controller and, in addition, the comfort information based on the match level between the current condition information and the occupants' preferences. The HMIs also provide a social networking mechanism for occupants and FMs, in order to support easy and efficient communications and collaborations as well as encourages energy saving behaviors (see Figure 14.8 and Figure 14.11). One highlight feature of the Occupants' HMI is its comfort arbitration through which occupants can provide their immediate feedback for building conditions. Through this Advanced Collaborative HMI, FMs can also easily create a DR event and effectively disseminate the information to all occupants.

A key feature to distinguish our cEMC technology from other *occupancy-based* control systems is that the proposed system is a pure software implementation. Occupants need to voluntarily submit their calendar and preference information. Even though we developed a user friendly HMI to expedite this process, occupants still need about 5 minutes to setup the system and several seconds to adjust the preferences for daily usage. This raises a question: What if the occupants are not cooperative? We address this fundamental challenge with two approaches:

(1) We developed an algorithm to *split the whole building energy bill* into individual zones. From the psychology perspective, we are more interested in knowing the information and feedback regarding our behavior. Therefore, our hypothesis is that occupants are curious to know about their energy impacts, as opposed to the total energy consumption of the whole building.

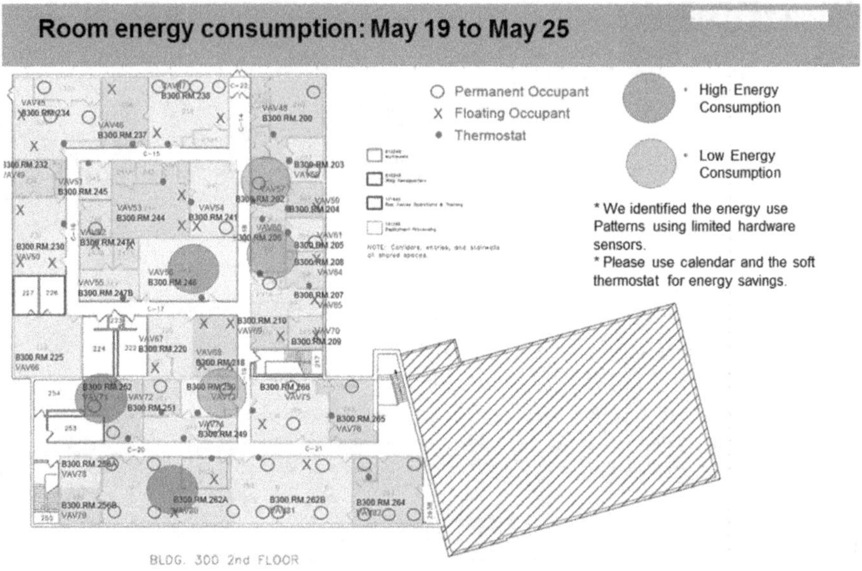

Figure 14.12 Zonal energy consumption

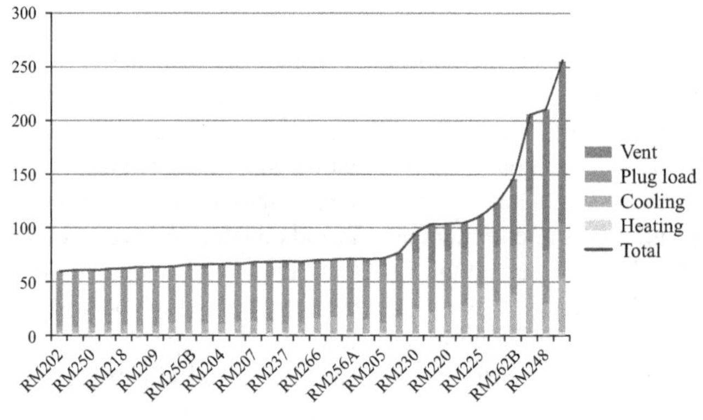

Figure 14.13 Energy consumption decomposition

(2) We establish an energy game, where occupants *compete and collaborate* for the best energy ranking. Each week, we present the energy consumption in a graphic chart as shown in Figure 14.12. Each zone has a different number of occupants and a different area. We normalized the differences and computed the energy intensity per person. The green zones are more energy efficient than the red zones. We also provided the ranking on individual zones with detailed energy consumption, as shown in Figure 14.13, where the room numbers are on the *x*-axis. Due to limited space, some room numbers are not visible. In this game, occupants need to collaborate with the colleagues in their zone, and compete with other zones. The suggested collaboration methods include:

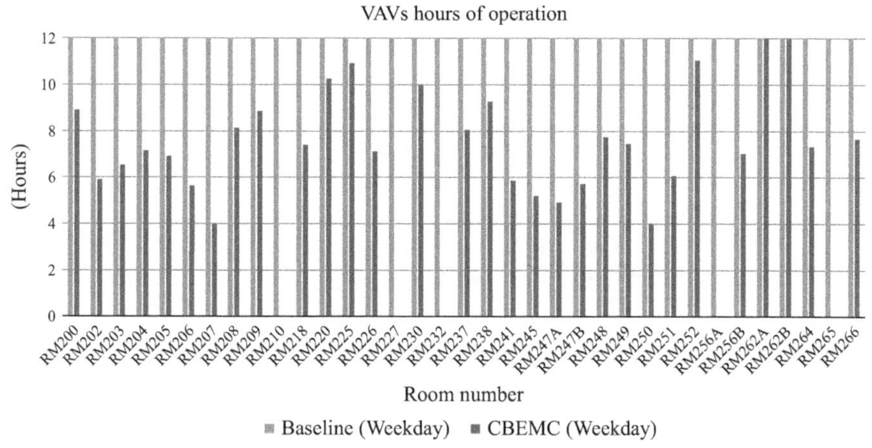

Figure 14.14 Comparisons of air-conditioning hours

(1) *Use the calendar*: Remind each other to adjust the calendar once per week. (2) *Control natural light*: For example, avoid additional thermal load by closing curtains to avoid too much sunshine during summer afternoons. (3) *Ecofriendly dressing code*: In summer, wearing clothes with less thermal insulations, e.g., short sleeves, in order to keep higher temperature set points. Of course, take the reverse actions in winters.

In summary, we designed the engaging HMI based on the following hypotheses:

Hypothesis 1: Occupants are more engaged if the visualized data are directly related to their behaviors.
Hypothesis 2: Occupants are more engaged if the energy game has both competitive and collaborative factors.

We have a quantitative metric to evaluate the hypothesis. When we deployed the software to an office building at Pittsburgh, we observed significant reduction during the office hours, which means that occupants used the system. From September 15 to September 21, 2014, we successfully reduced the air-conditioning hours from 12 hours per day per zone to 4–12 hours per zone, averaged about 8 hours per weekday per zone, as shown in Figure 14.14. Due to the limited experiment time, we did not conduct experiments to evaluate the impacts of each hypothesis. However, the experimental data indicates that occupants used the HMI and reduced conditioning hours by more than 50%. The HMI designed based on those hypothesis has fulfilled its goal.

14.2.3 Temperature and ventilation arbitrations using convex optimization

We first present our *temperature arbitration* algorithm based on convex optimization. Then, we briefly describe the principles of the *ventilation arbitration* method.

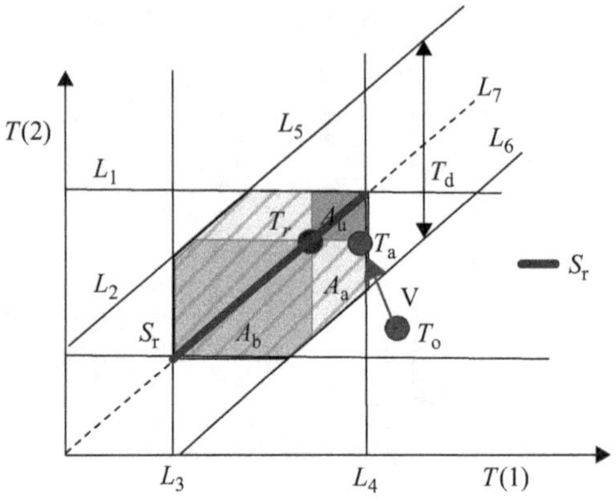

Figure 14.15 Temperature set point correlation diagram for two adjacent zones

One key design philosophy is to provide ensured energy savings and load shedding via arbitration algorithms, based on solid physics and social regulations. Today, control parameters in commercial building are hardcoded during the commissioning phase. For example, the air flow set point shall be calculated based on the number of occupants in each thermal zone. In theory, the flow set points shall be adjusted when occupants arrive or leave their desks. Siemens' VAV box controller, namely Actuating Terminal Equipment Controller (ATEC) controller, has 99 parameters and about 30 of them can be changed during run time, via the BACnet connections. For the experimental site, there are 30 ATEC controllers to serve about 60 occupants. In other words, there are around 3,000 parameters for the FM to monitor and about 900 of them should be adjusted every 15 minutes. In theory, every time an occupant leaves his/her desk for dozens of minutes, the FM shall adjust about 3–6 parameters out of the 3,000 parameters, which is an unaffordable effort. In commercial buildings, these parameters are usually established before occupants move in, and remain fixed for several years. Our cEMC platform collects occupants' preferences and optimizes HVAC parameters automatically. In the following paragraphs, we present our temperature arbitration algorithm in details. Due to limited space, we only include a brief overview of the ventilation arbitration algorithm.

The motivation of the temperature arbitration algorithm is to *minimize discomfort*. In order to quantitatively analyze the discomfort, we propose a novel "set point correlation diagram" as shown in Figure 14.15. The *y*-axis depicts the lower and upper temperature limits of one zone, i.e., L_2 and L_1. The *x*-axis shows the temperature set points of another zone. The graphics can be generalized to show the correlations of any number of zones, as shown in Figure 14.16b. As shown in Figure 14.15, the *solution of zone temperature arbitration is not unique*. Given the initial occupants' temperature preference, $\mathbf{T_o}$, not within the *acceptable temperature area*, $\mathbf{A_a}$, there are an infinite number of possible methods to map occupants' preference to *arbitrated temperature set point*, $\mathbf{T}_a$, such that $\mathbf{T}_a$ is within $\mathbf{A_a}$. Therefore, it is valuable to have an

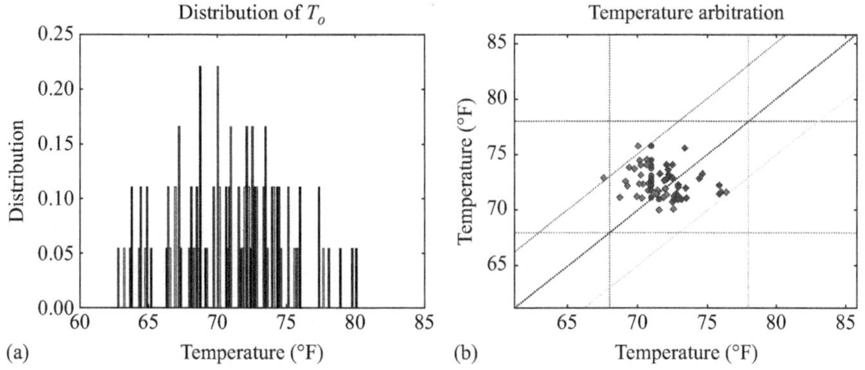

Figure 14.16 Optimal temperature arbitration. (a) Simulated occupants' preference. (b) Temperature arbitration. Gray diamonds: before arbitration. Black circles: after arbitration

optimization method, which finds the best solution based on certain criterion defined by the user. We present a function f, which maps a given $\mathbf{T_o}$ to a $\mathbf{T}_a$ inside $\mathbf{A_a}$ with the least variation $\mathbf{v}$, which is defined as $\mathbf{v} := \mathbf{T}_a - \mathbf{T_o}$. In other words, we implemented a function $\mathbf{T}_a = f(\mathbf{T_o})$, such that $\mathbf{T}_a \in \mathbf{A_a}$ and $\|\mathbf{v}\|$ is minimized. For more energy saving, we introduce a parameter, T_r, the *reference temperature* from the FM, where T_r is a parameter of f, i.e., $\mathbf{T}_a = f(\mathbf{T_o}; T_r)$. To illustrate the concept, we draw a trivial two adjacent zone scenario shown in Figure 14.6, where each zone has one occupant. The occupants' temperature reference is a vector $\mathbf{T_o}$, defined by $\mathbf{T_o} = [T_o[1] \;\; T_o[2]]^T$, where $T_o[1]$ and $T_o[2]$ are the preference from occupants 1 and 2, respectively. The zone temperature set points after arbitration are $\mathbf{T_a} = [T_a\mathbf{[1]} \;\; T_a\mathbf{[2]}\,]^T$, $T_a[1]$ and $T_a[2]$ are the zone set points after arbitration. Line L_1 is the upper limit on $T_a[1]$, such as 78°F (25.6°C). The acceptable region is the area below this line, i.e., $T_a\mathbf{[1]} \leq l_u$, where l_u is the upper limit and $l_u = 78$ in this case. Line L_2 is the lower limit, such as 68°F (20°C). The acceptable region is above this line, i.e., $T_a[1] \geq l_l$, where l_l is the lower limit and $l_l = 68$ in this case. Lines L_3 and L_4 are the lower and upper limits on $T_a[2]$ and they can be posed as a constraint of the same inequality form. The maximum temperature difference T_d imposes two inequality conditions, i.e., $\mathbf{T}_a[1] - \mathbf{T}_a[2] \leq T_d$ (i.e., line L_6) and $\mathbf{T}_a[2] - \mathbf{T}_a[1] \leq T_d$ (i.e., line L_5). In the figure, the area that satisfies all the *fixed temperature constraints* is $\mathbf{A_a}$. In order to give FMs a tuning knob to balance between further energy saving and comfort, we allow FM to change T_r based on a schedule. In summer, the additional constraints are $\mathbf{T}_a[1] \geq T_r$ and $\mathbf{T}_a[2] \geq T_r$, which is the convex upper area, $\mathbf{A_u}$. In winter, the constraints are $\mathbf{T}_a[1] \leq T_r$ and $\mathbf{T}_a[2] \leq T_r$, where the convex area $\mathbf{A_b}$, at the bottom of $\mathbf{A_a}$, is the valid area for $\mathbf{T}_a$. FMs can pick any T_r within the range of S_r.

Since the notations are written in linear algebra, we can generalize the problem for an arbitrary number of zones and occupants and formulate it as a mathematical optimization problem, specifically, a convex quadratic programming problem [22], which is formulated as follows:

$$\text{Min } \|\mathbf{v}\|_2$$

$$\text{Subject to } \mathbf{G}\mathbf{v} \leq \mathbf{h}$$

where $\mathbf{v} := \mathbf{T}_a - \mathbf{T_o}$; $\mathbf{v} \in \Re^n$; $\|\mathbf{v}\|_2$ is the 2-norm of vector $\mathbf{v}$; n is the number of zones; $\mathbf{T}_a \in \mathbf{A}_u$; $\mathbf{T_o}$ is the single occupants temperature set points in a zone for Figure 14.15. We construct matrix $\mathbf{G}$ and vector $\mathbf{h}$ to capture the constraints applied to $\mathbf{A}_u$ or $\mathbf{A}_b$in Figure 14.6. The following equations are immediate:

$$\begin{bmatrix} sI \\ I \\ -I \\ \mathbf{Z}_1 \\ \mathbf{Z}_2 \\ \vdots \\ \mathbf{Z}_{\mathbf{n}-1} \end{bmatrix} \mathbf{T_a} \leq \begin{bmatrix} sT_r \cdot \mathbf{1} \\ l_u \cdot \mathbf{1} \\ -l_l \cdot \mathbf{1} \\ T_d \cdot \mathbf{1} \\ T_d \cdot \mathbf{1} \\ \vdots \\ T_d \cdot \mathbf{1} \end{bmatrix}, \text{ so that } \begin{bmatrix} sI \\ I \\ -I \\ \mathbf{Z}_1 \\ \mathbf{Z}_2 \\ \vdots \\ \mathbf{Z}_{\mathbf{n}-1} \end{bmatrix} \mathbf{v} \leq - \begin{bmatrix} sI \\ I \\ -I \\ \mathbf{Z}_1 \\ \mathbf{Z}_2 \\ \vdots \\ \mathbf{Z}_{\mathbf{n}-1} \end{bmatrix} \mathbf{T_o} + \begin{bmatrix} sT_r \cdot \mathbf{1} \\ l_u \cdot \mathbf{1} \\ -l_l \cdot \mathbf{1} \\ T_d \cdot \mathbf{1} \\ T_d \cdot \mathbf{1} \\ \vdots \\ T_d \cdot \mathbf{1} \end{bmatrix},$$

where s is 1 for winter and -1 for summer; $\mathbf{1}$ is an all one column vector; We also define a *zone permutation matrix* $\mathbf{Z_i}$ as

$$\mathbf{Z_i} = \begin{bmatrix} 0 & \dots & 1 & -1 & 0 & \dots & 0 \\ 0 & \dots & -1 & 1 & 0 & \dots & 0 \\ 0 & \dots & 1 & 0 & -1 & \dots & 0 \\ 0 & \dots & -1 & 0 & 1 & \dots & 0 \\ \vdots & \vdots & \vdots & \vdots & \vdots & \vdots & \vdots \\ 0 & \dots & -1 & 0 & 0 & \dots & 1 \end{bmatrix}, \\ \qquad \uparrow i - th\ column$$

where the non-zero entries start from the i-th column.

So, we have

$$\mathbf{G} = \begin{bmatrix} I & I & -I & \mathbf{Z}_1 & \mathbf{Z}_2 & \dots & \mathbf{Z}_{\mathbf{n}-1} \end{bmatrix}^T$$

$$\mathbf{h} = \mathbf{G} \cdot \mathbf{T_o} + \begin{bmatrix} T_r \cdot \mathbf{1} & l_u \cdot \mathbf{1} & -l_l \cdot \mathbf{1} & T_d \cdot \mathbf{1} & T_d \cdot \mathbf{1} & \dots & T_d \cdot \mathbf{1} \end{bmatrix}^T.$$

To solve this convex optimization problem, we use an open source Python library called CVXOPT [22]. Without cEMC, occupants are allowed to change the temperature set points by about 1–2°F (0.6–1.1°C). After cEMC installation, each occupant can type in their preference from 68 to 78°F (20.0–25.6°C). The maximum temperature difference after arbitration is about 5°F (2.8°C). If the occupant's preferred temperatures are subject to a Gaussian distribution with the expectation at 72°F (22.2°C) (Figure 14.16a), $\mathbf{T}_r$ is 71°F (21.7°C), then the arbitrated zone temperature set points

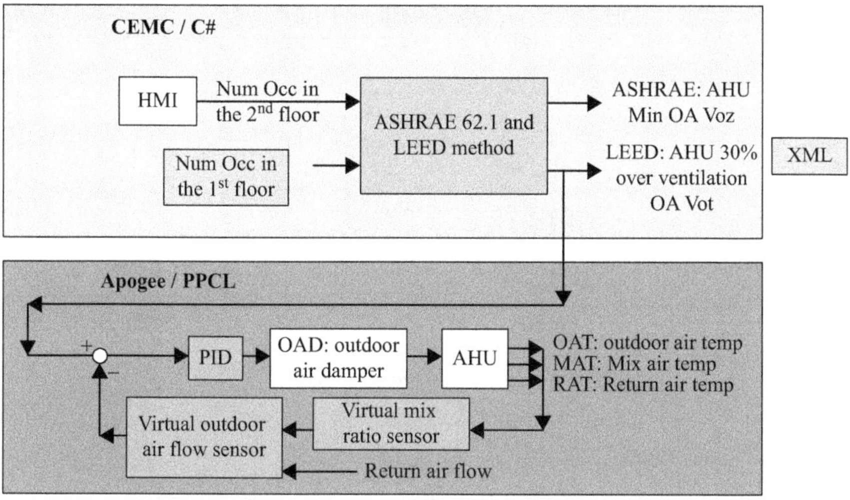

Figure 14.17 Ventilation arbitration system diagram

are shown in red dots in Figure 14.16b. The norm, max, and mean of temperature variation are the following:

$$\|\mathbf{v}\|_2 = 2.6, \max(|\mathbf{v}|) = 1.8, \mathit{mean}(|\mathbf{v}|) = 0.45.$$

Under this case, with relatively small shifts on occupants' temperature preferences, all temperatures are proposed within the area defined by facility manager, i.e., $\mathbf{A_a}$ in Figure 14.15.

The mentioned occupants' preference is hypothetical, *but it shows an important difference between arbitrated software thermostat and a hardware thermostat*: the software thermostat allows a wider temperature band. (1) When the FM allows 2°F (1.1°C) adjustment ranges on hardware thermostat, the maximal variation among, for instance, 10 terminal box temperature set points is also 2°F (1.1°C). (2) With optimal set point arbitration, each occupant experiences *similar* temperature variations, i.e., max 2°F (1.1°C) change and 0.45°F (0.25°C) shift in average, but the temperature set points difference among 10 terminal boxes *can be much larger*. In this example, it is 5°F (2.8°C). The wider temperature band allows extra energy saving opportunities.

Due to the limited space, we will only describe the principle of the proposed *ventilation arbitration* algorithm. The ventilation arbitration diagram is shown in Figure 14.17. The cEMC calculates the ventilation related set points, such as static pressure, damper positions, etc., and sends the points to individual ATECs and RTUs. It is a hierarchical control system with two layers. Both ASHRAE 62.1 and LEED standards specify the minimal fresh air required by each occupant, i.e., 5 cubic feet per minute (CFM) of outdoor air. The LEED standard is more conservative and demands 30% more fresh air than the ASHRAE 62.1 standard. Based on occupant's calendar, the *ventilation arbitration* algorithm calculates the proper ventilation rates

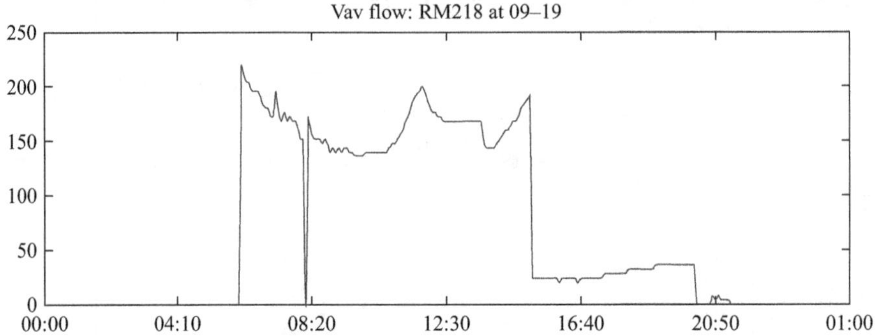

Figure 14.18 Room 218 VAV flow

Table 14.1 DR modes

DR mode	Summary	Occupant engagement	Temperature arbitration	Ventilation arbitration	Manual plug load control
1	Email	No	No	No	No
2	Ensure comfort	Yes	Yes	Yes	No
3	Less comfort	Yes	Yes	Yes	Yes
4	Loss of comfort	Yes	No	No	Yes

for breathing air (fresh air) and for the conditioning air (circulated air for temperature and humidity control). Since there are intrinsic couplings among the breathing air and the conditioning air, the ventilation arbitration algorithm needs to find a trade-off between energy saving and hygiene, i.e., fresh air. Figure 14.18 is the real flow meter reading that we collected from an office building at Pittsburgh. The flow rate is adjusted in real time according to the occupancy rate in the zone. After the occupants left the zone, the damper was almost closed, only allowing a small infiltration flow.

14.2.4 Collaboration for DR scenarios

In this section, we introduce four DR modes, with mode 1 the baseline, i.e., standard email-based DR scenario. DR modes 2, 3, and 4 are cDR scenarios with the following differences. In DR mode 2, the temperature and ventilation are automatically adjusted to the extreme set points within the comfort zone, e.g., the zone set points are set to the upper limit of the occupants' preferences in summer; provided that the temperature would not trigger VAV reheat. In mode 2, occupants are not required to shut down their plug loads. In DR mode 3, yet occupants received a list of action items to shut down plug loads, on top of behaviors of mode 2. In mode 4, the RTU is shut down temporarily, with minimal ventilation available. Thanks to thermal inertia, the indoor temperature will not change dramatically. Therefore, the comfort loss is still limited. The comparisons of the different DR modes are shown in Table 14.1.

The detailed settings of each mode are summarized in the following bullet points:

- DR mode 1
  Scenario: traditional email-based notification. Not controlled by cEMC. On occupants HMI, there are optional choices for turning off the plug loads.

  DR actions:
  1. Increase VAV values
  2. Shutdown plug loads (optional)
- DR mode 2
  Scenario: cEMC takes control. The RTU is optimized to follow occupants' settings. The room temp slightly increased. Optional choice for turning off the plug loads.

  DR actions:
  1. Increase VAV values
  2. Shutdown plug loads (optional)
- DR mode 3
  Scenario: Unified setback. The cEMC switches VAV boxes (terminal units) to "night mode" for the duration of the DR event. Optional choice for turning off the plug loads.

  DR actions:
  1. Switch VAV values to night mode
  2. Increase AHU: supply air temperature
  3. Decrease AHU: static air pressure
  4. Shutdown plug loads (optional)
- DR test mode 4
  Scenario: Unified setback. The cEMC turns the RTU to shut down mode during the DR event. Mandatory plug loads turnoff by occupants.

  DR actions:
  1. Increase VAV values
  2. Shutdown AHU
  3. Shutdown plug loads (mandatory)

14.2.5 Energy split algorithm for energy games

Due to the limited space, we present the principles of the energy split algorithm and the game-theoretical Optimal Incentive Design (OID) algorithm, without technical details.

Our HVAC energy split algorithm calculates RTU heating, cooling, and ventilation to the level of individual VAV boxes. Technically, we can even split the energy consumption to an individual person, based on his/her calendar. However, due to our hypothesis 2, we assume it might be more effective to engage occupants, if we just stop at the VAV box level. Each week during the experiment, the FM announced the most and least energy-efficient zones and the ranking of each zone. The peer pressure motivated the occupants to keep their calendar updated each week. This is a simple energy competition game. Many researchers proposed numerous energy games, such as Energy Chickens [23], in order to motivate occupants to adjust their behaviors. It is certainly interesting to validate hypothesis 2 in future work and improve our energy game design.

Calculating the RTU heating and cooling energy is a standard practice. We need to estimate the thermal energy injected to the individual zone through each VAV box. Now, we provide the algorithm outline for the VAV box thermal energy flow calculation. Assuming all the heat is transferred as sensible heat, we can estimate the energy flow, $\dot{q}$ in unit of BTU/hr, due to temperature set point increase, ΔT. It is trivial to convert BTU/hr to kW.

$$\dot{q} = \left[\frac{0.075\ lb_m}{ft^3}\right] \times V\left[\frac{ft^3}{\min}\right] \times \left[\frac{60\ \min}{hr}\right] \times \left[\frac{0.24\ BTU}{lb_m^{\circ}F}\right]$$
$$\times \Delta T = 1.08 \times V\left[\frac{ft^3}{\min}\right] \times \Delta T$$

Notice that the aforementioned equation does not consider sensor error. In practice, air flow meters have a rather large drift. A closed-loop observer is required to calibrate the VAV box energy flow against the total RTU energy consumption for the whole building. We will present the closed-loop observer in our future work.

In the hardware experiment, we presented weekly zonal energy consumption plots to occupants each week. We announced the top three energy-efficient teams. It is natural to extend this psychological incentive to real *financial incentive*, such as cash, gifts, gift cards, coupons, or energy credits. Due to corporate policies, we could not evaluate financial incentives in the hardware experiment. However, we investigated the feasibility in simulations [24].

It is easy to observe that FMs need a tool to facilitate the incentive split, as shown in Figure 14.16. The figure presents the incentive split challenge from the perspective of an FM. The FM should promise to award "top contributing occupants" if the DR target is achieved. If the occupants collaborate and shed enough peak loads, the FM receives financial incentive from the utility company. The FM needs to consider two conflicting factors:

1. *Too little incentive:* From the economic perspective, occupants do not have long motivation to help the FM, if the FM does not share the incentive with occupants. Without supports from the occupants, the FM is not likely to reach the DR target and is not likely to receive any incentive from the utility.
2. *Too much incentive:* If the FM promised too much incentive to occupants, the FM may end up with significant financial loss. The income from the utility may not cover the costs to pay the occupants.

From the FM's perspective, he or she shall define Nm, the *maximum number of winning zone*, and Pm, the *ratio of incentive against zonal energy consumption.* If the top 3 load shedding zones get a reimbursement of 60% of the energy bill during the DR period, then Nm is 3 and Pm is 60%. There are, of course, other methods to parameterize the incentive design policy.

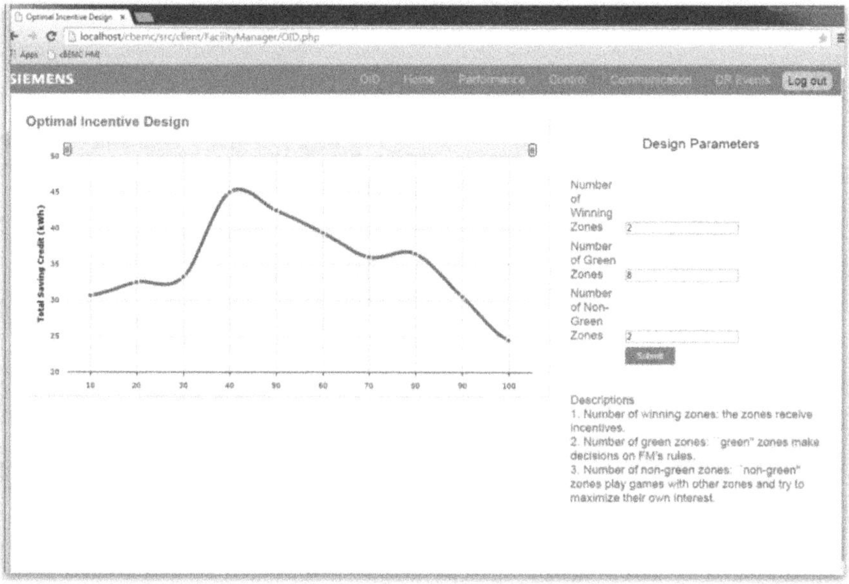

Figure 14.19 Optimal incentive design dashboard

We simplified the mental set of occupants into two categories: Green and Non-Green. Those Green occupants are willing to take more efforts when the FM promises more total incentives. The Green occupants continue environmental aware behaviors, even if the chance of winning is insignificant. The Non-Green occupants appreciate incentive more than the possible environmental impacts. The Non-Green occupants calculate the possibility for them to receive the incentive, and quit the competition if the winning chance is lower than their expectation. A given number of Green and Non-Green occupants, Nm, and Pm, we can simulate the FM's financial gain when the whole building reaches a Nash Equilibrium in the Game Theory. When there are a large number of occupants, proof or disproof of the existence of a Nash Equilibrium point is impossible for manual calculations.

In this chapter, we define the OID problem as *maximizing financial gain* for the FM with temperature and ventilation arbitration and energy split. The relationship between Pm and shedable load is called *DR Incentive Curve,* which is shown in Figure 14.19. Due to the limited space, we will not present the calculation method. The DR incentive curve is highly non-linear and not intuitive to predict. Comparing to other work on lighting systems, the game-theoretical incentive curve [24] is much smoother. It may due to that fact that thermal systems have complicated dynamics while lighting systems are static without complicated coupling.

We argue that an OID tool is *necessary,* for collaborative DR application, due to the complicity of the DR incentive curve, as well as the significant economic and environmental impacts (as mentioned in Section 14.1).

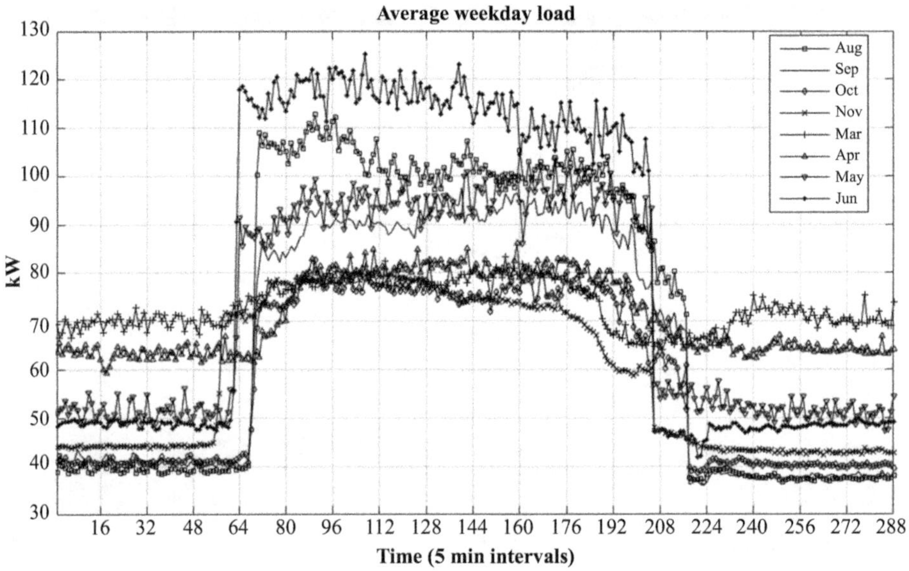

Figure 14.20 Average monthly weekday demand curves for the site

14.3 Experimental study

14.3.1 Deployment site

We deployed our cEMC system to an office building at Pittsburgh and conducted cDR experiments for more than 6 months. There is one RTU and 30 VAV boxes to serve 30 full-time employees and about 30 part-time employees. Before the project, there was one thermal zone, i.e., the whole floor was air conditioned even if there was only one occupant. The HVAC operation time was 14 hours per day, including weekends. The building has more than 3,100 software configurable points in its VAV terminal box controllers. Although we did not utilize every point in our software, it is important to understand each point, in order to ensure that the system is functioning properly. A large set of these points were established at the initial commissioning, when the field engineer tuned the parameters based on the worst-case scenario. Although it is well known that these points affect both occupant comfort and building energy usage and that they should be re-tuned with respect to changes in dynamic occupancy, manual adjustment of these parameters by the FM on a daily basis, or whenever an occupant enters or leaves a particular thermal zone, is impractical. This one-way, top-down process for building control leads to energy waste in some cases, comfort complaints in other cases, and most often both.

14.3.2 Baseline setup

The baseline was established based on historical monthly demand curves. Figure 14.20 shows some average monthly demand curves for the site. Through examination

Table 14.2 DR results summary

DR modes	Load shedding (%)
DR Mode 2	11.2
DR Mode 3	15.1
DR Mode 4	56.7

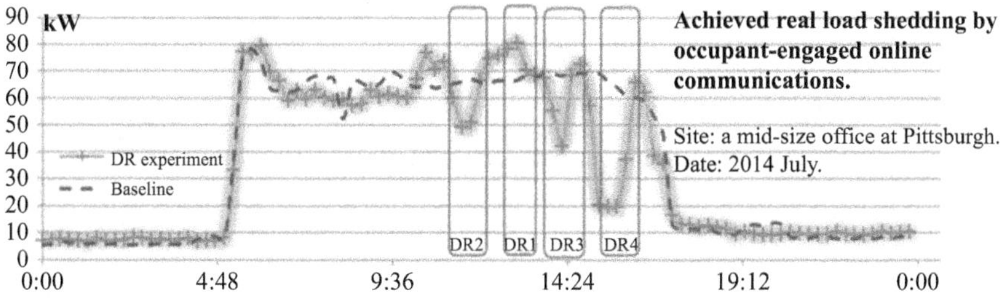

Figure 14.21 DR experimental results

of such curves, it was clear that the building peak load generally occurs between 7:30 a.m. and 8:30 a.m.. In some cases, it also between 2:00 and 3:00 p.m. Based on the metered data, it was determined that the building daily peak load varied between 39 and 180 kW during the period between August 7, 2013 and June 21, 2014. Another observation is that there is no prominent peak load time. The demand, in general, varies a lot between and these variations do not occur at any specific times. Therefore, the peak demand reduction objective of this project was reformulated as the reduction of the aggregated demand with reference to the demand at 15 minutes prior to the start of the DR period. The DR strategies were designed to reduce the aggregated electrical demand at the start of the DR period (pre-DR) by 15–30%.

In this approach, in order to determine the electrical load for a future day the weather forecast for that day is used along with historical weather (temperature, humidity, and wind speed) and trended electrical load data. The weather forecast of the target day is then compared to obtain "N" closest matches from the historical weather data. Then, the electrical loads of these "N" closest matches are combined to compute the electrical load forecast for the target day. This electrical load forecast can then be used as the baseline consumption for that day.

14.3.3 Field experiment

During field experiments on the testing site in 2014 summer, we observed up to 1.6% load reduction for the traditional email-based DR method, i.e., DR mode 1 scenario, and 11–15% load reduction using our cDR system without compromising on comfort. For the specific experiment presented in Figure 14.21, the measured load is compared with the predicted load baseline, which is fitted based on historical data.

Table 14.3 Results summary for DR2

Aggregate kWh consumption during DR period	58.40 kWh
Aggregate kWh for baseline during DR period	65.76 kWh
kWh savings during DR period	7.36 kWh (11.2%)
Average kW reduction during DR period	7.36 kW (11.2%)

Table 14.4 Results summary for DR3

Aggregate kWh consumption during DR period	58.31 kWh
Aggregate kWh for baseline during DR period	68.70 kWh
kWh savings during DR period	10.39 kWh (15.1%)
Average kW reduction during DR period	10.39 kW (15.1%)

Table 14.5 Results summary for DR4

Aggregate kWh consumption during DR period	29.13 kWh
Aggregate kWh for baseline during DR period	67.33 kWh
kWh savings during DR period	38.20 kWh (56.7%)
Average kW reduction during DR period	38.20 kW (56.7%)

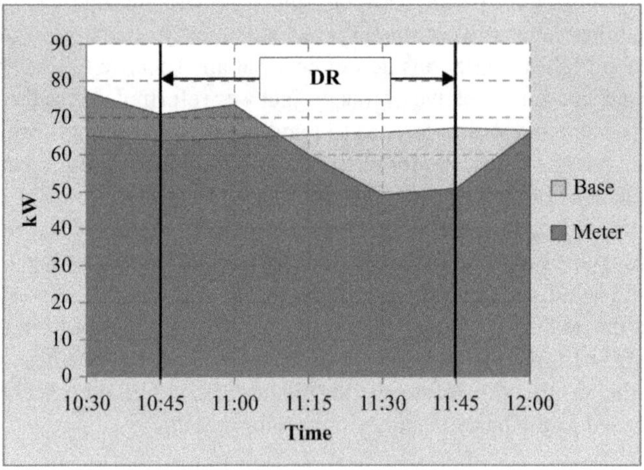

Figure 14.22 Baseline and actual demand for DR2

DR2 is the automatic cDR scenario where the room temperatures are automatically adjusted during the DR periods to reduce load yet still maintain the comfort level. For the experiment in Figure 14.21, DR2 achieved 11.2% load reduction. In the scenario of DR3, occupants are notified via the web portal to reduce plug load. The

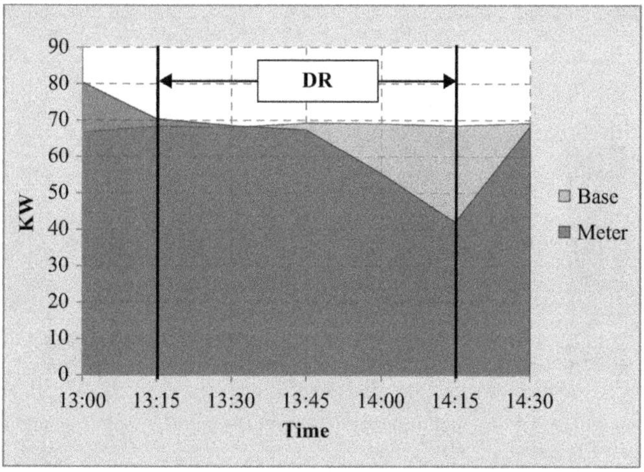

Figure 14.23 Baseline and actual demand for DR3

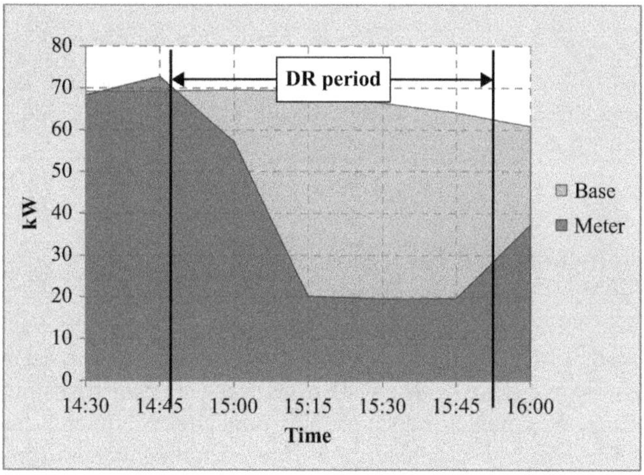

Figure 14.24 Baseline and actual demand for DR4

room temperature is also automatically adjusted within occupants' comfort range. The result is 15.1% load reduction. DR4 is the most aggressive DR strategy, where the AHU is temporarily shut down and the occupants' comfort is compromised. Up to 56.7% load is shed for this scenario. DR2 is a non-intrusive short DR strategy for daily usage. The extra load shedding from DR3 to DR2 illustrates the impact of interactive web portal.

The load shedding results are summarized in Table 14.2 and shown in Figure 14.21. The detailed data for individual experiments are shown in Tables 14.3–14.5 and the associated plots figures are shown in Figures 14.22, 14.24. Figure 14.25 shows

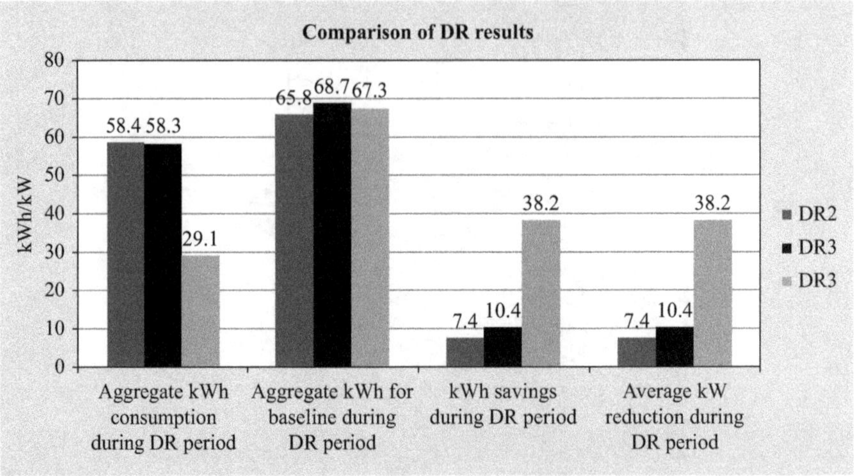

Figure 14.25 Comparison of DR results

the comparison of various results from three DR events. In the end, we suggest DR 2 and 3 for standard cDR events, when 11% and 15% peak load shedding is feasible without comfort loss.

14.4 Conclusions and future work

In this chapter, we presented an occupant-engaged, collaborative building control platform named cEMC. The DR module of this platform is called cDR, stands for *collaborative demand response.* The system has been deployed in a commercial building in Pittsburgh. We conducted a set of short-time DR experiments during a period of 6 months. The experiments achieved 15% load shedding without comfort loss and up to 56% load shedding with certain comfort loss.

Specifically, the cDR features the following novel functions: (1) An *Energy split* algorithm and *zonal* energy dashboard: Based on the VAV and RTU sensor readings, the whole building energy bill is split into individual zones at real time. (2) *Temperature and ventilation arbitration*: Based on convex optimization, our software automatically minimizes the discomfort of an occupant subject to other occupants' preferences, constrains of the HVAC hardware, and the energy policy from FMs. In order to achieve ensured results for energy savings and load shedding, we developed *arbitration* algorithms based on convex optimization theory. (3) *Pure software solution*: In order to minimize the initial investment, our method is a pure software approach that does not demand hardware occupancy sensors. (4) *Engage occupants via energy games*: The performance of the pure software approach depends on occupants' participation. Built upon our energy split algorithm, we introduced a zonal energy dashboard to encourage occupants to participate energy games. (5) *Optimal incentive design*: We demonstrated a prototype of game-theoretical incentive design tool. Based on the collaboration model of the cDR scenario, the tool can plot DR incentive curves and help FMs to find the maximum financial gain.

Different from traditional HVAC engineering, we designed the system from the perspective of CPSS and considered *human as sensors and actuators*. While the human factor is one of the drivers of our work, our foundation is the physics. The core functions, such as energy savings and load shedding, are built on convex optimization and HVAC thermal dynamics. We developed an energy split algorithm, also based on physics, to engage occupants. With integrated social and human factors, we provided a pure software solution with minimal installation costs, which are infeasible without cooperation from the occupants and FMs. In the end, we demonstrated an interactive CPSS design methodology, where optimization algorithms based on hard physics were developed to serve the social factors, i.e., minimize discomfort. Similarly, a game-theoretical method was used to motivate human subjects for quantitative improvement of physics systems.

Acronyms

BAS: Building Automation System.
cEMC: Collaborative Energy Management and Control.
cDR: Collaborative Demand Response.
DR: Demand Response.
FM: Facility Manager.

Acknowledgments

This work is partially sponsored by DOD ESTCP (EW-201336) program. We appreciate valuable discussions with Dr Yan Lu, Raphael Imhof, Paul Sedlock, Vivian Loftness, Bertrand Lasternas. We thank Kevin Sankatumvong for creating the user interface design. We thank all reviewers and the editor's helpful comments and feedback.

References

[1] M. H. Albadi and E. F. El-Saadany, "Demand response in electricity markets: An overview," In *Power Engineering Society General Meeting*, Tempa, FL, pp. 1–5, June 2007. [Online]. Available: http://dx.doi.org/10.1109/pes.2007.385728, accessed August 2016.

[2] New York Independent System Operator (NYISO), "Power trends 2014 evolution of the grid," 2014. [Online]. Available http://tinyurl.com/zgpjy8e, accessed August 2016.

[3] U.S. Department of Energy, "Benefits of demand response in electricity markets and recommendations for achieving them – a report to the United States Congress pursuant to section 1252 of the energy policy act of 2005," February 2006. [Online]. http://tinyurl.com/ztbxyqm, accessed August 2016.

[4] Siemens Building Technologies, "APOGEE building automation." [Online]. Available: http://www.siemens.com/bt/file?soi=A6V10301530, accessed August 2016.

[5] P. De Martini, "DR 2.0: a future of customer response. prepared for department of energy and federal energy regulation commissions," Newport Consulting, Tech. Rep., 2013. [Online]. Available: http://tinyurl.com/okbhox5, accessed August 2016.

[6] U.S. Federal Energy Regulatory Commission (FERC), "A national assessment of demand response potentials," 2009. [Online]. Available: http://www.ferc.gov/legal/staff-reports/06-09-demand-response.pdf, accessed August 2016.

[7] Consolidated Edison (ConEdison) Inc., "Service classification," 2012. [Online]. Available: http:// www.coned.com/rates/elec.asp, accessed August 2016.

[8] U.S. Energy Information Administration (EIA), "Commercial Buildings Energy Consumption Survey (CBECS)," 2012. [Online]. Available: http://www.eia.gov/consumption/commercial/reports/2012/preliminary/, accessed August 2016.

[9] E. Lee, "Cyber-physical systems – are computing foundations adequate?," Position Paper for *NSF Workshop on Cyber-Physical Systems*, October 16–17, 2006, Austin, TX.

[10] Consolidated Edison (ConEdison) Inc., "Demand response programs details." [Online]. Available: http://www.coned.com/energyefficiency/demand_response_program_details.asp, accessed August 2016.

[11] California ISO, "What the duck curve tells us about managing a green grid." [Online]. Available: http://tinyurl.com/ofky69q, accessed August 2016.

[12] Thermal Environmental Conditions for Human Occupancy, ANSI/ASHRAE Standard 55-2010, 2010.

[13] P. O. Fanger, *Analysis and Applications in Environmental Engineering*. McGraw-Hill Book Company New York, 1970.

[14] F. Nicol and M. Humphreys, "Adaptive thermal comfort and sustainable thermal standards for buildings," *Energy and Buildings*, 2002, vol. 34, no. 6, pp. 563–572. doi:10.1016/S0378-7788(02)00006-3.

[15] V. Erickson and A. Cerpa, "Thermovote: participatory sensing for efficient building HVAC conditioning," in *Proceedings of the Fourth ACM Workshop on Embedded Sensing Systems for Energy-Efficiency in Buildings (BuildSys 2012)*, pp. 9–16, 2012.

[16] Y. Lu, Z. Song, V. Loftness, K. Ji, S. Zheng, B. Lasternas, F. Marion, and Y. Yu, "Advanced, integrated control for building operations to achieve 40% energy saving," 2012. [Online]. Available: http://www.osti.gov/scitech/servlets/purl/1059657, accessed August 2016.

[17] Z. Song, L. Wang and Y. Lu, "Social Learning soft Thermostat for Commercial Buildings," in *Proceedings of the Fourth ACM Workshop on Embedded Sensing Systems for Energy-efficiency in Buildings*, Toronto, Canada, 2012.

[18] Z. Song, S. Zheng and Y. Lu, "Collaborative building control to optimize energy saving and improve occupants' experience," *ASHRAE Transactions*, 2013, vol. 119, no. 1, pp. AA1–AA8.

[19] L. J. Ratliff, J. Ming, I. C. Konstantakopoulos, C. Spanos, and S. S Sastry, "Social game for building energy efficiency: incentive design," In *52nd Annual Allerton Conference on Communication, Control, and Computing (Allerton 2014)*, pp. 1011–1018, 2014.

[20] T. G. Holmes, "Eco-visualization: combining art and technology to reduce energy consumption," in *Proceedings of the sixth ACM SIGCHI Conference on Creativity & Cognition* (C&C '07), New York, NY, pp. 153–162, 2007.

[21] P. Mansourifard, F. Jazizadeh, B. Krishnamachari, and B. Becerik-Gerber, "Online learning for personalized room-level thermal control: a multi-armed bandit framework," in *Proceedings of the Fifth ACM Workshop on Embedded Sensing Systems for Energy-Efficiency in Buildings (BuildSys 2013)*, pp. 1–8, 2013.

[22] J. Dahl and L. Vandenberghe, "CVXOPT: a python package for convex optimization," 2008. [Online]. Available: http://abel.ee.ucla.edu/cvxopt/, accessed August 2016.

[23] B. Orland, N. Ram, D. Lang, K. Houser, N. Kling, and M. Coccia, "Saving energy in an office environment: A serious game intervention," Energy and Buildings, vol. 74, pp. 43–52, May 2014. [Online]. Available: http://dx.doi.org/10.1016/j.enbuild.2014.01.036.

[24] Z. Song and K. Ji, "Optimal energy incentive design for occupant-engaged commercial building control," in *Proceedings of the 1st ACM Conference on Embedded Systems for Energy-Efficient Buildings (BuildSys'14),* Memphis, TN, pp. 178–179, 2014.

Part 5

Testbeds for validation of the research concepts

Chapter 15

A testbed for closed-loop cyber-physical-social system simulation and security analysis

Ren Liu[1], *Ryan Goodfellow*[2], *and Anurag K. Srivastava*[1]

Abstract

To analyze the complex interdependencies of power, social, and cyber domains, it is necessary to develop a cohesive modeling and simulation architecture for cyber-power-social systems. A real-time cyber-physical testbed with abilities to perform unified simulation, emulation, and interface with hardware devices provides a great platform to analyze the impact of cyber events on the power grid. In this chapter, development of a real-time cyber-physical testbed and applications have been discussed. The developed testbed has four different layers: (1) power system layer, (2) sensor and control layer, (3) communication layer, and (4) application layer. In the power system layer, Real-Time Digital Simulator (RTDS) and related software RSCAD are used to simulate the power system. In the sensor and control layer, real physical hardware devices such as phasor measurement units (PMU) and phasor data concentrator (PDC) have been used as sensors and supporting infrastructure. Real-Time Automation Controller (RTAC) and Synchrophasor Vector Processor (SVP) are used as controllers in this layer. Satellite-synchronized clock provides the accurate time signal to all the synchrophasor devices and simulator. In the communication layer, DeterLab have been used to emulate the communication network. With the unique features available in the DeterLab, cyber attack generation, cyber protection control, and communication traffic collection and analysis can be performed to analyze the impact of cyber events on the power grid. In the application layer, Real-Time Voltage Stability Monitoring and Control (RT-VSMAC) Tool is used to demonstrate the wide-area monitoring and close-loop control concepts.

[1]School of Electrical Engineering and Computer Science, Washington State University, Pullman, WA, USA
[2]USC Information Sciences Institute, Internet and Networked Systems Division, Marina Del Rey, CA, USA

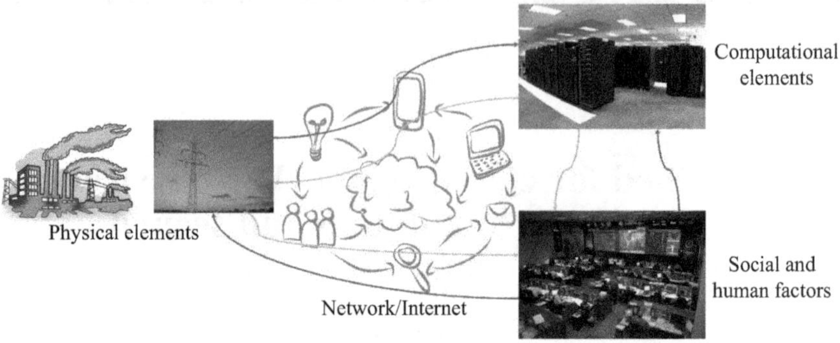

Figure 15.1 Architecture of smart grid

15.1 Introduction

15.1.1 Smart grid: cyber-physical-social system

Smart grid generally refers to a class of technologies to bring utility electricity delivery systems into the twenty-first century, using computer-based remote control and automation. These systems are made possible by two-way communication technology and computer processing that has been used for decades in other industries [1]. The smart grid is truly a cyber-physical-social system, since it requires integration of communication and computational technologies into the electric grid to provide enhanced decision support to human operators enabling reliable and resilient energy delivery service to our society. The architecture of smart grid is shown is Figure 15.1. Integration of advanced technologies also allow the smooth integration of the renewable energy and distributed generation.

Consumers are more engaged through demand response program with deployment of smart grid technologies. Humans play a critical role in adoption of demand response program as an important stakeholder in success of smart grid. With two-way communication between consumers and utilities, consumers can receive the real-time price signal, and automatically or manually shift or lower their power consumption. Additionally, human operators play a critical role in operating smart electric grid based on situational awareness and decision support provided by enhanced utilization of digital technologies. Even though these new digital technologies results in higher efficiency and reliability, it also brings new challenges, such as need for analyzing complex interdependency between physical and cyber systems, and providing solutions for cybersecurity caused by new vulnerabilities of electrical power grid from advanced communication, computer technologies, and human activities [2,3].

15.1.2 Need and challenges for cyber-physical-social testbeds

Power grid is the most complex large-scale networked systems engineered by man. Electric power systems are undergoing an architectural transformation necessitated by a societal-level shift in the way energy is procured and consumed. Integration of relatively small, intermittent, and typically asynchronous sources of energy into a grid designed for large, synchronous, and reliable generators brings some additional problems. Consumption patterns are also evolving. The introduction of the so called prosumer devices that can either produce or consume electricity, the prime example being electric cars, change the equations and management strategy for energy dispatch. Demand management technologies such as transactive energy systems have also introduced new degrees of freedom into overall grid operations and energy management. These changes are, at the very least, increasing the complexity of operating modern power systems. To help in dealing with this operational complexity, we are increasingly relying on computers and communication infrastructure [4,5].

Advanced control systems designed for wide-area monitoring and control of large-scale power grid requires hard timeline in completing the tasks of sensing, communicating, computing, and receiving the control signal. This will be increasingly difficult to achieve with increasing size of control variables and parameters involved in implementing such control architecture. To achieve coordinated operation in such a large-scale shared infrastructure environment using computers, we may rely on decentralized distributed computing systems. Moreover, the Internet is the only network infrastructure that can accommodate grid-scale interconnected computing systems. By virtue of being a shared infrastructure, use of the Internet for energy systems operations that may be operationally critical or privacy sensitive in nature brings on significant challenges. Recent history has shown that networked computer systems are the targets of many forms of attack. The criticality of our reliance of electric power as a cornerstone of our society suggests that building secure, resilient, and dependable energy systems should be a subject of careful and rigorous study.

The use of operational power systems is not the place for developing and testing new monitoring and control systems. Thus, we rely on testbed systems to design and test such systems. Testbeds allows modeling and simulating most of possible scenarios to test and validate the new algorithms and implementation architecture being field implementation. Testbeds also offer educational opportunities to demonstrate the operation of the sensors, controllers, or physical system operation. Additionally, based on the emulated, simulated, or hardware-in-the-loop (HIL) architecture, testbeds help with real-life modeling and simulation. Testbeds for cyber-physical systems (CPSs) in the power grid are comprised of five primary components: a power system simulator, sensor and actuator layer, a network environment, an applications layer, and interfaces to connect the different layers.

There are several big challenges faced while developing cyber-physical testbed. They are synchronization, interface problems, and scalability:

i. Synchronization: Synchronization of the testbed is important and difficult, since there are different layers in the cyber-physical testbed. All the layers are closely coupled and should be managed in time for integrated co-simulation. The whole testbed will lose the reliability and the accuracy without synchronization.
ii. Interface: Since the cyber-physical testbed is a multi-layer co-simulation addressing the interface problems between different layers is another issue. For example,

if the HIL feature needs to be added in the cyber-physical testbed, how to configure the interface between the real hardware device (such as phasor measurement units (PMU)) and other parts of the testbed (such as power system simulator and communication network simulator) is a challenge which needs to be solved.

iii. Scalability: Generally, due to the capability of hardware power system simulator, the size of the simulated power system is limited, which also restricts the scalability of the cyber-physical testbed. In order to simulate the larger power system, addressing the scalability problem becomes important.

In this chapter, a specific example of testbed for real-time closed-loop monitoring and control systems with HIL has been discussed. The power system simulation components, the networked system emulation components, and their composition as a usable testbed system are discussed next. The social components are not directly considered into this testbed, since the focus of this testbed example is to analyze the interdependency of the cyber and physical assets. Additionally, this testbed environment is compared with other approaches to cyber-physical power grid.

15.1.3 Types of power grid cyber-physical testbeds

There are quite a few options when creating a CPS testbed for the power grid. The choices are most concretely reflected in the type of simulator employed to represent power grid phenomena and the type of network emulator chosen to represent communications phenomena and house monitoring and control components. There are a few key parameters that define the type of physical simulator. For power systems, the first distinction that is typically made is electromechanical versus electromagnetic simulation. Electromechanical simulators are primarily concerned with capturing the dynamic behavior of large-scale synchronous machines interacting with electrical loads across transmission networks. These simulators typically focus on transmission networks and use dynamic load aggregation models such as the ZIP [6] model. The primary benefit in conducting electromechanical simulations is that the equations are simple relative to their electromagnetic counterparts and the timescale of phenomena of interest is much slower than interesting electromagnetic phenomena. Taken together, these two properties allow for very large (albeit simplified) systems to be simulated, controlled, and analyzed in a testbed environment. A further distinction that can be made here is single phase versus multi-phase. For many power system phenomena, it is sufficient to consider a single phase for a monitoring and control algorithm. For others it may be necessary to consider all three phases. Furthermore, there are simplifying transformations that also may be used given some simplifying assumptions, the most popular being the DQ0 transformation [6]. It is also worth noting that in the electromechanical space, the transmission system traveling wave differential equations may be reduced to non-linear algebraic equations.

The electromechanical models of power system components employed within electromechanical simulation environments are derived simplifications of electromagnetic models. Indeed, the defining characteristic of a simulator is the type of model it can simulate. In the electromagnetic space, the simulation of the models is much more computationally intensive. Electromagnetic models have a higher overall

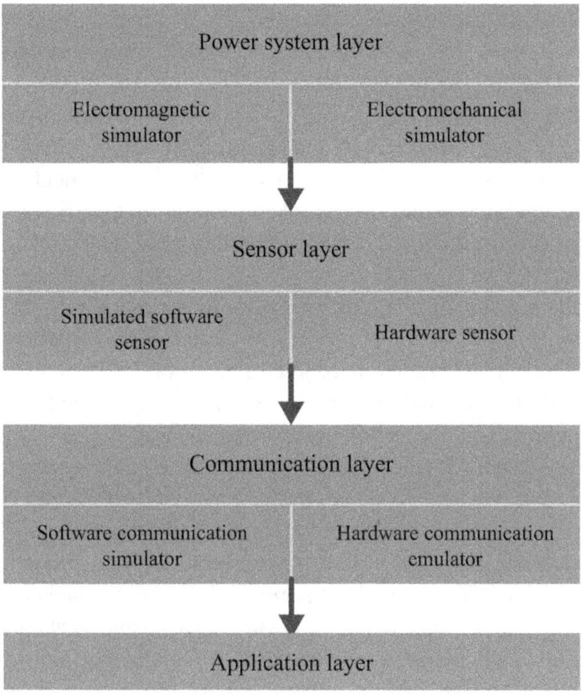

Figure 15.2 Electromechanical and electromagnetic testbed environment

level of complexity in terms of variable space, but the real challenge for simulators is in the timescale at which phenomena of interest takes place. The electromagnetic transient response to a fault in a transmission line across one or several phases can exist at the sub-microsecond timescale. For the simulation environment this extremely small timescale means extremely small time steps which amounts to more computation. If real-time simulation is required, this becomes a substantial challenge. This discussion highlights the important fact that there are many types of phenomena, that only electromagnetic simulation environments can capture. The trade-off is that achieving any kind of system-level scale in the electromagnetic space is highly elusive. Typically, a different type of simplification is performed. Instead of simplifying to electromechanics, systems engineers and researchers will take large models, say in the tens of thousands of buses and reduce them to an equivalent electromagnetic model in the tens or hundreds of buses, such that the aggregated model preserves some particular behavior of the system under study.

Figure 15.2 gives a generic diagrammatic representation of a power systems CPS testbed design. This design incorporates both an electromagnetic and electromechanical simulator. Typically, the electromagnetic simulators are based on high performance computation or custom hardware-based simulators, that may be capable of producing both analog and digital signals as output as well as control input. The diagram highlights the fact that the analog signals emanating from electromagnetic transients program (EMTP) type simulators need to be translated into digital format for consumption within a cyber environment. The diagram shows two such options for accomplishing this. To the far right are dedicated relays that can produced derived digital signals, IEC-61850-9-2 sampled values or IEEE-C37-118 synchrophasor values for example. The alternative is having a specialized component called a merging unit whose sole purpose is analog-to-digital conversion for a particular digital format. Such units are readily available in COTS format for a range of digital communications protocols.

When digital signals enter a cyber testbed environment they do so through the switching fabric of that environment. When an experiment is designed to run within the testbed, an embedded virtual environment, commonly called a slice, is allocated within the virtual environment. Say for example, that the experiment calls for 47 computers interconnected over a wide-area network (WAN) composed of 11 routers and a handful of switches. It is the job of the cyber testbed environment to allocate these resources within the physical hardware that comprises the testbed. This embedding process is a combination of computer allocation and virtual network construction through techniques such as VLAN partitioning. In addition to allocating the network and compute resources for a testbed experiment, the cyber testbed environment must also provide coherent and isolated communication. This means that if two experiments are running on the testbed simultaneously they are guaranteed not to interfere with each other, whether through communication leakage or resource starvation.

Sensor and actuator layers can be modeled and embedded in physical system simulation or interfaced as an actual HIL. Sensors and actuators generally require low voltage/current signals and relatively easier to interface compared to the power devices integrated in power-hardware-in-the-loop (PHIL). Integrating actual devices provide real-life simulation instead of modeling these.

There are also choices to be made in terms of the cyber environment. We typically call a computing environment that houses computers and programs together with infrastructure for emulating network phenomena as network testbed. The range of network testbed environments goes from a single computer emulating a complex network through routing matrix mathematics and forwarding traffic accordingly to the embedding of networks across the actual Internet. There is also a range of options in between. An example of the former is the NS3 [7] network simulator. This is an event-driven simulation engine in which data networks may be described and simulated on a single machine or a host of machines in a LAN environment. A network description file is handed off to the NS3 machine and the host computers described in the file are hooked up to the same LAN network. Then the host computers communicate with each other using the addresses specified in the network description file, communication is routed through the NS3 simulation engine that gives it the specified behavior in terms of things like delay, drop rates, and link failures.

At the other end of the spectrum are network testbeds that are themselves distributed across the Internet. Examples of these types of networks are GENI/PlanetLab [8]. The workflow for these network testbeds is quite similar in concept to the network in a box type of systems. The user provides a network description file specifying the properties of the desired network and the runtime infrastructure of the testbed goes about creating the network. In the distributed network testbed, this process involves allocating distributed resources across the underlying testbed network and creating an embedded network called a slice. Being a multiuser system, the testbed resources are shared among multiple slices at any given time.

In between the network-in-a-box option and full-blown WAN embedding testbeds, are emulated testbed environments. Like the WAN testbeds, emulated network environments provide both computing and network resources. For example, given a description file of a network with 47 computers, 47 computers are automatically allocated and interconnected according to the network topology presented in the description file. The difference between emulated testbeds and WAN testbeds is how WAN behavior is represented. In a WAN testbed, the WAN behavior is captured by virtue of being on a WAN itself. In the emulated environments the behavior is computed similar to the network-in-a-box approach. In fact, the simplest approach to creating an emulated environment would be to use a boxed network environment and build a resource allocation mechanism on top. In Section 15.2.3.1 we will be presenting the emulated network testbed used for closed-loop CPS testbed in more detail, DeterLab.

The interaction between the simulator and the network testbed environment is very important. If not done correctly, artifacts may be produced that invalidate experimental results. There are a number of issues in play here, most of which are matters of timescale and fidelity. The timescale issues are straightforward in concept but can be incredibly difficult to deal with in practice. If the cyber components in the CPS testbed environment are real-time, then the simulator must also be real time. If the simulator is not real-time, then the network environment must manage time in such a way that is opaque to applications yet produces a temporally consistent space across the entire testbed.

Applications layers generally include software-based computation using data delivered by the communication systems. Application layers may give control signals back to physical simulators going through communication layer. The environment presented here is entirely real time, thus eliding the host of difficult time management issues.

15.1.4 Existing cyber-physical testbed

There are lots of different cyber-physical testbeds developed by other researchers, but they all have their own advantages and limitations. In Reference 9, authors develop a platform called EPOCHS for agent-based electric power and communication simulation. In this platform, PSCAD/EMTDC is used as electromagnetic simulator, PSLF is used as electromechanical transient simulator, and Network Simulator 2 (NS 2) is used as the communication simulator. In order to analyze and evaluate the security of the current cyber-physical environment, an remotely configurable and community-accessible experimental testbed facility design is presented in Reference 10. In Reference 11, authors introduce a cyber-physical testbed called Experimentation Platform for Internet Contingencies (EPIC), which contains two main components: Matlab as power system simulator and Emulab as communication network emulator. EPIC is used to provide assessment of the impact from cyber-attack on both cyber and power domain. In order to create a large-scale control system test environment, researchers utilize OPNET as the communication network simulator and PowerWorld as the power system simulator to build the Virtual Control System Environment (VCSE) [12]. In Reference 13, a real-time cyber-physical platform "GridSim" is created to develop and test wide-area application. TSAT is used to simulate the power system in real time for electromechanical dynamics, and cyber aspect is simulated by a high-bandwidth networked communication system called GridStat. In order to make the assessment of power grid wide-area application, researchers in Virginia Tech develop a Global Event-driven Co-simulation framework (GECO) [14]. NS2 is used as the communication network simulator and PSLF is utilized to simulate the power system. Another testbed which includes Real-Time Digital Simulator (RTDS) and the Internet-scale event and attack generation environment (ISEAGE) has been developed by Iowa State University [15]. In order to analyze the performance and the security of SCADA system and corporate with other testbeds, the Virtual Power System Testbed (VPST) is developed at the University of Illinois at Urbana-Champaign. PowerWorld is utilized to handle electrical simulation and network simulation is based on RINSE [16]. Researchers from Florida State University and Missouri University of Science and Technology develop the smart grid CPSs testbed which combining RTDS as power system simulator, Mamba, and TS-7800 single board computer as computing platform [17]. In Reference 18, a cyber-physical co-simulation platform called INSPIRE is developed using DIgSILENT Power Factory and OPNET. In order to synchronize the time in this platform, substation data processing unit has also been implemented into this platform.

15.2 Developed cyber-physical testbed

A developed cyber-physical testbed will be introduced in this section. There are four different layers in this cyber-physical testbed: (1) the power system layer, (2) the monitoring layer, (3) the communication layer, and (4) the application and control layer.

In the power system layer, a RTDS and related software RSCAD [19] are used to simulate the power system. In the monitoring layer, real physical hardware devices such as PMU and phasor data concentrators (PDCs) have been used as sensors and supporting infrastructure. A satellite-synchronized clock provides the accurate time signal to all the synchrophasor devices and the simulator. In the communication layer, DeterLab [20] is used to emulate communication network. In the application and control layer, Real-Time Voltage Stability Monitoring and Control (RT-VSMAC) Tool and Remedial Action Schemes (RAS) will be used as an example to demonstrate the wide-area monitoring and close-loop control concepts. A Real-Time Automation Controller (RTAC) and Synchrophasor Vector Processor (SVP) are used as controllers in this layer. The architecture of this four layered cyber-physical testbed is shown in the Figure 15.3.

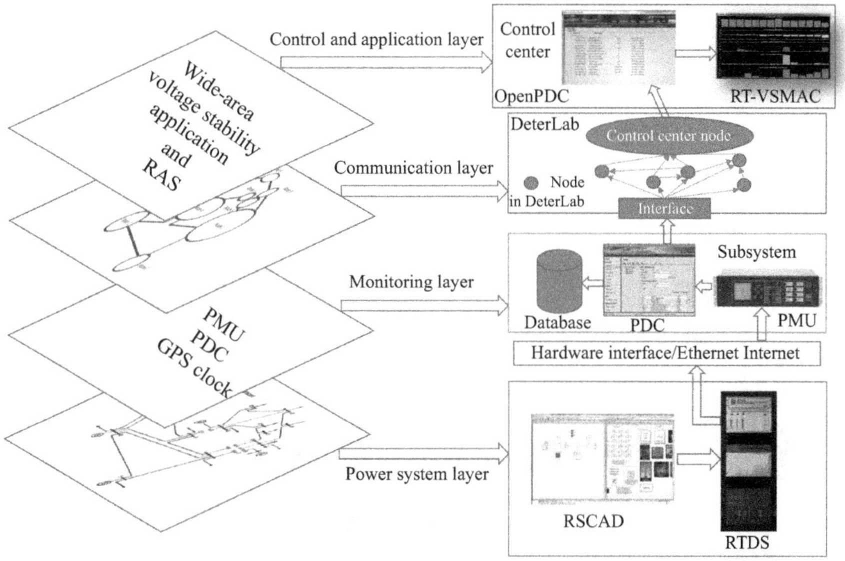

Figure 15.3 The architecture of the developed cyber-physical testbed

15.2.1 Power system layer

In this developed cyber-physical testbed, RTDS and RSCAD are used to model and simulate the power system and associated control components in real time. RTDS is designed to simulate power system and to interface with others devices into the HIL simulation. RTDS is capable of performing continuous real-time simulation, since its simulation time-step is 50 μs. With the various Giga-Transceiver Input/Output cards, RTDS could exchange the analog or digital signal with the external devices, such as monitoring, control, and protection devices. The Giga-Transceiver Network Communication Card (GTNET) provides a real-time communication channel between simulator and external devices through Ethernet. GTNET supports different protocols for the different applications, such as TCP/UDP transport protocol, IEEE C37.118 data stream, IEC 61850-9-2 sampled values, and DNP3 communication [21]. RSCAD is an interface software which is designed specifically for the RTDS. In the RSCAD, there are different modules to help users create, execute, control, and analyze the real-time simulation. RSCAD also offers the ComponentBuilder (CBuilder) to let user create the user-defined components, including both power system components and control components. By using CBuilder, it is easy to implement the self-designed algorithm into the real-time HIL simulation. The RTDS used in this chapter is capable of running IEEE Standard 14-bus and 30-bus transmission system with all the control components. The steady state values of real-time simulation are verified to validate the accuracy of the model.

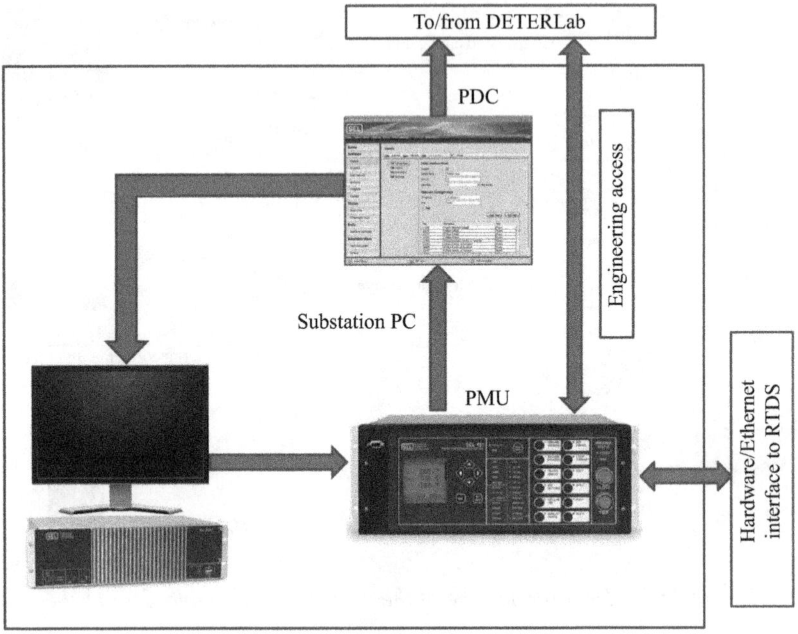

Figure 15.4 The substation view of the monitoring layer

15.2.2 Monitoring layer

The purpose of the monitoring layer of the testbed is to monitor the system status. To achieve this goal, SCADA system and/or PMU can be used. In this testbed, the monitoring layer consists of simulated potential transformers and current transformers in the RTDS and real industry hardware devices, such as PMU, PDC, GPS clock, and Substation PC.

Figure 15.4 shows the substation view of the monitoring layer. In each of the substation, PMU receives the analog signals from the RTDS hardware interface. After estimating the phasor data, PMU sends the phasor data to substation PDC followed by C37.118 standard. After archiving the data from the PMU, substation PDC sends the data to control center node or the controller in other substation through the emulated communication network. In case of the data loss caused by the communication failures, all the data from the substation PDC are also stored into a temporary database in the substation PC. When the communication connections recover, the substation PC can resend the temporarily stored data to their original destination. In order to synchronize the testbed, the high resolution time-stamp signal is offered by the SEL GPS clock to RTDS and all the synchronized devices, such as PMU and PDC.

In the developed testbed, there are four SEL hardware PMUs connecting with RTDS through Giga-Transceiver Analogue (GTAO) card, and eight software PMUs simulated by the GTNET card in the RTDS. The hardware PMUs also have the relay function, which can send the control signal back to RTDS through Giga-Transceiver

Digital Input (GTDI) card for utilizing the control action in the simulated power system. All the PMUs communicate with substation PDC based on PMU ID, port number, IP address, and communication protocol.

15.2.3 Communication layer

15.2.3.1 DeterLab introduction

DeterLab is a network testbed designed to conduct scientific cybersecurity experimentation for wide-area networked systems. A networked system is simply a group of computers interconnected by data networking equipment, such as switches and routers. At a very literal level, a networked system is considered to be wide-area when it is a geographically dispersed. However, most often when a network is referred to as wide-area, it is also assumed to be a shared infrastructure not under the control of any one entity and also that it is a routed (also called Layer-3) network. These characteristics distinguish WANs from LANs which are typically under unilateral control and do not typically contain Internet-protocol (IP) routers. DeterLab allows users to conduct cybersecurity experimentation by designing and instantiating an experimental WAN environment, deploy distributed software systems in the experimental WAN environment, subject the environment to attack, and analyze the response of the distributed software system to the attack.

The workflow of designing a cybersecurity experiment in DeterLab is quite simple. The first step is to describe a network environment using a scripting language called NS2 which is purposely built for describing network environments. The core components that make up a network environment description are computers and substrates. Each of these components comes in a variety of types and configurations. For example, a computer can be anything from a server grade bare-metal machine running a webserver to something as simple as a barebones operating system running a few processes as a lightweight container or virtual machine. A substrate is an abstraction of a network connection. In the simplest case, a substrate may be a point-to-point link connecting two computers, it could be a LAN interconnecting multiple machines, or it could represent an entire WAN itself exhibiting complex network behavior such as probabilistic latency, throughput, and packet loss. Taken together a collection of computers and substrates model a network environment, which may or may not be wide area.

When the description of a network environment is complete, it may be submitted to DeterLab for compilation and instantiation. The DeterLab runtime performs basic syntax and semantic checks on network descriptions, giving users feedback when needed for debugging. Once a network description passes these checks, the environment, it represents is instantiated by the DeterLab runtime. The terminology used by DeterLab for instantiating and experiment is called swapping-in. Swapping-in is the process by which the DeterLab runtime allocates the necessary resources to create the environment specified by the description file, configures those resources according to the supplied specification and ultimately provides the user with access to the environment.

To make this all a bit more concrete, consider the example NS2 code and corresponding visualization in Figure 15.5. This is a very simple representation of a WAN system that is composed of two LAN systems interconnected by a WAN link connecting nodes 2 and 5. Taken within a power systems context, this environment could model two substations each with two devices and a gateway. This design represents a subtle and simple yet important concept when modeling WAN systems with DeterLab. The WAN link between nodes 2 and 5 is not meant to be representative of the implementation of a real-world WAN interconnection, it is meant to be representative of the behavior or a WAN interconnection between two computers. While this may seem like a shortcoming from a design perspective, it is a more realistic representation of our ability to observe real WAN interconnections whose design is outside our control. Consider for example if this mini-model did represent two substations and their interconnection, and these two substations were located in different cities or even different states. We can get a good idea of the network interconnect performance between the two sites by simply running a ping test between the two sites. This will give us a reasonable measurement of the latency between the two sites. More sophisticated tests could also establish bandwidth capacity and drop rates between the two sites. Hence we have an immediately tractable way to establish the aggregate link behavior between the two sites. On the contrary consider trying to model the actual physical network infrastructure that connects the two sites is a much more difficult task. One that involves proprietary information that is unlikely to be accessible and will likely provide very little additional experimental benefit with respect to distributed monitoring and control software.

When this NS2 file is submitted to DeterLab, the depiction in the visualization is precisely the environment that the user gains access to once swap-in is complete. The default operating system that is loaded onto experiment nodes is Ubuntu Linux, however there are several operating systems to choose for experimental designs. The user also has the option to load, install, and run software at node startup. DeterLab provides a network file system (NFS) that is mounted on each experimental node at boot time. This is the preferred method to install custom software. Users can place the software installation packages in their projects NFS directory along with scripts to install the software locally on an experiment node. These installation scripts may then be specified with special NS2 commands in the experiment description file for the nodes on which the software installation is needed. Standard Linux software packages are also made available through distribution-specific repositories such as Debian aptitude servers.

DeterLab experiments are carefully isolated from the Internet and subsequently the rest of the world. When an experiment is instantiated, it is created behind a special firewall specifically designed to keep the phenomena taking place within the experiment from escaping to the outside world. It is because of this isolation that it is possible to execute malware, viruses, distributed denial of service attacks (DDOS), and other forms of malicious software and attack mechanisms within the testbed. In fact, this was the original motivation for creating the testbed, to study cybersecurity experimentally.

```
set ns [new Simulator]
source tb_compat.tcl

for {set j 0}{$j < 2}{incr j}{
    set lanstr($j)""
    for {set i 0}{$i < 2}{incr i}{
        set n [expr $j * 3 + $i]
        set node($n) [$ns node]
        append lanstr($j) "$node($n) "
    }
    set lan($j) [$ns make-lan "$lanstr($j)" 100Mb 0ms]
}
set link25 [$ns duplex-link $node(2) $node(5) 100Mb 33ms DropTail]

$ns rtproto Static
$ns run
```

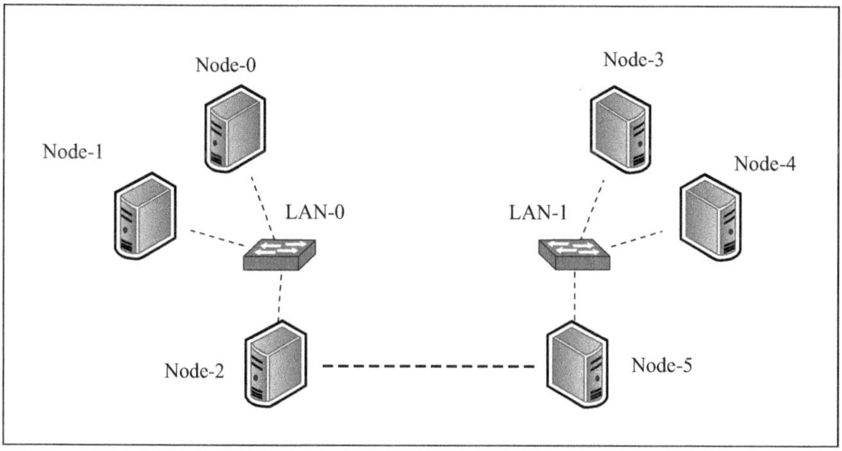

Figure 15.5 Sample NS2 file and topology

Finally, DeterLab experiments are sharable and repeatable. When users create experiments they are stored in the experiment database. This way, if results are published about a particular experiment, other researchers from the community may run the experiment for themselves and verify results, copy the experiment to make tweaks or modifications for follow on research, or use parts of the experiment as building blocks for an entirely different experiment.

15.2.3.2 DeterLab power systems experiments

Creating power systems experiments in DeterLab is a bit more involved than designing pure cyber experiments. Power systems experiments in real time typically involve specialized real-time simulators, high-performance computing resources, and/or hardware-based monitoring and control equipment. Thus additional infrastructure and capability is needed beyond the pure node, substrate, and software facilities that have been described thus far. The traditional LAN-based power systems testbed environment is depicted in Figure 15.6.

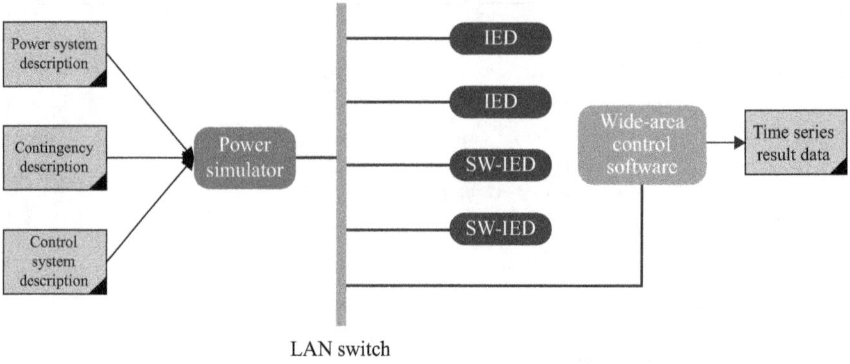

Figure 15.6 LAN-based testbed environment

Figure 15.6 shows a typical setup for a HIL testbed. A real-time simulator is connected to intelligent electronic devices (IED) through a switch which may also be connected to a computer running system-level monitoring and control software. This type of setup has been accomplished with a range of different types of simulator. At one extreme of the spectrum, real-time EMPT type simulators may be employed for conducting experimentation on protection type systems which have a relatively small time and system scale. At the other extreme are electromechanical simulators which have a much larger time and system scale. Typically the latter form of simulator is not designed for real-time operation but may be augmented to perform in such a manner as in Reference13. The simulator is fed a power system description, a contingency event timeline containing some sort of destabilizing phenomena such as a transmission system fault and there can optionally be control component descriptions for simulators that support running control routines as a part of the simulation itself.

When a simulation is executed by the simulator, the time series data produced by the execution is pushed out onto the testbed LAN by the simulators networking facilities and consumed by the IEDs. IEDs can then conduct independent or collaborative control or, create aggregated data to be pushed higher up the control chain to the system-level control software. This is very useful setup for prototyping and evaluating small-scale control systems. By integrating this environment with DeterLab, we gain the ability to model WAN behavior as well as to introduce cyber attacks. Consider Figure 15.7. The situation is quite similar to that of Figure 15.6, however, this time the controllers are attached to a WAN environment in addition to the LAN.

The WAN environment which is depicted by the cloud containing nodes of its own is a DeterLab hosted WAN. The key technology that allows DeterLab to extend itself into other environments is called federation. As the name suggests federation technology allows DeterLab to federate with other testbeds, effectively forming a cooperative relationship between DeterLab and one or more other testbed environments. Federation is achieved by installing an experiment controller at the federated testbed site. The experiment controller establishes a common communication protocol between the federated testbed and DeterLab. An experiment controller extends DeterLab like

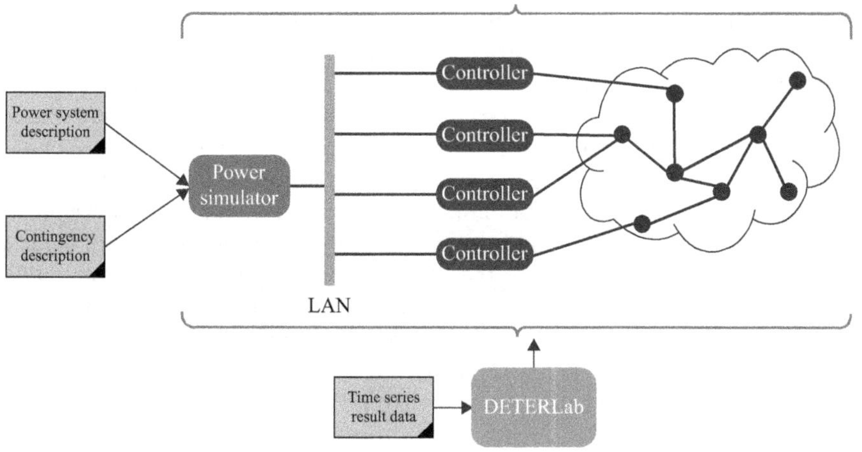

Figure 15.7 WAN-based testbed environment

functionality into the federated testbed environment. For example, once an experiment controller is installed at a federation site, uniform resource allocation across testbed sites becomes possible. Essentially, the user of the testbed environment may be unaware that resources are being allocated across different testbed environments, and is only aware that the resources themselves are available.

The experiment controller software also provides routing functionality to facilitate communication between DeterLab and federated sites. The link that is established between DeterLab and a federated site is a layer-2 tunnel, so layer-2 functionalities like ARP still work across federated links. The routing facilities are provided through a Quagga [22] installation that accompanies the experiment controller installation. The Quagga router running on the experiment controller node is aware of the routes within a DeterLab experiment and can advertise these routes within a federation site. This way, sites with a different IP addressing scheme than a DeterLab experiment can still effectively communicate as if they are a part of the experiment. Depending on the configuration of networked devices in the federated lab, some manual configuration and static routing may be required. However, the presence of the routing facilities in the experiment controller greatly reduce the need to take on this sort of minutia.

By installing a federation endpoint in the lab that houses the LAN-based power system testbed, duplex communication access to a DeterLab experiment is achieved. Hardware-based controllers in the power system testbed can talk to WAN nodes in the DeterLab environment and vice versa. The typical way to develop WAN power systems experiments in DeterLab is to first create a mapping relationship between hardware IEDs and experiment nodes. This is depicted in Figure 15.7 where each controller is mapped to one of the nodes in the DeterLab environment. The communication scheme for this setup is as follows. When sampled physical values are emitted from the simulator, they are carried by the power system testbed LAN to the hardware controllers. As these sampled values streams are received by the controllers, they are either handled locally by the controller, for example if the stream triggers a protection function, or the controller engages in some sort of distributed communication protocol. The distributed communication protocol can be as simple as augmenting the sampled values with additional information and passing this information upstream as PMU do, or it can be as complex as the control units engaging in a multi-phase iterative distributed decision algorithm. In either case this communication takes place through the WAN environment provided by DeterLab. This way the wide-area nature and behavior of the communication is captured. For example, if two controllers are engaged in a distributed coordination protocol, but are modeled as highly geographically disparate, the latency and underlying communication reliability protocol overhead will manifest themselves in the performance of the protocol. Communication from hardware controllers back to the simulator is still conducted over the power systems testbed LAN. This is reflective of the real-world scenario, as controllers in a substation will typically use layer-2 communications for control using protocols such as GOOSE.

15.2.3.3 Cybersecurity experimentation

There are many potential avenues of attack for degrading the performance of or completely disabling a power grid control system. In cybersecurity parlance, these avenues of attack are called attack vectors. The DeterLab environment provides and supports a plethora of attack vectors, ranging from malware and viruses to botnets and DDOS attacks. In the remainder of this section, we present how a DDOS attack may be leveraged against a power system RAS. This is based on an experiment produced for the US National Institute of Standards and Technology (NIST) Smart America challenge. A RAS is the power grid's last line of defense against localized and possibly cascading blackouts. In a nutshell, IED cooperatively detect and perform control actions to mitigate destabilizing phenomena before critical system equipment is damaged or a singular point in system state is passed where drastic action such as massive load shedding becomes necessary.

The DeterLab experiment that was constructed to test the effect of DDOS on RASs is a federated experiment between RTDS facility and the main DeterLab cluster. RTDS and industry standard IEDs were used for implementing RAS schemes. To build the experiment, an experiment controller was installed in RTDS facility and control system and substation automation software was installed on experiment nodes hosted on DeterLab. In addition to the layer-2 tunnel provided by the experiment controllers interconnecting DeterLab and RTDS facility, some customized static routing was put in place to allow control center and substation-level RAS software elements operating within the DeterLab testbed to communicate with the RAS hardware elements.

For the experiment itself, there were several control center and substation nodes set up in DeterLab as well as attacker and defender nodes. Using the Magi orchestration framework which is a distributed event coordination framework that comes as a part of DeterLab, a two-stage attack on the power system was created. In the first stage, a malicious trip signal was sent to a protective relay to take a transmission line out of service in the real-time simulation. The result of this trip is that a particular critical transmission path becomes overloaded if the appropriate RAS action is not taken within a specified time threshold. Immediately following the first stage of the attack, a DDOS is launched from DeterLab on the RAS elements. This attack in most cases prevents the RAS elements from enacting the appropriate scheme within an acceptable time frame and a localized blackout occurs. A key lesson that was learned in this experiment was the attack only succeeds most of the time, which is in contrast to the attack succeeding all the time when it originates on the same LAN and easily overwhelms the RAS elements. The effect of having an emulated WAN in place actually showed the relative difficulty in effectively mounting such an attack in the wide area and provided a path for some new defensive strategies to capitalize on the properties of WAN's themselves.

15.2.4 Control and application layer

The application and control layer in the testbed considered as the control center or the local controller based on the different applications.

For the wide-area application, the application and control layer is represented by the control center. Figure 15.8 shows the control center view of the application and control layer. Super PDC archives the phasor data from all the substation PDC through emulated communication network, aligns this data based on the time stamp, and stores this data into the control center database. The data in the control center database is used in both real-time application and non-real-time application. Human–Machine Interface (HMI) and visualization tools may be utilized based on the requirements of applications which are implemented in the cyber-physical testbed. A wide-area monitoring and control application is implemented in this cyber-physical testbed as an example.

For the local control application, the application and control layer is represented by the local controller, such as SVP and RTAC, located in the specific substation. For example, Figure 15.9 shows the substation view of the application and control layer

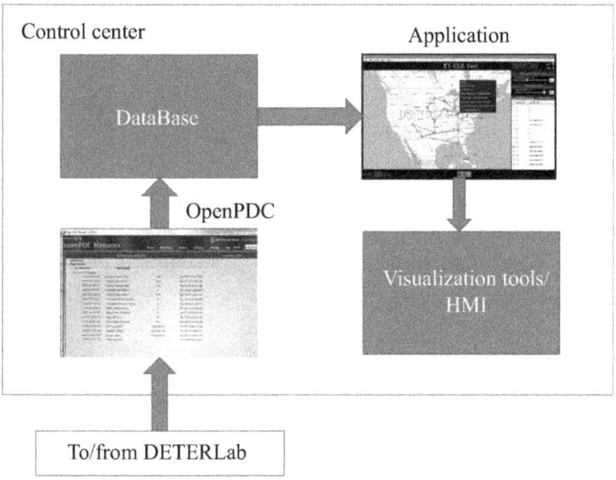

Figure 15.8 Control center view of the application and control layer

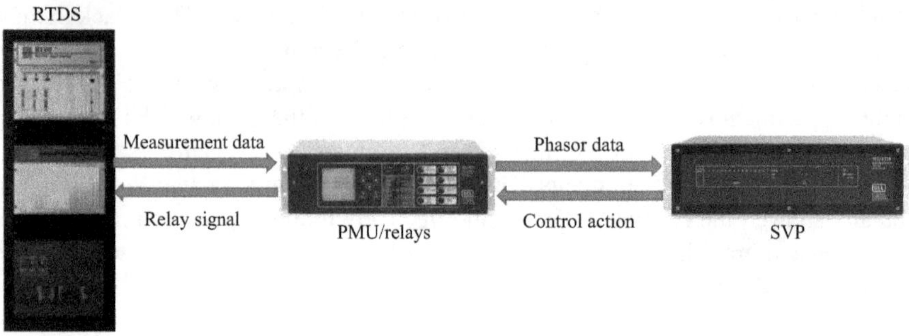

Figure 15.9 Substation view of the application and control layer for RAS

for RAS. The local controller archives the phasor data from the PMU in the same substation. Based on the real-time phasor measurement, the controller generates the control action and sends it back to the relay. The relay finally sends the relay signal back to RTDS through GTDI card for circuit breaker tripping.

15.2.4.1 Wide-area voltage stability monitoring and control

Wide-area voltage stability monitoring and control (RT-VSMAC) application [23] in the application and control layer is utilized as one of the energy management system (EMS) applications. It is based on a non-iterative method that makes use of wide-area voltage phasor measurements and system topological information to calculate terms of Voltage Stability Assessment Index (VSAI), which ranges from 0 to 1, where values near 0 indicate highly voltage stable load bus and values near 1 indicate that the load bus is near the point of voltage collapse. The system VSAI is given by the VSAI of the weakest bus in the system. Since the developed algorithm is completely non-iterative in nature, the computation speed is very fast and the run time for each loop is in the range of a few milliseconds, which allows this application to run in real time.

Once the system VSAI goes above the threshold, RT-VSMAC generates the appropriate control action based on the available control resource and the VSAI. The control action is sent back to the specific substation relay through the emulated communication network.

RT-VSMAC is developed on C# Language and XAML. To demonstrate the ability of the cyber-physical testbed, a modified IEEE 14-bus system shown in Figure 15.10 is simulated in the cyber-physical testbed with the full observability using PMUs at all the substations. Normal stressed situation and UDP flood attack situation are simulated. Simulation results come from RT-VSMAC and RTDS. Figures 15.11 and 15.12 show the real-time visualization window and monitoring visualization window of RT-VSMAC.

Normal stressed situation

In the normal stressed situation, the sequence of events which leads to a possible voltage collapse are shown in Table 15.1. When the RT-VSMAC detects the system

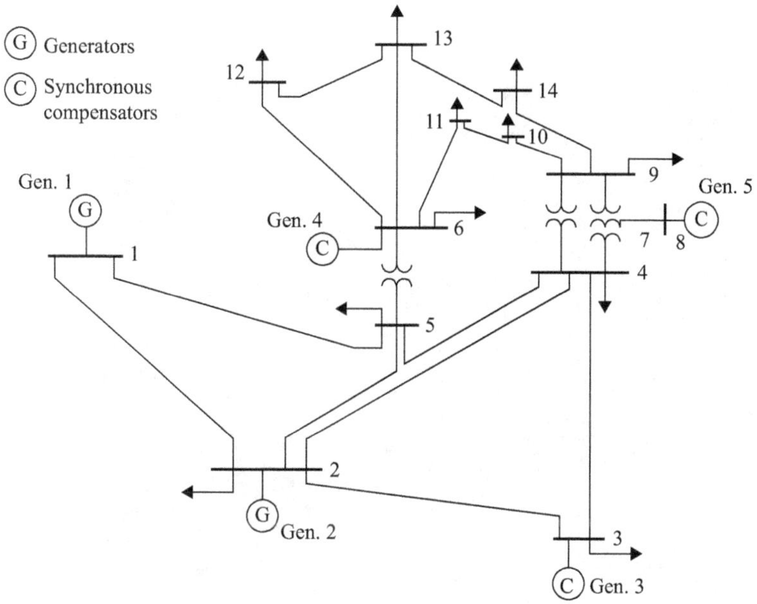

Figure 15.10 Modified 14-bus system

VSAI goes over the threshold 0.75, it generates the appropriate control actions [24]. All the control actions given by the RT-VSMAC are shown in Table 15.2. The VSAI and related voltage phasor data are shown in Figure 15.13. The first four control actions are inserting the shunt capacitor bank at bus-14 and bus-9. After these control actions, the system VSAI does not go below the threshold. Then, the next two control actions are load shedding at bus-14. From Figure 15.13, it is clear that system VSAI drops below threshold after these two control actions. To keep the availability of the control devices, RT-VSMAC automatically revokes the ineffective inserting shunt capacitor bank control actions.

UDP flood attack

With the near realistic environment supported by DeterLab, many different real-life cyber events can be implemented into the cyber-physical testbed, such as communication delay, packet drop, Man-in-the-Middle Attack, and Denial of Service Attack. UDP flood attack is used as the example to show the ability of cyber-physical testbed to analyze the cyber-physical impact of cyber attacks on power grid.

UDP flood attack is one kind of bandwidth consumption attack. The goal of the UDP flood attack is to congest the network bandwidth so that legal data flow cannot reach the attacked server. Since UDP is a connectionless computer networking protocol, there is no handshake validation required to send or receive UDP data. An attack can forge a fake IP address and send a large number of UDP packets to random ports on the targeted server, which may lead to the saturation of the bandwidth so that other legal services cannot access the targeted server. Meanwhile, the targeted server

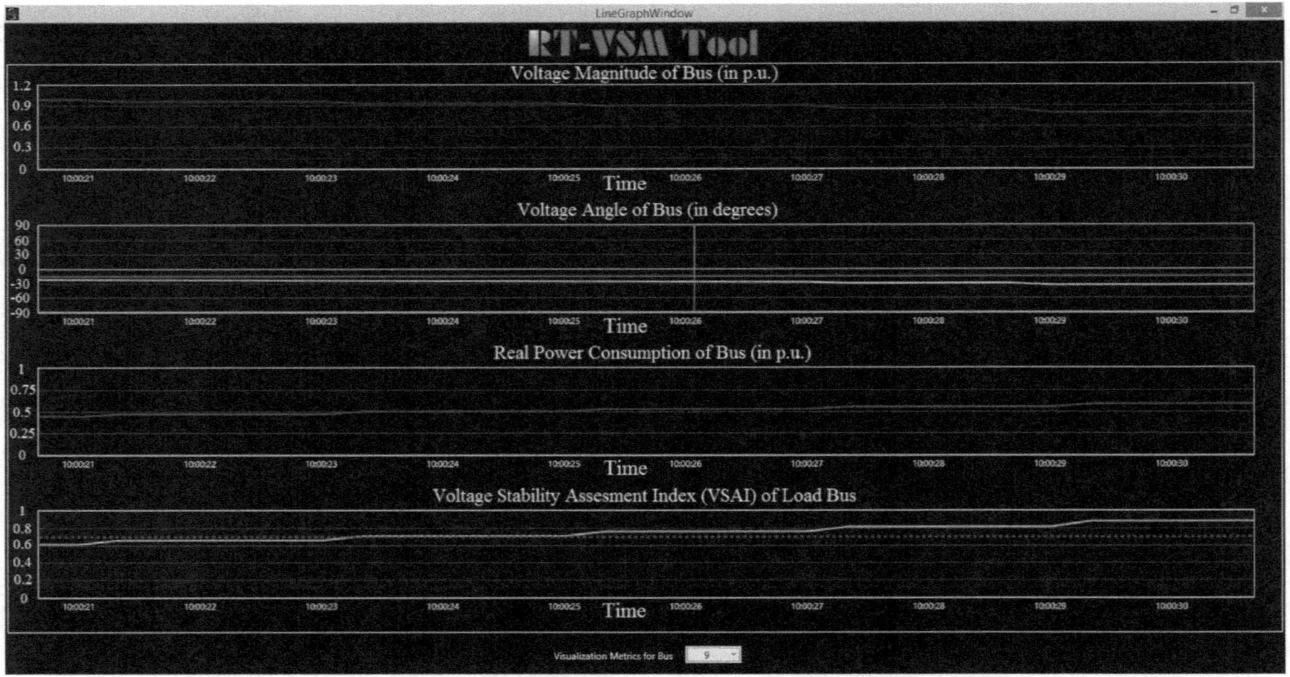

Figure 15.11 Real-time visualization window of the RT-VSMAC tool showing change of critical system metrics of the selected load bus

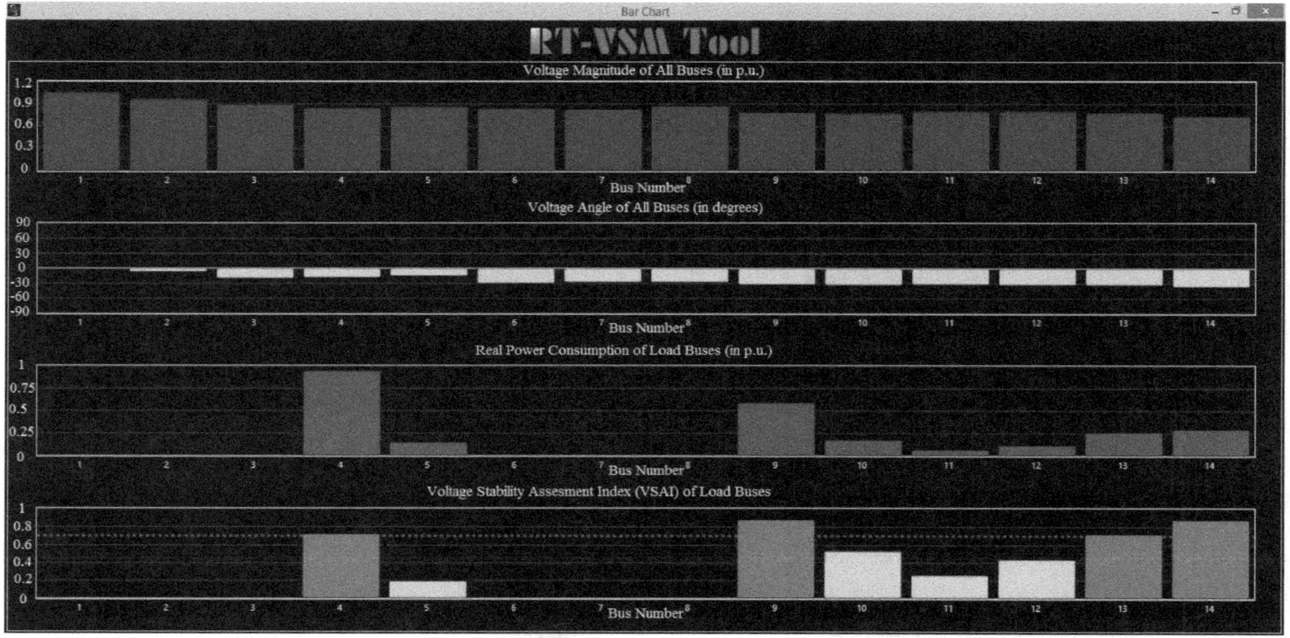

Figure 15.12 Wide-area visualization window of the RT-VSMAC tool showing critical system metrics of all load buses in IEEE 14-bus test case during the base case

Table 15.1 Sequence of events leading to a possible voltage collapse

Event	Time stamp	Event description
1	t = 5 s	Base case
2	t = 10 s	Load at bus_9 increase to real power consumption 59 MW and reactive power consumption is 33.2 MVAR. (base case: 29.5 MW and 16.6 MVAR)
3	t = 15 s	Load at bus_{14} increase to real power consumption 29.8 MW and reactive power consumption is 10 MVAR. (base case: 14.9 MW and 5 MVAR)
4	t = 20 s	Load at bus_{14} increase to real power consumption 44.7 MW and reactive power consumption is 15 MVAR.
5	t = 25 s	Load at bus_{13} increase to real power consumption 27.6 MW and reactive power consumption is 11.6 MVAR. (base case: 13.5 MW and 5.8 MVAR)
6	t = 30 s	Load at bus_{14} increase to real power consumption 59.6 MW and reactive power consumption is 20 MVAR.
7	t = 35 s	Load at bus_{14} increase to real power consumption 89.4 MW and reactive power consumption is 30 MVAR.

Table 15.2 Sequence of events with related control actions for normal condition

Event	Time	Event description
8	t = 40 s	Control action by RT-VSMAC tool: Capacitor rated 1 MVAR connected at Bus-14
9	t = 45 s	Control action by RT-VSMAC tool: Capacitor rated 2 MVAR connected at Bus-14
10	t = 50 s	Control action by RT-VSMAC tool: Capacitor rated 1 MVAR connected at Bus-9
11	t = 55 s	Control action by RT-VSMAC tool: Capacitor rated 2 MVAR connected at Bus-9
12	t = 60 s	Control action by RT-VSMAC tool: Load-shedding at Bus-14: Real power loading at Bus-14 to 70.775 MW Reactive power loading at Bus-14 to 21.8706 MVAR
13	t = 65 s	Control action by RT-VSMAC tool: Load-shedding at Bus-14: Real power loading at Bus-14 to 56.8063 MW Reactive power loading at Bus-14 to 17.1831 MVAR
14	t = 70 s	Control action by RT-VSMAC tool: Capacitor rated 1 MVAR disconnected at Bus-9
15	t = 75 s	Control action by RT-VSMAC tool: Capacitor rated 2 MVAR disconnected at Bus-9
16	t = 80 s	Control action by RT-VSMAC tool: Capacitor rated 1 MVAR disconnected at Bus-14
17	t = 85 s	Control action by RT-VSMAC tool: Capacitor rated 2 MVAR disconnected at Bus-14

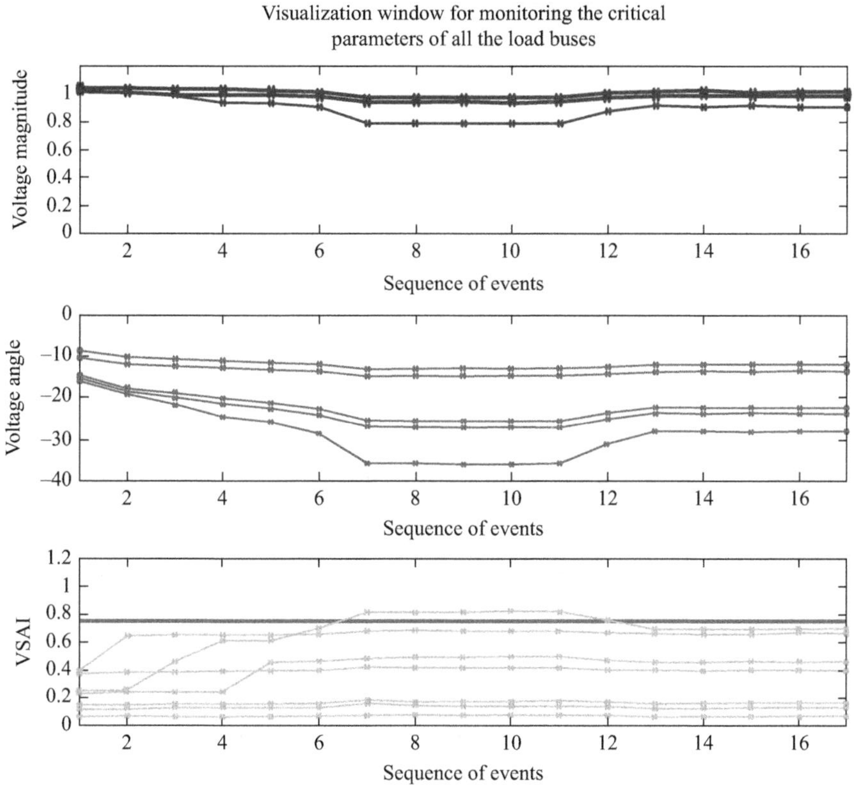

Figure 15.13 Wide-area VSAI and voltage phasor data for normal condition

checks whether there are applications listening to the attacked ports. Since there is no application listening to the attacked ports, the targeted server replies with Internet Control Message Protocol (ICMP) destination unreachable packets to the fake IP address. If the attacker forges the fake IP address as the IP address of other clients, the targeted server is unreachable for those clients.

In the UDP flood attack simulation, the attacker floods UDP packets to substation 9. Due to UDP attack, the bandwidth of substation 9 is occupied by attack data. Communication delay and packet loss appear on the data streams from substation 9. Figure 15.14 shows the impact of UDP flood attack on the communication delay of substation 9 data stream from four situations. X-axis represents the length of the attack and Y-axis represents the communication delay on substation 9 data stream. From the results, for the 40 Mbps attack, the communication delay is still at the normal level. When the attack increases to 45 Mbps, huge communication delay appears on the node 9 data stream. The bandwidth of the substation 9 is congested by the attack data at this time. With the increase of attack rate, the communication delay also keeps increasing. When the attack rate is over 50 Mbps, RT-VSMAC detects that

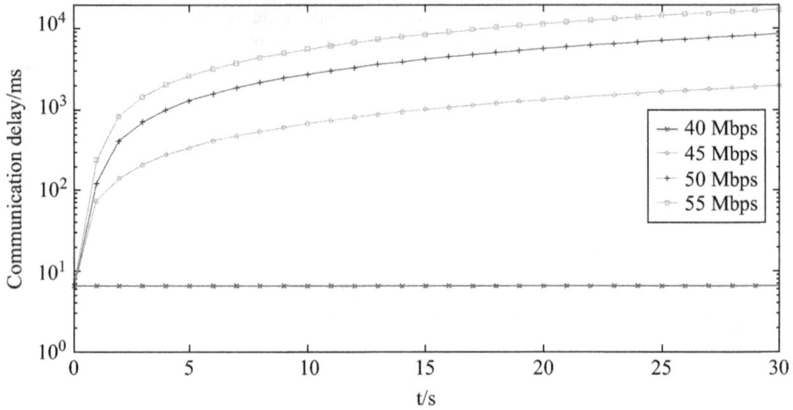

Figure 15.14 The impact of UDP flood attack on the communication delay of substation 9 data stream for four possible scenarios

one synchrophasor data stream has large communication delay compared to other data streams. This situation may be caused by the occurrence of communication failure or power system failure. In order to avoid incorrect control action based on the bad data, RT-VSMAC keeps using previous VSAI data until all the phasor data streams come back to the normal condition. In the UDP flood attack, since the application enters the safety mode caused by the bad input data, it cannot observe that the system is operating near the instability condition, and there is no coordinated control action being sent back to power system, which may finally lead to a blackout.

15.3 Summary

In this chapter, motivation, requirements and example of a cyber-physical testbed for the smart grid is discussed. Different kinds of existing power grid cyber-physical testbed are also discussed. The developed cyber-physical testbed, using RTDS, DeterLab, various sensor devices, and local controller, is presented in detail. RT-VSMAC and RAS are implemented in real time using the cyber-physical testbed as possible examples of the smart grid applications. UDP flood cyber attack is emulated in the cyber-physical testbed to show the impact of the real-life cyber attack on the power system.

Acknowledgments

The authors would like to thank US Department of Energy (USDOE), Power Systems Engineering and Research center (PSERC), Schweitzer Engineering Lab (SEL) to financially support development of this testbed. We would also like to thank

Dr Saugata Baswis and Ceeman Brightson Vellathurai for developing the RT-VSMAC tool and helping with development of the real-time cyber-physical testbed.

References

[1] Office of electricity delivery and energy reliability. [Online]. Available: http://energy.gov/oe/services/technology-development/smart-grid
[2] R. Liu, C. Vellaithurai, S. S. Biswas, T. T. Gamage, and A. K. Srivastava, "Analyzing the cyber-physical impact of cyber events on the power grid," *IEEE Trans. Smart Grid*, Sep 2015, vol. 6, no. 5, pp. 2444–2453.
[3] A. K. Srivastava, T. Morris, T. Ernster, C. Vellaithurai, P. Shengyi, and U. Adhikari, "Modeling cyber-physical vulnerability of the smart grid with incomplete information," *IEEE Trans. Smart Grid*, Mar 2013, vol. 4, no. 1, pp. 235–244.
[4] S. M. Amin and B. F. Wollenberg, "Toward a smart grid: power delivery for the 21st century," *IEEE Power and Energy. Mag.*, Sep 2005, vol. 3, no. 5, pp. 34–41.
[5] F. Shariatzadeh, S. Chanda, A. K. Srivastava, and A. Bose, "Real time benefit computation for electric distribution system automation and control," in *Proc. IEEE Industry Applications Society Annual Meeting*, Oct 2014, pp. 1–8.
[6] P. Kundur, N. J. Balu, and M. G. Lauby, *Power System Stability and Control*. McGraw-hill New York, 1994, vol. 7.
[7] G. F. Riley and T. R. Henderson, "The NS-3 network simulator," in *Modeling and Tools for Network Simulation*. Springer, Berlin, 2010, pp. 15–34.
[8] M. Berman, J. S. Chase, L. Landweber, *et al.*, "GENI: a federated testbed for innovative network experiments," *Computer Networks*, 2014, vol. 61, pp. 5–23.
[9] K. Hopkinson, X. Wang, R. Giovanini, J. Thorp, K. Birman, and D. Coury, "EPOCHS: a platform for agent-based electric power and communication simulation built from commercial off-the-shelf components," *IEEE Trans. Power System*, May 2006, vol. 21, no. 2, pp. 548–558.
[10] T. Edgar, D. Manz, and T. Carroll, "Towards an experimental testbed facility for cyber-physical security research," in *Proc. The Seventh Annual Workshop on Cyber Security and Information Intelligence Research*, ACM, 2011, p. 53.
[11] C. Siaterlis, B. Genge, and M. Hohenadel, "EPIC: a testbed for scientifically rigorous cyber-physical security experimentation," *IEEE Trans. Emerging Topics in Computing*, Dec 2013, vol. 1, no. 2, pp. 319–330.
[12] M. J. McDonald and G. N. Conrad, "TC Service, and RH Cassidy. Cyber Effects Analysis Using VCSE," Tech. Rep. SAND2008-5954, Sandia National Laboratories, Tech. Rep., 2008.
[13] D. Anderson, C. Zhao, C. H. Hauser, V. Venkatasubramanian, D. E. Bakken, and A. Bose, "'Intelligent design' real-time simulation for smart grid control and communications design," *IEEE Power and Energy. Mag.*, 2012, vol. 10, no. 1, pp. 49–57.

[14] H. Lin, S. S. Veda, S. S. Shukla, L. Mili, and J. Thorp, "GECO: global event-driven co-simulation framework for interconnected power system and communication network," *IEEE Trans. Smart Grid*, Sep 2012, vol. 3, no. 3, pp. 1444–1456.

[15] A. Hahn, A. Ashok, S. Sridhar, and M. Govindarasu, "Cyber-physical security testbeds: architecture, application, and evaluation for smart grid," *IEEE Trans. Smart Grid*, Jun 2013, vol. 4, no. 2, pp. 847–855.

[16] D. C. Bergman, D. K. Jin, D. M. Nicol, and T. Yardley, "The virtual power system testbed and inter-testbed integration," in *Proc. CSET*, 2009.

[17] M. J. Stanovich, I. Leonard, K. Sanjeev, *et al.*, "Development of a smart-grid cyber-physical systems testbed," in *Proc. IEEE PES Innovative Smart Grid Technologies (ISGT)*, Feb 2013, pp. 1–6.

[18] H. Georg, S. C. Muller, N. Dorsch, C. Rehtanz, and C. Wietfeld, "INSPIRE: integrated co-simulation of power and ICT systems for real-time evaluation," in *Proc. IEEE International Conference on Smart Grid Communications (SmartGridComm)*, Oct 2013, pp. 576–581.

[19] A. K. Srivastava and N. Schulz, "Applications of real time digital simulator in power system education and research," in *Proc. American Society for Engineering Education*, 2009.

[20] The DETER Project. [Online]. Available: http://www.deter-project.org

[21] RTDS Technologies. [Online]. Available: https://www.rtds.com/the-simulator/our-hardware/gtnet-card/

[22] Quagga Routing Suite. [Online]. Available: http://www.nongnu.org/quagga/

[23] S. S. Biswas and A. K. Srivastava, "RT-VSMAC tool: a real time voltage stability monitoring and adaptive control tool for electric power grids," US patent pending, Washington State University, Pullman, 2014.

[24] R. Liu and A. K. Srivastava, "Integrated simulation to analyze the impact of cyber-attacks on the power grid," in *Proc. Modeling and Simulation of Cyber-Physical Energy Systems (MSCPES) Workshop, IEEE*, 2015, pp. 1–6.

Chapter 16

Cyber-physical-social system security testbeds for an attack-resilient smart grid

Aditya Ashok[1], *Pengyuan Wang*[2], *and Manimaran Govindarasu*[2]

Abstract

With the ever increasing cyber security concerns for critical infrastructures such as the power grid against sophisticated and advanced persistent threats, the need for realistic cyber-physical-social system (CPSS) security testbed environments is compelling. CPSS security testbeds capture the complex interactions between the various domains and their interdependencies appropriately, while also providing necessary system abstractions for limiting experimental scale and yet maintaining higher fidelity than traditional simulation-based tools. In this chapter, we motivate the need for CPSS security testbeds and provide a quick overview of various objectives in the design of such testbeds. We then provide a comprehensive survey of state-of-the-art research in the area of CPSS security testbeds. We also identify key research areas enabled by CPSS security testbeds and highlight certain limitations of individual testbeds, which motivate the need to federate CPSS security testbeds. As part of experimental case studies, we present a brief proof-of-concept case study to illustrate the feasibility of CPSS security testbed federation. We also describe a detailed case study of coordinated cyber attacks on Wide-Area Monitoring, Protection, and Control (WAMPAC) applications for the Smart Grid using the PowerCyber CPSS security testbed at Iowa State University. We conclude this chapter with a brief overview of how CPSS security testbeds could be leveraged beyond experimental research for educational and outreach activities to enhance the overall cyber security awareness of the current and future workforce in this area.

16.1 Need for testbeds

Cyber security of critical infrastructures (CIs) like the power grid is becoming extremely important and is currently one of the top R&D priorities globally. It has

[1]Pacific Northwest National Laboratory, Richland, Washington, USA
[2]Iowa State University, Ames, IA, USA

been well documented in several government reports, and technical literature that there has been an increasing trend of highly sophisticated and targeted cyber attacks on the systems that monitor, protect, and control the power grid [1–7]. Several of the government reports highlight the urgent need to assess risk due to the cyber threats, identify existing system vulnerabilities and their potential impacts, and the roadmaps to incorporate defense measures that could mitigate the potential threats and also provide a level of resiliency to the future smart grid [1,6,7].

It is quite obvious that existing operational systems cannot be used to perform various types of cyber security experiments to discover vulnerabilities and fix them. Cyber-physical-social system (CPSS) security testbeds serve as test environments, which can be used to perform various types of cyber attack and defense experimentation using a hybrid combination of real, simulated, and emulated components that represent the real world as closely as possible. The real utility of testbeds over traditional standalone simulation tools is the ability to model complex cyber-physical interdependencies and their relevant interactions across multiple domains such as computation, communication, and physical systems in a single coherent environment. One of the distinguishing features of CPSS testbeds is the ability to map real hardware devices seamlessly into the experimental framework to perform real-time, hardware-in-the-loop experiments, which allow the ability to observe computation, communication, and physical system dynamics together, some of which otherwise would have been abstracted away due to shortcomings in the modeling platform. Also, CPSS testbeds provide a platform to conduct various attack/defense training exercises, such as red team vs. blue team exercises, which enable operator training in real time on situational awareness and on effective response strategies based on attacker actions. Through such training exercises, not only the system security features can be fully learned by the system operators, but also the weaknesses of their habitual online social activities. Therefore, CPSS testbeds would also effectively improve awareness of cyber security issues for power system operators and engineers.

16.2 Testbed design objectives and challenges

The key design objectives [8] for a CPSS security testbed include the following:

- *Modularity* – being composed of different functional units, which do not affect each other's functionality and could be abstracted depending on experimental needs.
- *Scalability* – the ability to able to increase/decrease the size of the system to be modeled without loosing out on the ability to integrate real hardware.
- *Repeatability* – the ability of the testbed platform to be able to consistently reproduce identical results for repeated trials of an experimental scenario.
- *Re-configurability* – the testbed should be able to support multiple configurations that allow a variety of experimental scenarios using the same hardware or tools.

- *Interoperability* – components in the testbed should be able to be replaced seamlessly, provided they have similar functionalities and interfaces such as communication protocols and services.

While CPSS testbeds provide modeling of computation, communication, and physical system dynamics in a single environment, there are several design challenges because of the necessity to have real and emulated devices interacting with simulated components. Some of the key challenges in designing a CPSS security testbed are: accuracy, scalability, and cost. The modeling accuracy is driven by the interdependencies that are to be modeled. The scalability of the system is limited by the amount of physical hardware that is available to be mapped into the experiment and also mostly by the capability of the real-time simulation tools that model the physical system. It is straightforward to understand how these two factors interact with cost – a system with all real hardware would be extremely expensive, but would have a high-modeling accuracy, whereas a system with all simulated components would be cheap, but would loose out on the modeling accuracy.

Optimal CPSS testbed design involves the identification of the right level of balance between these three competing factors using a hybrid combination of simulated components, emulated components, and real physical components. The design of the CPSS testbed should carefully consider the type of experimental use-case scenario that would be implemented on it. For example, a testbed that is built to test/prototype a new protection device would require a real-time power system simulator with the capability to interface real devices through its analog and digital I/O's along with the associated communication protocols. Similarly, a testbed that is expected to test market algorithms would require a network of multiple market agents that are deployed on computing nodes distributed realistically over an emulated wide-area communication network.

16.3 Conceptual testbed architecture

A typical CPSS testbed conceptual architecture can be explained with the help of a layered abstraction approach, based on each layer's functionality [8]. This method of functionally organizing the testbed helps to vary models that are used for different use-case scenarios and abstract the unnecessary details appropriately. Figure 16.1 shows the conceptual testbed architecture with the three different layers, namely, information/control layer, communication layer, and physical layer.

- *Information/control layer* – This layer consists of devices/software components that act on the various analog and status measurements coming from the substations to perform monitoring and control of the physical system components. Specifically, this layer consists of the supervisory control and data acquisition (SCADA) energy management systems (EMS), that perform various analytics at the control center like state estimation, automatic generation control (AGC), contingency analysis, etc., substation automation system hardware and software including remote terminal units (RTU) that collect data from the different

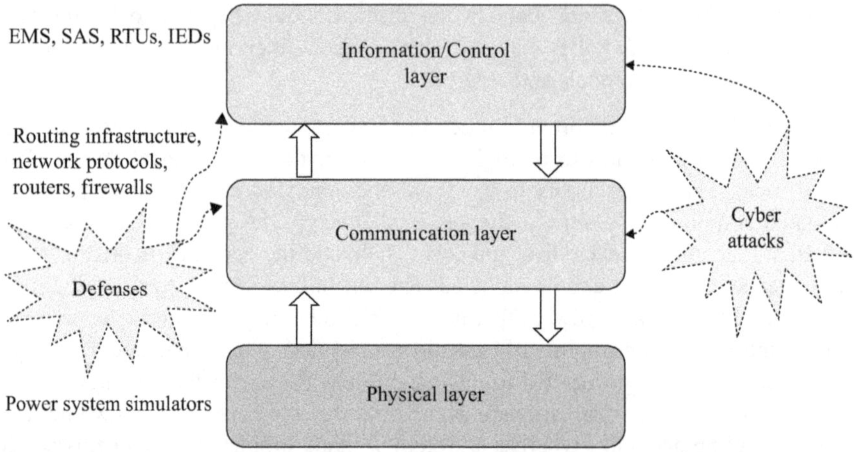

Figure 16.1 Conceptual testbed architecture ©2011 ACM. Reprinted, with permission, from [8]

intelligent electronic devices deployed in the field like protective relays and phasor measurement units.

- *Communication layer* – This layer consists of the network infrastructure that interfaces the control layer with the physical layer. Specifically, this layer consists of the network infrastructure that emulates wide-area network routing, network routers and switches, security devices such as firewalls, Intrusion Detection Systems, and the communication protocols that are used to enable intra- and inter-substation communications (e.g. distributed network protocol (DNP3), IEC 61850 Manufacturing Message Specification (MMS), IEC 61850 Generic Object-Oriented Event (GOOSE), IEC 61850 Sampled Measured Values), and wide-area communications with the control center (e.g. DNP3, Modbus).
- *Physical layer* – This layer consists of the components/devices that model the physical system. Typically, this layer consists of real-time power system simulators with the ability to interface real devices to perform hardware-in-the-loop experiments. The capability to interface with real devices with real communication protocols allows a realistic testing of the various attack/defense scenarios. Also, this helps to study the impacts of malicious events in the SCADA control layer (say the opening or closing of a breaker) on the physical layer (in the power grid) and the communication layer (in the wide-area network) in real time and vice versa.

16.4 State-of-the-art – literature review

The development of CPSS security testbeds and cyber security experimentation leveraging CPSS testbeds has been an active research topic over the last few years. Several educational institutions and research laboratories are setting up individual CPSS security testbeds for validating their research and are also exploring possibilities for

CPSS testbed federation to perform large-scale, high-fidelity, hardware-in-the-loop cyber attack/defense experiments.

The National SCADA Testbed, at Idaho National Laboratory, is one of the earliest collaborative testbed efforts across several US national laboratories where actual power system resources (generation and transmission lines), and the associated cyber system components were used to perform vulnerability assessments and impact analysis studies [9]. Sandia National Laboratory houses the Virtual Control System Environment, which is a cyber-physical testbed with a hybrid mix of simulated, emulated, and physical system components for attack/defense experimentation [10]. The University of Illinois at Urbana-Champaign hosts the virtual power system testbed, consisting of a mix of simulated, emulated, and physical components to perform cyber security research across a wide range of topics including vulnerability assessments, communication protocol, and security product testing and research validation [11]. The powerNET testbed at Pacific Northwest National Laboratory (PNNL) leverages a combination of cloud-based computing resources, simulators, and a variety of real field equipment to provide a highly customizable and configurable environment for cyber security experimentation [12]. The PowerCyber CPSS security testbed at Iowa State University (ISU) consists of a hybrid mix of industry-grade SCADA hardware and software, emulators for wide-area network routing, and real-time power system simulators, which enable a high-fidelity, hardware-in-the-loop, cyber-physical testbed environment for cyber-physical security research and experimental validation studies [13]. The SCADA testbed at the University College, Dublin, is used to perform experiments on intrusion detection and anomaly detection related to power grid cyber security [14]. Mississippi State University consists of a SCADA security testbed, which is used to perform intrusion detection and security testing of phasor measurement units and data concentrators [15].

The University of Southern California's Information Sciences Institute hosts the DETER cyber security testbed, which is a large-scale testbed for cyber security experimentation, but it does not contain physical system resources, such as relays and phasor measurement units, which some of the earlier testbeds mentioned have for hardware-in-the-loop experimentation [16]. The Smart Grid testbed at Washington State University consists of several hardware resources such as phasor measurement units and phasor data concentrators interfaced with a real-time power system simulator, but is primarily focused on developing new power system monitoring and control algorithms. Other notable research efforts that leverage CPSS security testbeds for experimentation on telecommunication, networking, and intrusion detection include the European CRUTIAL project [17], Testbed for Analyzing Security of SCADA Control Systems (TASSCS) testbed at the University of Arizona [18], SCADASim testbed at the Royal Melbourne Institute of Technology [19].

16.5 Testbed research areas

As mentioned previously, CPSS security testbeds capture the cyber-physical-social interdependencies accurately and enable flexibility in the type of experimental scenario that could be evaluated by leveraging a mix of simulated, emulated, and physical devices appropriately using abstraction and virtualization. Therefore, they enable research and validation of a broad spectrum of research areas [13]. Some of the major research areas are:

1. *Vulnerability assessment* – This allows for the systematic inspection of industry standard software and hardware platforms, network protocols and configurations to detect unknown vulnerabilities, which could be exploited, and responsibly disclosing them to vendors and appropriate regulatory authorities.
2. *Impact analysis* – CPSS security testbeds allow the study and quantification of the physical system impacts from various cyber attack scenarios on the cyber system components such as relays, phasor measurement units, and also on the communication network in terms of network latencies, and throughput. The development of metrics, which combine cyber-physical properties, is a critical research gap that could be addressed through the use of CPSS testbeds.
3. *Mitigation research* – The integration of industry grade software/hardware allows the implementation of different types of defense measures such as firewalls, intrusion detection systems, and software patches, and evaluates the performance of various mitigation strategies against different attack vectors, system topologies, and configurations.
4. *Data and model development* – The availability of standardized models and datasets often are a limiting factor in performing realistic experimentation in the area of CPSS security. Testbeds could be used to develop and share standard models and datasets that provide researchers with the information required to evaluate and test innovative security approaches.
5. *Conformance testing* – CPSS security testbeds provide environments where new security devices or algorithms could be tested for compliance requirements, and also for the purposes of interoperability/standardization.
6. *Operator training* – CPSS security testbeds serve as a realistic training environment for system operators to be trained under several types of cyber attack scenarios through orchestrated cyber attack/defense exercises. These improve the situational awareness and response capabilities of system operators and cyber first responders during actual cyber contingencies.

16.6 Testbed Federation

One of the factors that limit the utility of CPSS security testbeds is the scalability of the system that can be used for a particular type of experiment. The cost of scaling a single CPSS testbed in order to perform a large-scale, real-time, hardware-in-the-loop experiment becomes prohibitively expensive and therefore renders such scenarios impractical. By exploiting the concepts of abstraction and virtualization, we can simplify the system model under study to a certain extent and scale the cyber components; however, overall system scalability is still limited by the availability of real hardware devices.

Therefore, it is critical to federate CPSS testbeds, which is the idea of pooling together testbed/simulation resources across organizations in a distributed manner to share resources and create an integrated environment to perform large-scale, high-fidelity, real-time, closed-loop cyber attack/defense experimentation.

Conceptually, there are two possible approaches to testbed federation: federation at the information/control layer and federation at the physical and communication layers. The first approach involves the pooling of resources at a high level such as control areas, exchanging limited data as part of the experiment. Although this approach cannot support use-case scenarios such as protection and control, it provides an environment to study higher level EMS applications such as automatic generation control. The latter approach could potentially support real-time, wide-area protection, and control applications but are governed by the strict application latency constraints which may not be satisfied, and/or modeling accuracy when two simulators are coupled over a wide-area communication network. Although there are several challenges that exist in ensuring the tight integration of physical components at various locations, CPSS testbed federation provides a sustainable approach to scalability and its potential definitely merits further exploration.

16.6.1 Proof-of-concept federation architecture – Smart America challenge

This section briefly describes the CPSS testbed federation that was demonstrated at the Smart America Challenge Expo held in Washington, D.C., on June 1, 2014 [20]. As part of this effort, the PowerCyber cyber-physical testbed at ISU was successfully federated with University of Southern California's DETER testbed to perform realistic, scalable, and high-fidelity attack/defense studies on wide-area protection application. Figure 16.2 shows the implementation architecture of the testbed federation between ISU and DETER. The two testbeds were interconnected using a virtual private network (VPN) tunnel to establish seamless network connectivity among control center, substations, and field devices that were distributed on the federated testbed.

The control center and a few substations were implemented inside the DETER testbed. The DETER testbed environment allows the configurability of the topology of the cyber network and therefore, the connectivity between control center and the substations were setup corresponding to realistic configurations. Some of the substations were hosted inside the ISU PowerCyber testbed. Both the substations running inside DETER and the PowerCyber testbed shared the physical relays Relay 1 and Relay 2 inside PowerCyber testbed. The gateway nodes ensured that all network traffic was tunneled across the testbeds. The relays were interfaced to the real-time digital simulator (RTDS) at PowerCyber through the GOOSE protocol, which is a part of the IEC 61850 standards. The control center communications with the substations used the industry standard DNP3 protocol. The substations used the MMS protocol, which is also a part of IEC 61850 standards, for communicating with the relays. In the experimental scenarios, the attack originated from inside the DETER testbed and was tunneled into the PowerCyber testbed to target the relays as shown by the dotted arrows in Figure 16.2.

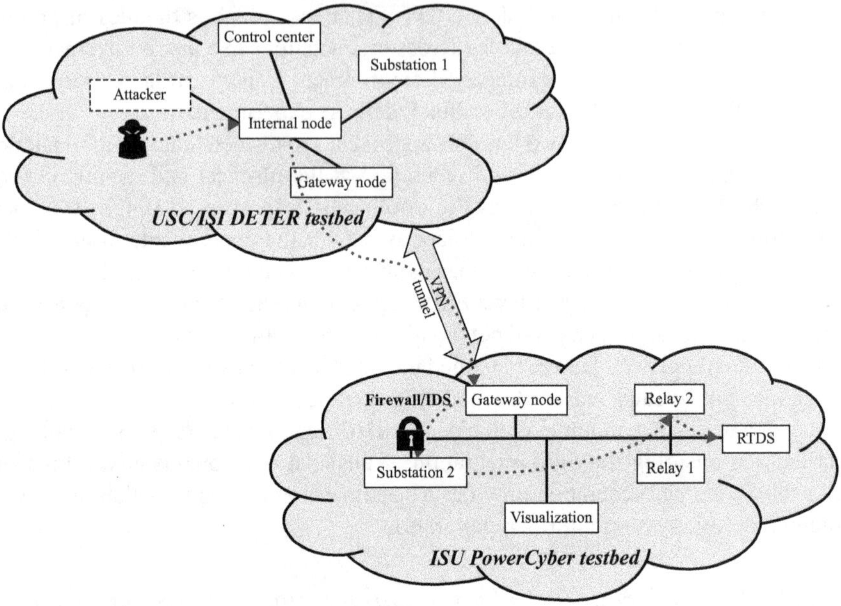

Figure 16.2 Proof-of-concept implementation architecture of CPSS testbed federation

As part of the defense strategy, perimeter defense was employed and therefore each substation network had an intrusion detection system, which filtered malicious network traffic. The visualization node was running an open platform communications (OPC) server, which talked to the substations to gather data about the status of the relays and fed it to a visualization engine. The visualization engine overlaid relay status information of various lines and attack/defense dynamically on a map as they occurred. As part of the experimentation, the latencies across the two end points on the VPN tunnel were measured and the use-case scenario was chosen appropriately. This work was a proof-of-concept to show the value of CPSS security testbed federation. A more systematic exploration on the feasibility of other architectures and use-case scenarios are being explored actively by leveraging and expanding on this foundation.

16.7 Case study application – Coordinated cyber attacks on WAMPAC

In this section, we present a detailed case study of a coordinated cyber attack performed on Wide-Area Monitoring, Protection and Control (WAMPAC) applications, namely the AGC, and a generator rejection remedial action scheme on ISU's Power-Cyber CPSS security testbed. Figure 16.3 shows the implementation architecture for the use-case scenario in this case study.

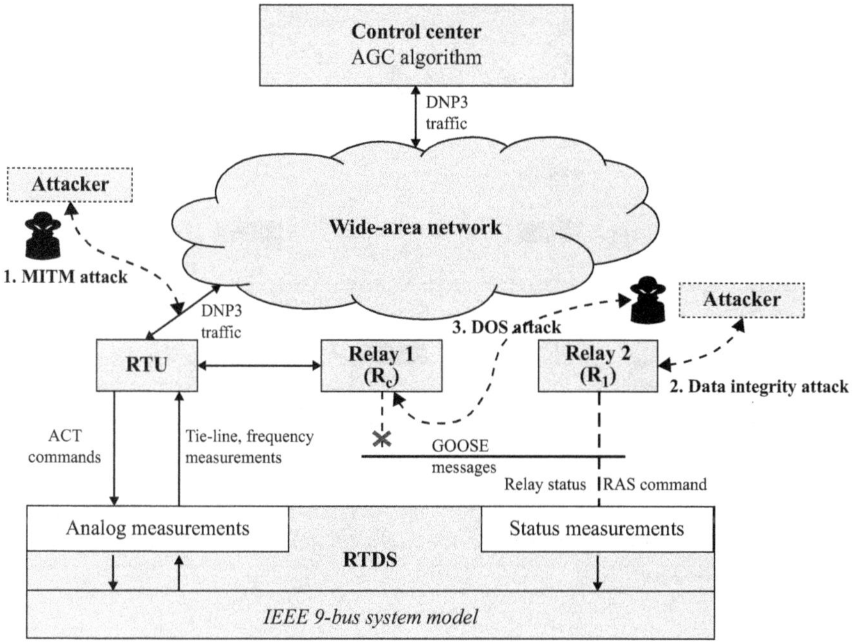

Figure 16.3 Implementation architecture of use-case scenario on the PowerCyber testbed

The standard IEEE 9-bus power system model is implemented in the RTDS. The analog measurements and control signals corresponding to the AGC algorithm are communicated from/to the RTDS to/from the control center via the substation RTU over the DNP3 communication protocol. Figure 16.3 shows only one RTU for simplicity. Two physical relays Relay 1 and Relay 2 in the testbed are utilized for implementing the RAS scheme along with the RTDS. The relays and RTDS exchange GOOSE messages similar to real substation implementations to communicate the status signals when a relay trips or in order to send generation rejection commands to the generating stations in the RTDS model. The relays in the testbed are mapped to control breakers in the RTDS power system model enabling real-time, hardware-in-the-loop experimentation. Therefore, actions such as data integrity attacks on the cyber layer and/or denial of service (DOS) attacks on the communication layer can be observed in the power system model instantaneously.

Figure 16.4 shows the IEEE power system model along with the control areas for AGC algorithm, and also the location of the generator rejection RAS that has been implemented. The relays Relay 1 and Relay 2 on the testbed are mapped to R_c, and R_1 in the power system model, respectively. As part of the experimental scenario, we have also implemented line overload protection on the critical transmission lines, and under frequency load shedding (UFLS) similar to real-world implementations. The relay R_2 corresponding to overload protection in the power system model is modeled

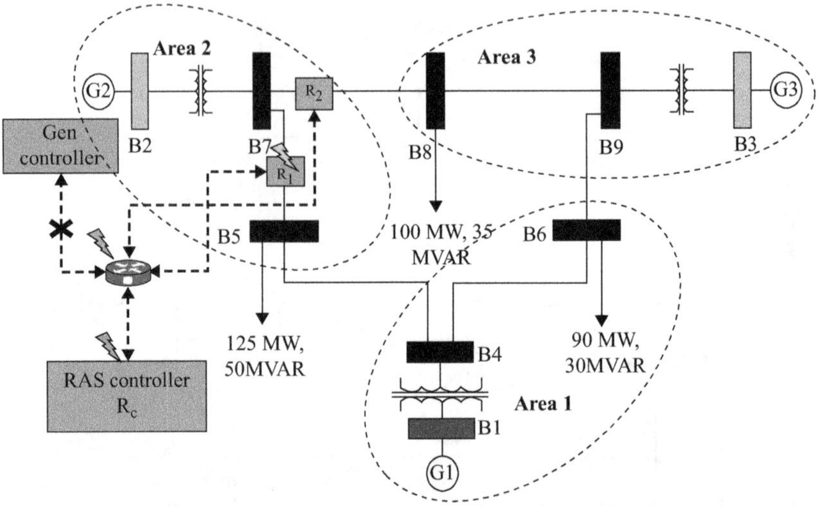

Figure 16.4 IEEE 9-bus power system model with AGC control areas and RAS location

inside RTDS. We leveraged the communication modules in RTDS for exchanging the real-time simulation data using the DNP3 protocol (Tie-line power flow and frequency measurements), and the GOOSE protocol (relay status, RAS generator rejection command). This enables a realistic implementation of AGC and RAS functions similar to real-world settings.

In this case study, we only consider the AGC functioning with respect to Area 1 as shown in Figure 16.4 for simplicity. The associated tie-line flows for Area 1 are flows on transmission lines 4-5 and 6-9. According to AGC operation, any load change on Bus 6 would be addressed by correspondingly adjusting the mechanical turbine input power for generator G1 corresponding to the area control error (ACE) value.

We have implemented the generator rejection RAS to decrease the generation level of G2 when either line 7-8 or line 7-5 is out of service and the power output from G2 is over a certain threshold. For example, if line 7-5 is tripped due to a natural or malicious event, the relay corresponding to line 7-5 (R_1) will send out a status change alert to the RAS controller (R_c). On the condition that the RAS is currently armed (when G2 is over a pre-determined threshold), the RAS controller will send a command to G2 to ramp it down by a preset value. After the ramping down command is successfully carried out, the overload on line 7-8 due to the tripping of line 7-5 can be alleviated.

As line overload protection is often armed for critical transmission lines, when the line flow goes beyond the secure level for a certain amount of time, the relay associated with line 7-8 (R_2) will trip the line for self-protection. In this case study, overload protection is installed only for line 7-8 and line 7-5. As a last defense measure to prevent system frequency from causing system islanding, UFLS is configured as to shed

Table 16.1 IEEE 9 bus system base case

Line flow (MW)		Generation level (MW)		Load level (MW)	
Line 7-8	42.34	Gen1	150.3	Load1	90
Line 7-5	37.71				
Line 4-5	87.65	Gen2	80.1	Load2	125
Line 4-6	62.61				
Line 9-6	27.53	Gen3	85.4	Load3	100
Line 9-8	57.84				

Table 16.2 RAS and UFLS configuration

Scheme	Arming condition	Trigger condition	Action
RAS for G2	P2 > 85 MW	Either line 7-8 or line 7-5 is out of service	Ramp down G2 by 40 MW
UFLS in area 2	Always	Frequency < 59.7 Hz	Shed local load by 70 MW with 2 s delay
		Frequency < 59.6 Hz	Shed local load by 40 MW with 0 s delay
Overload protection for line 7-8/line 7-5	Always	Flow on the remaining single line > 70 MW for more than 20 s	Trip the overloaded line

70 MW load from Bus 5 with 2 s delay when system frequency goes below 59.7 Hz and to further shed another 40 MW more immediately if the frequency is below 59.6 Hz.

The base case of the IEEE 9 bus system is listed in Table 16.1, and the configuration of line overload protection, RAS and UFLS in Table 16.2. The coordinated attack vector that was implemented for the case study has been described in the following paragraphs.

16.7.1 Coordinated attack vector

In practice, RAS are seldom used, as the condition to arm the RAS is satisfied only under heavy loading conditions such as peak summer loads. But when under peak load, RAS would be enabled to maximize transmission line utilization. In our case study, the generation level of G2 is very close the arming threshold in the base case. As shown in Table 16.1, the generation level of G2 has only a 5 MW margin for the RAS to be armed. The coordinated attack vector consists of the following actions:

Step 1: Man-in-the-middle (MITM) attack

The attacker compromises the tie-line flow and frequency measurements of Area 1 as they are in transit between the RTU and the control center. This MITM attack sends spoofed measurements to the control center resulting in ACE values that decrease the generation of Area 1, i.e., G1 will decrease its power output gradually. This causes the other generators to increase their output due to the governor actions to counteract the drop in system frequency. This increases generator G2's output above its threshold, effectively arming the RAS.

Step 2: Data integrity attack

The attacker trips line 7-5 (R_1) by sending a trip command to Relay 2 in the testbed after the RAS gets armed. This initiates the RAS sequence, where R_1 sends a status update to the RAS controller R_c. R_c checks if the RAS is armed, and sends a generator rejection command to the generation controller for G2 (modeled inside RTDS). If this command does not reduce the generation within a certain amount of time, overload protection relay (R_2) will trip line 7-8 and isolate generator G2 from the rest of the system.

Step 3: Denial of service (DoS) attack

The attacker blocks the RAS generation rejection command sent from the RAS controller R_c, which should ramp down G2. This is achieved by carrying out a DoS attack on the relay R_c. As G2 does not ramp down due to the DoS attack, overload protection relay R_2 times out and trips the line 7-8.

The dotted arrows in Figure 16.3 show the implementation of the attack vector on the PowerCyber testbed.

16.7.2 Impact analysis

Figures 16.5 and 16.6 show the various impacts of the coordinated attack on AGC and RAS. The top subplot in Figure 16.5 shows the variation of system frequency with time and the bottom subplot shows the corresponding variation of critical tie-line power flows on lines 7-8, 7-5, and 4-5. Similarly, the top subplot in Figure 16.6 shows the variation of voltages at buses 4, 5, 7, and 8, respectively, as they are involved in the attack vector. The bottom subplot in Figure 16.6 shows the output of the three generators and the load in area 2 during the entire scenario.

The timeline of all the major events in this attack vector is as explained below:

1. AGC measurement spoofing attack is carried out 8 s after simulation begins, and this causes G1's output to reduce slowly. Consequently, G2's output increases due to governor action and eventually crosses the threshold to arm the RAS for G2 at around 28 s.
2. R_1 (Relay 2) is tripped by the attacker at around 28 s almost instantaneously after the RAS is armed. At the same time, the attacker also floods the RAS controller R_c (Relay 1).
3. Due to the DoS attack, the generator G2 does not ramp down and therefore around 48 s, line 7-8 is tripped by the overload protection (R_2) and G2 is completely separated from the system.
4. System frequency decreases sharply due to a significant loss in generation, which triggers both the UFLS stages. After 110 MW load out of the total 125 MW being tripped, the system frequency finally recovers to 60 Hz.

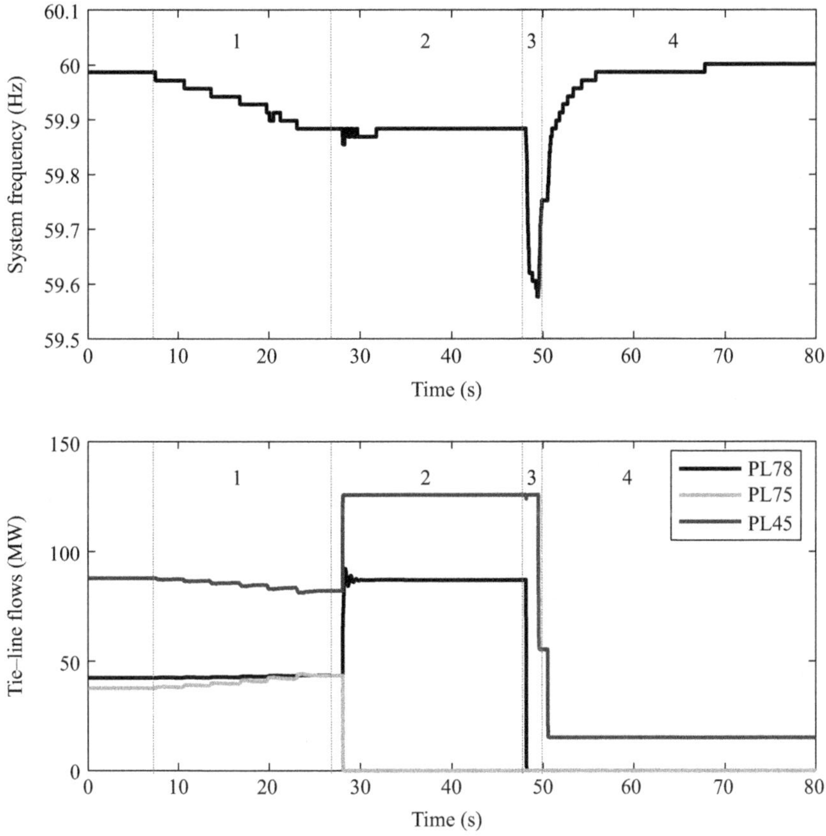

Figure 16.5 System frequency (Hz) and tie-line flows (MW) during the coordinated attack

All the subplots in Figures 16.5 and 16.6 have been annotated with these four major phases as mentioned in the timeline to aid the reader's understanding.

Frequency impacts – From the top subplot in Figure 16.5 we can see that the initial AGC attack (Phase 1) causes a steady decline in system frequency. The triggering of the RAS by tripping R_1 (Phase 2) causes a minor disturbance, but the governor actions restore the frequency to a small extent. After the overload protection trips the line 7-8 (Phase 3), there is a sharp drop in frequency, as G2 is isolated, thereby, causing a huge deficit in generation. Finally, UFLS schemes kick in (Phase 4) to shed loads in two major chunks to restore the frequency back to normal.

Tie-line flow impacts – From the bottom subplot of Figure 16.5, we can see the impacts on the tie-line flows during the attack. As mentioned previously, the AGC attack (Phase 1) causes a reduction of generation in Area 1, which decreases the flow on line 4-5 steadily. After the line 7-5 is tripped (Phase 2), the power flow on

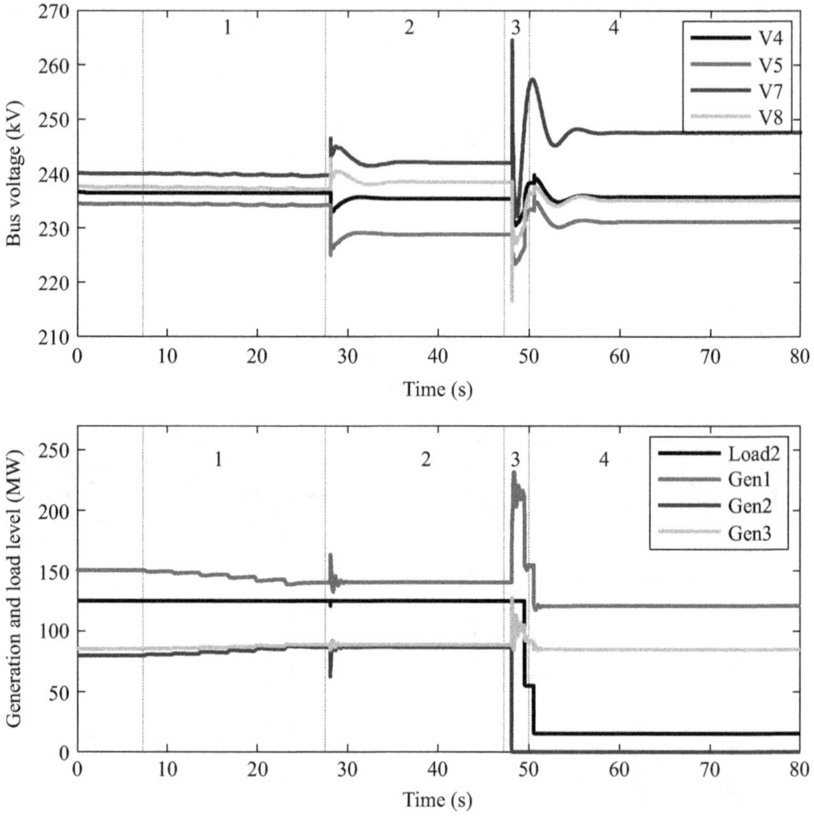

Figure 16.6 Bus voltages, generation, and load levels during the coordinated attack

line 4-5 increases again due to governor action to supply the load at bus 5. This is also accompanied by a loss in frequency. As a result of the DoS attack, the overload on line 7-8 (Phase 3) does not reduce until it is tripped out by thermal overload protection. Finally, the power flow on line 4-5 is reduced after the UFLS sheds the load (Phase 4).

Generation and load impacts – The bottom subplot of Figure 16.6 shows the impact of the generation outputs and the load during the coordinated attack. The AGC attack clearly can be observed as a steady decrease in the output of G1, corresponding to a steady increase in generators G2 and G3 due to the governor actions (Phase 1). All generators experience a minor fluctuation when the RAS is triggered (Phase 2). Due to the DoS attack, generation output of G2 does not decrease until it is isolated completely due to tripping of line 7-8 (Phase 3). UFLS kicks in to shed a bulk of the load at bus 5, which reduces the output of the generator G1 that was feeding the load (Phase 4). Overall, this coordinated attack isolated G2 from rest of the system, and caused a forced load shedding of 110 MW in area 2.

Voltage impacts – From the top subplot of Figure 16.6 we can observe that the impacts on system voltages are much more severe than generation outputs. The line trip (Phase 2) creates a temporary voltage fluctuation, which damps out quickly. However, after the UFLS kicks in (Phase 4), there is a huge voltage fluctuation as the magnitude of load shedding is significant. In this case study, we have not modeled any voltage sensitive loads on the system. However, the evaluation of voltage impacts on voltage sensitive loads such as induction motors would reveal further insights into the stability of the system during such attacks. These would be explored as part of our future work.

16.8 Educational and outreach aspects

CPSS security testbeds also serve as excellent educational and outreach platforms and could be leveraged as part of undergraduate/graduate research curriculum to impart critical inter-disciplinary knowledge, and hone practical tool-based skills, thereby increasing awareness of cyber security among students, and also practicing engineers/industry professionals through short courses, hands-on workshops and cyber attack/defense training exercises. As part of the curriculum CPSS testbeds could be leveraged for laboratory exercises on power system simulation, communication, and networking aspects relating to cyber security, and security concepts such as port scanning, vulnerability assessment, and malware. Also, they provide platforms for students to work on graduate course projects or undergraduate capstone senior design projects under the guidance of other graduate students. Another potential utility of CPSS security testbeds is that they could be potentially used as a training ground for transforming the current passive, table-top grid security exercise such as the NERC GridEx [21,22] into an active, real-time cyber attack/defense exercises to improve cyber security preparedness and response capabilities against various cyber threats of utility and industry stakeholders.

16.9 Conclusion

This chapter motivated the need for CPSS security testbeds, and briefly described the various testbed design objectives and challenges for CPSS security experimentation. A conceptual three-layer testbed architecture was introduced and along with a review of state-of-the-art research in CPSS security testbeds. Then, the chapter described the broad topical research areas that CPSS security testbeds enable.

The chapter also addressed some of the limitations of current testbed efforts and motivated the need for CPSS testbed federation. The chapter briefly introduced general architectures for testbed federation and identified their strengths and weakness for enabling realistic, large-scale, high-fidelity attack/defense experimentation on the Smart Grid. A proof-of-concept case study of CPSS security testbed federation for coordinated attack/defense experiment at the Smart America Challenge was also presented. Then, a detailed case study of coordinated cyber attacks on WAMPAC using the PowerCyber CPSS security testbed at ISU was presented. Finally, the chapter described how CPSS testbeds could potentially be leveraged for educational and outreach activities to promote workforce training and development in the critical area of CPSS for the smart grid.

References

[1] Government Accountability Office (U.S.), "Critical Infrastructure Protection Challenges and Efforts to Secure Control Systems," GAO-04-354, March 2004.

[2] Idaho National Laboratory (U.S.), "Common Cyber Security Vulnerabilities Observed in Control System Assessments by the INL NSTB Program," November 2008.

[3] North American Electric Reliability Corporation and the Department of Energy (U.S.), "High-Impact, Low-Frequency Event Risk to the North American Bulk Power System," November 2009.

[4] S. Baker, S. Waterman, and G. Ivanov, "In the Crossfire: Critical Infrastructure in the Age of Cyber War," McAfee, 2009.

[5] N. Falliere, L. Murchu, and E. Chien, "W32.Stuxnet Dossier," Version 1.3, Symantec, November 2010.

[6] The U.S. White House – Office of Press Secretary, "Presidential Policy Directive – Critical Infrastructure Security and Resilience," February 2013. [Online]. Available: https://www.whitehouse.gov/the-press- office/2013/02/12/presidential- policy-directive-critical-infrastructure-security-and-resil, last accessed: 10 October 2015.

[7] Government Accountability Office (U.S.), "More Comprehensive Planning Would Enhance the Cybersecurity of Public Safety Entities' Emerging Technology," GAO-14-125, January 2014.

[8] A. Ashok, A. Hahn, and M. Govindarasu, "A cyber-physical security testbed for smart grid: System architecture and studies," in *Proceedings of the Seventh Annual Workshop on Cyber Security and Information Intelligence Research Series* CSIIRW '11, ACM, New York, NY, 2011.

[9] Idaho National Laboratory (U.S.), "National SCADA Test Bed: fact sheet," 2007.

[10] M. J. McDonald, "Modeling and Simulation for Cyber-Physical System Security Research, Development and Applications," SAND2010-0568, Sandia National Laboratories, 2010.

[11] D. C. Bergman, D. Jin, D. M. Nicol and T. Yardley, "The virtual power system testbed and inter-testbed integration," Presented at *Second Workshop on Cyber Security Experimentation and Test*, 2009.

[12] T. Edgar, D. Manz, and T. Carroll, "Towards an experimental testbed facility for cyber-physical security research," In *Proceedings of the Seventh Annual Workshop on Cyber Security and Information Intelligence Research (CSIIRW '11)*, 2011, ACM, New York, NY.

[13] A. Hahn, A. Ashok, S. Sridhar, and M. Govindarasu, "Cyber-physical security testbeds: Architecture, application, and evaluation for smart grid," *IEEE Transactions of Smart Grid*, June 2013, vol. 3, pp. 847–855.

[14] J. Hong, S.-S. Wu, A. Stefano, et al., "An intrusion and defense testbed in a cyber-power system environment," in Presented at *Power and Energy Society General Meeting*, Detroit, MI, 2011.

[15] B. Reaves and T. Morris, "An open virtual testbed for industrial control system security research International Journal of Information Security," 2012. [Online] Available: http://dx.doi.org/10.1007/s10207-012-0164-7

[16] T. Benzel, "The science of cyber security experimentation: The deter project," Proceedings of the *27th Annual Computer Security Applications Conference Series* ACSAC'11, ACM, New York, NY, 2011, pp. 137–148.

[17] G. Dondossola, G. Deconinck, F. Garrone, and H. Beitollahi, "Testbeds for Assessing Critical Scenarios in Power Control Systems," Berlin: Springer, 2009, pp. 223–234

[18] M. Mallouhi, Y. Al-Nashif, D. Cox, T. Chadaga, and S. Hariri, "A testbed for analyzing security of SCADA control systems (TASSCS)," Presented at IEEE PES Innovative Smart Grid Technologies (ISGT), Anaheim, CA, 2011.

[19] C. Queiroz, A. Mahmood, and Z. Tari, "SCADASim – a framework for building SCADA simulations," *IEEE Transactions of Smart Grid*, December 2011, vol. 2, pp. 589–597.

[20] G. Manimaran, T. Benzel, and A. Hahn, "Smart Energy CPS – CPS Security Testbed Federation for Coordinated Cyber Attack/Defense Experimentation," [Online] Available: http://smartamerica.org/news/iowa-state-researchers-to-demonstrate-cyber-physical-security-testbed-for- power-grid-at-smartamerica-challenge-expo/, last accessed: 10 October 2015.

[21] North American Electric Reliability Corporation (U.S. and Canada), "2011 NERC Grid Security Exercise – After Action Report," March 2012. [Online]. Available: http://www.nerc.com/pa/CI/CIPOutreach/GridEX/NERC_GridEx_AAR_16Mar2012_Final.pdf, last accessed: 10 October 2015.

[22] North American Electric Reliability Corporation (U.S. and Canada), "Grid Security Exercise (GridEx II) – After Action Report," March 2014. [Online]. Available: http://www.nerc.com/pa/CI/CIPOutreach/GridEX/GridEx%20II%20Public%20Report.pdf, last accessed: 10 October 2015.

Chapter 17

Distributed real-time simulations for electric power engineering

Manish Mohanpurkar[1], *Mayank Panwar*[2], *Sayonsom Chanda*[1], *Marija Stevic*[3], *Rob Hovsapian*[1], *Vahan Gevorgian*[4], *Siddharth Suryanarayanan*[2], *and Antonello Monti*[3]

Abstract

Real-Time Simulations (RTS) are increasingly being used to understand the complex device and system level interactions in power grids. RTS provides the capability to create detailed, highly accurate, and diverse set of power and control system components at low time steps (order of microseconds) that are based on "real-world clock-time." RT simulator is a unique architecture with specialized processors and communication boards that allow time synchronization of simulations and the clock-time. Lean operating systems, specialized processors, faster communications, etc. are the typical attributes of RT simulators. RT simulators provide a unique capability of interfacing with power and control components via analog and digital interfaces. However, RT simulators have limited computational capability that constrains the size of power and control systems that can be simulated. Multiple simulators connected locally is typically used to increase the computation capability, however this is not always economical. Additionally, RT simulators are used at facilities with unique test infrastructure in the form of grid emulators, inverters, photovoltaic, wind turbine, microgrids, etc. Performing distributed RTS via Internet can augment simulation capacity and leverage unique infrastructure that is dispersed in academia and research laboratories. Research related to distributed RTS and its application in electric power engineering is discussed in this chapter.

[1]Power and Energy Systems Department, Idaho National Laboratory, Idaho Falls, ID, USA
[2]Department of Electrical and Computer Engineering, Colorado State University, Fort Collins, CO, USA
[3]Institute for Automation of Complex Power Systems, E.ON Energy Research Center, RWTH Aachen University, Aachen, Germany
[4]Energy Systems Integration Facility, National Renewable Energy Laboratory, Golden, CO, USA

17.1 Introduction

Rapid evolution and modernization of the electric grid in myriad ways, necessitates a holistic systems approach to understanding power and energy systems. The complexity of architecture and operation of electric grids is increasing exponentially. Renewable energy based Distributed Energy Resources (DERs) are one of the prime architectural and operational changes in distribution networks. DERs are increasingly utilized to compensate carbon footprint of generation, enhance resiliency, defer investment in large power plants, and maintain critical loads with greater reliability. Advanced power electronics based devices are being added to the transmission networks for optimal scheduling and operations. Switching frequencies of such devices are in the range of a few kilohertz (kHz) to megahertz (MHz) introducing transients and harmonics to the grid [1,2]. Thus, there is a need for component as well as system level analysis under steady-state and dynamic conditions to fully interpret the impacts of DER penetration and advanced power electronic devices. Real-Time (RT) simulators are increasingly being utilized to analyze the dynamic and transient response of electric power systems with DERs and power electronic converters. Additionally, RT simulators provide a suitable platform for performing Hardware-in-the-Loop (HIL) testing of power and control components to characterize functionality [3]. A single RT simulator may not always provide the necessary computational capability for analyzing a detailed power network of practical size. A locally connected bank of RT simulators locally connected is typically used for solving such larger cases [4]. However, in most cases the RT simulators may be distributed at different locations that are not directly connected.

An historical account of efforts related to distributed RTS is summarized in Section 17.3. An approach to perform geographically distributed RT simulations, based on standard communication protocols, between two power systems that are loosely coupled is discussed in the Sections 17.4 through 17.6. Section 17.7 summarizes few innovative applications of distributed RTS. HIL using the geographically distributed RT simulators, at Idaho National Laboratory (INL), Idaho Falls, Idaho and National Renewable Energy Laboratory (NREL), Golden, Colorado for characterizing devices is also provided in this chapter. Furthermore, an approach to leveraging assets of power and energy test-beds located across the globe at various research centers in universities and research laboratories is presented in Section 17.8. A futuristic vision of expanding the connectivity between RT simulators based on distributed Real-Time Simulations (RTS) between different research centers is elaborated under concluding remarks. Facilitating greater collaboration between power and energy research community based on distributed RTS and HIL is one of the long-term objectives of the research discussed in this chapter.

17.2 Distributed real-time simulations

RTS is a unique approach that allows a thorough assessment of a component in a suitable test system based on actual "clock-time." The independent time variable in RTS is programmed to generate simulation conditions that are very close to real-world conditions. RTS environments or RT simulators are based on special hardware and software architectures that enable the distinctive feature of RT. At the core, RT simulators consist of highly advanced and powerful processors, high speed communication between processors, specialized operating systems, and interface to hardware [4]. These commonly observed components of RT simulators may be either vendor specific or commercially off the shelf. Such design characteristics determine the capabilities and critical performance specifications of the simulators. The development of processors and their computation capability has significantly influenced the overall RT simulator evolution.

RTS have numerous applications in the world of engineering and technology. In power systems, RT simulators have been historically utilized for rapid controller prototyping via verification and validation as Controller-Hardware-In-the-Loop (CHIL), HIL testing of power devices, protection scheme design and testing, and renewable energy related impact studies [5–8]. Model-Based Design (MBD) and Automatic Code Generation (ACG) are salient features of RT simulators of the current generation which greatly facilitates research methodologies [9]. MBD is a widely adopted technique of performing complex system design and analysis based on mathematical and graphical methods [9]. Inclusion of ACG and its allied features in MBD-based RT simulators has multifold advantages leading to script creations and automation of lengthy experiments.

RT simulators are utilized for a multitude of applications varying from DER and power generation, automotive design, protective schemes, electric ship design and testing, aviation, and power engineering education and research [4,8]. RT simulators developed for such applications are based on specific hardware and software architectures and design principles. Parallel computing and low overheads for operating systems are two such important design principles. Power systems network partitioning into subsystems to load the multiple processor cards in the RT simulators is a critical design process, as each processor card needs to be optimally utilized [4]. Commercially available RT simulators have a mixed trend of either manually portioning or algorithmically optimizing the processor loadings. Multiple RT simulators are available and they possess diverse sets of capabilities and functionalities. A holistic comparison of commercially available RT simulators based on architecture, user interface, simulation capacity, capabilities, and modeling platform is provided in Reference 10. Research and simulation objectives govern the choice of RT simulators as its inherent characteristics will influence the outcome.

17.2.1 Philosophy of distributed real-time simulation

RT simulators are specialized digital simulators that are capable of performing specialized computations within 'real clock' constraints. They also allow a user-friendly modeling environment and ability to process hardware-related information simultaneously. As described earlier, high-end processors at the core of RT simulators provide the necessary hardware to simulate systems as necessary. The number of processors utilized in the design of an individual RT simulator or a bank of such simulators

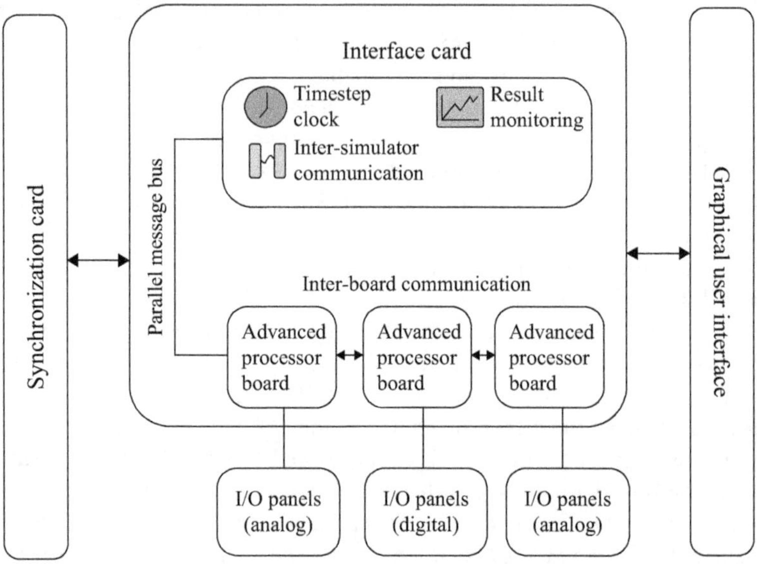

Figure 17.1 A sample architecture of utilizing synchrophasors with a power system for model validation and co-simulation of tightly coupled power systems

determines the RTS capacity. Other typical components of RT simulators are shown in Figure 17.1. Components of RTS can be classified under control systems, power electronics, power systems components, sensors and measurements, and mechanical systems. The size of any unified system representation modeled for analysis in an RT simulator is primarily constrained by the capacity of processors. The evolution of the computing capability of processors has increased drastically and it has reflected in the overall simulation capacity of RT simulators. Limitation on the size of systems simulated in RT is a function of numerous factors such as hardware, software, processor architectures, and type of environment. Multiple units of such simulators may be configured together to operate as a single unit to provide larger simulation capacity and perform system analysis. Design of experiments for RTS involving either large-scale or greater granularity of power systems may become challenging, as larger simulation capacity is needed, which may not be available at the same location. This may be a hindrance to obtaining highly accurate results from the RT simulators for large-scale power systems. Larger system simulations require more units of RT simulators that may be cost-prohibitive in many cases.

A potential solution to address the lack of access to larger cluster of locally available RT simulators for larger system simulations is the fast paced evolution of the processor capacity and related design of the RT simulator. This development requires equally critical development of robust communication mechanism between the processors to support the increased bandwidth. The cost of processor-based computations is decreasing, and will eventually lead to a lower cost of RT simulators as well. Thus, RT simulators with a greater computation capacity and comparatively lower costs may make locally available clusters of RT simulators more economically viable. This progression is well driven by the commercial processor businesses as well as commercial RT simulator vendors.

Another potential solution is to directly connect geographically distributed RT simulators by point-to-point cables. This serves as an even more cost-prohibitive solution than creating larger RT simulator capabilities and is not a feasible solution for most research centers. However, custom point-to-point solutions may have applications in remote physical systems that are inaccessible, especially after field commissioning. These custom network solutions have to be incorporated during the design phase of complex systems. Numerous versions of point-to-point interconnections that resemble intended applications of distributed RTS for troubleshooting and firmware upgrades exist in practice as described in Section 17.3.6. Lack of flexibility and high cost are the two major drawbacks in such communications between systems and simulators. However, advantages include a controlled environment for performing mission critical tasks and limited cyber issues and interferences.

One of the most cost-effective and technically feasible ways to connect geographically distributed RT simulators is the utilization of standard communication protocols. RT simulators are equipped with network interface cards suitable for implementing standard communication protocols that are typically used in power systems operation and controls. Numerous communication protocols, for both Wide Area Network (WAN) and Local Area Network (LAN), are typically available in commercial RT simulators for various applications. Utilizing an existing WAN along with the Internet Protocol (IP) for communication is a feasible way to connect distributed RTS [11,12].

17.2.2 Transmission-distribution-communication co-simulation

Transmission-Distribution-Communication (TDC) co-simulation refers to a concerted approach of modeling and simulation of coupled critical power system components, events, and scenarios in a common framework, so that interaction between the constituent models can be investigated with closest resemblance to real-world phenomena. Geographically distributed RTS derives basis from TDC co-simulations in regard to "loose-coupling" of RT simulators. In this regard, loose-coupling of RT simulators implies that two environments are coupled with WAN-based communication without direct changes to the environment itself. In general, experiences from TDC co-simulation, parallel processing, and High-Performance Computing (HPC) clusters may be leveraged to rapidly develop the concept of distributed RTS and HIL.

Prior to the availability of RT simulators, utilities assembled a scaled-down model of their physical infrastructure for comprehensive HIL testing of control systems, protection equipment, and inter-device communication commonly referred to as Transient Network Analyzers (TNA). However, utilities required significant time, effort, and financial investments to build scalable, scenario-specific TNAs. With advancements in parallel processing capabilities of the digital computer in the early 1990s, there was an industry-wide paradigm shift toward testing using RT simulators.

Unlike TNAs, RT simulator capabilities are not limited to closed-loop HIL testing or studying traveling wave phenomena in large transmission lines. They can be used to co-simulate communication between power system components, which are critical to proper coordination of control and systematic operation of the power system. Thus, providing a complete platform to recreate the engineering ecosystem required to sustain reliable operation of power systems, and geographically distributed RT simulations further increases the possibilities of simulating larger power systems scenarios. In order to proactively identify gaps and protect the electric grid from systemic failures or external attacks [13], the importance of co-simulation of TDC is acknowledged and emphasized. Domain-specific simulators – RT simulators e.g., RTDS® and Opal-RT®, steady-state simulators for transmission system e.g., MATPOWER, simulators for distribution systems e.g., Synergi® and GridLAB-D, power systems transients simulators e.g., PSS®E and ETAP®, and simulators for communication networks e.g., NS3, already exist. In order to study power system events, porting a model from one software to another is currently the adopted practice. Incompatibility, missing modules, and license limitations frequently challenge co-simulation workflow, and affect the accuracy of the module. With the increase in wide-area monitoring of power systems, utilities are implementing control algorithms that require extensive communication for feedback from distant sensors and functioning. TDC co-simulation enables testing of different communication protocols used in the industry. TDC is commonly employed for observing data latency of communications between components and systems and its impact on the operations. TDC co-simulation will also assist the innovation of communication protocols specifically for smart devices and grid management systems for modernization.

Geographically distributed RT simulation can leverage and benefit from the developments in TDC. Minimization of data latency can improve geographically distributed RTS. HIL testing (e.g., synchrophasors) with RTS leads to investigation of actual time delay between communication packets and the consequential control actions between the two hardware under close to real-world conditions. Figure 17.2 pictorially depicts the interaction of HIL and RTS for TDC co-simulation of tightly coupled power systems.

Federal research laboratories in the United States are extensively engaged in TDC co-simulation research. Pacific Northwest National Laboratory's (PNNL) Framework for Network Co-Simulation (FNCS) platform includes support for mixed discrete events (e.g., communications) and continuous time (e.g., power flow) simulations over a ZeroMQ messaging bus [14]. NREL's Integrated Grid Modeling System (IGMS) [15] is characterized by a highly scalable hierarchical architecture using Message-Passing Interface (MPI) parallelisms to link ZeroMQ message buses at multiple levels. In recent times, this framework was used to link distribution and Home Energy Management System (HEMS) simulations [16,17]. Lawrence Livermore National Laboratory's (LLNL) Federated Simulator Toolkit (FSKIT) [18] provides a coupling interface for both continuous and discrete time simulations for large-scale distributed memory high-power computing systems using MPI. Oakridge National Laboratory's (ORNL) Toolkit for Hybrid Modeling of Energy Systems (THYME) uses an event-driven simulation of the electrical power system equations to naturally

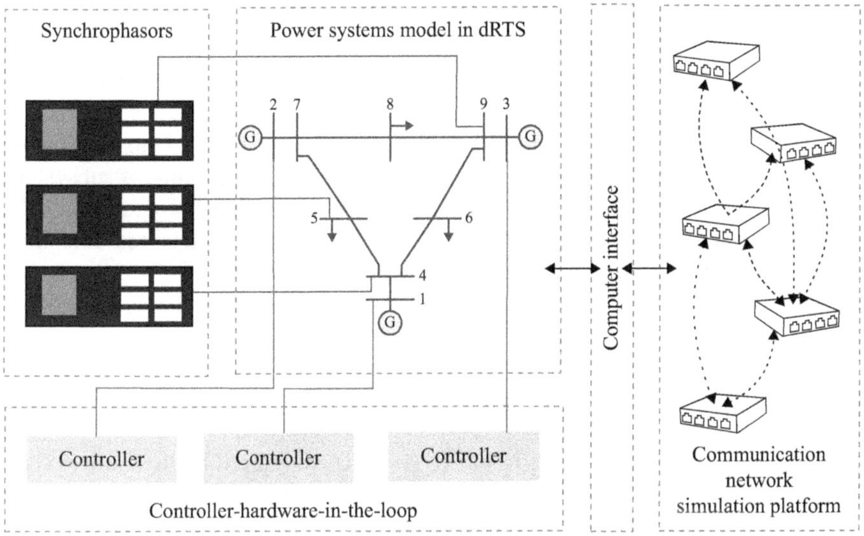

Figure 17.2 A sample architecture of utilizing synchrophasors with a power system for model validation and co-simulation of tightly coupled power systems

and precisely integrate its model of electrical-mechanical dynamics with dynamic models of event-driven control and communication systems [19].

17.3 Historical efforts in distributed RTS

Research based on infrastructure placed at different geographical location was first reported in the automotive industry in the 1990s. This concept was later on adopted in power engineering research and has progressed through numerous initiatives by research agencies, universities, and industry. Examples of prior work in distributed RTS are summarized in this section.

17.3.1 Remote testing and distributed simulation based on the virtual test-bed

One of the first realizations of a virtual environment for remote testing of devices has been reported in Reference 20. A virtual environment based on the Virtual Test-Bed (VTB) integrated with a LabVIEW virtual instrument for remote signal processing was designed and implemented. The platform was experimentally verified and applications for remote model validation of a transformer and remote design verification of a fuel cell supplying a dc motor drive were demonstrated. In both cases, the physical devices were located at the Polytechnic University of Milan, Italy, while VTB-based virtual environment was located at the University of South Carolina, USA.

RT extension of the VTB simulation platform, VTB-RT, was enhanced by a synchronization module to meet requirements for geographically distributed simulation applications [21]. An FPGA-based module that utilizes a common time base derived from a Global Positioning System (GPS) is developed to ensure synchronization of distributed RT simulators [22]. Furthermore, efforts have been made to develop and implement interface algorithms for both signal and natural coupling of subsystems for simulation of power and power electronics systems. Results of geographically distributed RTS demonstrate relatively high fidelity with respect to integrated simulation. However, systematic assessment of the impact of the dynamics of a communication medium on performance and stability of geographically distributed RTS was not part of the work described earlier.

17.3.2 Multiple university research initiative

A Multiple University Research Initiative (MURI), supported by the United States Department of Defense (US DoD), investigated the concept of a virtual testing environment for enabling interconnection of power system laboratories located at five universities. Namely, Drexel University, Northeastern University, Iowa State University, Texas A&M University (TAMU), and Mississippi State University (MSU), participated in the MURI initiative. The initiative focused on the development of remote testing and measurements models and methods for coupling hardware assets over the Internet with main application related to the US Navy's all-electric ship [23]. A generalized coupling method based on overlapped decoupling between subsystems was proposed and implemented in VTB-RT [24].

Distributed RTS based on transmission line models was pursued jointly by MSU and TAMU [12]. This research was based on creating custom interfaces for RT simulators based on National Instruments (NI) interface and external communication cards. Internet-based protocols were used to create a connection between the RT simulators at MSU and TAMU. Test-bed validation and accuracy analysis is also presented based on comparisons of steady-state and dynamic test results obtained from both distributed and localized RTS. Latency associated with the Internet-based communication between the two research centers is not considered significant. However, in our experience, we have found that Internet-based latency plays a critical role in geographically distributed simulations. Solutions to address latency in the form of prediction techniques are one of the current directions of research in this field.

17.3.3 Cyber-security test-bed and testing

An early research effort reported on distributed RTS was a project between Center for Advanced Power Systems (CAPS) at Florida State University (FSU) and Sandia National Laboratory (SNL) [11]. The experiment connected the control center at SNL with the power systems designed in the RT simulator at FSU. Data exchanged between the two facilities included the control set points and protection settings associated with the power systems model in RT simulator. The two research facilities were connected via a Virtual Private Network (VPN) on WAN, leveraging the Florida LambdaRail infrastructure which is a part of the National LambdaRail network. Florida LambdaRail is an independent research and education network that connects universities in the state of Florida with a cost-effective, ultra-high speed, broadband service for innovative and cutting-edge scientific research [25]. A cyber-security exercise was performed using this experimental setup for understanding the impacts of "successful attacks" on a critical infrastructure and the potential impacts. This experimental test-bed allowed impact analysis, risk assessment, and generated inputs for business case of cyber-security impacts on power systems.

17.3.4 Geographically distributed thermo-electric co-simulation

The main motivation behind interconnecting RT simulators over a long distance is to leverage individual competence and resources of various research facilities and enable a strong collaboration among them. This is particularly advantageous in the context of multi-domain co-simulation. Two commercial RT simulators were interconnected for RT thermo-electric co-simulation studies with the focus on achieving a high-fidelity simulation model of an all-electric ship [26]. RTDS® located at CAPS at FSU was utilized for the simulation of an electrical model; while an Opal-RT® located at the University of Alberta performed the simulation of a thermal system. A comprehensive hardware setup was developed as two heterogeneous RT simulators were interfaced over the Internet [27]. The work demonstrates the applicability of geographically distributed setup for thermo-electric co-simulation as communication delay is not a critical aspect in co-simulation applications with widely different time constants of subsystems, such as electric and thermal domains.

17.3.5 A modular architecture for virtually interconnected laboratories

A modular framework for interconnection of RT simulators located at geographically dispersed locations is being developed at the Institute for Automation of Complex Power Systems (ACS) at the Rheinisch-Westfaelische Technische Hochschule (RWTH) in Aachen, Germany. The Virtually Interconnected Laboratories for LArge systems Simulation/emulation (VILLAS) architecture includes a generic VILLAS node at each facility to enable a harmonized integration of diverse assets into a joint experiment. Experiments have demonstrated flexibility and applicability of the framework for interconnection of Opal-RT® and RTDS® simulators located at various research facilities [28,29]. Namely, Opal-RT® and RTDS® simulators located at RWTH are interconnected with Opal-RT® simulators hosted at the facilities of SINTEF Energy Research, University of South Carolina, and the Polytechnic University of Turin. Section 17.6 provides a brief overview of the results and includes details of the VILLAS architecture. A hybrid approach to the design of a simulator-to-simulator interface for coupling geographically distributed RT simulators is also presented.

17.3.6 Automotive engineering application

The concept of integrating distributed hardware assets over the Internet has been applied in the field of automotive engineering with motivation to allow simultaneous engineering design and testing. A set of comprehensive practical experiments as well as theoretical contributions have been reported in literature with applications of distributed RTS in automotive engineering [30–35]. RT interconnection of distributed HIL facilities over the Internet is presented in Reference 30 where the Engine-in-the-Loop simulation setup at the University of Michigan and the Driver-in-the-Loop ride motion simulator at the US Army Tank-Automotive Research, Development and Engineering Center (TARDEC) are coupled. Thus, the fidelity of experiments is improved as the real hardware engine is interfaced to the ride motion simulator with a human operator driving a virtual vehicle. Different methods are implemented to ensure stability and achieve high fidelity of geographically distributed HIL setups. Namely, methods based on observers [31], model-free prediction approach [32], and learning control concept [33] are developed. Various studies, such as variation-based and statistical transparency analysis and analysis of the effect of coupling point selection on fidelity are performed based on a simplified and linear analytical framework [34,35]. Furthermore, conclusions were verified based on realistic experiments of coupling geographically distributed HIL setups. Direct application of these methods for the development of distributed electrical system simulation is not feasible as the time constants of mechanical and electrical systems differ in range. However, the formalization of the interface design and systematic assessment of inherent issues in geographically distributed RTS may be relevant to power systems engineering research.

17.4 Systematic approach toward distributed real-time simulations

Distributed RTS is in a nascent research stage of its applications for electric power engineering. Although distributed RTS is considered a matured research methodology in many other industries, the adoption to power engineering research is not straightforward. Several challenges, and hence research opportunities, exist in the application of distributed RTS for power engineering approach and a systematic approach to performing distributed RTS is proposed in this section.

17.4.1 Objectives and assumptions

Interconnected RT simulators must be capable of maintaining a synthetic electrical coupling by exchanging subsystem data at the interface that reflects the system state accurately over a range of frequencies of interest. Each simulator involved acts as both a server and receiver for transformed data to be exchanged via WAN-based communication. An interface is designed for the two RT simulators in distributed RTS to perform transformations of signals from the time-domain to phasor-domain and vice versa. At the sending end, the interface layer performs a transformation of time-domain signals to phasor-domain, whereas the receiving end interface layer transforms the phasor data back to time-domain for RTS. For example, a simulation where the two subsystems are a distribution network and a transmission network, interconnected via a sub-transmission line and a substation transformer. The decoupling interface is modeled as an equivalent Frequency Dependent Voltage Source (FDVS) connected at the interconnection point of distribution grid or primary side of the substation transformer and as a current injection source at transmission grid side. The data exchange in this case will be Root Mean Square (RMS) voltage, current, and phase angle measured on the interface bus at fundamental frequency. The baud rate of data communication is determined based on the phenomenon under investigation. This will be constrained by the sampling function and data transceiver hardware used at the interfaces across the communication channels. In case the simulators are at distant locations, the issue of latency in data communication can result in inaccurate state representation, or even lead the simulation into numerical instability. Most commercial RT simulators provide measures to ensure numerical stability in decoupled simulations with delay which is typically within each time step of simulation. As shown in Figure 17.3 power system phenomena and their simulations vary on a wide timescale, namely a few microseconds to several hours or even days. The simulation phenomena investigated using the distributed RTS setup are stationary events such as steady-state and harmonic power flow, frequency variations, transients due to switching and protection events, dynamic events such as power oscillations, voltage recovery, and ancillary services in regulation and future fast-response markets. Transient propagation in local network cannot be investigated using a WAN-based distributed RTS setup over large distances. Investigation of inertial response, wide area control and measurement system, adaptive wide area protection, and fast-response ancillary services are some of the most promising phenomena that can be beneficial for the future electric grid. Studying the coupling to and the effect of emerging markets on system stability, device certification and interoperability, and investigation of stationary power quality events provides an avenue for longer timescale experimentation (ranging from minutes to hours) using distributed RTS environment.

The latency in data communication presents constraints on the events of interest investigated by distributed RTS. If the event of interest has a time response which is orders of magnitude larger compared to the latency, then the latency does not severely impact the accuracy of the result. For large-scale power systems, frequency events take seconds to propagate across the large systems [36–38]. The exact severity of the event depends on the magnitude of disturbance and the origin of the event. Some low frequency and high inertia disturbances such as power swings can impact large interconnections but high frequency transients can decay locally without a wide area impact. While slow events can be studied accurately under delayed connectivity, fast transient events must be studied for smaller regional grid with closely coupled, low-latency, detailed Electro-magnetic Transient (EMT) models. Transient studies are not appropriate use cases for distributed RTS based on WAN, unless latency mitigation techniques are used to manage the integrity of data exchange.

Large systems require large simulation and processing capacity, which is not always feasible due to limited simulation hardware. Such systems can be simulated by

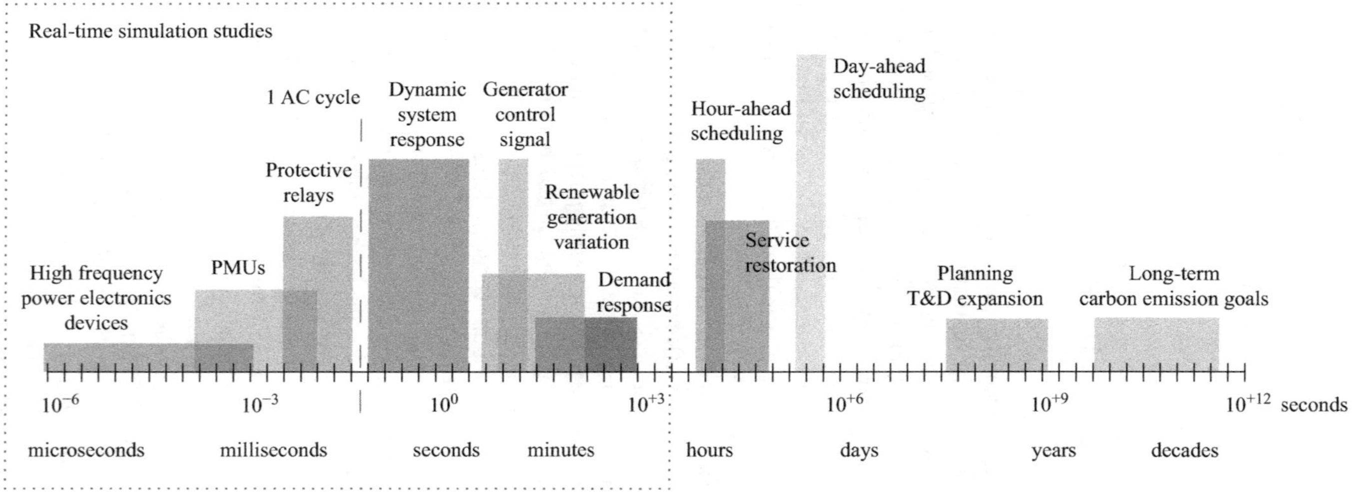

Figure 17.3 Summary of power systems phenomenon with their typical durations. Reproduced from Reference 1

using network reduction and partitioning techniques. Network reduction techniques usually provide equivalent representations that result in loss of information at the compressed band of the simulation model. Frequency-Dependent Network Equivalents (FDNE) are one such category of EMT models used for hybrid Transient Stability Analysis embedded inside systems [39–41]. Network partitioning is used to retain the essential characteristics of the network by dividing it into smaller subsystems. The subsystems are coupled to each other with interfaces capable of state information exchange through connecting communication channels or ports [42]. Network partitioning for distributed RTS can be performed based on multiple criteria. Graph theory based partition to divide systems into several subsystems based on connectivity is feasible. System partition based on voltage levels within transmission and distribution systems is another way. Accurate performance of RTS is essential for any HIL experimentation of interest which is even more sensitive toward latency. HIL simulations for two-way partitions can be categorized as follows:

1. Simulated power systems partitioned at node of interconnection of Power-Hardware-in-the-Loop (PHIL)/CHIL.
2. Simulated power systems partitioned into two subsystems with one or more PHIL/CHIL connected on one side.
3. Simulated power systems partitioned into two subsystems with at least one PHIL/CHIL connected on both ends.
4. Simulated power systems with two or more PHIL/CHIL physically present at two locations but electrically connected on the same node.

These four modular configurations extend the distributed RTS and HIL to formulate more complex scenarios between multiple geographically separate locations. Each RT simulator location will present a slightly different challenge for emulating the communication layer.

17.4.2 Impacts of data latency

Data latency associated with WAN-based communications is a key research area in network engineering, especially to gauge its impacts on various applications. Latency associated between any two locations of communication network is a stochastic variable which makes WAN-based distributed RTS as a challenging field. Latency is a function of multiple factors, such as time of the day, data traffic, communication protocols, physical distances, and network configuration [43]. Data latency of any WAN-based communication protocol is the summation of signal propagation delay, network processing delay, transmission delay, and queuing delay [43]. Signal propagation delay is the time associated with the signal to travel from the source to destination, which is approximately 70% of the speed of light in vacuum. Network propagation delay is the time utilized by the network components such as routers, firewalls, and gateways to direct the data. Transmission delay is the time necessary to push the data into the physical layer and is a function of the bandwidth of the channels. Queuing delay is the congestion-related delay associated with multiple packets to be exited from the same "exit port." Typical values of data latency vary from a few to several hundred milliseconds. The four components of data latency functions of several uncontrollable factors, and hence cannot be predicted easily; described as a stochastic variable.

The impact of data latency varies and is significantly dependent on end-user applications. For example, traditional applications such as e-mail can typically tolerate a latency of few hundred milliseconds, whereas advanced financial trading applications are sensitive to even microseconds of delays [43]. “Low latency networks” are preferred in cases where the stakes are too high and dependent on variable data latencies. The latency aspects from the perspective of distributed RTS are found via empirical results. The time steps associated with RTS may be as low as 50 microseconds to a few hundred microseconds depending on the type of analysis. For distributed RTS based on WAN, variable data latency may lead to inaccuracies as the synchronization between RT simulators will be affected.

It is critical to understand the limitations that may be imposed on the type of signals that are transmitted over the WAN for performing accurate distributed RTS. A theoretical foundation and experimental verification is an acceptable way of determining the most suitable scenarios for distributed RTS within reasonable accuracy.

17.5 Distributed RTS between INL and NREL

The following section describes the experiences of INL and NREL in performing distributed RTS based on the background discussed in Section 17.4. Simulation setup and results are included to indicate the effectiveness of this approach.

17.5.1 Experimental setup

RTS at different locations need to be started in step else numerical errors may be introduced in unsynchronized simulations. RT simulators can be synchronized to external clock signals from multiple sources i.e., workstation time, Internet-based time, or external device time. The approach adopted for performing distributed RTS between INL and NREL (approximately 600 miles apart) is the utilization of an external clock. Both the simulators were synchronized to a commercially available GPS-based clock, providing reliable time reference value for both RT simulators. Distributed RTS performed under this research activity is synchronized and programmed to start at the same clock-time. This ensures that there is no out-of-step RTS, leading to numerical instability. An example of a two subsystem simulation is shown in Figure 17.4, resembling simulations performed between INL and NREL. The 2-Area 4-Generator system, [44], was simulated at INL and the IEEE 13-node test feeder, [45], was simulated at NREL. The reader can obtain more information about the 2-Area 4-Generator system and the IEEE-13 node test feeder from References 44 and 45, respectively. This partition of the power systems network on the two RT simulators located at INL and NREL is based on voltage levels i.e., transmission and distribution voltages, and can allow HIL connections on either side. RMS values of voltage, frequency, and phase angle measured at bus 9 are transferred over the IP secured tunnel from INL

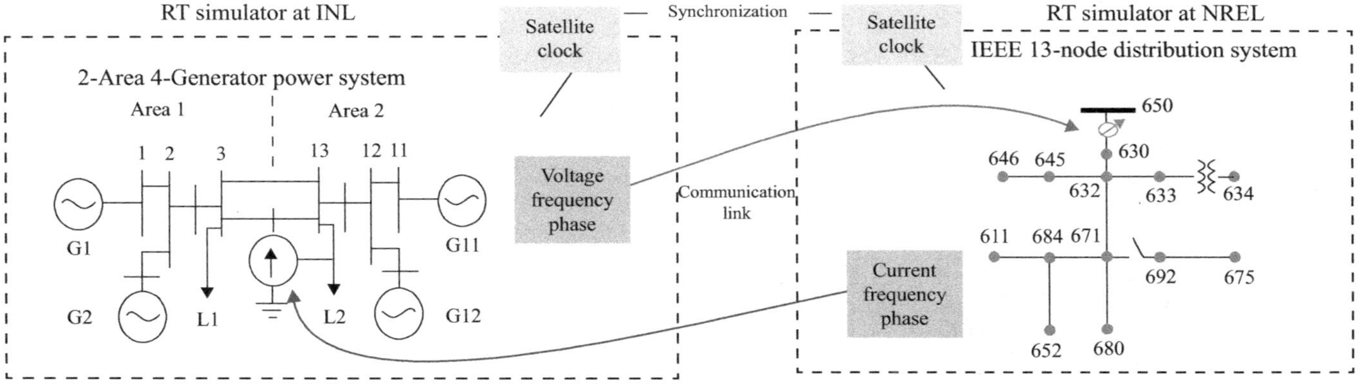

Figure 17.4 INL-NREL distributed RTS setup

to NREL, thus providing the states for the grid supply at the distribution network. RMS values of current, frequency, and phase angle measured at node 650 in the distribution system are sent as a response from NREL to INL, thus acting as a load in the transmission network.

17.5.2 Latency analysis between INL and NREL

An empirical approach to understand the WAN connection between INL and NREL was chosen. The WAN connection between INL and NREL partly utilizes both the Energy Sciences Network (ESNET) and the public network. This involved performing data latency measurement experiments with observations related to the time taken for data exchange between communication cards of RT simulators located at INL and NREL. A secured IP tunnel was established between the two facilities to provide a VPN connection. This provided a relatively controlled environment to perform geographically distributed RTS. Time stamped data, using the reference GPS clock, was exchanged between the two centers. A set of bidirectional flow of data was initiated at different times of the day and the data captured was analyzed to compute statistical indices of the data latency and other network characteristics. Figure 17.5

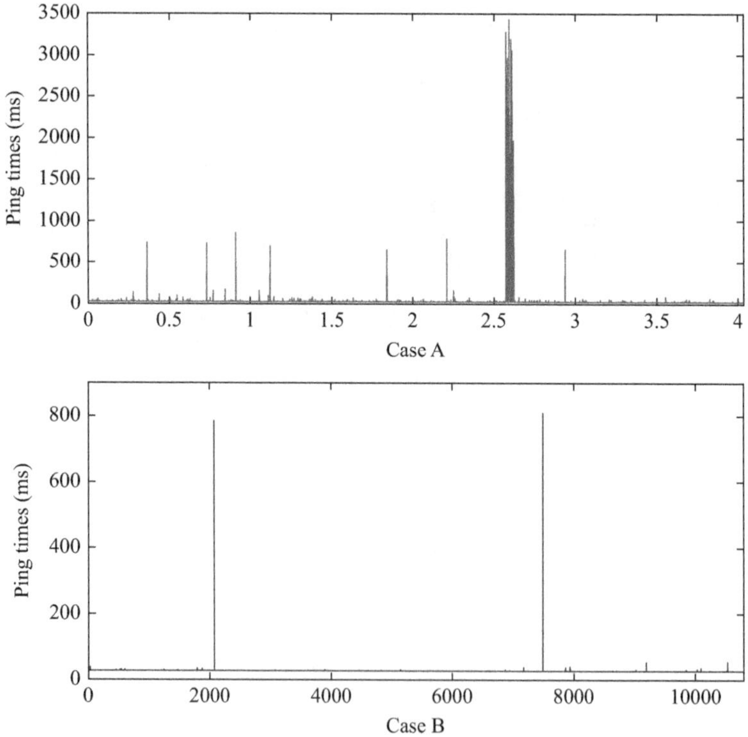

Figure 17.5 Data latency plots of sample ping between the two RT simulators at INL and NREL

shows two such data latency plots for data exchanged between the RT simulators at INL and NREL.

The delays, jitters (deviation of high frequency signals from the norm leading to loss of data), and drops observed for the two samples are significantly different, although the same network cards were used. Two sets of latency data samples Case A and B representing pre- and post-communication network upgrade, respectively, are discussed. The minimum latency observed for both the cases is 26 ms, corresponding to the best case scenario. The maximum delay observed for the Case A is approximately 3.4 s, whereas for Case B the maximum delay observed is approximately 810 ms. The average latency observed for Cases A and B are 34.78 and 27.20 ms, respectively. Additionally, the jitter, observed in Case A is higher compared to Case B. This indicates that the data latency has a greater variability for Case A implying network conditions and its impact on a point-to-point channel such as the one discussed here is significant. During the initial phase of this project, RT simulators at INL were connected to the public WAN via a cable of 1 Gbps capacity, whereas the RT simulators at NREL had a 10 Gbps connection. Several data loss and jitter issues were observed with this setting attributable to the bottleneck 1 Gbps channel. The network connection for RT simulators at INL was upgraded to 10 Gbps leading to both lower jitter and data loss.

For the experiments intended WAN-based distributed RT simulators such data latency characteristics will have a significant impact as data packets may be delayed by significantly longer time than expected and may arrive at the remote location thousands of steps later. The experimental setup of geographically distributed RTS must have the capability to detect and accommodate such abrupt delays in data packets and maintain simulation stability. This is a nontrivial prediction process as the set of variables of interest i.e., data latency and simulation variables, are stochastic and are an outcome of the interaction of multidimensional communication and power systems networks, respectively.

17.5.3 Simulation results

Utilizing the setup described earlier and the User Datagram Protocol (UDP) for data exchange, grid scenarios such as single phase to ground fault, two phase to ground fault, three phase to ground fault, and load drop in the transmission network were simulated. The details of the test conditions are provided in Table 17.1. Controlled

Table 17.1 Test conditions for distributed RTS

No.	Event type	Conditions	Location
1	Load drop condition	L1	Bus 3
2	Single phase to ground fault	3 cycles	Bus 13
3	Two phase to ground fault	3 cycles	Bus 13
4	Three phase to ground fault	3 cycles	Bus 13

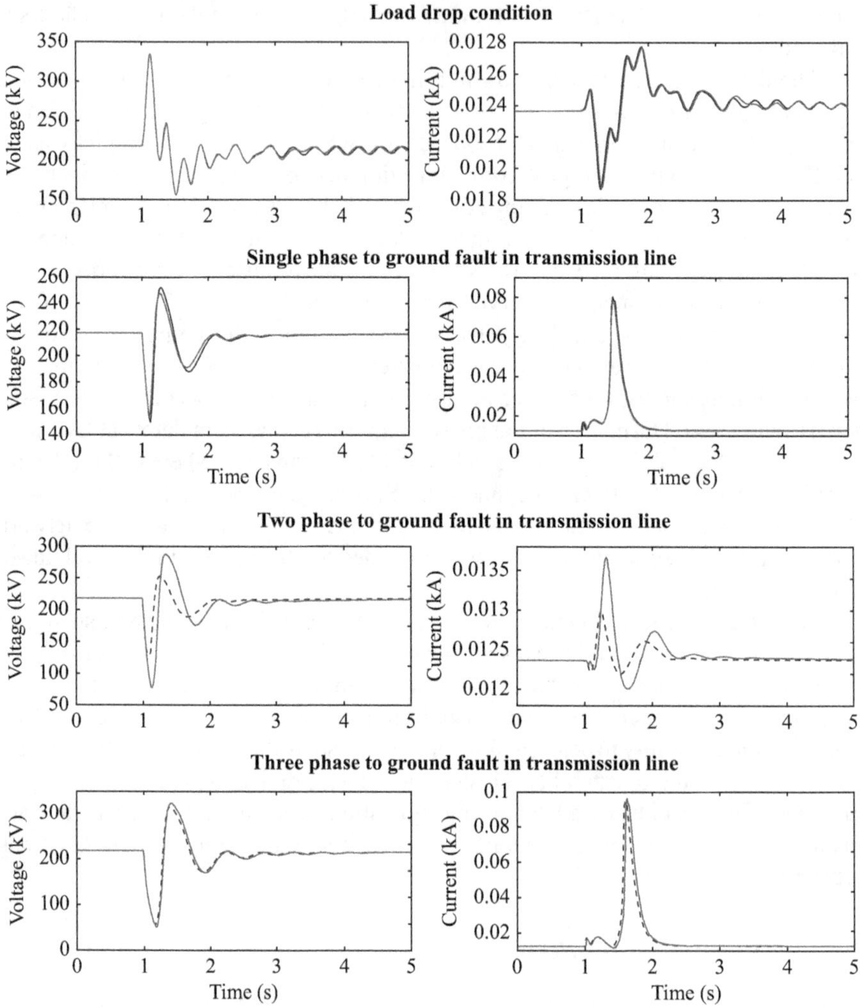

Figure 17.6 RMS voltage and current waveforms at the point of interconnection for different fault conditions (The dashed line represents results from zero delay case, and the solid line are the results from the distributed RTS

voltage and current sources are designed to account for the response of transmission and distribution networks, respectively.

Firstly, a scenario of a load drop at transmission network bus 3 is simulated, leading to voltage transient at bus 13 (distribution network connection). The impact of subjecting a voltage transient on a distribution network at its source is shown in Figure 17.6. All the simulations are repeated for two RT simulators with no delay in exchanging data. In other words, the zero delay case represents two RT simulators are connected locally at INL, and hence serves as the best case scenario. Comparison of results obtained from the simulations between INL and NREL are compared with those obtained from the zero latency case, to assess the impact of data latency on distributed RTS. The impact of the data latency in this case is observed to be low, as the two RMS waveforms (voltage and current) for both zero delay case and distributed RTS case are identical. Fault scenarios (balanced and unbalanced) are applied to bus 13 in the transmission network and the resulting impacts are recorded. It can be observed from the two phase to ground fault that the impact of data latency is not trivial as the two RMS waveforms (voltage and current) for both zero delay case and distributed RTS are significantly different. Numerical analysis using indices to formalize the impact of latency on specific simulations. One of the future challenges in this research is the generalization of the impact of latency on distributed RTS and eventually mitigating it through compensation techniques.

17.6 Modular architecture for interconnected laboratories

Interconnection of RT simulators over WAN imposes many challenges to be addressed, as described in the previous sections. In addition, a holistic virtual framework with a modular and generic architecture is required to enable flexible cooperation and integration of diverse assets. To achieve this goal, a platform for VILLAS is being developed by the Juelich Aachen Research Alliance (JARA), a collaboration between RWTH Aachen University and Juelich Research Center in Germany.

17.6.1 Generic design approach

The platform is designed with the goal of enabling integration of the resources available at each laboratory, and providing a flexible utilization of an infrastructure as a whole. A conceptual architecture diagram of the VILLAS platform is illustrated in Figure 17.7.

Three integration layers, namely, hard RT, soft RT, and offline, are included with respect to the timing requirements for data exchange and the nature of interconnected assets. Hard RT layer is utilized for the interconnection of various RT simulators, actual hardware devices, and equipment into a joint experiment. This layer addresses requirements for interconnection of multiple HIL enabling unique experiments that would not be feasible otherwise due to diverse locations of hardware assets. High-level user interfaces, such as an interface for online interaction, coordination of an experiment, and an interface for data logging, are managed by a soft RT layer. This layer enables integration and testing of system level tools. In this context, "Simulation as a Service" is foreseen as an on-demand concept to enable access to third parties for remote testing of monitoring and control systems. The third VILLAS integration layer is an offline layer. Offline layer allows for local analyses based on domain-specific methods and tools. These analyses refer to pre-experiment as well as post-experiment phases. The offline layer is of great importance for embedding specific know-how of individual research groups. Offline services are envisioned to enable utilization of

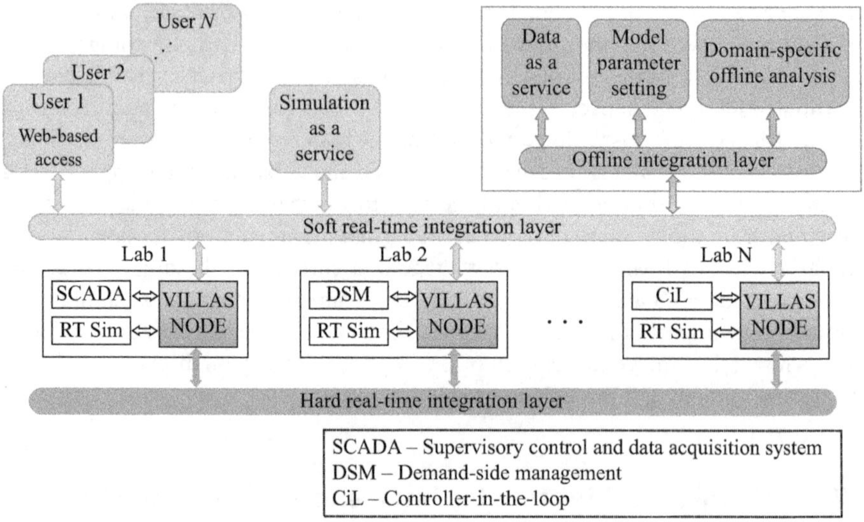

Figure 17.7 Virtually interconnected laboratories for LArge systems Simulation (VILLAS)

the infrastructure beyond RTS and PHIL testing. Integration of diverse assets hosted at dispersed geographical locations provides an environment for pursuing unique and holistic experiments, resulting in data sets that are of great value. In this context, "Data as a Service" is foreseen as an offline service that provides Internet-based access to a historical database of experiments. An advanced realization of such service includes domain-specific analysis and visualization tools for data interpretation.

Every facility is equipped with a generic VILLAS node [28] to address different interconnection requirements arising from diversity of applications and involved facilities. VILLAS node, an in-house software developed at RWTH Aachen University, is designed to provide a flexible interconnection of geographically distributed assets into different integration layers. At the time of writing this chapter, a public domain for VILLAS software is in the preparation process for release [46]. The VILLAS node includes a set of VILLAS interfaces allowing for a harmonized integration of diverse assets into VILLAS architecture. Namely, VILLAS node at each laboratory acts as a gateway between the local RT simulators, devices, and the WAN. The software runs on an RT-patched version of Linux® with a minimal Fedora® installation to reduce the impact of background tasks on data exchange between the simulators. Independent processes take care of different interfaces. Furthermore, processes for sending and receiving data between laboratories are asynchronous and independent, thus the additional delay that might be caused by waiting for a response from a VILLAS node instance is avoided. VILLAS node forwards data between simulators and remote VILLAS node instances, drops invalid packets, collects statistics of the communication link, and runs arbitrary operations on the forwarded data if required. VILLAS nodes communicate via UDP for the purpose of data exchange between the simulators. To provide data security a VPN based on the open source software Tinc [47], which runs directly on the VILLAS node, is utilized. This VPN domain establishes a completely decentralized and fully meshed virtual network on top of the Internet and enables point-to-point data exchange, which is not typical for other VPN solutions.

A modular design and features of the VILLAS platform have been exploited in various projects to enhance collaboration and leverage assets at participating facilities. As an example, interconnection of laboratories within the context of the European Real-time Integrated Co-simulation Laboratory (ERIC-Lab) is based on the described platform [48]. ERIC-Lab is a federation of laboratories within the EU that is supported by the European Commission Joint Research Centre. It is initiated with the main goal to enable coordinated and integrated approach for research and technology innovation within the EU for a sustainable energy system. Applicability of the ERIC-Lab environment was demonstrated based on RT co-simulation of the transmission system of the Piedmont region and the distribution system of the city of Turin, Italy. RT simulators located at RWTH Aachen University, Germany and at Polytechnic University of Turin, were utilized for the simulation of the transmission and distribution systems, respectively. The described scenario represents a realistic situation in which confidentiality issues of sharing data and models among operators might limit wide area and interoperability studies necessary for the identification of possible unacceptable interactions within interconnected systems.

17.6.2 Hybrid interface design

Fidelity and stability of a distributed simulation is highly affected by dynamics of a communication medium that is utilized for the data exchange between different subsystems. A common characteristic of RT simulators is the utilization of multiple computing units operating in parallel to achieve the required simulation capabilities. Although a hard RT data exchange between units is ensured by dedicated communication buses or shared memory, time delay between subsystems is a critical aspect in designing a scalable RT simulator. Communication bottleneck in case of geographically distributed simulation becomes even more critical as a larger time delay is introduced and a hard RT communication link cannot be guaranteed typically. While limited bandwidth is not considered as a critical factor, time-varying delay, packet reordering, and packet loss introduced between subsystems violate the fundamental characteristics of RT simulators, such as hard RT computation and data exchange. However, it is not envisioned to design a simulator-to-simulator interface for geographically distributed simulation of high frequency transients as described in Section 17.4.1. A realistic goal is to ensure interface transparency and simulation fidelity for low frequency transients in case of RTS distributed over WAN.

RT simulators perform simulations based on instantaneous values of current and voltage waveforms. However, interface quantities between geographically distributed RT simulators are typically not exchanged in the form of instantaneous values of waveforms. Intrinsic characteristics of WAN would significantly distort the waveforms and introduce additional issues apart from time delay [28]. The simulation setup illustrated

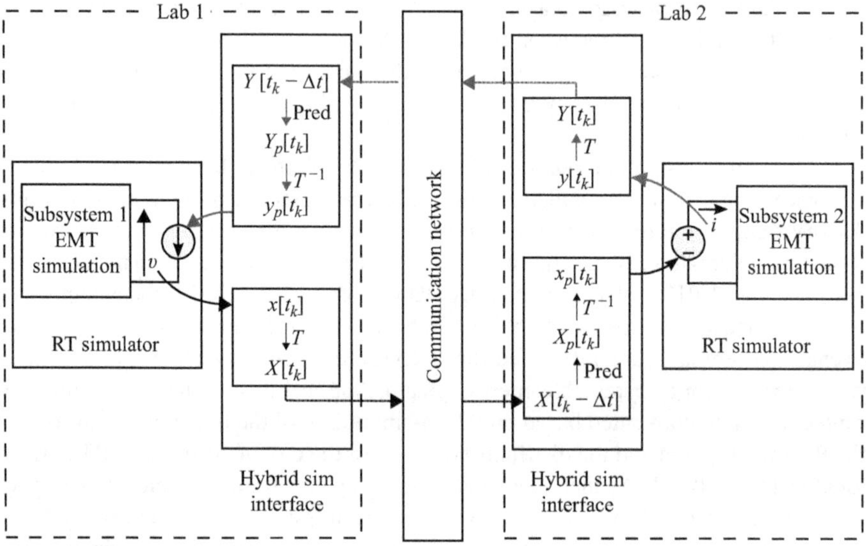

Figure 17.8 A real-time representation of the hybrid approach to the interface design for performing geographically distributed RTS

in Figure 17.4 is based on the representation of each interface quantity using its RMS value, estimated frequency, and phase, as described in Section 17.5.1. An approach adopted in Reference 29 is based on the estimation of time-varying Fourier coefficients of interface quantities. Both representations average current and voltage waveforms and thus filter high frequency transients through the interface. This is considered advantageous for coupling systems over WAN in the field of bilateral teleoperation that is closely related to the problem of geographically distributed simulation [49]. A simulator-to-simulator interface that is based on transforming current and voltage waveforms to the form that is not directly used for computation in RT simulators is referred to as hybrid simulation interface. This section provides an analysis of the proposed approach with emphasis on the advantages with respect to communication bottlenecks.

The hybrid simulation interface for geographically distributed simulation is illustrated in Figure 17.8. The original system is decoupled based on controlled current and voltage sources. While each subsystem is simulated in RT simulator in the time domain based on signal waveforms, interface quantities are transformed to other suitable domains before being sent over WAN to the remote RT simulator. An adequate representation of interface quantities is characterized by quantities that do not change rapidly with respect to the communication limitations. Thus, a lower sending rate can be used and coefficients can be utilized for the extraction of a time signal at the receiving side over a time window that is comparable to communication limits. At the same time, interface transparency over a frequency range of interest must be ensured. Time-varying Fourier coefficients, known as dynamic phasors [50], are considered here for representation of interface quantities. Namely, a time-domain waveform $x(\tau)$ can be represented based on the time-varying complex Fourier coefficients $X_i(t)$ on the interval $\tau \in (t-T, t]$: $x(\tau) = \sum_{i=-\infty}^{\infty} X_i(t)e^{ji(2\pi/T)\tau}$, where T represents a fundamental period of the system. These Fourier coefficients are represented in the time-frequency domain and the following averaging function determines the i-th coefficient: $X_i(t) = \int_{t-T}^{t} x(\tau)e^{-ji(2\pi/T)\tau}\,d\tau$. Under certain conditions, a few coefficients provide an adequate approximation and vary slower than instantaneous quantities. Approaches such as dynamic phasor based simulation and coupling an electromagnetic transient RT simulator and dynamic phasor based system equivalents benefit from these characteristics to address large-scale simulation scenarios [51]. In the context of geographically distributed simulation, number of coefficients that are necessary for an adequate approximation is not a critical aspect, but coefficients should vary slowly with respect to communication limitations among RT simulators.

The hybrid simulation interface is described below in more detail for the case of a dynamic phasor based representation of interface quantities. The original simulation model is decoupled into two subsystems, where each subsystem is simulated in an RT Simulator with a time step that is suitable for a particular study. With respect to the Subsystem 1, voltage measurement v across the decoupling point is provided as an input $x[t_k]$ at the time instant t_k to the hybrid simulation interface block at every time step in order to enhance the accuracy of the interface. Instantaneous values of a waveform $x[t_k]$ are converted to a n-set of time-varying Fourier coefficients: $X[t_k] = \{X_1[t_k], X_2[t_k], \ldots, X_n[t_k]\}$ based on the transformation T. A simple approach that is utilized here is to determine the n-set of the Fourier coefficients $X[t_k]$ based on the discrete short-time Fourier transform over the running time window of the length equal to the system period. To improve the accuracy of the transformation T, advanced methods for RT estimation of dynamic phasors, such as algorithms based on Taylor–Fourier transform [52], could be considered and applied. After obtaining the n-set of Fourier coefficients $X[t_k]$, the complete data set is sent to the remote simulator according to the selected sending rate. RT simulator that performs simulation of the Subsystem 2 receives the n-set of the complex Fourier coefficients with a time delay $X[t_k - \Delta t]$ where Δt varies in time. An adequate prediction method is expected to be applied and compensated for the time delay. However, in this analysis delay impact on simulation stability and fidelity is of interest to obtain first insights into the performance of the proposed interface. Thus, hybrid simulation interface for Subsystem 2 assumes that the n-set of complex Fourier coefficients for the predicted signal are equal to the values of the last received data set: $X_p[t_k] = X[t_k - \Delta t]$. At this point, instantaneous values of the time-domain waveform $x_p[t_k]$ are reconstructed based on the n-set of complex Fourier coefficients $X_p[t_k]$. At each time step, a new value $x_p[t_k]$ is calculated that represents an input to the controlled voltage source in the Subsystem 2. The procedure for sending the current measurements from Subsystem 2 to the controlled current source in the Subsystem 1 is analogous to the described procedure for sending the voltage measurements from Subsystem 1 to the controlled voltage source in Subsystem 2.

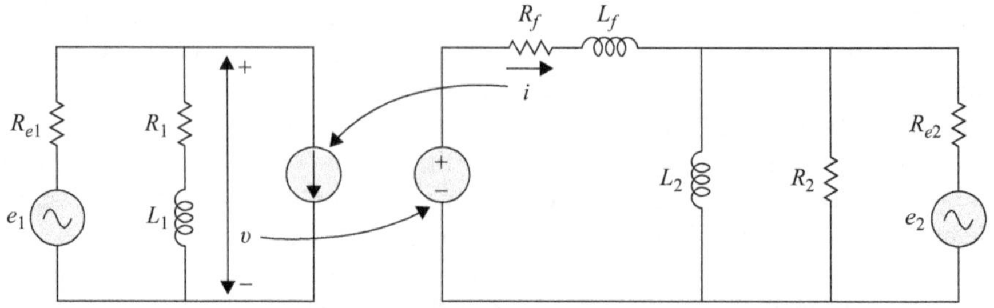

Figure 17.9 Decoupled simple electrical circuit

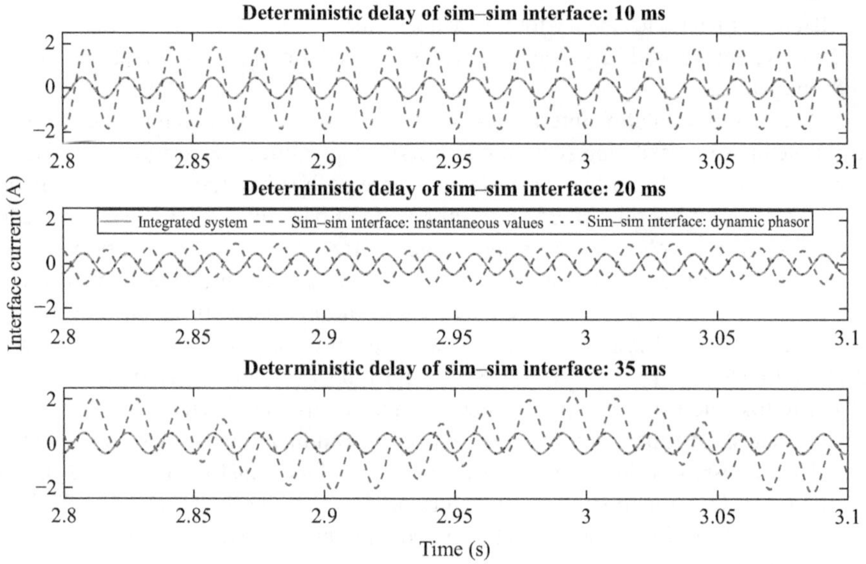

Figure 17.10 Stability limit with respect to deterministic interface delay: the simulation stability is preserved for delay of 35 ms in case of the interface based on dynamic phasors, while the interface based on instantaneous values of waveforms becomes unstable. Simulation results for the interface based on dynamic phasors coincides with simulation results obtained based on the integrated system.

Benefits of a hybrid approach to the interface design for geographically distributed simulation based on dynamic phasors are demonstrated based on a simulation performed for a simple electrical circuit that is decoupled as depicted in Figure 17.9. Namely, the impact of a fixed communication delay on system stability is analyzed.

Simulation results illustrated in Figure 17.10 show that the stability of the system with respect to interface delay is preserved for a larger time delay in case of the dynamic phasor based interface. The obtained results demonstrate an appealing

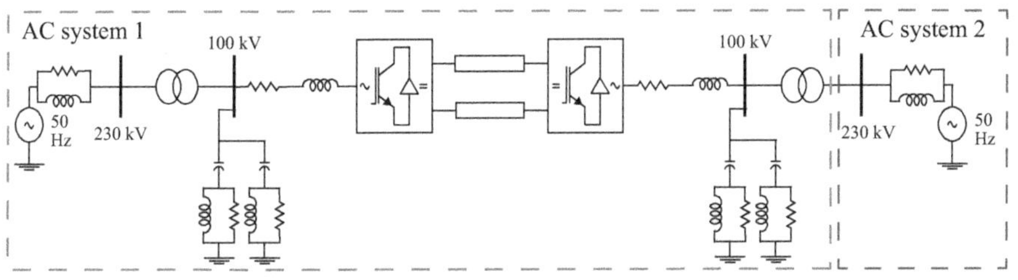

Figure 17.11 Decoupling of the system for geographically distributed simulation: VSC-based HVDC point-to-point link connecting two AC systems. The figure is created based on the data available in Reference 54

characteristic of a hybrid simulation interface that gives flexibility for further interface design such as improving accuracy based on prediction methods as mentioned in Section 17.5.2. Namely, the system that is decoupled based on the dynamic phasor interface is characterized by a larger delay margin. As depicted in Figure 17.10, the system with the interface based on instantaneous values of waveforms is unstable for the delay of 35*ms* while the system with the hybrid simulation interface preserves stability. Representation of interface quantities adopted here does not guarantee delay independent stability in an arbitrary case, but it encourages further research in this direction. Interface approaches that are insensitive to time delay have been designed in other fields, such as the scattering transformation that is applied for the data exchange between master and slave units in bilateral teleoperation [53].

The proposed interface is verified in a realistic framework that includes WAN as a communication medium and with a nonlinear system as a simulation use case. Namely, two RT simulators located at RWTH Aachen University, Germany, and at the University of South Carolina, USA, are interconnected over the Internet. A system under study is illustrated in Figure 17.11 and represents a voltage-source converter (VSC) based high-voltage DC (HVDC) point-to-point link that connects two AC systems. As reference cases, an integrated system and a system with deterministic delay equal to an average time delay of the communication link between the two locations are adopted. The hybrid simulation interface described above is utilized with n-set of Fourier coefficients where $n = 10$ is selected based on a heuristic approach to accommodate for faster transients and nonlinearities.

Figure 17.12 illustrates results of a simulation that is carried out for a dynamic response to a three-phase voltage sag applied in AC system 2 in the time interval $t = [6.0, 6.15]$ s. Results obtained for a metric that represents a two-norm of the error relative to the reference response of an integrated system are included. Simulation fidelity in case of deterministic delay and Internet delay can be compared based on the described metric. Obtained results indicate that time-varying characteristics of the Internet connection jeopardize simulation fidelity in addition to the deterministic delay. However, based on the results of two-norm of the error it can be concluded that the main resource of fidelity degradation during voltage sag transient is averaged delay.

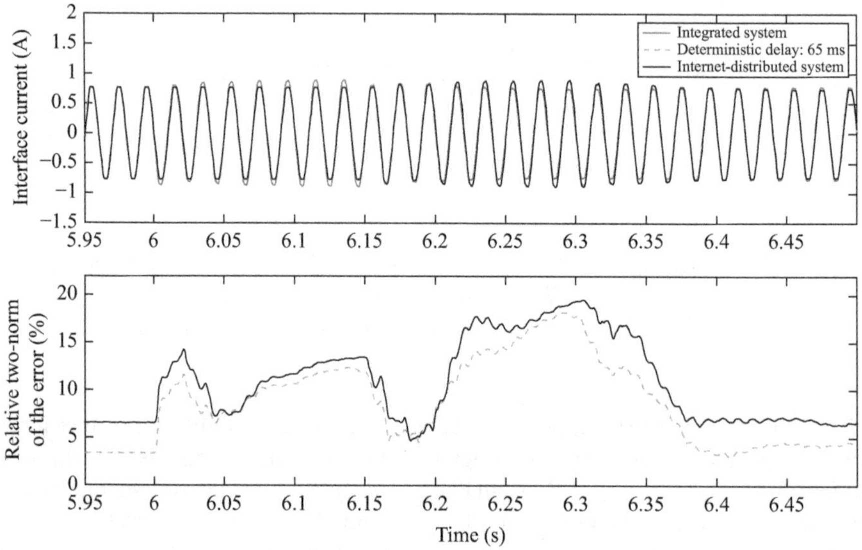

Figure 17.12 Geographically distributed simulation results of VSC-based HVDC point-to-point link connecting two AC systems: a three-phase voltage sag is applied in AC system 2 at time interval t = [6.0, 6.15] s

Analyses presented in this section show the importance of an adequate representation of interface quantities toward development of an interface for geographically distributed simulation. Relatively high level of fidelity can be achieved in case of slow transients and relatively slow variations of interface quantities with respect to communication limitations. Furthermore, feasibility of improving stability limit with respect to communication delay is presented based on a simulation of a simple circuit as illustrated in Figure 17.10.

17.7 Applications of distributed real-time simulations

INL and NREL are leading the development of distributed RTS and HIL integration by pursuing a wide range of power and energy systems projects. The communication link that exists between INL and NREL is used to perform HIL experiments for leveraging unique, mutual assets. Following are examples of some projects currently under progress:

17.7.1 Wind and hydropower research

Geographically distributed RT simulators are utilized by researchers at INL and NREL, to explore new technological developments in wind and hydropower plants. This includes current projects investigating scientific advancements to enhance the participation of wind turbines and Run-of-the-River (ROR) Hydroelectric Power Plants (HPPs) in ancillary service markets. In hydropower research, the INL-NREL link is being used to develop controls and integrate Energy Storage Systems (ESS) based on advanced power electronics, cohesive response to network events, interaction of multiple ROR HPPs with grid, and its equivalence to a large HPP. Distributed RTS with CHIL and PHIL (as shown in Figure 17.13) will provide a verification of the proposed ESS solution by capturing the RT dynamic and transient interactions between the rotational inertia of the HPP and the virtual inertia of the ESS due to power converters. Unique facilities and infrastructure at INL and NREL have been integrated for this project using a communication link between the respective RT simulators in this case. INL is leading the development of a front-end controller to connect a hardware supercapacitor to the RTDS®, while the Controllable Grid Interface (CGI) at NREL will be used to tie a wind turbine as PHIL, or hydro models to a loosely coupled power system model, emulating a real generator. Using the communication link between INL and NREL, electric grid models and the hardware connected to them can be co-simulated despite being part of different facilities. The communication link allows RT data exchange (with a 13 ms data latency) between power systems modeled in RTDS® at both laboratories.

INL-NREL distributed RTS setup (shown in Figure 17.13) helps develop and test coordinated controls of active power by wind turbines, short-term energy storage, and large industrial motor drives for providing various types of ancillary services to the grid. INL RTS facilities include CHIL and PHIL capabilities and modeling of large power systems. NREL's infrastructure includes the National Wind Technology Center (NWTC), a combination of multi-MW utility scale wind turbine generators, variable-frequency motor drives (VFD), new 8 MW energy storage testing facility, 1 MW solar PV array, and 7 MVA CGI. This wind power related research project will produce test data for understanding the impacts of a single turbine and wind power plant level short-term energy storage on various types of ancillary service controls (inertial control, power factor correction, automatic generation control, variability smoothing, and mechanical loads). The results of this project can also help understand the trade-offs between better gearbox design and predictive control on inflow for mechanical loads minimization and responsive control on grid side with electrical storage. In addition, the project is expected to produce data to demonstrate coordinated control of wind power and industrial motor loads for providing optimized ancillary services to the grid.

In order to achieve the technical goals, the project will utilize PHIL of a wind turbine and a variable frequency drive hardware at NWTC. This is connected to the RTDS® racks running the grid models and multiple CHIL for wind turbines at INL. This will provide an opportunity to perform grid level simulations using the characterized data as the inputs and controller as hardware in RT. The existing INL-NREL communication link will facilitate data exchange between the synchronized digital RT simulation between both the labs. INL will use a CHIL bank, at INL's Real Time Power and Energy Systems Innovation Test-Bed, consisting of multiple hardware controller cards (up to 50) where each controller card will act as an adaptive controller for an individual wind turbine. The CHIL bank setup will emulate the interaction

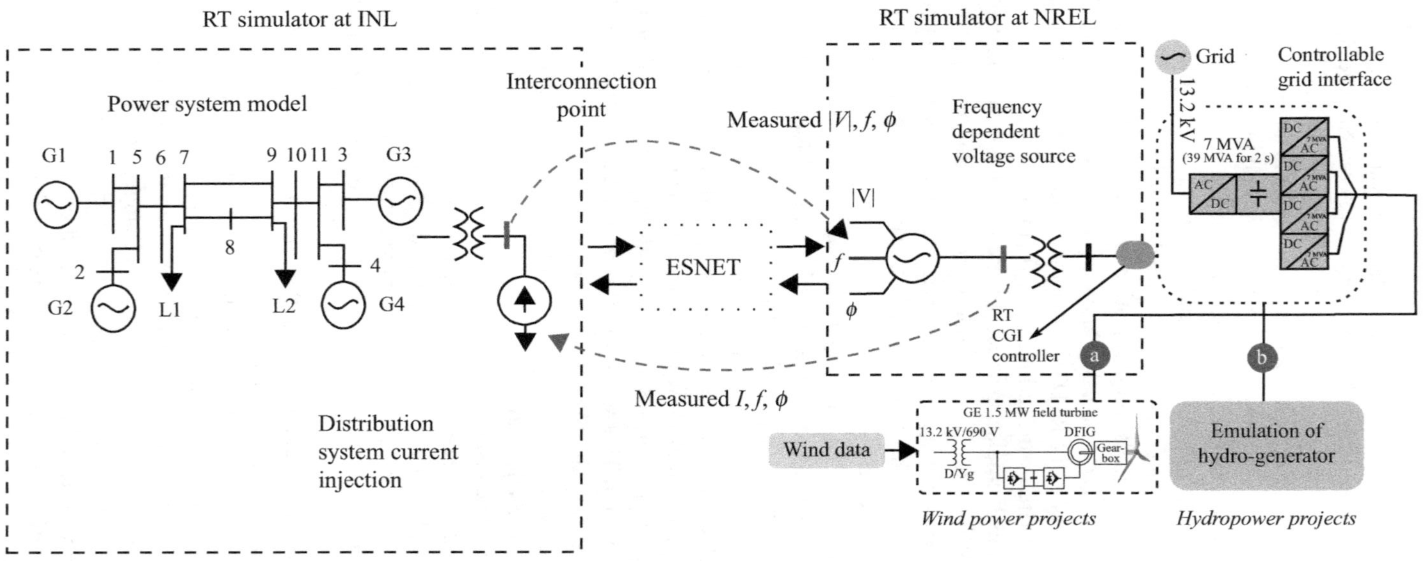

Figure 17.13 Schematic representation of ongoing wind and hydropower research projects carried out through geographically distributed real-time simulation between INL and NREL

and coordinated control action of multiple adaptive controllers for individual wind turbine in a wind power plant for large-scale grid-level simulations using DRTS racks. This task will help in understanding the actual RT interaction of wind power plants, controllable loads, and coordinated controllers in a large power grid.

The setup used for the wind power plant project has been extended to hydropower plants (shown by a branch (b) in Figure 17.13). The overarching objective of the hydropower plant project is to investigate and demonstrate the potential for ROR HPPs to participate in ancillary service markets.

17.7.2 Hydrogen applications in power systems research

A 120-kW electrolyzer is interfaced with an RTDS® at NREL's Energy Systems Innovation Facility (ESIF), which is capable of communication with RTDS® located in INL's Energy System Laboratory (ESL) via ESNET. The project is assessing the potential of electrolyzers for demand response (DR) and grid stability services in regions with high consumer demand for hydrogen. A control system developed at INL can be used to control the electrolyzer output at the distant facility, and its impacts can be recorded. Analysis of the data obtained from research also indicates that the electrolyzer connected to a power grid can also be used to increase stability of the power grid [55]. This setup is being used to perform 200-h PHIL testing and characterization of the electrolyzer, remotely (Figure 17.14).

Thus, from the above discussion of the projects currently being pursued (or planned) at INL and NREL, it is evident how scientific resources and infrastructure in two laboratories – separated by hundreds of miles – can be combined to solve larger problems. In the absence of distributed RTS, each laboratory would have no choice but to duplicate efforts, require multiplicity of labor and funds in order to study similar problems in the power system industry.

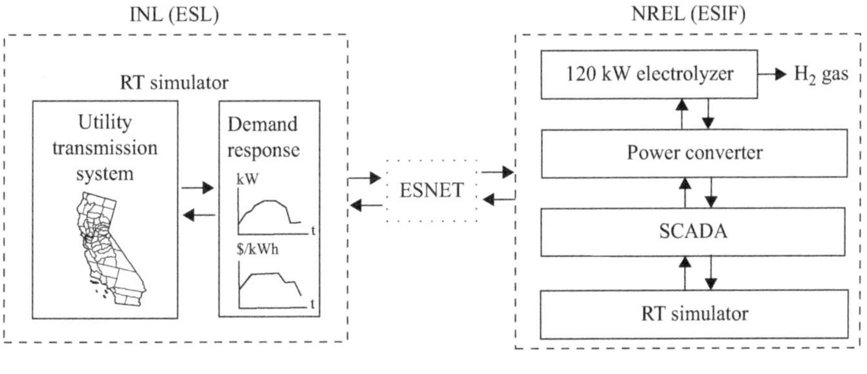

Figure 17.14 Schematic representation of INL and NREL joint research project on characterization of electrolyzers for participation in demand response and ancillary grid services market

17.8 Concluding remarks and future work

Research centers under both academic and research agencies are working toward addressing the future challenges of the electric grid. However, greater research challenges can be addressed, via an active collaboration that leverages spatially distributed assets such as the ones described earlier. RT simulators and power and energy systems testing and research centers are dispersed across the globe. Interconnecting these simulators using WAN is proposed as one such feasible approach to link research facilities. The major application of geographically distributed RT simulation is modeling large power networks, integrating component models or hardware to the model, and subsequently, their validation and testing.

HIL testing using RTS also follows the testing principles of MBD for rapid prototyping, verification, and deployment on-site. This design principle can be used for development of large power networks for geographically distributed simulation, where an RT model can be used as a critical input for the design and development of rest of the devices and power system under study. Distributed RTS combined with HIL testing has the potential of serving tremendous advantages. Testing and troubleshooting of controllers and systems that are not easily accessible can be facilitated greatly by distributed RTS and HIL. Research facilities located in different national laboratories and universities have unique blend of equipment and systems which can be integrated. RT simulators at INL and NREL are connected via WAN-based IP that allows leveraging unique test assets such as electrolyzers, wind turbines, power converters. Multitude of efforts of linking physical test-beds at different locations with and without RT simulators has also been discussed in this article. In the past, the automotive industry has led the way in terms of testing equipment and systems remotely. Power and energy systems research can immensely benefit from formalizing and creating novel testing networks (such as the one proposed earlier) to create a more tightly linked research network. Utilities may also benefit greatly from a well-connected testing network with the ability to simulate, test, and analyze, and hence mitigate a wider range of grid issues and challenges. Establishing a formal mathematical base of distributed RTS is a critical component of such research allowing its widespread adoption.

A concept of "RT-Super Lab" is envisioned by the virtue of unification of distributed RTS and HIL extension to numerous facilities. As shown in Figure 17.15, the RT-Super Lab can allow the simulation of large-scale power systems networks in RT that are vendor neutral and have diverse capabilities of HIL. It will allow a greater access to test facilities and infrastructure that may be available at only select locations such as the national laboratory complex in the United States. The RT simulator connectivity between locations in the United States with European research and university centers is also being pursued at INL with an objective of creating the RT-Super Lab. The team pursing this project comprises two national laboratories (INL and NREL) and five universities (Florida State University, Colorado State University, University of South Carolina, RWTH Aachen, and Polytechnic University of Turin). A joint collaboration across the power and energy community can be enhanced to greater levels when such projects are pursued. Perhaps the most important potential

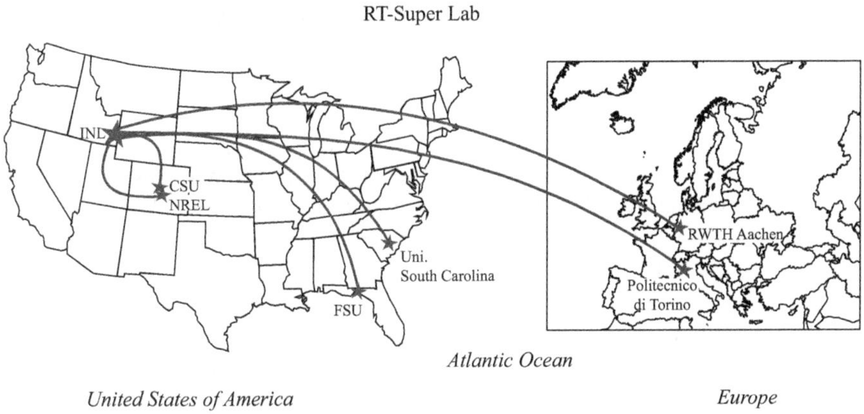

Figure 17.15 RT-Super Lab concept utilizing geographically distributed RTS based on WAN

outcome of pursuing distributed RTS and HIL is enhanced collaboration and leveraging research personnel. Tremendous amount of work happens related to RTS and device characterization around the world with mostly passive exchange of findings. We foresee this growing into a stimulus for research centers to build extensive testing networks to address greater challenges in power and energy systems.

References

[1] P. Kundur, N. J. Balu, and M. G. Lauby, *Power system stability and control.* McGraw-hill New York, 1994, vol. 7.

[2] J. M. Carrasco, L. G. Franquelo, J. T. Bialasiewicz, *et al.*, "Power-electronic systems for the grid integration of renewable energy sources: a survey," *IEEE Transactions on Industrial Electronics*, Jun. 2006, vol. 53, no. 4, pp. 1002–1016.

[3] P. Venne, J.-N. Paquin, and J. Bélanger, "The what, where and why of real-time simulation," in *IEEE Power and Energy Society General Meeting*, 2010, pp. 37–49.

[4] L.-F. Pak, M. O. Faruque, X. Nie, and V. Dinavahi, "A versatile cluster-based real-time digital simulator for power engineering research," *IEEE Transactions on Power Systems*, May 2006, vol. 21, no. 2, pp. 455–465.

[5] S. Biswas, C. Vellaithurai, and A. Srivastava, "Development and real time implementation of a synchrophasor based fast voltage stability monitoring algorithm with consideration of load models," in *2013 IEEE Industry Applications Society Annual Meeting*, Oct. 2013, pp. 1–9.

[6] P. Crolla, A. Roscoe, A. Dysko, and G. M. Burt, "Methodology for testing loss of mains detection algorithms for microgrids and distributed generation using real-time power hardware-in-the-loop based technique," in *2011 IEEE*

8th International Conference on Power Electronics and ECCE Asia (ICPE ECCE), May 2011, pp. 833–838.
[7] X. Guo, X. You, and Y. Song, "Real-time digital simulation of high-power electrical traction system," in *Power Electronics and Motion Control Conference (EPE/PEMC), 15th International*, Sep. 2012, pp. 1–5.
[8] M. Steurer, F. Bogdan, W. Ren, M. Sloderbeck, and S. Woodruff, "Controller and power hardware-in-loop methods for accelerating renewable energy integration," in *IEEE Power Engineering Society General Meeting, 2007*, Jun. 2007, pp. 1–4.
[9] C. Dufour, A. Andrade, and J. Belanger, "Real-time simulation technologies in education: a link to modern engineering methods and practices," in *Proceedings of the 11th International Conference on Engineering and Technology Education, INTERTECH*, Brazil, Mar. 2011.
[10] M. D. O. Faruque, T. Strasser, G. Lauss, *et al.*, "Real-time simulation technologies for power systems design, testing, and analysis," *IEEE Power and Energy Technology Systems Journal*, Jun. 2015, vol. 2, no. 2, pp. 63–73.
[11] J. Tang, R. Hovsapian, M. Sloderbeck, *et al.*, "The CAPS-SNL power system security test bed," in *Proceedings of 3rd CRIS*, 2006.
[12] K. Ravikumar, N. Schulz, and A. Srivastava, "Distributed simulation of power systems using real-time digital simulator," in *Power Systems Conference and Exposition*, Mar. 2009, pp. 1–6.
[13] T. Godfrey, S. Mullen, R. C. Dugan, C. Rodine, D. W. Griffith, and N. Golmie, "Modeling smart grid applications with co-simulation," in *IEEE International Conference on Smart Grid Communications*, 2010, pp. 291–296.
[14] S. Ciraci, J. Daily, J. Fuller, A. Fisher, L. Marinovici, and K. Agarwal, "FNCS: a framework for power system and communication networks co-simulation," in *Symposium on Theory of Modeling & Simulation – DEVS Integrative*. Society for Computer Simulation International, 2014, p. 36.
[15] B. Palmintier and A. Pratt, "Evaluation of the Netherlands' International Test Facility for Smart Grids," *NREL White Paper*, 2015.
[16] S. Mittal, B. P. Zeigler, J. L. R. Martín, F. Sahin, and M. Jamshidi, "Modeling and simulation for systems of systems engineering," *Systems of Systems – Innovations for the 21st Century*, 2008.
[17] M. Ruth, A. Pratt, M. Lunacek, S. Mittal, H. Wu, and W. Jones, "Effects of home energy management systems on distribution utilities and feeders under various market structures," in *International Conference on Electricity Distribution*, 2015.
[18] B. M. Kelley, P. Top, S. G. Smith, C. S. Woodward, and L. Min, "A federated simulation toolkit for electric power grid and communication network co-simulation," in *IEEE Workshop on Modeling and Simulation of Cyber-Physical Energy Systems*, 2015, pp. 1–6.
[19] S. Chouikhi, I. E. Korbi, Y. Ghamri-Doudane, and L. A. Saidane, "A comparison of wireless sensor networks co-simulation platforms for smart grid

applications," *International Journal of Digital Information and Wireless Communications*, 2013, vol. 3, no. 1, pp. 93–102.

[20] L. Cristaldi, A. Ferrero, A. Monti, F. Ponci, W. McKay, and R. Dougal, "A virtual environment for remote testing of complex systems," *IEEE Transactions on Instrumentation and Measurement*, 2005, vol. 54, no. 1, pp. 123–133.

[21] J. Bastos, J. Wu, N. Schulz, R. Liu, and A. Monti, "Distributed simulation using the virtual test bed and its real-time extension," in *Proceedings of the 2007 Summer Computer Simulation Conference*. Society for Computer Simulation International, 2007, pp. 757–765.

[22] H. Figueroa, J. Bastos, A. Monti, and R. Dougal, "A modular real-time simulation platform based on the Virtual Test Bed," in *IEEE International Symposium on Industrial Electronics*, 2006, pp. 1537–1541.

[23] K. Miu, V. Ajjarapu, K. Butler-Purry, *et al.*, "Testing of shipboard power systems: a case for remote testing and measurement," in *IEEE Electric Ship Technologies Symposium*, 2005, pp. 195–201.

[24] J. Wu, N. N. Schulz, and W. Gao, "Distributed simulation for power system analysis including shipboard systems," *Electric Power Systems Research*, 2007, vol. 77, no. 8, pp. 1124–1131.

[25] M. O. Faruque, M. Sloderbeck, M. Steurer, and V. Dinavahi, "Thermo-electric co-simulation on geographically distributed real-time simulators," in *IEEE Power & Energy Society General Meeting*, 2009, pp. 1–7.

[26] M. O. Faruque, V. Dinavahi, M. Sloderbeck, and M. Steurer, "Geographically distributed thermo-electric co-simulation of all-electric ship," in *IEEE Electric Ship Technologies Symposium*, 2009, pp. 36–43.

[27] M. O. Faruque, M. Sloderbeck, M. Steurer, and V. Dinavahi, "Thermo-electric co-simulation on geographically distributed real-time simulators," in *IEEE Power and Energy Society General Meeting*, 2009, pp. 1–7.

[28] M. Stevic, S. Vogel, A. Monti, and S. D'Arco, "Feasibility of geographically distributed real-time simulation of HVDC system interconnected with AC networks," in *IEEE PowerTech*, 2015, pp. 1–5.

[29] M. Stevic, A. Monti, and B. Andrea, "Development of a simulator-to-simulator interface for geographically distributed simulation of power systems in real time," in *IECON - Annual Conference of the IEEE Industrial Electronics Society*, 2015, pp. 20–25.

[30] T. Ersal, M. Brudnak, A. Salvi, J. L. Stein, Z. Filipi, and H. K. Fathy, "Development of an Internet-distributed Hardware-in-the-Loop simulation platform for an automotive application," in *ASME Dynamic Systems and Control Conference*, 2009, pp. 73–80.

[31] A. Tandon, M. J. Brudnak, J. L. Stein, and T. Ersal, "An observer based framework to improve fidelity in Internet-distributed Hardware-in-the-Loop simulations," in *ASME Dynamic Systems and Control Conference*, 2013.

[32] X. Ge, M. J. Brudnak, P. Jayakumar, J. L. Stein, and T. Ersal, "A model-free predictor framework for tele-operated vehicles," in *American Control Conference*, 2015, pp. 4573–4578.

[33] T. Ersal, M. Brudnak, A. Salvi, Y. Kim, J. B. Siegel, and J. L. Stein, "An iterative learning control approach to improving fidelity in Internet-distributed Hardware-in-the-Loop simulation," *Journal of Dynamic Systems, Measurement, and Control*, 2014, vol. 136, no. 6, pp. 061012-061012-8.

[34] T. Ersal, M. Brudnak, J. L. Stein, and H. K. Fathy, "Statistical transparency analysis in Internet-distributed Hardware-in-the-Loop simulation," *IEEE/ASME Transactions on Mechatronics*, 2012, vol. 17, no. 2, pp. 228–238.

[35] T. Ersal, R. B. Gillespie, M. J. Brudnak, J. L. Stein, and H. K. Fathy, "Effect of coupling point selection on distortion in Internet-distributed Hardware-in-the-Loop simulation," *International Journal of Vehicle Design*, 2012, vol. 61, pp. 67–85.

[36] J. S. Thorp, C. E. Seyler, and A. G. Phadke, "Electromechanical wave propagation in large electric power systems," *IEEE Transactions on Circuits and Systems I: Fundamental Theory and Applications*, 1998, vol. 45, no. 6, pp. 614–622.

[37] S.-J. Tsai, L. Zhang, A. G. Phadke, *et al.*, "Frequency sensitivity and electromechanical propagation simulation study in large power systems," *IEEE Transactions on Circuits and Systems I: Regular Papers*, 2007, vol. 54, no. 8, pp. 1819–1828.

[38] J. Yu, M. Zhang, and W. Li, "Static equivalent method based on component particularity representation and sensitivity consistency," *IEEE Transactions on Power Systems*, 2014, vol. 29, no. 5, pp. 2400–2408.

[39] Y. Liang, X. Lin, A. M. Gole, and M. Yu, "Improved coherency-based wide-band equivalents for real-time digital simulators," *IEEE Transactions on Power Systems*, 2011, vol. 26, no. 3, pp. 1410–1417.

[40] X. Lin, A. Gole, and M. Yu, "A wide-band multi-port system equivalent for real-time digital power system simulators," *IEEE Transactions on Power Systems*, 2009, vol. 24, no. 1, pp. 237–249.

[41] U. D. Annakkage, N. K. C. Nair, A. M. Gole, *et al.*, "Dynamic system equivalents: a survey of available techniques," in *IEEE Power and Energy Society General Meeting*, 2009, pp. 1–5.

[42] F. Ma and V. Vittal, "Right-sized power system dynamic equivalents for power system operation," *IEEE Transactions on Power Systems*, 2011, vol. 26, no. 4, pp. 1998–2005.

[43] R. Kay, "Pragmatic network latency engineering fundamental facts and analysis," cPacket Networks, Tech. Rep., 2009. [Online]. Available: https://www.network-visibility.com/downloads/cPacket_Whitepaper_Introduction-to-network-latency-engineering.pdf

[44] M. Klein, G. J. Rogers, and P. Kundur, "A fundamental study of inter-area oscillations in power systems," *IEEE Transactions on Power Systems*, 1991, vol. 6, no. 3, pp. 914–921.

[45] W. H. Kersting, "Radial distribution test feeders," *IEEE Power Engineering Society Winter Meeting*, 2001, vol. 2, pp. 908–912.

[46] "RWTH Aachen University – Institute for Automation of Complex Power Systems – Homepage," (2016, Mar. 3). [Online]. Available: http://www.acs.eonerc.rwth-aachen.de

[47] "Tinc VPN Project Homepage," (2016, Mar. 3). [Online]. Available: http://tinc.nl.linux.org

[48] "ERIC-lab | European Real-time Integrated Co-simulation Laboratory," (2016, Mar. 3). [Online]. Available: http://www.eric-lab.eu

[49] G. Niemeyer and J.-J. Slotine, "Stable Adaptive Teleoperation," *IEEE Journal of Oceanic Engineering*, 1991, vol. 16, no. 1, pp. 152–162.

[50] Sebastian Henschel, "Analysis of electromagnetic and electromechanical power system transients with dynamic phasors," PhD dissertation, The University of British Columbia, Vancouver, Canada, 1999.

[51] H. Konara, "Interfacing dynamic phasor equivalent to an electromagnetic transient simulation," MS thesis, University of Manitoba, Winnipeg, Canada, 2015.

[52] P. Castello, J. Liu, C. Muscas, P. A. Pegoraro, F. Ponci, and A. Monti, "A fast and accurate PMU algorithm for P+M class measurement of synchrophasor and frequency," *IEEE Transactions on Instrumentation and Measurement*, 2014, vol. 63, no. 12, pp. 2837–2845.

[53] R. J. Anderson and M. W. Spong, "Bilateral control of teleoperators with time delay," *IEEE Transactions on Automatic Control*, 1989, vol. 34, no. 5, pp. 494–501.

[54] S. Casoria, "VSC–based HVDC transmission system (detailed model)," *Matlab, SimPowerSystems*, 2010.

[55] J. Eichman, K. W. Harrison, and M. Peters, *Novel electrolyzer applications: providing more than just hydrogen*. National Renewable Energy Laboratory, 2014.

Index